harr

Springer Series in
OPTICAL SCIENCES 98

founded by H.K.V. Lotsch

Editor-in-Chief: W. T. Rhodes, Atlanta

Editorial Board: T. Asakura, Sapporo
K.-H. Brenner, Mannheim
T. W. Hänsch, Garching
T. Kamiya, Tokyo
F. Krausz, Wien and Garching
B. Monemar, Lingköping
H. Venghaus, Berlin
H. Weber, Berlin
H. Weinfurter, München

Springer
Berlin
Heidelberg
New York
Hong Kong
London
Milan
Paris
Tokyo

Springer Series in
OPTICAL SCIENCES

The Springer Series in Optical Sciences, under the leadership of Editor-in-Chief *William T. Rhodes*, Georgia Institute of Technology, USA, provides an expanding selection of research monographs in all major areas of optics: lasers and quantum optics, ultrafast phenomena, optical spectroscopy techniques, optoelectronics, quantum information, information optics, applied laser technology, industrial applications, and other topics of contemporary interest.

With this broad coverage of topics, the series is of use to all research scientists and engineers who need up-to-date reference books.

The editors encourage prospective authors to correspond with them in advance of submitting a manuscript. Submission of manuscripts should be made to the Editor-in-Chief or one of the Editors. See also http://www.springer.de/phys/books/optical_science/

Editor-in-Chief

William T. Rhodes
Georgia Institute of Technology
School of Electrical and Computer Engineering
Atlanta, GA 30332-0250, USA
E-mail: bill.rhodes@ece.gatech.edu

Editorial Board

Toshimitsu Asakura
Hokkai-Gakuen University
Faculty of Engineering
1-1, Minami-26, Nishi 11, Chuo-ku
Sapporo, Hokkaido 064-0926, Japan
E-mail: asakura@eli.hokkai-s-u.ac.jp

Karl-Heinz Brenner
Chair of Optoelectronics
University of Mannheim
Institute of Computer Engineering
B6, 26
68131 Mannheim, Germany
E-mail: brenner@uni-mannheim.de

Theodor W. Hänsch
Max-Planck-Institut für Quantenoptik
Hans-Kopfermann-Strasse 1
85748 Garching, Germany
E-mail: t.w.haensch@physik.uni-muenchen.de

Takeshi Kamiya
Ministry of Education, Culture, Sports
Science and Technology
National Institution for Academic Degrees
3-29-1 Otsuka, Bunkyo-ku
Tokyo 112-0012, Japan
E-mail: kamiyatk@niad.ac.jp

Ferenc Krausz
Vienna University of Technology
Photonics Institute
Gusshausstrasse 27/387
1040 Wien, Austria
E-mail: ferenc.krausz@tuwien.ac.at
and
Max-Planck-Institut für Quantenoptik
Hans-Kopfermann-Strasse 1
85748 Garching, Germany

Bo Monemar
Department of Physics
and Measurement Technology
Materials Science Division
Linköping University
58183 Linköping, Sweden
E-mail: bom@ifm.liu.se

Herbert Venghaus
Heinrich-Hertz-Institut
für Nachrichtentechnik Berlin GmbH
Einsteinufer 37
10587 Berlin, Germany
E-mail: venghaus@hhi.de

Horst Weber
Technische Universität Berlin
Optisches Institut
Strasse des 17. Juni 135
10623 Berlin, Germany
E-mail: weber@physik.tu-berlin.de

Harald Weinfurter
Ludwig-Maximilians-Universität München
Sektion Physik
Schellingstrasse 4/III
80799 München, Germany
E-mail: harald.weinfurter@physik.uni-muenchen.de

Yuri Shvyd'ko

X-Ray Optics

High-Energy-Resolution Applications

With 181 Figures

Springer

Dr. habil. Yuri Shvyd'ko
Institut für Experimentalphysik
Universität Hamburg
Luruper Chaussee 149
22761 Hamburg, Germany
E-mail: yuri.shvydko@desy.de

ISSN 0342-4111

ISBN 3-540-21484-4 Springer-Verlag Berlin Heidelberg New York

Library of Congress Control Number: 2004103358

This work is subject to copyright. All rights are reserved, whether the whole or part of the material is concerned, specifically the rights of translation, reprinting, reuse of illustrations, recitation, broadcasting, reproduction on microfilm or in any other way, and storage in data banks. Duplication of this publication or parts thereof is permitted only under the provisions of the German Copyright Law of September 9, 1965, in its current version, and permission for use must always be obtained from Springer-Verlag. Violations are liable for prosecution under the German Copyright Law.

Springer-Verlag is a part of Springer Science+Business Media

springeronline.com

© Springer-Verlag Berlin Heidelberg 2004
Printed in Germany

The use of general descriptive names, registered names, trademarks, etc. in this publication does not imply, even in the absence of a specific statement, that such names are exempt from the relevant protective laws and regulations and therefore free for general use.

Camera-ready by the author using a Springer T_EX macropackage
Cover concept by eStudio Calamar Steinen using a background picture from The Optics Project. Courtesy of John T. Foley, Professor, Department of Physics and Astronomy, Mississippi State University, USA.
Cover production: *design & production* GmbH, Heidelberg

Printed on acid-free paper   57/3141/tr   5 4 3 2 1 0

To my wife Natasha

*To my parents
Anastasia and Vasily*

Preface

The use of x rays has moved in the forefront of science and technology in the second half of the 20th century. This progress has been greatly stimulated by the advent of synchrotron x-ray sources in the 1960s. The undulator-based synchrotron radiation sources which have appeared in the last decade of the 20th century gave a new impetus to such development. The brilliance of the x-ray sources has increased by 12 orders of magnitude in 40 years and this trend does not show any signs of stagnation.

The future x-ray sources of the 21th century based on free-electron lasers driven by linear accelerators will provide sub-picosecond radiation pulses with by many orders of magnitude higher brilliance and full transverse coherence. The x-ray sources of the newest generation offer a possibility to realize more than ever before the great potential of x-ray optics and, as a consequence, to elaborate new sophisticated instrumentation with unprecedented resolution and eventually to move in new directions of research in x-ray technology, materials science, fundamental physics, life sciences, etc.

The use of the radiation with a well defined wavelength, and the ability to measure precisely the radiation wavelengths and frequencies are of fundamental importance in physical and other natural sciences. X rays with a well defined wavelength enable precise structure determination. They can be used, for example, to observe such tiny changes in the crystal structure as those caused by replacement of constituent atoms with different isotopes. Measuring precisely the change of the x-ray wavelength in scattering from objects provides information on the dynamics of the nanoscopic parts of these objects. The information on vibrational dynamics of solids, liquids and macroscopic biological molecules can be accessed in this way.

Bragg backscattering is a tool for creating, manipulating, and analyzing x rays with highest spectral purity. Bragg backscattering has the unique feature of selecting x rays with a narrow spectral bandwidth. Bragg backscattering is essential for the realization of x-ray Fabry-Pérot resonators, devices which could provide access to x rays with even smaller spectral widths.

Theoretical foundations, experimental studies, applications of x-ray optics with high spectral resolution based on Bragg backscattering, as well as development of the instrumentation, which uses Bragg backscattering, are the topics discussed in this book. Special attention is payed to the results of the theoretical and experimental studies aimed at developing Fabry-Pérot resonators for x rays and Mössbauer radiation.

The book is organized as follows.

In Chap. 1 an overview of high spectral-resolution x-ray crystal optics is presented. In particular applications of high-resolution x-ray optics to different physical problems are discussed. The physical principles underlying high-resolution x-ray optics, and how these principles can be applied to design high-resolution x-ray instruments, such as monochromators, analyzers, resonators, interferometers, etc., are addressed in the rest of the book. In Chap. 1 also a historical background is given.

In Chap. 2 the dynamical theory of x-ray diffraction in perfect crystals is considered with emphasis on the points crucial to the backscattering geometry. The consideration is not restricted to backscattering alone. On the contrary, a general approach is used for the whole range of incidence angles without division into "standard" and "backscattering" regimes. Two-beam and multiple-beam diffraction are analyzed in detail.

Principles of x-ray diffraction involving an arbitrary number of crystals is considered in Chap. 3. X-ray beams with desired spectral and angular characteristics - monochromatic beams, beams with extremely narrow angular divergence, etc. - can be obtained by Bragg reflections off several crystals. Different multiple-crystal arrangements, dispersive, nondispersive, mixed, double-dispersive, etc., are considered theoretically, and graphically illustrated with the help of DuMond diagrams.

In Chap. 4 the theory of perfect and imperfect x-ray Fabry-Pérot resonators is propounded, and the technical challenges of their realization are discussed.

The fundamentals of the theory of x-ray diffraction in perfect crystals are applied in the following chapters to address the problem of designing highly accurate spectrometers for studying dynamics and structure of condensed matter with atomic resolution. Different design concepts of high-resolution, i.e., with a bandwidth of a few meV and less, x-ray monochromators are presented in Chap. 5, while designs of high-resolution x-ray analyzers are presented in Chap. 6. Their performance is illustrated with examples from recent experiments.

An account of first experiments on x-ray resonators is presented in Chap. 7.

The theories discussed in the book are very often accompanied by examples of numerical calculations of the reflectivity and transmissivity of crystals, monochromators, analyzers, and interferometers as a function of photon energy, angle of incidence, and time delay after excitation with radiation pulses, etc.. For this purposes original computer codes are used written by the author, unless otherwise stated.

It is the intention of the present book to help graduate and postgraduate students to enter the field of high-resolution x-ray optics. The author believes that also the scientists with experience in x-ray optics may find in this book in concentrated form the information useful for tackling new and sophisticated problems in x-ray high-resolution optics. The author hopes that this book will help in clearing up the confusion that many encounter when starting to study and apply dynamical theory of x-ray diffraction in single crystals. How many well-established scientists know and take advantage in praxis of the fact that the relative spectral width of the Bragg reflections is an invariant quantity? How many of them know and take advantage of the fact that in asymmetric Bragg scattering the crystal acts like a strongly dispersive optical prism? How many know that exact backscattering takes place *not necessarily* at normal incidence to the reflecting atomic planes? How many know that the disturbing reflections, which are accompanying Bragg reflections in backscattering, can be suppressed by using asymmetrically cut crystals? How do the multiple-crystal energy dispersive monochromators work? Is an x-ray Fabry-Pérot resonator feasible? If this book can help to answer some or even all of these questions, the author will feel that writing the book has been worthwhile.

Acknowledgments

Without help and support of my family, friends, and colleagues it would be impossible to accomplish this work.

I would like to express first of all my sincere gratitude to Prof. Erich Gerdau, Dr. Dierk Rüter, and Prof. Patrick McNally, who helped me write, edit, and proof this book.

Prof. Erich Gerdau of Hamburg University provided a first impetus to the work on this book. I wish to thank him for this initial impulse, subsequent stimulated interest, encouragement, and for reading and reviewing the whole manuscript.

Dr. Dierk Rüter of Hamburg University has reviewed the entire manuscript with greatest care and insight. A big THANKS for investing so much time and effort into this book project!

Prof. Patrick McNally of Dublin University has read the whole manuscript and improved its readability. For these and other invaluable contributions to this book project I wish to thank him very much.

I would like to express my gratitude to Prof. Roberto Colella of Purdue University for reading Chapters. 1, 2, and 4 and his encouraging appraisal. I am also grateful to Dr. Alexander Chumakov of European Synchrotron Radiation Facility (ESRF) and Dr. Thomas Toellner of Advanced Photon Source at Argonne National Laboratory (APS at ANL) for reading Sects. 1.2.1 and 5.4 of the manuscript, and for many valuable comments. I am indebted to Dr. Ralf Röhlsberger of HASYLAB at DESY for reading Chapt. 5, and Sect. 6.2, and for suggesting many improvements. Many thanks to Dr. Ercan Alp and Dr. Harald Sinn of APS at ANL, and to Dr. Michael Krisch of ESRF for reading Sects. 1.2.1 and 6.2, and for critical comments.

I would like to thank Dr. K. Achterhold (Technische Universität München), Dr. E.E. Alp (APS at ANL), Dr. A. Barla (ESRF), Dr. A.R.Q. Baron (SPring-8), Dr. C. Burns (Western Michigan University), Dr. A.I. Chumakov (ESRF), Dr. H. Franz (HASYLAB at DESY), Dr. J. Hill (Brookhaven National Laboratory), Dr. M. Krisch (ESRF), Prof. P. McNally (Dublin University), Dr. F. Sette (ESRF), Dr. D. Shu (APS at ANL), Dr. H. Sinn (APS at ANL), Dr. J. Sutter (SPring-8), Dr. T. Toellner (APS at ANL), Dr. J. Wong (Lawrence Livermore National Laboratory), and Dr. M. Yabashi (SPring-8) for placing original results of their research work at my disposal.

Finally, I wish to thank my dearest wife Natasha for her love, encouragement, and strength which she gave me – the most essential support during the years of working on this book.

Hamburg – Wedel, *Yuri Shvyd'ko*
April, 2004

Contents

1. **Overview of the Field** 1
   1.1 High-Resolution Bragg Optics and Backscattering 1
   1.2 Applications 9
       1.2.1 Atomic-Scale Dynamics 10
       1.2.2 Atomic-Scale Length Measurements 24
       1.2.3 Length Standard for Atomic-Scale Measurements ... 26
   1.3 Historical Background 31

2. **Dynamical Theory of X-Ray Diffraction (Focus on Backscattering)** 37
   2.1 Introduction 37
   2.2 General Solution of the Diffraction Problem 38
       2.2.1 Electric Susceptibility of Crystals 39
       2.2.2 Fundamental Equations 40
       2.2.3 Ewald Sphere 41
       2.2.4 Vacuum and in-Crystal Wave Vectors 43
       2.2.5 Angular Dispersion 45
       2.2.6 Scattering Geometries 46
       2.2.7 Exact Backscattering Geometry 53
       2.2.8 Fundamental Equations in Scalar Form 54
       2.2.9 Eigenvectors and Wave Fields in the Crystal 55
       2.2.10 Boundary Conditions 56
       2.2.11 Reflectivity and Transmissivity I. 57
   2.3 Single-Beam Case 58
   2.4 2-Beam Bragg Diffraction 59
       2.4.1 Radiation Fields in the Crystal and in Vacuum 60
       2.4.2 Reflectivity and Transmissivity II 65
       2.4.3 Reflectivity of Thick Non-Absorbing Crystals 65
       2.4.4 Deviation Parameter 67
       2.4.5 Bragg's Law 69
       2.4.6 Modified Bragg's Law for the Incident Wave 71
       2.4.7 Reflection Region. Symmetric Scattering Geometry . 73
       2.4.8 Reflection Region. Asymmetric Scattering Geometry 80
       2.4.9 Angular Dispersion 87

        2.4.10 Modified Bragg's Law for the Exit Wave 91
        2.4.11 Reflection Region in the Space of Exit Waves 93
        2.4.12 Reflectivity of Thin Crystals 103
        2.4.13 Reflectivity of Thick Absorbing Crystals 106
        2.4.14 Reflectivity of Crystals with Intermediate Thickness .. 109
        2.4.15 Time Dependence of Bragg Diffraction 110
  2.5 Multiple-Beam Diffraction in Backscattering:
      Kinematic Treatment 114
        2.5.1 Accompanying Reflections 115
        2.5.2 Examples of Si and α-Al_2O_3 Crystals 116
        2.5.3 Conjugate Pairs of Accompanying Reflections 117
        2.5.4 Umweganregung - Detour Excitation................ 121
        2.5.5 Polarization Effects 122
        2.5.6 Region of Multiple-Beam Excitation 123
  2.6 4-Beam Diffraction in Backscattering:
      Dynamical Treatment 126
        2.6.1 Fundamental Equations 127
        2.6.2 4-Beam Diffraction - "Degenerate" Case............. 131
        2.6.3 4-Beam Diffraction - General Case................. 134
  2.7 n-Beam Diffraction:
      Suppression of Accompanying Reflections 140

3. Principles of Multiple-Crystal X-Ray Diffraction 143
  3.1 Introduction 143
  3.2 Two-Crystal Configurations 145
  3.3 Diffraction from Crystals in $(+, +)$ Configuration 147
  3.4 Diffraction from Crystals in $(+, -)$ Configuration 151
  3.5 Diffraction in $(+, -, -, +)$ Configuration 155
  3.6 Diffraction in $(+, +, -)$ Configuration 159
  3.7 Exact Backscattering in $(+, +, +)$ Configuration 165

4. Theory of X-Ray Fabry-Pérot Resonators 171
  4.1 Introduction 171
  4.2 Multiple-Beam Interference in Optics 174
        4.2.1 Plane Resonators 174
        4.2.2 Wedge Resonators 179
  4.3 Perfect X-Ray Fabry-Pérot Resonator 180
        4.3.1 General Equations................................. 180
        4.3.2 Comparison with Optical Resonators 185
        4.3.3 Enhanced Phase Sensitivity....................... 186
        4.3.4 Influence of the Gap Width 188
        4.3.5 Spectral Dependence.............................. 189
        4.3.6 Time Dependence 197
        4.3.7 Angular Dependence.............................. 201
        4.3.8 Temperature Dependence......................... 207

		4.3.9 Finite Number of Multiple Reflections 208
		4.3.10 A Practical Expression for the Airy Phase 210
	4.4	Imperfect X-Ray Fabry-Pérot Resonator................... 210
		4.4.1 Roughness of Internal Surfaces 210
		4.4.2 Inhomogeneous Mirror Thickness 212
		4.4.3 Non-Parallel Mirrors 212
		4.4.4 Temperature Gradients 213

5. High-Resolution X-Ray Monochromators 215
  5.1 Introduction 215
  5.2 Principles of X-Ray Monochromatization 216
  5.3 Single-Bounce and $(+,-)$-Type Monochromators 218
    5.3.1 Spectral Bandpass................................ 218
    5.3.2 Choice of Crystals 223
    5.3.3 Access to Specific Photon Energies (Tunability) 225
    5.3.4 Reflectivity 231
    5.3.5 Figure of Merit 233
    5.3.6 Reflectivity in Exact Backscattering.
          Effects of Multiple-Beam Diffraction 235
  5.4 Multi-Bounce $(+,+)$-Type Monochromators 241
    5.4.1 Introduction 241
    5.4.2 Symmetric $(+,+)$ and $(+,-,-,+)$ Monochromators .. 242
    5.4.3 Nested $(+,+,-,-)$ Monochromators 244
    5.4.4 Asymmetric $(+,+)$ Two-Crystal Monochromators 255
    5.4.5 Asymmetric $(+,-,-,+)$ Monochromators 260
    5.4.6 Figure of Merit for $(+,+)$-Type Monochromators..... 268
  5.5 Multi-Bounce $(+,+,\pm)$-Type Monochromators 269
    5.5.1 Principle 269
    5.5.2 Spectral Bandpass................................ 272
    5.5.3 Angular Acceptance 277
    5.5.4 Energy Tuning 279
    5.5.5 Throughput, Figure of Merit, Spectral Function 280
    5.5.6 Technical Details 281
    5.5.7 Summary.. 283
  5.6 μeV-Resolution Fabry-Pérot Interference Filters 284

6. High-Resolution X-Ray Analyzers 287
  6.1 Introduction 287
  6.2 Spherical Crystal Analyzers............................... 288
    6.2.1 Historical Background 288
    6.2.2 Flat Crystal Analyzer............................. 290
    6.2.3 Focusing Array of Flat Crystals 297
    6.2.4 Exact Backscattering Analyzer and Ring Detector 300
    6.2.5 Inelastic X-Ray Scattering Spectrometers............ 302

XIV Contents

   6.3 Multi-Bounce $(+,+,\pm)$-Type
        Flat Crystal Analyzers 304
        6.3.1 High-Energy or Low-Energy Photons? 304
        6.3.2 Design Concept 306
        6.3.3 Monochromator Component 308
        6.3.4 Collimating Optics 309
        6.3.5 Inelastic X-Ray Scattering Spectrometers 313

7. **Towards Realizing X-Ray Resonators** 315
   7.1 Concept of the Test Device 315
   7.2 Technical Details 318
        7.2.1 Manufacturing Sapphire Mirrors................... 318
        7.2.2 Two-Axis Nanoradian Rotation Stage 319
        7.2.3 Temperature Control 321
   7.3 Experimental Setup..................................... 322
   7.4 Adjustment of Mirrors 325
   7.5 Experiments on X-Ray Resonators....................... 326
        7.5.1 Reflectivity of Mirrors 326
        7.5.2 Time Response.................................. 329
        7.5.3 Interference Effects 331

A. **Appendices** 335
   A.1 Si Crystal Data 335
   A.2 α-Al_2O_3 Crystal Data.................................... 337
   A.3 Bragg Back-Reflections in Si 342
   A.4 Two-Beam Bragg Backscattering Cases in α-Al_2O_3 350
   A.5 Low-Lying Levels of Stable Isotopes 355
   A.6 α-Al_2O_3 as a Universal meV-Monochromator 358
   A.7 Quality Assessment of α-Al_2O_3 Crystals.................. 360
   A.8 Radiation Wavelength and Angle of Incidence
        for Exact Bragg Backscattering 368

Bibliography.. 371

List of Symbols 387

Index 395

1. Overview of the Field

This introductory chapter is meant to give a brief overview of the general properties of x-ray Bragg crystal optics as opposed to Fresnel optics, and in particular, an overview of the special features of high-resolution Bragg crystal optics, which uses *backscattering*; an overview of applications of x-ray Bragg backscattering to different physical problems; and an overview of physical instruments which are based on Bragg backscattering, such as monochromators, analyzers, resonators, interferometers, etc. It is devoted also to giving a historical background of the field of high-resolution x-ray crystal optics.

The physical principles underlying high-resolution x-ray optics, and how these principles can be applied to the design of high-resolution monochromators, analyzers, resonators, interferometers, etc., are addressed in the following chapters of the book.

1.1 High-Resolution Bragg Optics and Backscattering

Scattering of electromagnetic radiation from single atoms is weak. In contrast, for an ensemble of atoms the weak scattering amplitudes can add up coherently to a macroscopic value proportional to the number of scatterers N_s and thus result in a high scattering intensity proportional to N_s^2. Coherently enhanced scattering from ensembles of atoms plays a fundamental role in optics, as it allows efficient transformations of radiation fields by reflection, refraction, rotation of polarization, etc.

If the radiation wavelength λ is larger than interatomic distances, the internal structure of the ensemble is practically insignificant. Coherent scattering is described by a macroscopic characteristic - the index of refraction n of a medium, see, e.g., (Born and Wolf 1999, Jackson 1967). In general, the index of refraction describes how much slower the radiation travels in a material than it does in vacuum. In the particular case of a medium composed of atoms having a single resonance energy E_a, the index of refraction is given by

$$n = 1 + \frac{h^2 c^2}{2\pi} \frac{N_0 Z\, r_\mathrm{e}}{E_\mathrm{a}^2 - E^2 + i\varGamma E}.$$

Here N_0 is the number density of atoms, Z is the number of electrons in each atom, $r_e = 2.81794 \times 10^{-15}$ m is the classical electron radius, c is the speed of light in vacuum, h is Planck's constant, Γ is the energy width of the atomic resonance, E is the energy of the radiation photon which relates to the wavelength λ in vacuum as $E = hc/\lambda$. The dependence of the index of refraction on the photon energy is generally called *dispersion*.

When the radiation passes through a boundary between regions of different refractive index, it can be reflected or its path can be changed. A well known example is the penetration through an optical prism, shown schematically in Fig. 1.1. Due to refraction and dispersion a collimated beam containing photons of different energies becomes a divergent beam with the propagation direction dependent on the photon energy.

Fig. 1.1. Refraction and dispersion in the prism converts a collimated beam into a divergent beam of photons with the propagation direction dependent on the photon energy.

The index of refraction enters the famous Fresnel formulae for reflection and refraction at the interface of two media with refraction indices n_1 and n_2. Visible light may be re-scattered (reflected) almost totally under certain conditions at any angle of incidence. This type of coherent scattering is called "Fresnel reflection". In particular, at normal incidence the reflectivity is given by

$$R = \left| \frac{n_1 - n_2}{n_1 + n_2} \right|^2.$$

For empty space $n_1 = 1$. Thus, one can obtain high reflectivity $R \simeq 1$ at normal incidence if $|n_2| \gg 1$. For the photons of visible light with $E \simeq 1$ eV this is achieved, for example, by using highly conducting metals as the reflecting medium, for which the index of refraction takes large imaginary values.

The situation changes dramatically for hard x rays. The energy is typically $E \gtrsim 5$ keV, and therefore very often much higher than the atomic resonance energies: $E \gg E_a$. As a result, the index of refraction becomes

$$n = 1 - w, \qquad w = \frac{N_0 Z r_e}{2\pi} \frac{h^2 c^2}{E^2}. \tag{1.1}$$

For $E \simeq 12.4$ keV ($\lambda \simeq 0.1$ nm), $N_0 \simeq 5 \times 10^{28}$ m^{-3}, and $Z = 14$ it differs from unity by only a very small negative value $-w$ ($w \simeq 3 \times 10^{-6}$). Due to this, the important boundary effect - Fresnel reflection - does not play such a significant role for hard x rays. No significant reflectivity can be achieved in this way [1].

The transition to the hard x-ray region also drastically changes the situation in another respect. Another type of coherent scattering – Bragg scattering (or Bragg diffraction) – starts playing a dominant role. This happens due to two important facts. First, the wavelengths of the hard x rays become smaller than interatomic distances. Secondly, atoms in crystals are arranged periodically in three dimensions. The periods of such three-dimensional crystal lattices are typically on the order of $1-0.1$ nm. Perfect crystals are natural three-dimensional optical gratings for x rays. Under these conditions, diffraction of x rays from the three-dimensional periodic ensemble of atoms takes place.

X-ray diffraction was observed for the first time by Friedrich, Knipping and von Laue in 1912 (Friedrich et al. 1913). Immediately after this discovery, W.H. Bragg and his son W.L. Bragg repeated the experiments on x-ray diffraction in crystals and interpreted the phenomenon as coherent reflection of x rays from sets of parallel atomic planes spaced by a period d_H apart, as shown in Fig. 1.2. They have first used systematically diffraction of x rays to measure interatomic distances and to analyze the geometrical arrangement of atoms in simple crystals. Diffraction of any kind of radiation in three-dimensional crystal lattices is often referred to as Bragg diffraction or Bragg scattering.

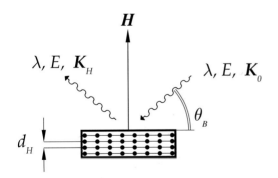

Fig. 1.2. Bragg reflection of x rays with wavelength λ (energy $E = hc/\lambda$, momentum $K_0 = K_H = 2\pi/\lambda$) from a set of parallel atomic planes in a crystal spaced by the interplanar distance d_H apart. Momentum conservation requires $\boldsymbol{K}_H = \boldsymbol{K}_0 + \boldsymbol{H}$, where $\boldsymbol{H}$ is the Bragg diffraction vector ($|\boldsymbol{H}| = 2\pi/d_H$).

---

[1] One can still achieve total reflectivity for x rays, however, at very small *glancing* angles of incidence to the interface. The Fresnel formulae predict that total reflectivity can be achieved at incidence angles $\Theta > \Theta_{cr}$, where Θ_{cr} is the critical angle of incidence defined as $\sin \Theta_{cr} = n_2/n_1$. This may happen if the incident wave propagates in an optically more dense medium ($n_1 > n_2$). Since the index of refraction is smaller than unity for x rays, the empty space is optically a more dense medium for them. The total reflectivity of x rays takes place at very small *glancing* angles of incidence $\theta \leq \theta_{cr} \simeq \sqrt{2w} \simeq 10^{-3}$. $\theta_{cr} = \pi/2 - \Theta_{cr}$ is called critical glancing angle of incidence, or critical angle of total reflection.

Intense diffraction maxima, which are called Bragg reflections, are observed in directions defined by Bragg's law:

$$\lambda = 2d_H \sin\theta_B, \qquad (1.2)$$

Bragg's law relates the radiation wavelength λ to the glancing angle of incidence, termed the Bragg angle θ_B, at which the radiation is reflected from the set of atomic planes. Unlike Fresnel reflection, Bragg reflection for a given wavelength takes place solely at the angle of incidence which satisfies Bragg's law. The glancing angle of incidence and x-ray photon energy $E = hc/\lambda$ are related similarly:

$$\frac{1}{E} = \frac{1}{E_H}\sin\theta_B, \qquad E_H = \frac{hc}{2d_H}. \qquad (1.3)$$

Here, E_H is termed Bragg energy. According to (1.3), x rays with this energy can be reflected at normal incidence ($\theta_B = \pi/2$) to the reflecting atomic planes, i.e., backwards. The Bragg energy is also the smallest energy of x-ray photons, which can be reflected from a given set of atomic planes. This is seen in Fig. 1.3. The dashed line reproduces the trace of the crystal reflectivity in the (θ, E) space as given by Bragg's law (1.3).

When Bragg's law is fulfilled, the waves scattered from atoms within the atomic plane and from different atomic planes interfere constructively, producing a strong diffracted wave thus giving rise to Bragg reflection. If only single-scattering from each atom is considered, i.e., each x-ray photon scatters only once before it is detected, the intensity of the Bragg reflection is proportional to the square of the total number of scattering atoms N_s^2. The single scattering treatment is commonly known as the kinematic or geometrical theory of x-ray diffraction. It predicts glancing angles of incidence and photon energies at which Bragg reflections occur - (1.3). The kinematic theory, however, fails to predict correctly the intensity of the Bragg reflection peaks, as with increasing N_s the reflected intensity grows unlimitedly.

The dynamical theory of x-ray diffraction in crystals, which takes into account multiple scattering of the radiation (as opposed to the kinematic theory, which assumes only single scattering), predicts reflection intensities in agreement with experiments. The foundation for the dynamical theory was laid in the works of Darwin (1914) and Ewald (1917) very soon after the discovery of the x-ray diffraction in crystals. von Laue (1931) has reformulated the dynamical theory of Ewald by solving the system of Maxwell's equations for a periodic electric susceptibility of crystals. Modern comprehensive treatments of the "standard" dynamical theory of x-ray diffraction can be found, for example, in the texts of Zachariasen (1945), von Laue (1960), Batterman and Cole (1964), Azaroff et al. (1974), Pinsker (1978), and Authier (2001). Its generalization to the backscattering regime will be discussed in detail in this book. Here we briefly summarize the basic predictions of the dynamical theory. They will be justified and discussed in detail in Chap. 2.

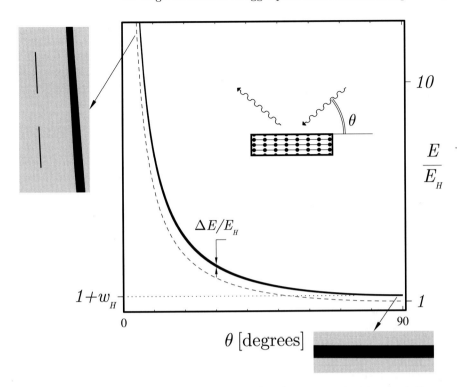

Fig. 1.3. Crystal reflectivity for x rays in the (E, θ) space, in the symmetric Bragg scattering geometry. The dashed line traces the position of the crystal peak reflectivity as given by Bragg's law $E = E_H / \sin \theta$. The solid line is the region of total reflection of x rays from perfect crystals, as given by the dynamical theory of x-ray Bragg diffraction; θ is the glancing angle of incidence to the reflecting atomic planes with interplanar distance d_H, E is the energy of the x-ray photons, $E_H = hc/2d_H$ is the Bragg energy, and w_H is the Bragg's law correction. The relative spectral width $\Delta E/E$ of the region is constant over the whole spectral range (but not $\Delta E/E_H$!). It is typically $\simeq 10^{-4} - 10^{-8}$ and smaller, and is greatly exaggerated here for clarity. The fragments of the diagram drawn to a larger scale display the advantages of backscattering: high reflectivity in a broad angular region with a narrow energy width.

The dynamical theory predicts total reflection, with reflectivity $R \simeq 1$, from crystals with negligible photo-absorption. Total reflection occurs within some region of photon energies ($E_c - \Delta E/2 \rightarrow E_c + \Delta E/2$) at a given glancing angle of incidence θ_c, and within some region of glancing angles of incidence ($\theta_c - \Delta\theta/2 \rightarrow \theta_c + \Delta\theta/2$) at a given energy E_c, about the values θ_c and E_c given by the modified (or dynamical) Bragg law:

$$E_c \sin \theta_c = E_H (1 + w_H). \tag{1.4}$$

Figure 1.3 shows the region of total reflection of x rays in the (E, θ) space, as given by the dynamical theory of x-ray Bragg diffraction.

The correction w_H in (1.4) to the kinematic Bragg's law (1.3) is to good accuracy, a constant for a given set of atomic reflecting planes. It is independent of photon energy and glancing angle of incidence. Under the so called symmetric diffraction conditions, when the atomic planes are parallel to the crystal surface, as in Fig. 1.2, it equals the deviation of the refractive index (1.1) from unity, taken at the Bragg energy:

$$w_H = 1 - n(E_H). \tag{1.5}$$

For a given glancing angle of incidence total reflection occurs at a photon energy, which is a factor of $1 + w_H$ larger than the photon energy predicted by the kinematic theory. This is attributed to the effect of refraction[2].

Though w_H is very small (typically 10^{-4}–10^{-6}), it is of great importance. First, it causes measurable shifts. Secondly, the angular width $\Delta\theta$ and the relative energy width $\Delta E/E_c$ of Bragg reflections are even smaller. $\Delta E/E_c$ is typically in the range from $\simeq 10^{-4}$ to $\simeq 10^{-10}$ for different Bragg reflections[3]. Importantly, the relative energy width $\Delta E/E_c$ of the region of total reflection is a quantity, which is *conserved* (to good accuracy) for the Bragg reflection with a given set of atomic planes. $\Delta E/E_c$ is independent of the glancing angle of incidence and of photon energy E_c. The relative spectral width of a Bragg reflection is inversely proportional to the *effective* number of atomic planes N_e contributing to Bragg diffraction[4]:

$$\frac{\Delta E}{E_c} = \frac{1}{\pi N_e}. \tag{1.6}$$

The relative energy width $\Delta E/E_c$ of Bragg reflections in Si, and in α-Al$_2$O$_3$ at $T = 295$ K is shown in Fig. 1.4.

The invariance of the relative energy width for a given Bragg reflection implies, in particular, that the smallest energy width of the Bragg reflection is achieved at the smallest photon energy. In other words, it is achieved in backscattering. Typically the energy bandwidths of Bragg reflections are on the order of 100 meV to sub-meV.

The independence of the relative spectral width on the glancing angle of incidence is valid for an incident *plane* wave. In practice, there are no plane waves. Finite beam divergence blurs the intrinsic spectral width, unless it is less than the intrinsic angular width $\Delta\theta$ of the Bragg reflection. For the performance of optical instruments, which use Bragg diffraction, it is thus

---

[2] The refractive index is slightly smaller than 1 for hard x rays. The wavelength in crystals is thus a little bit larger than in vacuum. To match Bragg's law (1.2) the wavelength of the incident radiation has to be smaller by the same factor and the photon energy respectively larger.

[3] Examples of Bragg reflections from different atomic planes in silicon and sapphire (α-Al$_2$O$_3$) single crystals and their parameters (including w_H, $\Delta E/E_c$, etc.) are presented in Appendices A.3 and A.4, respectively.

[4] N_e is the number of reflecting planes within the so-called extinction length.

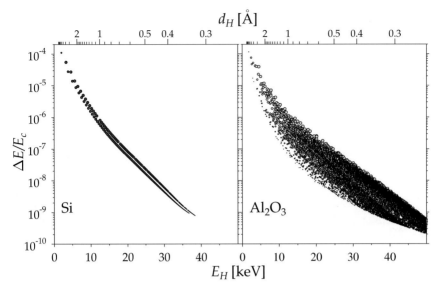

Fig. 1.4. Relative energy width $\Delta E/E_c$ of Bragg reflections in Si, and in α-Al$_2$O$_3$ (sapphire) at $T = 295$ K. Evaluations used the dynamical theory of x-ray diffraction in perfect thick crystals. Results for all allowed reflections are shown with Bragg energies E_H in the range from 0 to 50 keV, and reflectivity larger than 10%. The diameter of each point is proportional to the peak reflectivity of the corresponding back reflection. The smallest point corresponds to a reflectivity of 10% and the largest to 95%. For each Bragg reflection, the Bragg energy relates to the interplanar distance of the reflecting atomic planes as $E_H = hc/2d_H$.

always favorable to have the x-ray beam divergence much smaller than the angular width of Bragg reflections.

As predicted by the dynamical theory, the angular widths of Bragg reflections for glancing angles of incidence far from $\pi/2$ (far from normal incidence) scales with $\tan\theta$, whereas the proportionality constant is $\Delta E/E_c$:

$$\Delta\theta = \frac{\Delta E}{E_c} \tan\theta_c. \tag{1.7}$$

Since, $\Delta E/E_c < 10^{-4}$, see Fig. 1.4, this relation implies, that the angular width of Bragg reflections is very small. It is in the range of a few μrad and less. It increases, however, with increasing glancing angle of incidence θ.

The situation changes dramatically in backscattering geometry, i.e., in the vicinity of $\theta = \pi/2$, as can be seen from Fig. 1.3. The energy, at which the radiation is reflected, varies quadratically with incidence angle $\Theta = \pi/2 - \theta$ in backscattering, i.e., much slower than outside this region. As a result, the angular width of the region of total reflection becomes much broader. At exact normal incidence the dynamical theory predicts that the angular width is no more proportional to $\Delta E/E_c$, as in (1.7). Instead, it is proportional to the square root of the relative spectral width $\Delta E/E_c$:

$$\Delta\theta = 2\sqrt{\frac{\Delta E}{E_{\mathrm{c}}}}. \tag{1.8}$$

The μrad-broad Bragg reflections become mrad-broad in backscattering. This is a very important feature of Bragg back-reflections.

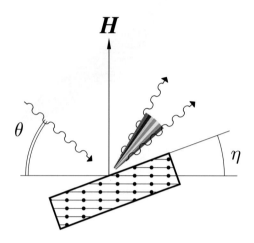

Fig. 1.5. Bragg reflection of x rays from an asymmetrically cut crystal with atomic planes at a nonzero asymmetry angle $\eta \neq 0$ to the crystal surface. The crystal acts like the optical prism, cf. Fig. 1.1, which disperses photons with different photon energies. The effect of angular dispersion is actually small, of an order of magnitude of the angular width of the Bragg reflection. It is shown here greatly exaggerated for clarity.

However small refraction may be, it plays a significant role in Bragg diffraction. Refraction at the crystal-vacuum interface is especially important if the crystal is cut asymmetrically, i.e., the atomic planes are at a nonzero asymmetry angle $\eta \neq 0$ to the crystal surface. Additional momentum transfer, due to refraction, disperses photons with different photon energies, as shown schematically in Fig. 1.5. An asymmetrically-cut crystal acts like an optical prism, cf. Fig. 1.1. The effect of angular dispersion is actually small, of an order of magnitude of the angular width of the Bragg reflection. It is shown in Fig. 1.5 greatly exaggerated for clarity. However, it is of great importance for high-resolution x-ray optics, as it offers additional possibilities in manipulating the spectral distribution of x rays on reflection.

One more distinctive feature should be mentioned typical for *exact* Bragg backscattering. It is often encountered in this book. Briefly formulated, it has to do with the so-called *multiple-beam* Bragg diffraction. Along with the beam backwards reflected from a selected set of atomic planes, accompanying beams appear, which originate from simultaneous reflection from other sets of atomic planes in the same crystal, as shown schematically in Fig. 1.6. This is an unfavorable situation for many applications of Bragg backscattering in x-ray crystal optics, as these simultaneously open reflection channels suppress reflectivity into the backscattering channel. Multiple-beam Bragg diffraction arises systematically if the condition for *exact* Bragg back-reflection is fulfilled. In crystals with a cubic Bravais lattice (like silicon, germanium, dia-

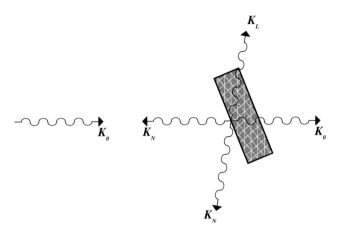

Fig. 1.6. Schematic drawing illustrating the four-beam coplanar Bragg diffraction case with the backscattering channel N. The three sets of parallel lines depict the three sets of reflecting atomic planes in the crystal. $\boldsymbol{K}_0$ and $\boldsymbol{K}_N$ are the wave vectors of the incident (or forward-transmitted) and Bragg backscattered waves, respectively. $\boldsymbol{K}_L$ and $\boldsymbol{K}_B$ are the wave vectors of a pair of the accompanying Bragg reflections. See, Sect. 2.5 for more details.

mond, etc.) all Bragg back-reflections with the exception of the two lowest order reflections suffer from the accompanying reflections. To avoid this problem, crystals with lower crystal symmetry should be used. For instance, many Bragg back-reflections in sapphire (α-Al$_2$O$_3$), a crystal with rhombohedral lattice, are multiple-beam-free reflections.

To summarize, Bragg back-reflections combine small spectral band-passes with comparatively large angular widths. It is a combination of these properties that makes Bragg diffraction in backscattering an effective means of obtaining x rays with high spectral purity even for poorly collimated beams. As will be discussed in this book, Bragg backscattering underlies the operation of high precision instruments with very small – meV and sub-meV – bandwidths, such as x-ray monochromators, analyzers, x-ray resonators, x-ray Fabry-Pérot interferometers, wavelength meters, etc. Achieving very high spectral resolution with the help of Bragg backscattering and designing highly accurate optical devices will be a primary focus of the discussion in the present book.

1.2 Applications

The use of radiation with sharp and well-defined frequencies and wavelengths, and the ability to measure these frequencies and wavelengths precisely are of fundamental importance to physical and other natural sciences. It is crucial for the realization of standards of time, and length, for the determination of fundamental constants, for fundamental research, etc. Monochromatic radiation enables both very precise structure determinations and studies of dynamics of living and non-living matter.

1.2.1 Atomic-Scale Dynamics

To understand the dynamics of matter, one needs *in situ* techniques, which can resolve both length (momentum) and time (energy). Techniques, which use visible or infrared light, such as Raman scattering, Brillouin scattering, and photon correlation techniques can probe the dynamics of matter over a very large range on the time scale from femtoseconds (10^{-15} s) to seconds. However, the resolution on the length scale is determined by visible light wavelengths and is limited therefore to micrometers. Radiation with shorter wavelengths of about 0.01 to 0.5 nm is required to access the atomic-scale dynamics.

X rays with energies $E \gtrsim 5$ keV and wavelengths $\lambda \lesssim 0.2$ nm, respectively, can be used to probe the dynamics of matter with the required atomic spatial resolution[5]. Thermal neutrons and $\simeq 100$ eV electrons have de Broglie wavelengths in the same range. However, consideration of these probes is beyond the scope of this book.

Figure 1.7 shows schematically some examples of collective excitations in condensed matter with their typical energy and time scales, such as electronic excitations, lattice vibrations (phonons), magnetic excitations, diffusion, fluctuations, tunneling, etc. Relaxation times of the non-equilibrium dynamical processes are measured directly in real time, if sufficiently short x-ray radiation pulses, and detectors with sufficient time resolution are available. Equilibrium dynamics is traditionally studied by measuring the energy spectra of excitations – the density of excited states $D(\varepsilon)$ – and (or) by measuring energy-momentum dispersion relations $\varepsilon(\boldsymbol{Q})$. Inelastic scattering of x rays, and measuring the energy and momentum of the incident and scattered x-ray photons, as schematically shown in Fig. 1.8, provides access to equilibrium atomic-scale dynamics of matter.

For this, well collimated and monochromatic x rays are needed with a spectral width smaller than the appropriate excitation energies. The available x-ray sources are very bright and well collimated, however, they are far from monochromatic. The modern undulator based x-ray sources at synchrotron radiation facilities of the third generation deliver hard x rays with photon energies $E \gtrsim 1$ keV in spectral bands of hundreds of electronvolts at each

---

[5] A recent extensive discussion of the advances in this field can be found in the materials of the Euroschool 2000 "Synchrotron radiation studies of new materials" published in a special issue of *Journal of Physics: Condensed Matter* **13** (2000) No. 34

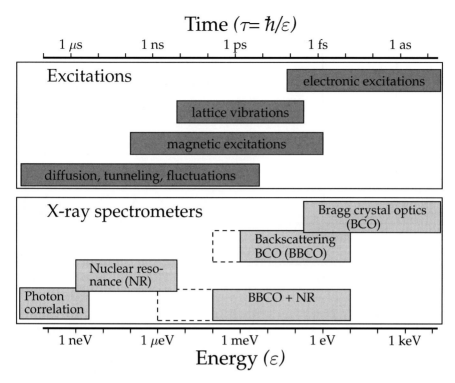

Fig. 1.7. Collective excitations in condensed matter with typical excitation energy and time scales, and corresponding x-ray spectrometers for probing these excitations. The low-energy edge of each box (in the lower part) indicates the achieved energy resolution and the high-energy edge indicates the tunability range of the appropriate x-ray spectrometer.

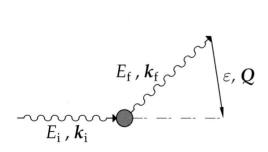

Fig. 1.8. Vector diagram for an inelastic scattering process of an x-ray photon with initial energy E_i, and momentum k_i, from a collective excitation in condensed matter with energy ε, and momentum Q. The energy and momentum conservation requires $\varepsilon = E_i - E_f$, and $Q = k_i - k_f$, respectively, where E_f, and k_f are the x-ray photon energy and momentum after the scattering process.

harmonic[6], see, e.g., recent reviews by Kunz (2001), Mülhaupt and Rüffer (1999). An example of an undulator spectrum is shown in Fig. 1.9.

---

[6] The relative spectral bandwidth of modern undulator based x-ray sources is approximately inversely proportional to the number of magnetic periods N_u of

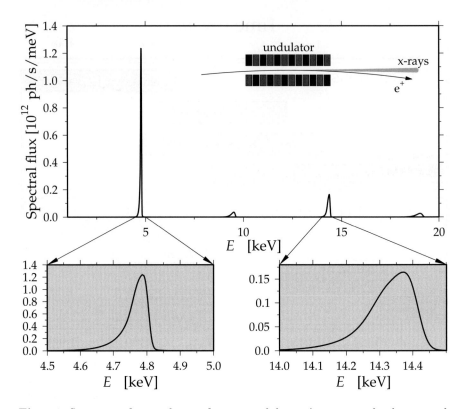

Fig. 1.9. Spectrum of x ray photons from an undulator. As an example, the spectral flux in the central cone of a prototype undulator for the PETRA-III 6 GeV positron storage ring (Hamburg, Germany) is shown. Undulator parameters: period 28.7 mm, number of periods $N_u = 696$, length 20 m, deflection parameter $K_u = 1.72$. Storage ring current 100 mA. Spectral flux through an aperture 1×1 mm² 40 m from the source. Calculations by Franz (2003) with the program SPECTRA (Tanaka and Kitamura 2001).

---

an undulator, and is typically 10^{-2}. The future x-ray sources, based on a single pass x-ray self-amplified spontaneous emission (SASE) free-electron lasers (FEL) driven by linacs, will provide sub-picosecond radiation pulses with many orders of magnitude higher brilliance and full transverse coherence (LCLS 1998, Materlik and Tschentscher 2001). Very long undulators with $N_u \simeq 5 \times 10^3$ magnetic periods, which are required to achieve self-amplified spontaneous emission in the SASE FELs (Murphy and Pellegrini 1990, Bonifacio et al. 1990, Saldin et al. 2000), will also provide x rays with superior spectral characteristics. X rays with a relative spectral width of $\simeq 1/N_u \lesssim 10^{-3}$ will be available. A two-stage SASE FEL consisting of two undulators and an x-ray monochromator located between them, may provide access to an x ray source with even smaller relative spectral width of $\simeq 10^{-6}$ (Saldin et al. 2001).

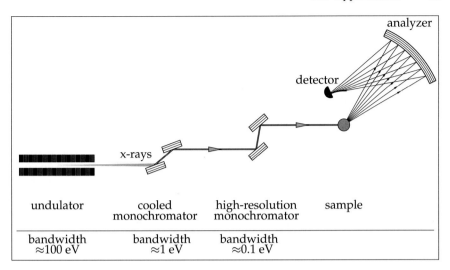

Fig. 1.10. Components of an inelastic x-ray scattering spectrometer for investigation of electronic excitations. Both, the two-bounce high-resolution crystal monochromator, and the spherically bent crystal analyzer have a bandpass ≈ 0.1 eV.

X-ray spectrometers have to be applied capable of monochromatizing the radiation from the x-ray sources and analyzing the scattered radiation with required resolution. Spectrometers should also have a sufficient tunability range allowing inspection of the whole excitation spectrum. Figure 1.7 shows in the lower part schematically the state-of-the-art x-ray spectroscopic techniques, which could be applied to probe excitations. The low-energy edge of each box in Fig. 1.7 indicates the achieved energy resolution and the high-energy edge indicates the tunability range of the appropriate x-ray spectrometer.

Electronic Excitations Investigated by Inelastic X-Ray Scattering Spectroscopy. Inelastic x-ray scattering offers a unique tool for studying electronic excitations in condensed matter because of its direct coupling to the electron charge. For an overview of the field see, e.g., (Schülke 2001), and (Hämäläinen and Manninen 2001). This technique was applied to study collective excitations, e.g., by Hill et al. (1996), Burns et al. (1999), single particle excitations by Caliebe et al. (2000), to name only few. Resonant enhancement of the charge transfer excitations, i.e., resonant inelastic x-ray scattering (RIXS) near the K-absorption edge, was demonstrated by Kao et al. (1996). RIXS was applied recently for high-resolution measurements of the dispersion relations of charge-transfer excitations in insulating La_2CuO_4 (Kim et al. 2002).

Electronic excitations, i.e., electron-hole pair creation, plasmons, core level excitations have energies greater than 0.1-1 eV. An energy resolution for

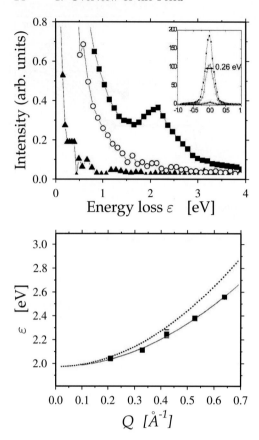

Fig. 1.11. Inelastic x-ray scattering signal from liquid metal Li(NH$_3$)$_4$ (lithium ammonia) (■), from pure liquid ammonia NH$_3$ (○), and from the empty sample cell (▲). The momentum transfer is $Q = 0.42$ Å^{-1}. The plasmon shows up as a peak in the liquid at an energy loss of 2 eV. The inset shows the elastic scattering peaks and energy resolution. (Burns et al. 1999).

Fig. 1.12. Dispersion of the plasmon energy. The solid squares are the data from the inelastic x-ray scattering measurements similar to those in Fig. 1.11, the solid line is the best fit to a quadratic dispersion, and the dotted line is the RPA (random phase approximation) fit. (Burns et al. 1999).

the inelastic x-ray scattering spectrometer of about 0.1-0.5 eV is, therefore, sufficient in most cases.

An x-ray spectrometer with such an energy resolution is in principle (although not always in practice) relatively easy to design, as the Bragg reflections from the atomic planes with largest interplanar distances, i.e., with the lowest Bragg energies, have a relative energy bandwidth of $\Delta E/E \simeq 10^{-4} - 10^{-5}$, see Fig. 1.4. Therefore, photons of energies $\simeq 5 - 30$ keV are filtered in an energy band of $\Delta E \simeq 0.1 - 1$ eV with these reflections. X-ray spectrometers based on crystal optics are most suitable for studies of electronic excitations through non-resonant or resonant inelastic x-ray scattering. A detailed optical design was described, e.g., by Kao et al. (1995). Figure 1.10 shows an example of an inelastic x-ray scattering spectrometer for investigation of electronic excitations. The spectrometers have a broadband tunability, which is achieved by incidence angle variation on the monochromator crystals.

Figure 1.11 shows an example of non-resonant inelastic x-ray scattering measurements (Burns et al. 1999). The plasmon in liquid metal Li(NH$_3$)$_4$

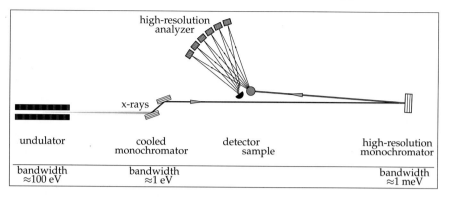

Fig. 1.13. Components of a modern inelastic x-ray scattering experiment with a very high energy resolution backscattering monochromator, and a very high energy resolution analyzer (two-dimensional array of $\simeq 10000$ flat crystals 1×1 mm^2 large mounted on a surface with spherical shape). The angular acceptance of the analyzer, which is typically $\approx 5-10$ mrad, and the size of the flat crystal elements are shown greatly exaggerated for clarity.

(lithium ammonia) was studied. Figure 1.11 shows the inelastic x-ray scattering spectrum at momentum transfer $Q = 0.42$ Å^{-1} in metal lithium ammonia (■), in pure liquid ammonia (○), and in the empty sample cell (▲). A feature at an energy loss of about 2 eV corresponds to the plasmon in lithium ammonia. Figure 1.12 shows the plasmon dispersion derived from the similar measurements performed at different momentum transfer Q.

Vibrational Dynamics Investigated by Inelastic X-Ray Scattering Spectroscopy. To study vibrational modes in solids, liquids, macroscopic biological molecules, or to study vibrations in magnetic lattices (the magnon modes), spectrometers with meV and sub-meV energy bandpass are required tunable in the $\simeq 0.1-1$ eV range, see Fig. 1.7.

A considerable amount of research on vibrational dynamics in matter has been performed with high-resolution inelastic scattering of thermal neutrons (Dorner 1982, Skold and Price 1986, Lovesy 1986, Izyumov et al. 1994). However, presently available neutron sources suffer from low brilliance, necessitating the use of large samples. Surface studies and use of small samples are difficult with neutrons.

X rays can be focused to a few micrometers in diameter, so that very small samples and surface studies become accessible. Furthermore, since photon energy and momentum transfers are uncoupled, x rays have no kinematic restriction in the low-momentum and high-energy transfer range, where the speed of sound exceeds the velocity of thermal neutrons. Inelastic x-ray scattering (IXS) with very high (meV) energy resolution applied to studies of vibrational modes in matter may be considered as a complementary technique to high-energy-resolution inelastic neutron scattering. Comprehensive reviews

16   1. Overview of the Field

covering the period 1983-2001 were published by Burkel (1991), Sette et al. (1998), Burkel (2000), and Sinn (2001).

Bragg reflections from single crystals can be used to monochromatize x rays to meV and sub-meV energy bandwidths, and to analyze the energy spectrum of scattered photons with the same energy resolution. Reflections with a very small relative bandwidth have to be used for this purpose. For example, a crystal spectrometer for 20 keV x-ray photons with a bandpass of 1 meV would require using Bragg reflections with $\Delta E/E \simeq 5 \times 10^{-8}$. Crystals with such Bragg reflections are available, see Fig. 1.4, and therefore make very high energy resolution spectrometers feasible. The backscattering geometry is preferable for design of both, the monochromator and the analyzer, for the two reasons mentioned in Sect. 1.1. First, the smallest absolute energy width of a selected Bragg reflection is achieved in backscattering. Second, the angular width of Bragg reflections becomes mrad-broad in backscattering. The latter is especially important for the design of analyzers, requiring a large acceptance solid angle.

The combination of an x-ray Bragg backscattering monochromator and a backscattering crystal analyzer composes a spectrometer for inelastic x-ray scattering measurements with meV energy resolution, as shown in Fig. 1.13. Energy tuning is achieved by changing the temperature in the crystal monochromator. The tunability range is limited to a few electronvolts.

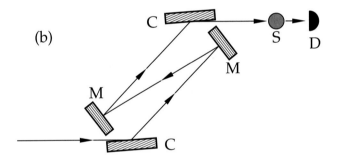

Fig. 1.14. Schematic arrangement of an experiment with a multi-bounce in-line monochromator. M: backscattering monochromator crystals, C: collimating crystals, S: sample, D: detector.

Along with the single-bounce backscattering monochromator there exist other design concepts for the high-resolution monochromators. Figure 1.14 shows one of the possible designs, the multi-bounce "nested" backscattering monochromator. The latter uses two crystals in close to back-reflection setting (M), serving as actual monochromator crystals, and two collimating crystals (C). In this scheme the direction of the primary beam is preserved. Therefore, they are also called in-line monochromators. The tunability range is $\simeq 100$ eV,

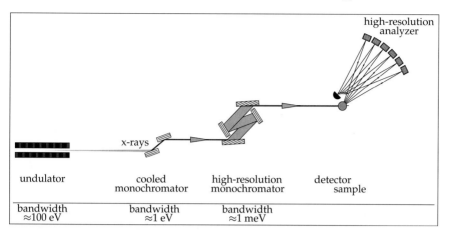

Fig. 1.15. Components of a modern inelastic x-ray scattering experiment with a very high energy resolution. The same as in Fig. 1.13, however, with a very high energy resolution in-line monochromator.

which is defined by the possible Bragg angle variation. The idea of nested monochromators was introduced by Ishikawa et al. (1992). The best energy resolution presently achieved with multi-bounce crystal monochromators is about $150 - 100$ μeV, as was reported by Toellner et al. (2001) and Yabashi et al. (2001). The physical principles of operation of this and other type of crystal monochromator will be addressed in this book. Figure 1.15 shows the scheme of an alternative spectrometer for IXS with an in-line multi-bounce monochromator.

To obtain a strong enough signal in the detector, the radiation emerging from the sample in a solid angle of $\simeq 10 \times 10$ mrad^2 has to be collected and analyzed. For this purpose, spherically bent crystal analyzers or arrays of flat analyzer crystals are applied. Although backscattering from flat crystals is capable of selecting radiation within sub-meV bandwidths, this is however valid only for plane incident radiation waves. Due to the large divergence of the radiation emerging from the sample, which has to be analyzed, the energy resolution of the crystal analyzer is inferior to the resolution of the backscattering monochromators. The total energy resolution of such spectrometers approaches, but is, still not better than 1 meV (Masciovecchio et al. 1996a,b, Sinn et al. 2002).

Figure 1.16(left) shows by way of example inelastic x-ray scattering spectra in large-grain polycrystalline specimens of face-centered cubic (fcc) δ-plutonium-gallium alloy (Wong et al. 2003). The spectra are characterized by an elastic contribution centered at zero energy transfer, and two inelastic contributions, corresponding to the creation (energy loss, Stokes component) and annihilation (energy gain, anti-Stokes component) of an acoustic phonon. Such data allow determination of the energy and momentum transfer asso-

18   1. Overview of the Field

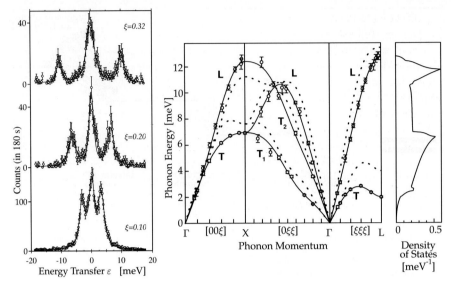

Fig. 1.16. Left: Representative spectra of inelastic x-ray scattering for the longitudinal acoustic phonon branches along [111] in a δ-Pu-Ga alloy. The experimental data are shown together with the results of the best-fit model of the phonon and elastic contributions. The corresponding reduced phonon wave vectors $[\xi\xi\xi]$ – $\xi = Q/(2\pi/a)$ – are indicated to the right of the spectra.
Center: Phonon dispersions along high-symmetry directions derived from the inelastic x-ray scattering measurements. The longitudinal and transverse modes are denoted as L and T, respectively. The experimental data are shown as circles. The solid curves are calculations using the fourth-nearest neighbor Born-von Kármán model. Dotted lines are calculations for pure Pu.
Right: The derived phonon density of states, normalized to three states per atom. (Wong et al. 2003)

ciated with phonons, and of the phonon dispersion relations. The phonon dispersion curves along the three principal symmetry directions in the fcc lattice are displayed in Fig. 1.16(center), together with a fit obtained by means of a standard Born-von Kármán model force constant model (Born and Huang 1954). The calculated phonon density of states derived from this model is shown in Fig. 1.16(right).

The crystal analyzer is a major bottleneck in the development of backscattering spectrometers with sub-meV resolution. Attempts to improve the resolution of backscattering crystal analyzers are ongoing. Alternatively, quite different techniques can be used, among them, e.g., nuclear resonant scattering. Since nuclear resonant scattering is much more than a tool for analyzing the phonon spectra, we shall present this technique in a general fashion.

Nuclear Resonant Scattering Techniques. Nuclear resonant scattering is scattering of photons from nuclei via their intermediate low-lying

($\lesssim 100$ keV) long-lived ($\gtrsim 10^{-11}$ s) excited states[7]. The long lifetime τ of the nuclear excited states determines the energy resolution $\simeq \hbar/\tau$ of the nuclear resonant scattering techniques, which is in the range of sub-neV to μeV[8].

The Mössbauer effect, which was discovered in 1958 (Mössbauer 1958, 1959), prevents nuclear resonance broadening from lattice vibrations and renders the precise and stable wavelength of the nuclear electromagnetic radiation named by Mössbauer. An excellent example is the Mössbauer radiation originating from the decay of the first excited state of ^{57}Fe nuclei. The energy of the excited state and the energy of the Mössbauer photons is $E_\mathrm{M} \simeq 14.4$ keV. The spectral width Γ of the ^{57}Fe Mössbauer radiation, determined by the lifetime of the excited state $\tau = 141$ ns via the uncertainty relation, is as small as $\Gamma = \hbar/\tau = 4.8 \times 10^{-9}$ eV.

Since the energy width of the nuclear resonance is so small, the nuclear resonance excitation spectra are sensitive to interactions of the nuclei with their environment, in particular to the hyperfine interactions of the electric and magnetic multipole moments with their own and neighboring atoms. The nuclear resonances are used as local probes for studying magnetic and electric structures in condensed matter and their dynamics such as electron spin relaxation, diffusion, etc. The ensemble of nuclei in crystals can be also used as a radiation filter with extremely high energy resolution in the range of neV to μeV, i.e., as high-resolution x-ray optical elements.

Radioactive sources of γ-radiation are used to perform nuclear resonance absorption spectroscopy in its traditional version, the Mössbauer spectroscopy, where a transmitted intensity is recorded vs. energy deviation from nuclear resonance. Energy tuning is achieved by Doppler shifting. Energy shifts, however, are typically not more than $\simeq 10$ μeV[9].

Since the pilot experiment of Gerdau et al. (1985), synchrotron radiation is being used as a source of the electromagnetic radiation in the nuclear resonant scattering experiments. The pulsed time structure of the synchrotron radiation is utilized to perform spectroscopic studies in the time domain, rather than in the energy domain, as is typical for Mössbauer spectroscopy. Figure 1.17 shows a schematic view of a modern nuclear resonant scattering experiment at a synchrotron radiation facility. Detector D_1 is used to measure the time dependence of the coherent response in the forward direction of the ensemble of resonant nuclei in the sample excited by a ≈ 100 ps long incident radiation pulse. The response time scales with the lifetime of the nuclear excited state τ (Kagan et al. 1979).

---

[7] Examples of low-lying excited states of stable isotopes with excitation energy less than 100 keV are listed in the table of Appendix A.5.

[8] In the first place, this statement is true for elastic scattering channels. For inelastic nuclear resonant scattering techniques, the energy resolution is very often determined by the monochromator.

[9] There is one important exception. Röhlsberger et al. (1997) have demonstrated a nuclear resonance μeV-monochromator for synchrotron radiation tunable up to about 1 meV. Doppler shifting in scattering from a rotating medium was used.

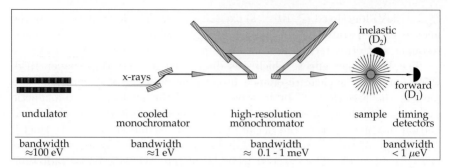

Fig. 1.17. Components of a modern nuclear resonant scattering experiment.

Figure 1.18 shows examples of such time spectra of resonant nuclear forward scattering (NFS) of 14.4 keV x rays from ^{57}Fe nuclei measured in a ^{57}FeBO$_3$ single crystal (a) with constant direction of the magnetic fields at the nuclear sites, as indicated in the figure; and (b) with the magnetic fields rotated by 90° within a few nanoseconds in the easy magnetization plane of the ^{57}FeBO$_3$ crystal at $t \simeq 18$ ns after excitation with the incident radiation pulse. As shown in the attached schemes the nuclear energy levels of ^{57}Fe split up into sublevels with magnetic quantum numbers m_g and m_e due to magnetic hyperfine interactions. Also shown are the transitions at which the nuclei radiate after their excitation at $t = 0$, and nuclear transitions at which they radiate after switching at $t \simeq 18$ ns, respectively. The superimposed oscillation structure in the time spectra, quantum beats (Gerdau et al. 1986), are due to the interference of the nuclear transitions, revealing hyperfine interactions and thus the structure and dynamics (in ps to µs regime) of local magnetic and electric fields at the nuclei in the sample.

The role of the monochromators in the time measurements of NFS, see Fig. 1.17, is to narrow the energy bandwidth of incident photons into the $10 - 0.1$ meV range, and thus to prevent the timing detector and electronics from overload. The center of the spectral bandpass of the monochromator is tuned to a fixed energy of the nuclear resonance.

Recoilless nuclear resonant absorption or emission of γ-radiation without creation or annihilation of phonons, the Mössbauer effect, takes place only with a certain probability, which is given by the Lamb-Mössbauer factor[10]. In the rest of the cases phonons or other collective excitations assist nuclear resonant scattering, which become inelastic.

The broad spectral bandwidth of the synchrotron radiation in combination with high-resolution tunable monochromators with ≈ 1 meV bandpass offer a possibility of studying vibrational dynamics in condensed matter in the meV regime by nuclear resonant inelastic scattering. This technique pioneered by Seto et al. (1995) and Sturhahn et al. (1995) is being intensively

---

[10] By its nature it is similar to the Debye-Waller factor, see Chap. 2

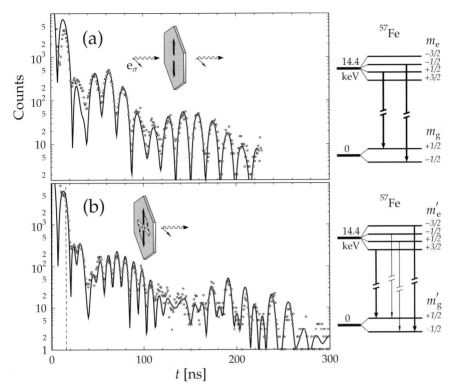

Fig. 1.18. Time spectra of resonant nuclear forward scattering (NFS) of 14.4 keV x rays from ^{57}Fe nuclei in a ^{57}FeBO$_3$ single crystal (a) with constant direction of the magnetic fields at the nuclear sites, as indicated in the figure; and (b) with the direction of the magnetic fields switched by 90° in the easy magnetization plane of the ^{57}FeBO$_3$ crystal at $t \simeq 18$ ns (Shvyd'ko et al. 1996, Shvyd'ko 1999). The excitation x-ray pulse of a $\simeq 100$ ps duration arrives at $t = 0$. It is too intense to be seen on this graph. The attached schemes (on the right) of nuclear energy levels of ^{57}Fe split up into sublevels with magnetic quantum numbers m_g and m_e due to magnetic hyperfine interactions show the transitions at which the nuclei radiate after their excitation at $t = 0$, and nuclear transitions at which they radiate after switching at $t \simeq 18$ ns, respectively. The energy splittings between the sublevels are ≈ 0.1 μeV. The solid line shows simulations with the theory of NFS.

developed. Reviews covering the period 1995-2001 were published by Chumakov and Rüffer (1998), Chumakov and Sturhahn (1999), Sturhahn and Kohn (1999), Rüffer and Chumakov (2000) and Alp et al. (2001b).

The energy dependence of inelastic incoherent scattering is measured with the detector D$_2$, as shown in Fig. 1.17, by changing the photon energy with the help of the high-energy-resolution monochromator. The resonant nuclei are used as built-in analyzers in the sample. Unlike other relevant methods like inelastic neutron and x-ray scattering, nuclear resonant inelastic

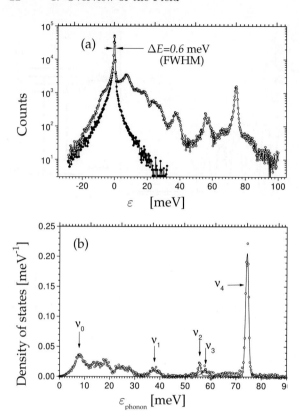

Fig. 1.19. (a) Spectra of nuclear resonant inelastic scattering in $(NH_4)_2Mg^{57}Fe(CN)_6$ measured at 30 K (open circles), as a function of the deviation $\varepsilon = E - E_M$ of the photon energy E from the Mössbauer resonance in ^{57}Fe at $E_M = 14.412$ keV. Solid circles show the instrumental function of the applied spectrometer with 0.65 meV (FWHM). Anti-Stokes components, with negative energy transfer ε, are practically not observed because of the low sample temperature. (b) Derived partial phonon densities of states of the iron atom in $(NH_4)_2Mg^{57}Fe(CN)_6$ at 30 K. Solid line shows the fit with the Gaussian functions of frequency distributions for several indicated modes. (Chumakov et al. 2003)

scattering provides direct access to the partial density of vibrational states of selected atoms containing resonant nuclei. Figure 1.19 shows as an example spectra of nuclear resonant inelastic scattering from ^{57}Fe nuclei in $(NH_4)_2Mg^{57}Fe(CN)_6$, measured as a function of the deviation $\varepsilon = E - E_M$ of the photon energy E from the nuclear resonance in ^{57}Fe at $E_M = 14.412$ keV. Also shown are the partial phonon densities of states of the iron atom, as derived from these measurements.

Versatile spectroscopic and optical techniques using nuclear resonant scattering of synchrotron radiation have recently been reviewed in (Gerdau and de Waard 1999/2000).

IXS and Nuclear Resonance. The energy analysis of x rays in IXS experiments could be performed alternatively by using nuclear resonant absorption. The nuclear resonances listed in the table of Appendix A.5 can be applied for this purpose. Again, a very good example is the 14.4 keV nuclear resonance in ^{57}Fe. The energy resolution of nuclear resonant analyzers is extremely sharp – in the sub-μeV range – as determined by the energy width of the nuclear resonances. The resolution is independent of the angular divergence of the radiation, which is to be analyzed. An analyzer based on nuclear resonant

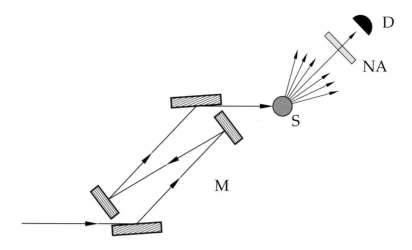

Fig. 1.20. Scheme of an experimental arrangement for measuring spectra of inelastic x-ray scattering by using a tunable multi-bounce in-line monochromator M, and a nuclear resonant analyzer NA. Other elements are: sample - S, detector - D.

absorption combined with a backscattering crystal monochromator capable of monochromatizing x-ray beams with meV to sub-meV bandwidths, and energy tuning over several eV, compose an instrument for inelastic x-ray scattering measurements with an energy resolution in the sub-meV range. A scheme of an experimental arrangement for measuring spectra of inelastic x-ray scattering by using nuclear resonant analyzer is shown in Fig. 1.20. The application of nuclear resonant analyzers to inelastic x-ray scattering was demonstrated by Chumakov et al. (1996a).

The nuclear resonant analyzers have much better energy resolution (0.01 – 1 μeV) than that of backscattering monochromators (0.1 – 10 meV). To further improve the energy resolution of spectrometers for inelastic x-ray scattering, monochromators with μeV resolution are required. Their tunability range should be at least $\simeq$ 100 meV to be able to cover the whole region of vibrational spectra. This could be achieved if x-ray resonators (or x-ray Fabry-Pérot interferometers) were used as interference filters. Steyerl and Steinhauser (1979) have pioneered the idea of a Fabry-Pérot-type interference filter for x rays. Bragg backscattering underlies the operation of these x-ray devices. The results of theoretical and experimental studies aimed at developing x-ray Fabry-Pérot resonators are presented in detail in Chapt. 4 and Chapt. 7, respectively. A scheme of a possible experimental arrangement for measuring spectra of inelastic x-ray scattering with the x-ray resonator used as an interference filter with μeV resolution is shown in Fig. 1.21.

X-ray Photon Correlation Spectroscopy. X-ray photon correlation spectroscopy (XPCS) covers the low energy range of excitations up to a few

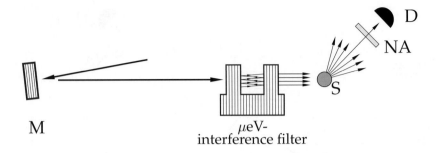

Fig. 1.21. Scheme of an experimental arrangement for measuring spectra of inelastic x-ray scattering with the x-ray resonator used as an interference filter with μeV resolution. Other elements are: meV backscattering monochromator - M, nuclear resonant analyzer - NA, sample - S, and detector - D. See Chap. 4 for details.

nanoelectronvolts, as shown in Fig. 1.7. Alternatively, it is also called intensity correlation spectroscopy. It is a novel technique for probing the dynamics of various equilibrium and non-equilibrium processes occurring in condensed matter systems in scattering experiments at large scattering vectors, i.e., on a nanometer length scale; see, e.g., (Dierker et al. 1995, Thurn-Albrecht et al. 1996, Riese et al. 2000). XPCS reveals sample dynamics via autocorrelation of the intensity scattered at different times t and $t + \delta t$ under particular coherent illumination. The second order correlation function $g_2(\delta t) = \langle I(t+\delta t)I(t)\rangle/\langle I(t)\rangle^2$ is measured as a function of momentum transfer $\boldsymbol{Q}$ to determine relaxation times $\tau(\boldsymbol{Q})$ of the system. The presently accessible relaxation times are in the milliseconds to seconds range, which is determined by the time resolution of the modern position sensitive detectors (CCD cameras). Fully or partly transverse coherent x-ray beams are required. Therefore, x-ray optics either has to be perfect or must not be present to maintain the coherent properties of the undulator based x-ray sources.

1.2.2 Atomic-Scale Length Measurements

Monochromatic electromagnetic radiation with sub-nanometer wavelength is required for distance measurements, and for precise structure determinations on atomic scales. Whatever x-ray scattering technique is applied, the result of the measurements is expressed in units of the wavelength of x rays used to probe the material. Therefore, the smaller the spectral bandwidth of the x rays, the more precise is the structural information, which can be obtained. This statement can be illustrated by the example of measuring interplanar distances d_H in crystals.

Bragg diffraction is very often used to measure interplanar distances d_H, as this parameter enters Bragg's law (1.4). For precise measurements of d_H,

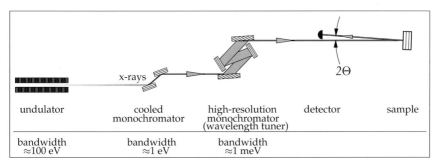

Fig. 1.22. Schematic of an experimental arrangement for measuring lattice spacings in crystals in almost exact backscattering geometry.

both the glancing angle θ and the energy E (wavelength λ) of incident photons must be well defined. In this respect, the backscattering geometry is most favorable. Indeed, the spectrum of the radiation reflected backwards from the atomic planes with interplanar distance d_H is centered at

$$\lambda_c = 2d_H \left[1 - \frac{\Theta^2}{2} - w_H \right].$$

This expression is derived from Bragg's law (1.4) under the assumption of a small angular deviation from normal incidence: $\Theta = \pi/2 - \theta \ll 1$. If $\Theta \leq \sqrt{2\epsilon}$, where ϵ is the required relative uncertainty of measurements, the simple relation

$$\lambda = 2d_H (1 - w_H)$$

holds even for a relatively coarse angular setting[11]. Thus in the almost exact backscattering geometry the incidence angle is no longer a parameter, which has to be determined, and which introduces some important uncertainty. As a result, a direct relation between the radiation wavelength and the interplanar distance can be established in backscattering. The accuracy of a measurement is eventually defined, firstly, by the spectral width of the Bragg back-reflection used in the sample, secondly, by the spectral purity of the incident x rays, and thirdly, by the accuracy of the x-ray wavelength measurements. Figure 1.22 shows the schematic of an experimental arrangement for measuring crystal lattice spacings in backscattering geometry. The wavelength measurements are performed with a wavelength meter, which is in fact a tunable high-resolution monochromator.

Figure 1.23 shows an example of precise lattice parameter measurements by using highly monochromatic x rays. It demonstrates the observation of such tiny changes in the crystal structure as those caused by the replacement

---

[11] E.g., for $\Theta \leq 100$ μrad it yields a relative uncertainty of better than $\epsilon \simeq 10^{-8}$.

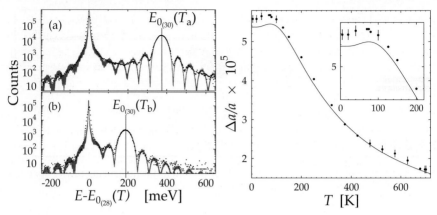

Fig. 1.23. Left: energy spectra of the (12 4 0) Bragg back-reflection from the layered epitaxial ^{30}Si and ^{28}Si crystal system at (a) $T_{\rm a} = 320$ K, and (b) $T_{\rm b} = 678$ K. The solid lines are fits with the dynamical theory of x-ray diffraction. The sharp peak at $E = E_{0_{(28)}}$ is due to the reflection from the single crystal substrate of ^{28}Si ($E_{0_{(28)}}(T_{\rm a}) - E_{0_{(28)}}(T_{\rm b}) = 18.09(8)$ eV). The broader peak at $E = E_{0_{(30)}}$ is due to the reflection from the thin single crystal epitaxial layer of ^{30}Si. From $E_{0_{(28)}} - E_{0_{(30)}}$ the difference $\Delta a = a_{(28)} - a_{(30)}$ in the lattice parameters of ^{30}Si and ^{28}Si crystals is determined.

Right: Temperature dependence of the relative difference $\Delta a/a$ of the lattice parameters of ^{30}Si and ^{28}Si, experiment and theory, demonstrating an anomalous behavior: the magnitude of the relative difference in lattice parameter first increases with temperature, reaches a maximum of $\Delta a/a$ at 75(10) K and only then starts to fall. (Wille et al. 2002)

of constituent atoms with different isotopes. The difference between the lattice parameters $\Delta a = a_{(30)} - a_{(28)}$ of silicon crystals composed of ^{28}Si and ^{30}Si isotopes is measured over the temperature range from 4.7 K to 700 K (Wille et al. 2002). The uncertainty of the measurements of the relative difference $\Delta a/a$ is 5×10^{-7}. An anomalous behavior of the isotopic effect on the lattice parameter in silicon is observed: the magnitude of the relative difference in the lattice parameter first increases with temperature, reaches a maximum of $|\Delta a/a|$ at 75(10) K, and only then decreases, i.e., starts to behave in the normal way.

1.2.3 Length Standard for Atomic-Scale Measurements

Length measurements require easy and precisely reproducible length standards. What length standards could be used for precise atomic-scale measurements? How do they relate to the standard meter? As will be discussed in this section, Bragg backscattering of x rays is an important tool for solving these problems as well.

Definition and Realization of the Meter. The definition of the meter as a standard for length measurements evolved with our growing knowledge of the basic laws of physics. In 1799 a platinum bar made equal as closely as possible to the one ten-millionth part of the polar quadrant of the earth passing through Paris was defined as one meter and deposited as a reference in the French National Archive. Replaced in 1899 by an 'International Prototype Meter' in the form of an alloy bar of platinum (90%) and iridium (10%), it served as a length standard until 1960, when the wavelength of the orange line from a ^{86}Kr standard lamp took over its role. Since 1983, the speed of light c is made a fixed constant, and the meter has been defined as the distance traveled by light in 1/299 792 458 of a second. This definition opens the way to major improvements in the precision with which the meter can be realized.

An approved way of realization of the meter is by a direct measurement of the frequency f of light with sharp spectral distribution, and calculating the wavelength: $\lambda = c/f$. Distances are measured then in terms of this wavelength with the help of, e.g., Michelson or Fabry-Pérot interferometers with movable mirrors.

Such definition and realization of the meter becomes possible due to the ability to measure laser frequencies relative to the atomic clock using ^{133}Cs atoms, due to the narrowness of the laser radiation line, and due to the ability to stabilize its frequency to a few parts in 10^{11}. Measuring frequencies and wavelengths of optical photons with even higher accuracy to a few parts in 10^{13} and better is in progress (Stenger et al. 2001, Udem et al. 2002).

In our days, an iodine stabilized He-Ne laser is routinely used as a source of light with well defined frequency f_s and wavelength $\lambda_s = c/f_s$. The values f_s and λ_s are *reproducible* to better than 3 parts in 10^{11}. However, the 633 nm wavelength of the He-Ne laser is too long to be used to measure interatomic distances, which are typically 10000 times smaller. Therefore, a secondary shorter length standard is necessary for atomic scale measurements.

Silicon Lattice as an "Ångström Ruler". Hart (1968) suggested using the periodic lattice of high quality silicon single crystals as "An Ångström ruler". The length of the lattice period, lattice parameter, is about 5.4 Å. A combination of Michelson visible-light and triple-Laue-case (LLL) x-ray interferometers (Bonse and Hart 1965, Bonse and Graeff 1977) can be used to link the silicon lattice parameter with the wavelength of the He-Ne laser, and thus to measure the lattice parameter of high quality silicon single crystals in terms of the wavelength of the He-Ne laser. The silicon lattice parameter was measured in this way in a series of experiments of Deslattes and Henins (1973) and Becker et al. (1981) with steadily increasing accuracy. In the experiments of Basile et al. (1994) a relative uncertainty of 2.9×10^{-8} was reached. The following value of the lattice parameter of silicon $a_{\mathrm{Si}} = 5.43102088(16) \times 10^{-10}$ m at 22.500 °C in vacuum is accepted nowadays as a length standard for interatomic measurements (Mohr and Taylor 2000).

One can measure the lattice parameter of silicon perhaps with even higher accuracy, as was discussed by Bergamin et al. (1999). However, to reproduce its value with the same accuracy is not easy, as many parameters such as crystal perfection, chemical purity, absolute temperature and pressure have to be known and precisely controlled during the experiment. Crystal defects are traceable even in silicon crystals of the best available quality (Tuomi et al. 2001, Tuomi 2002). The content and spatial homogeneity of residual carbon and oxygen impurities can be controlled only to a certain limit (Windisch and Becker 1990). As a result, the silicon lattice parameter may vary by about $\pm 5 \times 10^{-8}$ (in relative units) at different places of the same silicon sample of highest quality (Bergamin et al. 1999, Basile et al. 2000). At present, this number is most likely a limit of reproducibility of the silicon lattice parameter. It marks the precision which can be obtained by using the silicon lattice parameter as an absolute scale length reference, i.e., a secondary length standard.

γ-Ray Wavelength Standard for Atomic Scales. An easily reproducible and more accurate length standard for atomic scales is therefore desirable. In fact a solution to this problem is known. It is related to the Mössbauer effect and Mössbauer radiation, widely known since 1958 to have an unprecedentedly narrow spectral line and both precise and stable wavelength (Mössbauer 1958, 1959).

The most famous is the Mössbauer radiation originating from the decay of the first excited state of ^{57}Fe nuclei. The energy of the excited state and the energy of the Mössbauer photons is $E_M \simeq 14.4$ keV. The radiation wavelength $\lambda_M \simeq 0.86 \times 10^{-10}$ m is a perfectly suited wavelength standard for atomic scales.

The natural spectral linewidth Γ of the ^{57}Fe Mössbauer radiation, determined by the lifetime of the excited state $\tau = 141$ ns via the uncertainty relation, is as small as 4.8×10^{-9} eV. Thus the relative uncertainty of the radiation energy is $\Gamma/E_M \simeq 3 \times 10^{-13}$ and the same relative uncertainty corresponds to the wavelength λ_M of the Mössbauer radiation. The coherence length of the radiation is about 30 m. A potential source of line broadening originates from interaction of the nuclei with the environment, e.g., the electrons of the atomic shell. These hyperfine interactions may deteriorate the spectral purity of the Mössbauer radiation by a factor of about 100 at most. Thus, even if nothing is known about the hyperfine interactions, the relative uncertainty and reproducibility is still $\simeq 10^{-11}$. But this is a worst case scenario. It is not a problem at all to preserve the natural width of the radiation and to benefit from the very low relative uncertainty of $\simeq 3 \times 10^{-13}$.

The temperature dependent shift of the photon energy, i.e., the variation of the line position in the Mössbauer experiment is even smaller than the spectral width. For example, the relative temperature instability of the Mössbauer transition energy caused by the so-called second-order Doppler shift is only $\simeq 10^{-15}$ K^{-1} at room temperature. At lower temperatures it is still smaller. Since a temperature stability of $10 - 1$ mK is easy to achieve, the stability of the "nuclear oscillator" delivering Mössbauer radiation by far surpasses even that of the cesium fountain atomic clock (Sortais et al. 2001) - the most precise metrological instrument nowadays[12].

This outstandingly small uncertainty is a gift of nature and makes the wavelength of the Mössbauer radiation an attractive wavelength standard.

To use such a wavelength standard, one needs easily available sources of Mössbauer photons of high brilliance. In the mid 1960's Bearden considered the possibility of introducing a γ-ray wavelength standard, attempting to measure the wavelength of 14.4 keV Mössbauer radiation of ^{57}Fe nuclei with a relative accuracy of 10^{-5} (Bearden 1965). His conclusion was: to make the Mössbauer standard experimentally feasible the brightness of the radioactive source would have to be increased by a factor of 100 beyond his 200 mCi activity (Bearden 1965, 1967).

Until recently this was an unsolvable problem. The brightness of radioactive sources is limited by self-absorption. This limitation can be circumvented if one uses synchrotron radiation as a source. Mössbauer photons are filtered from the synchrotron radiation spectrum by coherent nuclear resonant scattering, see Sect. 1.2.1, Fig. 1.17, and Gerdau and de Waard (1999/2000) for a review. Any ^{57}Fe-containing substance exposed to synchrotron radiation in the 14.4 keV spectral range becomes a strong source of Mössbauer photons preserving the collimation of the incident beam. Such a source is free of radioactive contaminations. Since the pilot experiment of Gerdau et al. (1985) in 1984 the rate of the 14.4 keV Mössbauer photons available has increased from 0.1 Hz to $\simeq 10^4 - 10^5$ Hz within a solid angle of 20×20 μrad^2 at modern third generation synchrotron radiation facilities. The synchrotron-based sources of Mössbauer photons are already $\simeq 10^6$ times brighter than the 200 mCi radioactive source used by Bearden and will become another three to six orders of magnitude brighter with forthcoming x-ray free-electron laser facilities (Materlik and Tschentscher 2001, Saldin et al. 2001). The availability of such sources has stimulated attempts to establish this new length standard (Shvyd'ko et al. 2000).

After realizing the potential of the Mössbauer wavelength standard, an important first step leading to practical use is the precise measurement of the absolute value of the wavelength.

There are at least two options. One possibility, is to use silicon as a transfer standard and to measure the wavelength λ_{M} of the Mössbauer radiation with respect to the silicon lattice parameter a_{Si}. The ratio $a_{\mathrm{Si}}/\lambda_{\mathrm{M}}$ was measured recently with a sub-ppm precision (Xiaowei et al. 2000, Shvyd'ko et al.

---

[12] The uncertainty of the NIST-F1 cesium fountain atomic clock developed in 1999 at the NIST laboratories in Boulder, Colorado (USA), is 1.7×10^{-15}. The "Mössbauer clock" has therefore a potential to take over in the future also the role of the time standard.

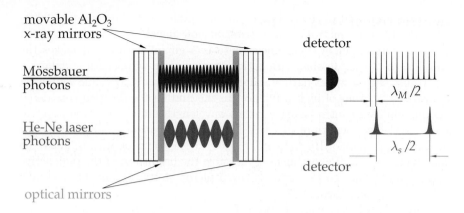

Fig. 1.24. Schematic of an experimental arrangement for direct measurement of the wavelength λ_M of Mössbauer radiation in terms of the wavelength of the He-Ne laser λ_s with the help of the combined Fabry-Pérot interferometer for He-Ne laser and Mössbauer radiation ($\lambda_\mathrm{s}/\lambda_\mathrm{M} \simeq 7 \times 10^3$). Sapphire ($\alpha$-Al$_2O_3$) single crystals are used as backscattering mirrors for Mössbauer radiation. Sapphire is transparent for visible light. Optical mirrors are fabricated, e.g., by thin film metalization of the sapphire x-ray mirrors.

2000)[13]. Backscattering of x rays from silicon single crystals was used for this purpose.

A more reliable and direct way is to measure the wavelength of the Mössbauer radiation in terms of the wavelength of the He-Ne laser. Figure 1.24 shows how this aim could be achieved with the help of a combined Fabry-Pérot interferometer for He-Ne laser- and Mössbauer radiation with movable backscattering mirrors. Sapphire (α-Al$_2$O$_3$) single crystals are used as backscattering mirrors for Mössbauer radiation. Sapphire is transparent for visible light. Optical mirrors can be fabricated, e.g., by thin film metalization of the sapphire x-ray mirrors. Unlike the visible spectral region where Fabry-Pérot interferometers are standard instruments, this is still a problem for x rays and Mössbauer radiation. Realization of the x-ray Fabry-Pérot interferometers requires backscattering mirrors of high reflectivity for x rays. The results of theoretical and experimental studies aimed at developing x-ray Fabry-Pérot resonators are presented in detail in Chap. 4 and Chap. 7, respectively.

---

[13] The wavelength, and thus the energy of the ^{57}Fe Mössbauer radiation was acknowledged to be known previously with a relative accuracy of only 10 ppm: $E_\mathrm{M} = 14413.00(15)$ eV (Firestone et al. 1996).

1.3 Historical Background

Sachs and Weerts (1930) were among the first, who have recognized the important role of x-ray Bragg backscattering in precision measurements. They have applied x-ray diffraction with scattering angles close to 180° for accurate determination of crystal lattice parameters and lattice strains. This technique was improved in the early 1970s by Bottom and Carvalho (1970), Sykora and Peisl (1970), Freund and Schneider (1972), and Okazaki and Kawaminami (1973), when x-ray backscattering instruments for measuring the relative changes of lattice parameters in single crystals with an accuracy approaching 10^{-6} (1 ppm) were built. Stetsko et al. (1988) have used exact backscattering geometry for precise determination of the crystal lattice parameters using the example of Ge. Tiny changes in the crystal structure caused by replacement of the constituent atoms with different isotopes is of fundamental interest for understanding properties of materials. Backscattering high-resolution monochromators have been applied as wavelength meters to study accurately the isotopic effect on the lattice parameter of silicon (Wille et al. 2002), and germanium (Hu et al. 2003). Absolute measurements of x-ray wavelengths and crystal lattice parameters with sub-ppm accuracy by using x-ray Bragg backscattering were carried out recently by Xiaowei et al. (2000), and Shvyd'ko et al. (2000, 2002).

Bragg backscattering from the viewpoint of the dynamical theory of diffraction in perfect crystals was first considered in the paper of Kohra and Matsushita (1972). Already in this publication two outstanding features of backscattering were pointed out: approaching the Bragg angle of $\pi/2$ the reflection width reaches its smallest value on the energy scale and its largest value on the angular scale. Backscattering thus promised to found a basis for x-ray optics with enhanced luminosity and high energy resolution. These predictions have stimulated further theoretical (Brümmer et al. 1979, Caticha and Caticha-Ellis 1982, Graeff and Materlik 1982, Caticha and Caticha-Ellis 1990a) and experimental (Graeff and Materlik 1982, Kushnir and Suvorov 1986, 1990) studies of backscattering. Though restricted to angles of incidence deviating from normal incidence to the reflecting atomic planes by $\delta\theta \geq 2$ mrad the experiments of Graeff and Materlik (1982) and of Kushnir and Suvorov (1986) have confirmed the two outstanding features of backscattering.

Bragg backscattering under grazing incidence conditions was studied both theoretically and experimentally by Stepanov et al. (1991). Along with the back-reflected beam the specularly reflected beam from the crystal surface is significant. A high sensitivity of the angular dependence to lattice spacing in a crystal surface layer was demonstrated both for the back-reflected and for the specularly reflected beams. Experimental measurements of specularly reflected and back-reflected intensities from (6 2 0) planes of a germanium crystal were carried out with CoK_{α_1} radiation.

Cusatis et al. (1996) were the first to observe x rays reflected exactly backwards from a Si(111) plate at an energy of 1.9 keV. The angular profiles in reflection and transmission were studied. Soon afterwards Shvyd'ko et al. (1998) have reported the observation of exact backscattering and comprehensive studies of its energy and angular dependence in sapphire (α-Al$_2$O$_3$) single crystals with well collimated and ideally monochromatic 14.4 keV Mössbauer radiation. High reflectivity at exact backscattering close to the theoretically predicted value was observed. A detailed comparison with the predictions of the dynamical theory of x-ray diffraction was made.

Steyerl and Steinhauser (1979) have pointed to the problem of multiple reflections at *exact* normal incidence Bragg diffraction. Multiple-beam Bragg diffraction occurs systematically in backscattering. For example, in crystals with cubic Bravais lattices (like silicon, germanium, diamond, etc.) all Bragg back-reflections with the exception of the two lowest oder reflections (1 1 1) and (2 2 0) are accompanied by additional Bragg reflections. The influence of the multiple-beam effects was first detected in the experiment of Giles and Cusatis (1991), in which the angular dependence of transmission at near-normal incidence to the (6 2 0) atomic planes in germanium was studied. Such an influence was also detected in the studies of the angular dependences of transmissivity and reflectivity at near-normal incidence to the (9 9 1) atomic planes in silicon by Kikuta et al. (1998). Shvyd'ko and Gerdau (1999), Sutter (2000), and Sutter et al. (2001) have studied the effects of multiple-beam diffraction on the angular and energy dependence of the reflectivity at near normal incidence to the atomic planes (0 4 12) in silicon crystals.

Multiple-beam diffraction effects in backscattering have also been investigated theoretically. The 4-beam diffraction case was briefly analyzed by Steyerl and Steinhauser (1979). Colella and Luccio (1984) have given a short analysis based on a general formalism developed by Colella (1974) for n-beam diffraction. Stepanov et al. (1991) have proposed the dynamical theory of 4-beam Bragg diffraction under grazing incidence conditions with the (0 2 6) Bragg back-reflection in germanium. Kohn et al. (1999) have considered the particular case of the 6-beam Bragg diffraction with the (1 9 9) Bragg back-reflection in silicon. Sutter (2000) has studied the 24-beam case with the (0 4 12) Bragg back-reflection in the same crystal.

To avoid multiple reflections, Shvyd'ko and Gerdau (1999) have suggested using crystals with lower crystal symmetry. For instance, sapphire (α-Al$_2$O$_3$), a crystal with a rhombohedral lattice, allows multiple-beam-free exact Bragg backscattering with high reflectivity for x rays. Crystals with lower symmetry are also favorable for Bragg backscattering optics from another point of view. Sapphire single crystals offer a greater diversity of possibilities compared to silicon crystals. The interplanar distances d_H and hence the Bragg energies $E_H = hc/d_H$ are much less degenerate in the rhombohedral lattice (e.g., α-Al$_2$O$_3$) than in the cubic lattice (e.g., Si). Sapphire single crystals allow exact backscattering at far more positions in a given photon energy interval, i.e., the linear density of back-reflections on the scale of photon energies is much higher than for silicon crystals.

The earliest proposals of inelastic x-ray scattering (IXS) experiments with meV and sub-meV energy resolution date back to the late 1970s and early 1980s (Steyerl and Steinhauser 1979, Moncton 1980, Dorner and Peisl 1983, Hastings et al. 1984). The feasibility of achieving meV energy resolution by using Bragg backscattering was demonstrated in the experiment of Graeff and Materlik (1982). The spectrometer with a single-bounce silicon crystal monochromator and a spherically bent silicon crystal analyzer operating in backscattering was proposed for IXS experiments by Moncton (1980), Dorner and Peisl (1983), and Dorner et al. (1986) - see Fig. 1.13. The first energy resolved observations of phonons with x rays were carried out by Burkel et al. (1987), see also Burkel (1991), at the synchrotron radiation facility Hamburger Synchrotronstrahlungslabor (HASYLAB) at DESY (Germany). An energy resolution of about 10 meV was achieved. The inelastic x-ray scattering technique was developed further with second generation instruments with an energy resolution between 1 and 3 meV at the European Synchrotron Radiation Facility (ESRF, France) by Sette et al. (1995), Masciovecchio et al. (1996b), Verbeni et al. (1996), and Sette et al. (1998); at the Advanced Photon Source (APS, USA) by Alp et al. (2000b) and Sinn et al. (2001); and at the Super Photon ring 8 GeV (SPring-8, Japan) by Baron et al. (2001b).

Spectroscopic techniques, which use nuclear resonant scattering of synchrotron radiation, have further stimulated the development of Bragg backscattering optics. The nuclear resonance scattering experiments require x-ray monochromators with meV and sub-meV energy-resolution operating at energies of low-lying nuclear resonances. High-resolution monochromators are needed to avoid overload in the x-ray detectors and timing electronics. More importantly, they are required for energy resolved measurements of phonon-assisted inelastic nuclear resonant scattering of synchrotron radiation.

Bragg energies in silicon, germanium and other high-quality semiconductor crystals of cubic lattice symmetry unfortunately do not match the nuclear resonance energies. It was proposed using crystals with lower crystal symmetry like sapphire for this purpose (Shvyd'ko and Gerdau 1999). One can always find a matching Bragg back-reflection in α-Al$_2$O$_3$ for any desired x-ray energy above 10 keV. The technique was demonstrated with a single-bounce sapphire backscattering monochromator for the 14.4 keV nuclear resonance of ^{57}Fe, for the 25.6 keV resonance of ^{161}Dy, and for some other nuclear resonances (Shvyd'ko et al. 1998, 2001, 2002). However, the lack of sapphire crystals with sufficient perfection, has not yet facilitated sub-meV resolution with sapphire backscattering monochromators.

Monochromators with several reflecting silicon crystals (multi-bounce monochromators) operating close to backscattering geometry have so far been the more popular solution. Faigel et al. (1987) have used two successively arranged channel-cut crystals in the $(+,+)$ setting[14]. Ishikawa et al. (1992) have proposed using a combination of an asymmetrically cut crystal, as an angular collimator reducing the beam divergence, and a backscattering crystal (with Bragg angle in the range from $\simeq 80°$ to $\simeq 89°$), as a high-resolution monochromator – see Fig. 1.14. This scheme, provides high energy resolution, reasonably good angular acceptance, and the possibility of matching the required energy of the nuclear resonance. First implemented by Toellner et al. (1992), Mooney et al. (1994), it is widely used at many synchrotron radiation facilities worldwide. Achieving energy resolution of a few meV became routine. This scheme was further perfected by Chumakov et al. (1996b), Toellner et al. (1997), Toellner (2000), and Baron et al. (2001a), to obtain ever improving energy resolution. Yabashi et al. (2001) and Toellner et al. (2001) have recently demonstrated multi-bounce monochromators with the relative energy resolution $\Delta E/E$ better than 10^{-8}. Such energy resolution is practically at the limit defined by the perfection of available silicon crystals.

A further improvement in the energy resolution can be achieved by using interference filters. Steyerl and Steinhauser (1979) have proposed a *Fabry-Perot-type interference filter for x rays* with the purpose of monochromatization of hard x rays to a bandwidth of less than 1 meV. Basically, they have proposed replacing the back reflection of the optical mirrors in the optical Fabry-Pérot interferometer by the Bragg back-reflection from parallel-sided silicon single crystals. Steyerl and Steinhauser (1979) have not only pioneered the idea of a Fabry-Pérot-type interference filter for x rays, they have also performed the first theoretical analysis, although restricted to *exact* normal incidence. Caticha and Caticha-Ellis (1990b) and Caticha et al. (1996) have extended the theoretical treatment to an arbitrary angle of incidence. A revised theory taking also into account possible imperfections of the x-ray Fabry-Pérot interferometer was given by Kohn et al. (2000), and Shvyd'ko (2002), see also Chap. 4 of this book.

Barbee and Underwood (1983), Lepetre et al. (1984), and Bartlett et al. (1985) have demonstrated a kind of soft x-ray Fabry-Pérot étalon. Tungsten-carbon multilayer mirrors *at grazing incidence* have been used. Because of low mirror reflectivity the observed resonances were not sharp. The performance of such devices has been analyzed by Guichet and Rasigni (1997).

Liss et al. (2000) have observed multiple reflections of x rays in backscattering from a system of two parallel silicon single crystals. The intensity of the successively reflected beams was, however, low, because of low reflectivity of silicon crystals in backscattering arising from accompanying Bragg reflections. Shvyd'ko et al. (2003a) have demonstrated that a system of two parallel sapphire (α-Al_2O_3) crystal mirrors in the backscattering geometry

---

[14] There are two ways of arranging two crystals with successive Bragg reflections. In the first setting, the second crystal deflects the beam in the same sense as the first. This is called $(+,+)$ configuration. In the second setting, the beam is deflected by the second crystal in the opposite sense from the first. This is called $(+,-)$ configuration.

produce a large number of intense overlapping beams, thus laying a foundation for future x-ray Fabry-Pérot interferometers. First direct evidence of the interference of the waves reflected from two separate sapphire crystals in backscattering was presented by Shvyd'ko et al. (2003b). An interferometer for hard x rays with two back-reflecting crystal mirrors - a prototype x-ray Fabry-Pérot interferometer - was built and demonstrated. A finesse of 15 and 0.8 μeV broad Fabry-Pérot transmission resonances were measured by the time response of the interferometer.

2. Dynamical Theory of X-Ray Diffraction (Focus on Backscattering)

2.1 Introduction

The dynamical theory of x-ray diffraction in perfect crystals is considered in the present chapter with an emphasis on the points crucial to the backscattering geometry. The consideration is not restricted to backscattering alone. On the contrary, a general approach is used for the whole range of incidence angles without division into "standard" and "backscattering" regimes.

The "standard" dynamical theory of two-beam x-ray diffraction, as presented in texts of von Laue (1931), Zachariasen (1945), von Laue (1960), Batterman and Cole (1964), Azaroff et al. (1974), Pinsker (1978), was extended to the backscattering geometry by Kohra and Matsushita (1972), Brümmer et al. (1979), Caticha and Caticha-Ellis (1982), and Graeff and Materlik (1982)[1]. Colella (1974) has considered a system of fundamental equations of the dynamical theory of x-ray Bragg diffraction for the general n-beam case including the extreme cases of grazing and normal incidence, and proposed a numerical procedure, by which the amplitudes of the diffracted waves can be determined. Particular cases of Bragg backscattering influenced by multiple-beam diffraction effects were addressed by Stepanov et al. (1991), Kohn et al. (1999), and Sutter (2000).

In Sect. 2.2 the fundamental equations of the dynamical theory are derived for a general multiple-beam diffraction case. The method of solving the multiple-beam diffraction problem including extreme cases of normal and grazing incidence is presented. In Sect. 2.4 the solution of the diffraction prob-

---

[1] In fact, there is no basic difference between the "standard" and the "backscattering" theory. The fundamental equations of the dynamical theory as derived by Ewald (1917) and von Laue (1931) are also valid in backscattering. However, the formal parameter of the theory - the so-called parameter of deviation from Bragg's condition denoted in the following as α - is used in the "standard" theory in an approximation, which fails for the backscattering geometry. Caticha and Caticha-Ellis (1982) have proposed using another approximation for the deviation parameter valid in backscattering. In this book we shall use an exact expression for α valid for any angle of incidence (Shvyd'ko et al. 1998, Shvyd'ko 2002). This general expression will allow us to formulate the results of the dynamical theory, e.g., the modified Bragg law, in a form valid for the whole range of glancing angles of incidence without division into "standard" and "backscattering" regimes.

lem in the two-beam diffraction case is discussed in detail, a regime where the incident wave excites only two waves inside the crystal: the forward-transmitted and the Bragg-reflected wave. Expressions are derived, which are used to further evaluate the angular, energy, time, and temperature dependence of the reflectivity and transmissivity of single crystal and multiple-crystal arrangements, as well as for the evaluation of the reflectivity and transmissivity of x-ray Fabry-Pérot resonators.

Much attention in the present discussion of x-ray diffraction in crystals is payed to establishing exact relations between the propagation direction of the incident wave and the reflected wave in the whole range of scattering angles including backscattering. The relations, which have been derived earlier in the existing versions of the dynamical theory, are extended here especially with respect to backscattering. The alteration in the propagation direction of the reflected wave with varying wavelength at a fixed direction of incidence - the effect of angular dispersion - is considered here in detail. The effect of angular dispersion underlying these relations is of great importance, in particular, for the theory of multiple-crystal diffraction addressed in Chap. 3, and for the practical purposes of designing high-resolution x-ray optics, such as x-ray monochromators and analyzers addressed in Chap. 5 and Chap. 6, respectively.

The two-beam-case Bragg reflection in backscattering is rather an exception than the rule. In most cases strong accompanying Bragg reflections can be excited simultaneously. The accompanying reflections can suppress the backscattered beam and thus reduce drastically the crystal reflectivity into the backscattering channel. Multiple-beam diffraction is unfavorable for many applications of Bragg backscattering in x-ray crystal optics. The kinematic treatment of the problem is given in Sect. 2.5. In particular, the systematics of multiple-beam diffraction cases appearing in crystals of different symmetry is addressed. In Sect. 2.6 the multiple-beam dynamical diffraction theory is used to analyze the influence of the accompanying reflections on the reflectivity into the backscattering channel. Four-beam diffraction with a back-reflection channel is analyzed in detail. Techniques of suppression of the accompanying reflections are addressed including the general n-beam diffraction case.

2.2 General Solution of the Diffraction Problem

We consider scattering of a plane monochromatic electromagnetic radiation wave

$$\mathcal{E}(\boldsymbol{r},t) = \mathcal{E}_\mathrm{i} \exp[\mathrm{i}(\boldsymbol{K}_0\boldsymbol{r} - \omega t)] \tag{2.1}$$

from a perfect crystal. The radiation frequency can be expressed here in the usual way as $\omega = E/\hbar$ in terms of the photon energy E. The magnitude of the wave vector in vacuum $\boldsymbol{K}_0$ is expressed via the wavelength λ and photon energy E as $|\boldsymbol{K}_0| = K = 2\pi/\lambda = E/\hbar c$. The incident wave is assumed to be linearly polarized. The crystal is assumed to be a plate of thickness d.

Our aim is to calculate the reflectivity and transmissivity of the crystal as a function of the radiation wavelength (or photon energy), incidence angle, polarization, crystal temperature, etc. For this, the radiation fields induced inside and outside the crystal by the incident wave, have to be calculated.

The incident wave excites a radiation field inside the crystal with electric vector $\boldsymbol{\mathcal{D}}(\boldsymbol{r},t) = \exp(-\mathrm{i}\omega t)\boldsymbol{\mathcal{D}}(\boldsymbol{r})$. The spatial part $\boldsymbol{\mathcal{D}}(\boldsymbol{r})$ is obtained as a solution of the wave equation

$$\left[-\nabla^2 - K^2\right] \boldsymbol{\mathcal{D}}(\boldsymbol{r}) = K^2 \chi(\boldsymbol{r}) \boldsymbol{\mathcal{D}}(\boldsymbol{r}), \tag{2.2}$$

which is derived from Maxwell's equations for a medium with electric susceptibility $\chi(\boldsymbol{r})$ (actually multiplied by 4π). In the derivation of (2.2) it is assumed that the electric field inside the crystal remains practically transverse[2].

2.2.1 Electric Susceptibility of Crystals

$\chi(\boldsymbol{r})$ is a continuous periodic function in space having the symmetry of the crystal lattice. It can be represented as a Fourier series:

$$\chi(\boldsymbol{r}) = \sum_H \chi_H \exp(\mathrm{i}\boldsymbol{H}\boldsymbol{r}). \tag{2.3}$$

Here $\boldsymbol{H}$ are the reciprocal lattice vectors of the crystal. The Fourier components of the susceptibility χ_0, χ_H, etc., are given by

$$\chi_H = -\frac{r_e F_H}{\pi V} \frac{h^2 c^2}{E^2} \equiv -\frac{r_e F_H}{\pi V} \lambda^2 \tag{2.4}$$

with

$$F_H = \sum_n f_n(\boldsymbol{H}) \exp\left[\mathrm{i}\boldsymbol{H}\boldsymbol{r}_n - W_n(\boldsymbol{H})\right] \tag{2.5}$$

as the structure factor of the crystal unit cell. The quantity

$$f_n(\boldsymbol{H}) = f_n^{(0)}(\boldsymbol{H}) + f_n'(E) + \mathrm{i} f_n''(E) \tag{2.6}$$

---

[2] In many texts, discussing the dynamical theory, the *electric displacement* field is calculated inside the crystal and not the *electric* field. This choice is dictated by the purely transverse character of the electric displacement field inside the crystal. However, since the electric field inside the crystal is practically transverse, which is justified by the smallness of χ, the distinction between the electric field and the electric displacement field representation inside the crystal in most cases is not essential. This problem is discussed in detail, e.g., by Pinsker (1978). In the following the electric field inside the crystal is denoted by $\mathcal{D}$, while the electric field in vacuum is denoted by $\mathcal{E}$.

is the atomic scattering amplitude, which is the sum of the atomic form factor $f_\mathrm{n}^{(0)}(\boldsymbol{H})$ and the anomalous scattering corrections $f_\mathrm{n}'(E)$ and $f_\mathrm{n}''(E)$ of an atom located in the unit cell at a point with radius vector $\boldsymbol{r}_\mathrm{n}$. The atomic form factor is the Fourier transform of the spatial electron density distribution within the atom. The corrections are energy (wavelength) dependent. Their values can be obtained, e.g., from the library of anomalous scattering factors computed using relativistic Hartree-Fock-Slater wavefunctions (Kissel and Pratt 1990, Kissel et al. 1995).

The reciprocal lattice vectors of a crystal are defined as $\boldsymbol{H} = h\boldsymbol{b}_1 + k\boldsymbol{b}_2 + l\boldsymbol{b}_3$, where h, k, and l are integer variables, the Miller indices, and $\boldsymbol{b}_1$, $\boldsymbol{b}_2$, and $\boldsymbol{b}_3$ is a set of three mutually non-coplanar vectors, called the basis vectors for the reciprocal crystal lattice. For each reciprocal vector $\boldsymbol{H}$ there is a set of atomic planes normal to $\boldsymbol{H}$ denoted as (hkl). The interplanar spacing between the atomic planes (hkl) is

$$d_H = 2\pi/H, \qquad H = |\boldsymbol{H}|. \tag{2.7}$$

The value $\exp(-W_\mathrm{n})$ in (2.5) is known as the square root of the Debye-Waller factor, cf., Ziman (1969), Ashcroft and Mermin (1976). The Debye-Waller factor measures thermal vibrations of the atoms. It is defined as

$$g_\mathrm{n}(\boldsymbol{H}) = \exp\left[-2W_\mathrm{n}(\boldsymbol{H})\right]$$

$$W_\mathrm{n} = \frac{B_\mathrm{n}(\boldsymbol{H})}{4\, d_H^2}, \qquad B_\mathrm{n}(\boldsymbol{H}) = 8\pi^2 \langle u_H^2 \rangle_\mathrm{n}, \tag{2.8}$$

where $\langle u_H^2 \rangle_\mathrm{n}$ is the mean square displacement of the atom n from the equilibrium position in the unit cell projected on the direction of $\boldsymbol{H}$.

The variable V in (2.4) is the unit cell volume and $r_\mathrm{e} = 2.81794 \times 10^{-5}$ Å is the classical electron radius. For hard x rays ($E \geq 5$ keV) the coefficients χ_H are negative with values typically 10^{-5} and less.

The classical derivation of (2.4)-(2.6) was given, e.g., by Zachariasen (1945), James (1950), von Laue (1960), Pinsker (1978), and others. The quantum mechanical derivation can be found, e.g., in the book of von Laue (1960). The quantum mechanical derivation of the atomic scattering amplitudes and structure factors with a very thorough account of the role of lattice vibrations was given by Afanas'ev and Kagan (1967).

The Fourier components of the susceptibility are complex values. In the following the real parts are denoted with primes as χ_0', χ_H', etc., and the imaginary parts - with double primes as χ_0'', χ_H'', etc. The same rule is applied for other complex values.

2.2.2 Fundamental Equations

The periodicity of $\chi(\boldsymbol{r})$ has far-reaching implications. It is due to this property that x-ray crystal optics is so different from Fresnel optics. Since $\chi(\boldsymbol{r})$ is periodic, the solution of equation (2.2) is a Bloch wave (see, e.g., Ziman (1969), Ashcroft and Mermin (1976))

$$\mathcal{D}(\boldsymbol{r}) = e^{i\boldsymbol{k}_0 \boldsymbol{r}} \sum_H \boldsymbol{D}_H e^{i\boldsymbol{H}\boldsymbol{r}} = \sum_H \boldsymbol{D}_H e^{i\boldsymbol{k}_H \boldsymbol{r}}, \qquad (2.9)$$

composed of an infinite number of plane waves with wave vectors

$$\boldsymbol{k}_H = \boldsymbol{k}_0 + \boldsymbol{H}. \qquad (2.10)$$

The in-crystal wave vector $\boldsymbol{k}_0$ is associated with the vacuum wave vector $\boldsymbol{K}_0$ of the incident wave. Their relation will be discussed soon. The plane wave component in the crystal with the wave vector $\boldsymbol{k}_0$ will be referred to as the forward scattered (or transmitted) wave.

Substituting (2.9) into (2.2) one obtains after transformations a system of linear algebraic equations for the vector amplitudes $\boldsymbol{D}_H$ of the plane wave components:

$$\frac{k_H^2 - K^2}{K^2} \boldsymbol{D}_H = \sum_{H'} \chi_{H-H'} \boldsymbol{D}_{H'}. \qquad (2.11)$$

They are called the fundamental equations of the dynamical theory. They have been originally introduced by Ewald (1917). The reciprocal vectors $\boldsymbol{H}$ and $\boldsymbol{H}'$ in (2.11) take all possible discrete values including zero. The system of fundamental equations has generally speaking a very large number of equations. To determine the radiation field inside the crystal one has to solve the system of fundamental equations for the unknowns $\boldsymbol{D}_H$ and $\boldsymbol{k}_0$.

2.2.3 Ewald Sphere

One can recast (2.11) as

$$\boldsymbol{D}_H = \frac{K^2}{k_H^2 - K^2(1+\chi_0)} \sum_{H' \neq H} \chi_{H-H'} \boldsymbol{D}_{H'}. \qquad (2.12)$$

As was already mentioned, the coefficients $\chi_{H-H'}$ are very small. Therefore, only those components $\boldsymbol{D}_H$ are significant for which also $|k_H^2 - K^2(1+\chi_0)|$ is small, i.e., for which

$$k_H \approx K \left|1 + \frac{\chi_0}{2}\right|. \qquad (2.13)$$

The latter is often referred to as the excitation condition. Among other things, it states that only those plane wave components are excited, whose wave vectors have approximately the magnitude $K|1+\chi_0/2|$. If we draw the end point of $\boldsymbol{k}_0$ at the reciprocal space origin 0, then the origin of $\boldsymbol{k}_0$ is at a

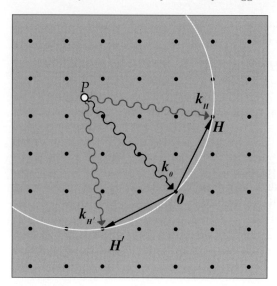

Fig. 2.1. Ewald sphere in the reciprocal space of the crystal. 0 is the reciprocal space origin. H, H' are reciprocal lattice vectors. P is the tie point - the common origin of the wave vectors $\boldsymbol{k}_0$, $\boldsymbol{k}_H$, $\boldsymbol{k}_{H'}$.

point P - often referred to as the tie point - Fig. 2.1. Within the approximation given by the excitation condition, the origins of the wave vectors $\boldsymbol{k}_H$ of the excited plane wave components being drawn also from the point P end on a sphere of radius $K|1 + \chi_0/2|$ - referred to as the Ewald sphere. For (2.10) to be fulfilled, the end point of $\boldsymbol{k}_H$ should coincide with the end point of the reciprocal vector $\boldsymbol{H}$ - Fig. 2.1. The same is applied to the component with wave vector $\boldsymbol{k}_{H'}$, etc. This implies that only those plane waves can be excited in the crystal whose wave vectors $\boldsymbol{k}_H$ and of course $\boldsymbol{H}$ end on the Ewald sphere. To be specific, in the following it is assumed that n plane wave components with different wave vectors are excited.

As follows from (2.11), the coefficients $\chi_{H-H'}$ are proportional to the probability amplitude of scattering the plane wave with wave vector $\boldsymbol{k}_{H'}$ into the plane wave with wave vector $\boldsymbol{k}_H$ with momentum transfer (scattering vector) $\boldsymbol{H} - \boldsymbol{H'}$. Scattering with momentum transfer equal to a reciprocal lattice vector is called Bragg scattering. Thus, (2.11) is a system of equations for the plane wave amplitudes $\boldsymbol{D}_H$ mutually coupled by multiple Bragg scattering in the crystal.

Bragg scattering with momentum transfer $\boldsymbol{H}$ of the wave with wave vector $\boldsymbol{k}_0$ can also be interpreted as reflection from the set of atomic planes (hkl) normal to $\boldsymbol{H}$ - (2.10). Because of the periodicity of the reflecting atomic planes this scattering process can be seen as diffraction from these planes.

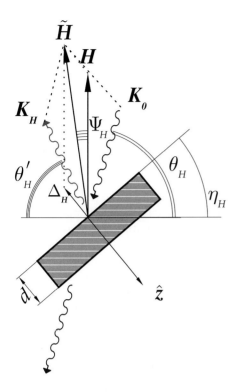

Fig. 2.2. Scattering geometry in direct space. Particular case is shown, in which the wave vector of the incident and reflected waves, K_0 and $K_H = K_0 + \tilde{H}$, respectively, are in the dispersion plane built of the diffraction vector H and the inward normal $\hat{z}$ to the crystal surface. White lines display the reflecting atomic planes normal to H.

$\tilde{H} = H + \Delta_H$ is the total momentum transfer. $\Delta_H = \Delta_H \hat{z}$ is the additional momentum transfer due to scattering from the vacuum-crystal interface. Δ_H is shown compared to H greatly exaggerated for clarity, resulting also in the exaggerated difference between the angles of incidence θ_H and reflection θ'_H. Other notations are explained in text.

Therefore it is often also called Bragg diffraction, and the momentum transfer H – as the diffraction vector.

2.2.4 Vacuum and in-Crystal Wave Vectors

Momentum Conservation. According to (2.9), the radiation field inside the crystal is a set of plane waves with wave vectors k_H. Their magnitudes are about $K|1 + \chi_0/2|$, i.e., they slightly differ from the magnitude K of the vacuum wave vector K_0 of the incident wave. Upon leaving the crystal the wave with wave vector k_H becomes an externally propagating wave with wave vector K_H - Fig. 2.2. The condition of continuity of the tangential component of the electric field at the vacuum-crystal interface (Jackson 1967) requires the continuity of the tangential component of the wave vectors:

$$k_0 = K_0 + \varkappa \hat{z}, \qquad k_H = K_H + \varkappa_H \hat{z}. \qquad (2.14)$$

Here $\hat{z}$ is the unit vector of an inward normal to the front crystal surface. The in-crystal and vacuum wave vectors may differ only by a component along the surface normal[3]. In the sense of the excitation condition (2.13), both $\varkappa$, and $\varkappa_H$ are very small, about $\simeq K|\chi_0|/2$. Equations (2.14) describe refraction of the incident and reflected waves at the crystal-vacuum interface. They can be interpreted also as the expression of momentum conservation in scattering from the crystal-vacuum interface.

By using (2.14), and (2.10), the wave vector of the diffracted wave in vacuum can be presented as

$$\boldsymbol{K}_H = \boldsymbol{K}_0 + \tilde{\boldsymbol{H}}$$
$$\tilde{\boldsymbol{H}} = \boldsymbol{H} + \boldsymbol{\Delta}_H, \qquad \boldsymbol{\Delta}_H = \Delta_H \hat{\boldsymbol{z}}, \qquad \Delta_H = \varkappa - \varkappa_H. \tag{2.15}$$

The relations (2.15) show that the momentum transfer for the radiation waves in vacuum is $\tilde{\boldsymbol{H}}$ and not $\boldsymbol{H}$ as for the radiation waves in the crystal - (2.10). The additional momentum transfer $\boldsymbol{\Delta}_H$ parallel to $\hat{\boldsymbol{z}}$ appears due to scattering from the vacuum-crystal interface, as shown in Fig. 2.2. If $\boldsymbol{H} \nparallel \hat{\boldsymbol{z}}$ then also $\boldsymbol{H} \nparallel \tilde{\boldsymbol{H}}$. From the viewpoint of vacuum waves this implies that the crystal reflects from virtual planes normal to $\tilde{\boldsymbol{H}}$, rather than from the real atomic planes normal to $\boldsymbol{H}$. The total momentum transfer is $\tilde{\boldsymbol{H}}$.

The angle Ψ_H between the virtual and the real atomic planes (the angle between $\tilde{\boldsymbol{H}}$ and $\boldsymbol{H}$) - Fig. 2.2 - is small. It will be calculated in the following section.

Energy Conservation. The wave vector corrections $\varkappa$ and $\varkappa_H$ are not independent. Since the vacuum is isotropic and Bragg scattering is elastic, energy conservation requires the magnitudes of the wave vectors of the incident and scattered radiation to be equal:

$$|\boldsymbol{K}_H| = |\boldsymbol{K}_0| = K. \tag{2.16}$$

This condition together with (2.15) leads to the following expression for the momentum transfer at the vacuum-crystal interface

$$\Delta_H = K\left(-\gamma_H \pm \sqrt{\gamma_H^2 - \alpha_H}\right), \tag{2.17}$$

where

$$\gamma_H = \frac{(\boldsymbol{K}_0 + \boldsymbol{H})\hat{\boldsymbol{z}}}{K}, \qquad \gamma_0 = \frac{\boldsymbol{K}_0 \hat{\boldsymbol{z}}}{K}, \tag{2.18}$$

are the direction cosines of the vector $\boldsymbol{K}_0 + \boldsymbol{H}$ and $\boldsymbol{K}_0$, respectively, with respect to the inward crystal surface normal[4], and

---

[3] We omit the index "0" in $\varkappa$ in order not to clutter up the following mathematical expressions.

[4] Equivalently, γ_0 equals the sine of an angle between $\boldsymbol{K}_0$ and the crystal surface, and γ_H equals the sine of an angle between $\boldsymbol{K}_0 + \boldsymbol{H}$ and the crystal surface. In this respect γ_0 determines the direction of propagation of the incident wave in vacuum with respect to the crystal surface.

$$\alpha_H = \frac{2\boldsymbol{K}_0\boldsymbol{H} + \boldsymbol{H}^2}{K^2}. \tag{2.19}$$

The dimensionless parameter α_H is a function of the magnitude and the direction of $\boldsymbol{K}_0$ relative to $\boldsymbol{H}$, i.e., it defines the scattering conditions. Due to reasons, which will be clarified in Sect. 2.4.4, it is termed "parameter of deviation from Bragg's condition" or simply "deviation parameter". The deviation parameters belonging to different reciprocal vectors are generally speaking not independent. The relation between them can be easily established. The deviation parameters play an important role in the subsequent discussion. The reflectivity and transmissivity of the crystal can be expressed as a universal function of these formal parameters, which in turn depend on the physical parameters of the scattering problem under discussion, such as the radiation wavelength, angle of incidence, crystal temperature, etc. The relation between the deviation parameter, a formal parameter of the problem, and physical parameters will be addressed in more detail in Sect. 2.4.4.

By using (2.14), (2.18), and (2.19) one may write (2.12) in the following form:

$$\boldsymbol{D}_H = \frac{1}{\alpha_H - \chi_0 + 2\gamma_H \varkappa/K + \varkappa^2/K^2} \sum_{H' \neq H} \chi_{H-H'} \boldsymbol{D}_{H'}. \tag{2.20}$$

With this, one may readily understand that a substantial excitation of the plane wave component with wave vector $\boldsymbol{k}_H$ only takes place if the correspondent deviation parameter α_H is small: $\alpha_H \approx \chi_0$. On the contrary, if the parameter is large ($|\alpha_H| \gg |\chi_0|$) the amplitude $\boldsymbol{D}_H$ is small and thus the plane wave component with wave vector $\boldsymbol{k}_H$ can be neglected.

For small α_H, and for $|\gamma_H|^2 \gg |\chi_0|$, i.e., no grazing emergence of the reflected wave takes place, the expression (2.17) can be closely approximated by a Taylor expansion in α_H:

$$\Delta_H = \Delta_H^{(1)} + \Delta_H^{(2)} + \ldots,$$

$$\Delta_H^{(1)} = -K\frac{\alpha_H}{2\gamma_H}, \qquad \Delta_H^{(2)} = -K\frac{\alpha_H^2}{8\gamma_H^3}, \qquad \text{etc.} \tag{2.21}$$

Very often, although not always, it is sufficient to consider only the leading term $\Delta_H^{(1)}$. In the above derivation of (2.21) the sign before the square root in (2.17) is chosen in such a way that the difference $(\varkappa_H - \varkappa)/K$, and thus also $\varkappa_H/K$ is small, as is required by the excitation condition (2.13).

2.2.5 Angular Dispersion

The wave vector correction $\varkappa$ and $\varkappa_H$, as well as the additional momentum transfer Δ_H due to scattering from the vacuum-crystal interface are material and wavelength dependent. For this reason also, the direction of the total momentum transfer $\tilde{\boldsymbol{H}}$, given by (2.15) and graphically shown in Fig. 2.2, is wavelength dependent. That means, the vacuum wave vector $\boldsymbol{K}_H$ of the reflection wave can change its direction with radiation wavelength $\lambda = 2\pi/K$ even if the direction of $\boldsymbol{K}_0$, the wave vector of the incident wave, is fixed. By using (2.15), one finds that the change of the direction of the unit vector $\boldsymbol{u}_H = \boldsymbol{K}_H/K$ in the most general case is

$$\delta \boldsymbol{u}_H = -\frac{\delta K}{K}\frac{\boldsymbol{H}}{K} + \frac{\delta \Delta_H}{K}\hat{\boldsymbol{z}}. \tag{2.22}$$

Here $\delta \Delta_H$ is the variation of Δ_H with wavelength. The alteration of the wave vector direction with radiation wavelength will be referred to as the *angular dispersion* of the reflected wave. As (2.22) demonstrates, angular dispersion may take place only in the plane composed of the diffraction vector $\boldsymbol{H}$ and inward normal $\hat{\boldsymbol{z}}$. The vector of the total momentum transfer $\tilde{\boldsymbol{H}}$ is clearly also lying in this plane. This plane will be termed the *dispersion plane* associated with diffraction vector $\boldsymbol{H}$. It is defined unequivocally only if $\boldsymbol{H} \nparallel \hat{\boldsymbol{z}}$.

2.2.6 Scattering Geometries

Classification. Examples of scattering geometries are shown in Fig. 2.3. Particular cases are presented in which the vacuum wave vectors $\boldsymbol{K}_0$ and $\boldsymbol{K}_H$ of the incident and the reflected waves, respectively, are in the dispersion plane built of vectors $\boldsymbol{H}$ and $\hat{\boldsymbol{z}}$. Figure 2.4 shows the general case, when the wave vectors are not lying in this plane.

If $\boldsymbol{K}_H$ is directed towards the outside of the crystal, so that $\boldsymbol{K}_0$ and $\boldsymbol{K}_H$ are on the same side of the crystal entrance surface, such scattering geometry is called Bragg geometry, and is shown schematically in Fig. 2.3a, and Fig. 2.4. The reflected wave with wave vector $\boldsymbol{K}_H$ is then called the Bragg-case wave.

If $\boldsymbol{K}_H$ is directed towards the inside of the crystal, so that $\boldsymbol{K}_H$ is on the other side of the crystal entrance surface with reference to $\boldsymbol{K}_0$, this is called Laue scattering geometry, as shown in Fig. 2.3b. The reflected wave is called the Laue-case wave.

The angle η_H between crystal surface normal $\hat{\boldsymbol{z}}$ and $-\boldsymbol{H}$ is nonzero in the general case. It is called the asymmetry angle. The same angle lies between the entrance crystal surface and the diffracting atomic planes perpendicular to $\boldsymbol{H}$. The asymmetry angle may change from 0 to π as is seen, e.g., in Fig. 2.3. One commonly speaks of the *symmetric Bragg* scattering geometry if $\boldsymbol{K}_0$ and $\boldsymbol{K}_H$ are on the same side of the crystal entrance surface, and the reflecting atomic planes are parallel to the crystal surface ($\eta_H = 0$). If $\boldsymbol{K}_H$ is on the other side of the crystal entrance surface with reference to $\boldsymbol{K}_0$, and the reflecting atomic planes are perpendicular to the crystal surface ($\eta_H = \pi/2$), one terms this as *symmetric Laue* scattering geometry. One speaks of asymmetric scattering geometry in all other cases.

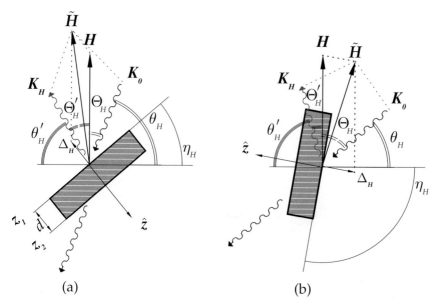

Fig. 2.3. Scattering geometries in direct space: (a) Bragg scattering geometry, (b) Laue scattering geometry. Particular cases are shown, in which the wave vector of the incident and reflected waves, $\boldsymbol{K}_0$ and $\boldsymbol{K}_H = \boldsymbol{K}_0 + \tilde{\boldsymbol{H}}$, respectively, are in the dispersion plane built of the diffraction vector $\boldsymbol{H}$ and the inward normal $\hat{\boldsymbol{z}}$ to the crystal surface. White lines display the reflecting atomic planes normal to $\boldsymbol{H}$. For other notations see text.

Reference System. In order to quantitatively describe the scattering geometry, we define a coordinate system $\{x', y', z'\}$ shown in Fig. 2.4 with the z'-axis along $\boldsymbol{H}$, the x'-axis in the plane composed of vectors $\boldsymbol{H}$ and $\hat{\boldsymbol{z}}$, and the y'-axis - perpendicular to the same plane: $\hat{\boldsymbol{z}}' = \boldsymbol{H}/H$, $\hat{\boldsymbol{y}}' = (\boldsymbol{H} \times \hat{\boldsymbol{z}})/|\boldsymbol{H} \times \hat{\boldsymbol{z}}|$, $\hat{\boldsymbol{x}}' = \hat{\boldsymbol{y}}' \times \boldsymbol{H}/H$. The magnitude of the vector product $|\boldsymbol{H} \times \hat{\boldsymbol{z}}| = H \sin \eta_H$. If $\eta_H \neq 0$ the unit basis vectors of the reference system are given by

$$\hat{\boldsymbol{z}}' = \frac{\boldsymbol{H}}{H}, \qquad \hat{\boldsymbol{y}}' = \frac{\boldsymbol{H} \times \hat{\boldsymbol{z}}}{H \sin \eta_H}, \qquad \hat{\boldsymbol{x}}' = \frac{1}{\sin \eta_H}\left(\hat{\boldsymbol{z}} + \frac{\boldsymbol{H}}{H} \cos \eta_H\right). \qquad (2.23)$$

If $\eta_H = 0$, then $\hat{\boldsymbol{x}}'$ and $\hat{\boldsymbol{y}}'$ are defined as two arbitrary mutually orthogonal vectors normal to $\boldsymbol{H}$.

In the $\{x', y', z'\}$ coordinate system the diffraction vector $\boldsymbol{H}$, and the surface normal $\hat{\boldsymbol{z}}$ can be represented as

$$\begin{aligned} \boldsymbol{H} &= H(0,\, 0,\, 1), \\ \hat{\boldsymbol{z}} &= (\sin \eta_H,\, 0,\, -\cos \eta_H). \end{aligned} \qquad (2.24)$$

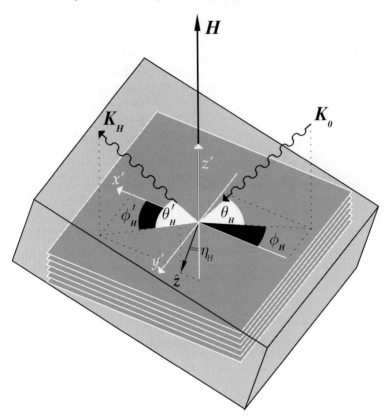

Fig. 2.4. Vacuum wave vectors $\boldsymbol{K}_0$ and $\boldsymbol{K}_H$ of the incident and reflected waves with respect to the reflecting atomic planes (dark grey) in Bragg scattering geometry. The crystal contour is shown with the solid lines. The general case is presented with $\boldsymbol{K}_0$ and $\boldsymbol{K}_H$ *not* lying in the dispersion plane (x', z') built of the diffraction vector $\boldsymbol{H}$ and the inward normal $\hat{\boldsymbol{z}}$ to the crystal surface, as opposed to the particular case of Fig. 2.3, in which $\phi_H = \phi'_H = 0$. For other notations see the text.

Direction of $\boldsymbol{K}_0$ and $\boldsymbol{K}_H$. The direction of the vacuum wave vector $\boldsymbol{K}_0$ or $\boldsymbol{K}_H$ we shall define by the set of two angles: the glancing angle, and azimuthal angle.

The glancing angle θ_H for the incident wave with wave vector $\boldsymbol{K}_0$ - the glancing angle of incidence - is defined as the smallest angle between $\boldsymbol{K}_0$ and the reflecting atomic planes normal to $\boldsymbol{H}$, as shown in Fig. 2.4 and 2.3. The glancing angle of reflection θ'_H for the reflected wave with wave vector $\boldsymbol{K}_H$ is defined similarly. The glancing angles may vary in the range from 0 to $\pi/2$. Also the angle $\Theta_H = \pi/2 - \theta_H$ shown in Fig. 2.3 will be used in the following, which is termed the incidence angle. It is a measure of deviation from normal incidence on the reflecting atomic planes. Similarly, Θ'_H is termed the reflection angle.

The azimuthal angle of incidence ϕ_H for the incident wave with wave vector $\boldsymbol{K}_0$ is defined as the angle between the plane composed of the vectors $(\boldsymbol{K}_0, \boldsymbol{H})$ and the dispersion plane (x', z'), as shown in Fig. 2.4. The azimuthal angle of reflection ϕ'_H for the reflected wave with wave vector $\boldsymbol{K}_H$ is defined in a similar way. The azimuthal angles may vary in the range from 0 to 2π. Figure 2.3 shows particular cases of scattering in the dispersion plane with $\phi_H = \phi'_H = 0$.

By using the above introduced definitions of the glancing and azimuthal angles, $\boldsymbol{K}_0$, and $\boldsymbol{K}_H$ can be presented in the $\{x', y', z'\}$ coordinate system as

$$\begin{aligned} \boldsymbol{K}_0 &= K(\cos\theta_H \cos\phi_H,\ \cos\theta_H \sin\phi_H,\ -\sin\theta_H), \\ \boldsymbol{K}_H &= K(\cos\theta'_H \cos\phi'_H,\ \cos\theta'_H \sin\phi'_H,\ \sin\theta'_H). \end{aligned} \quad (2.25)$$

Formal Parameters of the Scattering Geometry. Equations (2.24) and (2.25) can be used to express the direction cosines γ_0, γ_H, and the deviation parameter α_H, defined by (2.19), and (2.18), respectively, in terms of the glancing and azimuthal angles of incidence:

$$\gamma_0 = \cos\theta_H \sin\eta_H \cos\phi_H + \sin\theta_H \cos\eta_H, \quad (2.26)$$

$$\begin{aligned} \gamma_H &= \gamma_0 - \frac{H}{K}\cos\eta_H \\ &= \cos\theta_H \sin\eta_H \cos\phi_H - \sin\theta_H \cos\eta_H - \alpha_H \frac{K}{H}\cos\eta_H, \end{aligned} \quad (2.27)$$

$$\alpha_H = \frac{H}{K}\left(\frac{H}{K} - 2\sin\theta_H\right). \quad (2.28)$$

The crystal structure and symmetry relate the reciprocal lattice vectors $\boldsymbol{H}$, $\boldsymbol{H}'$, etc., to each other. Therefore, the pair of angles of incidence (θ_H, ϕ_H) defined with respect to $\boldsymbol{H}$ can be strictly related to another pair $(\theta_{H'}, \phi_{H'})$, defined with respect to the reciprocal vector $\boldsymbol{H}'$. For this reason, independent of the number of reciprocal vectors which contribute to the diffraction process, it is sufficient to define only one pair of angles of incidence with respect to a selected diffraction (reciprocal) vector.

At this point, we introduce also another formal parameter that characterizes the scattering geometry, the "asymmetry factor"

$$b_H = \frac{\gamma_0}{\gamma_H}. \quad (2.29)$$

It characterizes how far from the symmetric case the wave vector $\boldsymbol{K}_0$ of the incident wave and the wave vector $\boldsymbol{K}_0 + \boldsymbol{H}$ deviate with respect to the crystal surface normal $\hat{\boldsymbol{z}}$. If one neglects the small difference between $\boldsymbol{K}_0 + \boldsymbol{H}$ and $\boldsymbol{K}_H$, this statement also applies to $\boldsymbol{K}_0$ and $\boldsymbol{K}_H$.

Evidently, in the Bragg scattering geometry the parameters γ_H and b_H are negative. As follows from (2.26)-(2.27), and (2.29) in symmetric Bragg scattering ($\eta_H = 0$) the asymmetry factor $b_H = -[1 - \alpha_H K/(H\sin\theta_H)]$. Since, α_H is assumed to be very small, the asymmetry factor in the symmetric Bragg scattering geometry is commonly approximated with good accuracy as $b_H = -1$.

In the Laue scattering geometry the parameters γ_H and b_H are positive. If the scattering geometry is symmetric ($\eta_H = \pi/2$), then $\gamma_0 = \gamma_H$, see (2.26)-(2.27), and the asymmetry factor $b_H = 1$.

The Angle Between the Virtual Reflecting and Atomic Planes. By using (2.15), and (2.24), one may calculate the angle Ψ_H between the virtual reflecting and the atomic planes (the angle between $\tilde{\boldsymbol{H}}$ and $\boldsymbol{H}$), which was introduced in Sect. 2.2.4 and shown in Fig. 2.2: $\Psi_H = \sin\eta_H |\Delta_H|/H$. Applying (2.21) and leaving only the leading term $\Delta_H^{(1)}$, yields

$$\Psi_H = \frac{K|\alpha_H|}{2H|\gamma_H|}\sin\eta_H. \tag{2.30}$$

One should note that Ψ_H is not a universal parameter for a given Bragg reflection with diffraction vector $\boldsymbol{H}$, and asymmetry angle η_H. It depends on the photon energy (radiation wavelength) and angles of incidence, as a consequence of angular dispersion.

Reversibility of Bragg Scattering. Elastic scattering processes are reversible. In our case this implies that, if the incident wave with the wave vector $\boldsymbol{K}_0$ is reflected into the wave with wave vector $\boldsymbol{K}_H$, then the reverse situation can also take place: the incident wave with wave vector $-\boldsymbol{K}_H$ is reflected into the wave with wave vector $-\boldsymbol{K}_0$. Because of the rather complicated dependence of Δ_H on $\boldsymbol{K}_0$ given by (2.17)-(2.19), reversibility of (2.15) is, however, not evident. Reversibility requires that the momentum transfer Δ_H due to scattering from the vacuum-crystal interface appearing in (2.15) obeys the relation

$$\Delta_H(\boldsymbol{H},\hat{\boldsymbol{z}},\boldsymbol{K}_0) = \Delta_H(\boldsymbol{H},\hat{\boldsymbol{z}},-\boldsymbol{K}_H). \tag{2.31}$$

The above relation and thus the reversibility of Bragg scattering can be proved explicitly using definitions (2.17)-(2.19). The reversibility of Bragg scattering can be illustrated in the following example.

The vacuum wave vectors $-\boldsymbol{K}_0$ and $\boldsymbol{K}_H$ compose the same angle $\tilde{\Theta}$ with the total momentum transfer vector $\tilde{\boldsymbol{H}}$. This fact is illustrated schematically in Fig. 2.5. In contrast, $-\boldsymbol{K}_0$ and $\boldsymbol{K}_H$ compose different glancing angles of incidence θ_H and reflection θ'_H with respect to the reflecting atomic planes. In the particular case of $\phi_H = 0$ shown in Fig. 2.5(a_0) they are related to each other as $\theta'_H = \theta_H - 2\Psi_H$. Let us now reverse the scattering process as shown in Fig. 2.5(a_π). Now the azimuthal angle $\phi_H = \pi$ and the glancing angles of incidence are related as $\theta'_H = \theta_H + 2\Psi_H$. At this point it should be noted that due to reversibility $\Psi_H(a_0) = \Psi_H(a_\pi)$, and thus $\tilde{\Theta}(a_0) = \tilde{\Theta}(a_\pi)$. As a result $\theta_H(a_0) = \theta'_H(a_\pi)$, and $\theta'_H(a_0) = \theta_H(a_\pi)$.

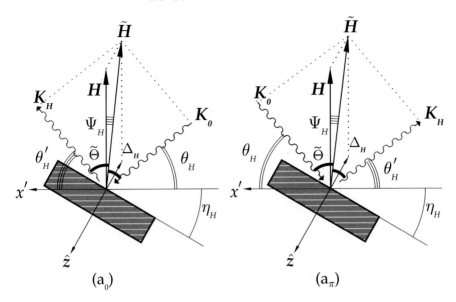

Fig. 2.5. Schematic drawing illustrating the reversibility of Bragg scattering in the asymmetric scattering geometry with $\eta_H \neq 0$.
(a_0) The direction of incidence is in the dispersion plane from right to left with $\phi_H = 0$ (compare with Fig. 2.4 showing a general case with $\phi_H \neq 0$).
(a_π) The direction of incidence in the dispersion plane from left to right with $\phi_H = \pi$.
The wave vectors $\boldsymbol{K}_H$ and $-\boldsymbol{K}_0$ compose the same angle $\tilde{\Theta}$ with the total momentum transfer vector $\tilde{\boldsymbol{H}}$. Note that $\Psi_H(a_0) = \Psi_H(a_\pi)$, and thus $\tilde{\Theta}(a_0) = \tilde{\Theta}(a_\pi)$, provided $\theta_H(a_0) = \theta'_H(a_\pi)$; and $\theta_H(a_\pi) = \theta'_H(a_0)$. Additional momentum transfer $\boldsymbol{\Delta}_H = \Delta_H \hat{z}$ due to scattering from the vacuum-crystal interface is shown compared to $\boldsymbol{H}$ greatly exaggerated for clarity, resulting also in the exaggerated difference between the angles of incidence and reflection.

All these results tell us that in general the angle of incidence is not equal to the angle of reflection. However, each angle is equal to its counterpart in the reversed scattering geometry.

Direction of $\boldsymbol{K}_0$, and $\boldsymbol{K}_H$ with Respect to the Entrance Crystal Surface. According to the definition, γ_0 is the cosine of the angle between the wave vector $\boldsymbol{K}_0$ of the vacuum incident wave and the crystal normal. The wave vector $\boldsymbol{K}_H$ differs slightly from $\boldsymbol{K}_0 + \boldsymbol{H}$. Therefore, strictly speaking, $\gamma_H = (\boldsymbol{K}_0 + \boldsymbol{H})\hat{z}/K$ does not have a meaning similar to that of γ_0. The quantity γ_H is rather a formal parameter of the problem. The cosine of the angle between the wave vector $\boldsymbol{K}_H$ of the reflected wave in vacuum and the crystal normal $\hat{z}$ can be calculated using (2.15), (2.17), and (2.18) as

$$\gamma'_H = \frac{\boldsymbol{K}_H \hat{z}}{K} = \pm\sqrt{\gamma_H^2 - \alpha_H}. \tag{2.32}$$

2. Dynamical Theory of X-Ray Bragg Diffraction

With the help of the relation between γ_0 and γ_H as given by (2.27), we obtain from (2.32) the general relation between γ_0 and γ'_H:

$$\gamma'_H = \pm\sqrt{\left(\gamma_0 - \frac{H}{K}\cos\eta_H\right)^2 - \alpha_H}. \tag{2.33}$$

In effect, (2.33) relates the glancing angles of incidence and reflection of the vacuum waves *with respect to the crystal surface*.

If $\boldsymbol{K}_0 + \boldsymbol{H}$ does not make a small angle with the crystal surface, so that $|\gamma_H|^2 \gg |\chi_0|$, and thus $|\gamma_H|^2 \gg |\alpha_H|$, one may use a simplified relation $\gamma'_H \simeq \gamma_H = \gamma_0 - (H/K)\cos\eta_H$, instead of (2.32)-(2.33). The application of the exact relations (2.32)-(2.33) becomes necessary under the condition of grazing emergence ($|\gamma_H|^2 \simeq |\chi_0|$) of the reflected wave. X-ray Bragg diffraction at grazing incidence and emergence was initially considered by Afanas'ev and Melkonyan (1983). For the rest of this chapter *no* grazing emergence of the reflected wave will be assumed.

Direction of $\boldsymbol{K}_0$, and $\boldsymbol{K}_H$ with Respect to the Reflecting Atomic Planes. The glancing and azimuthal angles of incidence are determined by the scattering geometry. The glancing and azimuthal angles of reflection on the contrary are unknowns, which have to be determined. The next step is to obtain general relations between the angles of incidence and reflection *with respect to the reflecting atomic planes*. For this, we insert (2.24) and (2.25) into (2.15) and obtain a system of equations

$$\cos\theta'_H \cos\phi'_H = \cos\theta_H \cos\phi_H + \frac{\Delta_H}{K}\sin\eta_H, \tag{2.34}$$

$$\cos\theta'_H \sin\phi'_H = \cos\theta_H \sin\phi_H, \tag{2.34'}$$

$$\sin\theta'_H = -\sin\theta_H + \frac{H}{K} - \frac{\Delta_H}{K}\cos\eta_H, \tag{2.34''}$$

which is a consequence of the conservation of energy and momentum for the incident and reflected vacuum waves. The three equations are not independent. One may readily check, that the sum of squares of the left- and righthand sides of the equations are equal, which is in agreement with the requirement of energy conservation expressed by (2.16). Equations (2.34)-(2.34'') allow one to determine the angles of reflection θ'_H, ϕ'_H for the given angles of incidence θ_H, ϕ_H.

Equations (2.34)-(2.34'') prove what was earlier stated: in the general case the angles of incidence and reflection are not equal: $\theta_H \neq \theta'_H$ and $\phi_H \neq \phi'_H$. The angles are equal only in some particular cases. For example, $\theta_H = \theta'_H$ and $\phi_H = \phi'_H$ if the asymmetry angle is zero: $\eta_H = 0$ (symmetric Bragg scattering geometry).

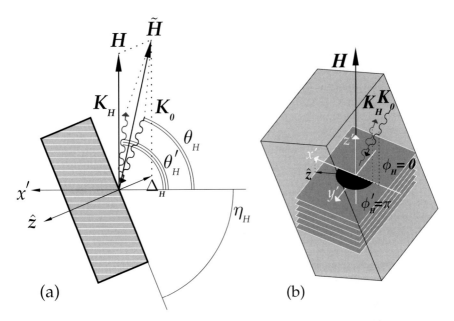

Fig. 2.6. Schematic drawing illustrating Bragg diffraction backscattering geometry from the asymmetrically cut crystal with $\eta_H \neq 0$: (a) a cross section through the dispersion plane $(\hat{z}, H)$, (b) a perspective view. Direction of incidence is in the dispersion plane from right to left with $\phi_H = 0$. It is similar to Fig. 2.5($a_\circ$), however: the wave vector K_H of the reflected wave is in the same quadrant as the wave vector K_0 of the incident wave, and therefore, the azimuthal angle of reflection $\phi'_H = \pi$. Additional momentum transfer $\Delta_H = \Delta_H \hat{z}$ due to scattering from the vacuum-crystal interface is shown in (a) greatly exaggerated for clarity.

2.2.7 Exact Backscattering Geometry

Let us consider the important and representative particular case: the scattering geometry in which the incident wave is reflected into itself. One commonly speaks in this case of *exact* Bragg backscattering. The wave being reflected into itself has a wave vector $K_H = -K_0$. This condition together with (2.15) yields for the wave vector of the incident wave

$$K_0 = -\frac{1}{2}\tilde{H} = -\frac{1}{2}(H + \Delta_H \hat{z}). \tag{2.35}$$

That is, to achieve exact back-reflection the wave vector of the incident radiation has to be directed opposite to the total momentum transfer vector $\tilde{H}$ rather than to the diffraction vector H. This fact can be easily understood by inspection of Fig. 2.6 showing the scheme of Bragg diffraction in backscattering geometry.

As follows from (2.35), only the waves with wave vector K_0 lying in the dispersion plane composed of the diffraction vector H and the inward normal $\hat{z}$ can be reflected into itself. Incident waves with $\boldsymbol{K}_0$ nonparallel to the dispersion plane, cannot be scattered by exactly 180°. While analyzing (2.20) we have already ascertained that the wave with the wave vector $\boldsymbol{K}_H$ is excited substantially if α_H is close to χ_0, i.e., if α_H is negative. If α_H is negative, the direction cosine γ_H is then also negative. Therefore $\boldsymbol{\Delta}_H$ (2.17) is negative under these conditions, and $\boldsymbol{\Delta}_H$ is directed opposite to $\hat{z}$, as in Figs. 2.5 and 2.6. Thus, exact backscattering is achieved only if the azimuthal angle of incidence is $\phi_H = 0$, and the azimuthal angle of reflection is $\phi'_H = \pi$, i.e., as shown in Fig. 2.6 (compare also with Fig. 2.5(a_0)).

Inserting (2.24) and (2.25) into (2.35), and using $\phi_H = 0$ yields

$$K \cos \theta_H = -\frac{\Delta_H}{2} \sin \eta_H, \qquad (2.36)$$

$$-K \sin \theta_H = -H/2 + \frac{\Delta_H}{2} \cos \eta_H.$$

Eliminating Δ_H from the above expressions, results in the following relation between the glancing angle of incidence θ_H and the wavelength $\lambda = 2\pi/K$ of the wave, which can be reflected into itself:

$$2d_H \cos(\theta_H + \eta_H) = -\lambda \sin \eta_H, \qquad (2.37)$$
$$\phi_H = 0.$$

This relation (as well as (2.35)) is solely a consequence of conservation of energy and momentum in Bragg scattering. It does not tell us yet how high or low is the reflectivity of the crystal for the wave of wavelength λ impinging at a glancing angle of incidence θ_H to the reflecting atomic planes.

It tells us only, that for $\eta_H = 0$ the wave of any wavelength λ may be reflected into itself at normal incidence, i.e., when $\theta_H = \pi/2$. If $\eta_H \neq 0$, this is no more the case. The direction of $\boldsymbol{K}_0$ has to be turned off-normal incidence by the offset angle $\Theta_H = \pi/2 - \theta_H \neq 0$, with θ_H given by (2.37). The offset angle is greater the closer is the asymmetry angle η_H to $\pi/2$.

The preceding considerations demonstrate that exact Bragg backscattering in particular, and Bragg scattering in general, is strongly influenced not only by the direction of $\boldsymbol{K}_0$ relative to the reflecting atomic planes, or equivalently to diffraction vector $\boldsymbol{H}$, but also the orientation of the crystal surface to the atomic planes, given by the asymmetry angle η_H, is important. The reason for this is refraction at the vacuum-crystal interface.

2.2.8 Fundamental Equations in Scalar Form

It was assumed at the beginning that the electric field inside the crystal is practically transverse. That means, each vector amplitude can be presented as a sum of two orthogonal linear polarization components

$$\boldsymbol{D}_H = D_H^\sigma \boldsymbol{\sigma}_H + D_H^\pi \boldsymbol{\pi}_H \qquad (2.38)$$

with two mutually orthogonal unit polarization vectors $\boldsymbol{\pi}_H$ and $\boldsymbol{\sigma}_H$. Both vectors are normal to $\boldsymbol{k}_H$. The unit vectors of the polarization components can be defined in different ways. Most common, is to define them as

$$\boldsymbol{\sigma}_H = \frac{\boldsymbol{k}_H \times \boldsymbol{k}_o}{|\boldsymbol{k}_H \times \boldsymbol{k}_o|}, \qquad \boldsymbol{\pi}_H = \frac{\boldsymbol{k}_H \times \boldsymbol{\sigma}_H}{|\boldsymbol{k}_H|}, \qquad (2.39)$$

for all vectors except $\boldsymbol{\sigma}_0$. The unit vector $\boldsymbol{\sigma}_0$ can be arbitrarily chosen parallel to a particular direction perpendicular to $\boldsymbol{k}_0$. For example, it can be chosen parallel to the entrance crystal surface (Colella 1974). If necessary, another definition can be used.

By using (2.38), one can rewrite the system of fundamental equations (2.11) for the polarization components of the plane wave amplitudes in the following form:

$$\sum_{H',s'} G^{ss'}_{HH'} D^{s'}_{H'} - \left(2\gamma_H \frac{\varkappa}{K} + \frac{\varkappa^2}{K^2}\right) D^s_H = 0, \qquad (2.40)$$

$$G^{ss'}_{HH'} = \chi_{H-H'} P^{ss'}_{HH'} - \alpha_H \delta^{ss'}_{HH'}, \qquad (2.41)$$

$$P^{ss'}_{HH'} = (\boldsymbol{s}_H \boldsymbol{s}'_{H'}). \qquad (2.42)$$

Here s (s') is a generic notation for the unit polarization vectors $\boldsymbol{\sigma}$ and $\boldsymbol{\pi}$; index s (s') denotes the two polarization components π or σ; $\delta^{ss'}_{HH'}$ is the Kronecker delta symbol for the two pairs of indices ss' and HH'. $P^{ss'}_{HH'}$ is the polarization factor, which describes possible polarization mixing $s \to s'$ in scattering of the plane wave with wave vector $\boldsymbol{k}_H$ and polarization vector $\boldsymbol{s}_H$ into the wave with wave vector $\boldsymbol{k}_{H'}$ and polarization vector $\boldsymbol{s}'_{H'}$. If $P^{ss'}_{HH'} = 0$ for $s \neq s'$, no polarization mixing occurs. The quantities $G^{ss'}_{HH'}$ are the elements of the so-called scattering matrix, which is of rank $2n$.

The fundamental equations (2.40) form a system of $2n$ homogeneous equations for the unknown amplitudes D^s_H. The solution is non-trivial only if the determinant of the system is zero. The determinant of the system will be in general a polynomial of degree $4n$ in the quantity $\varkappa$ (a characteristic polynomial of the system), giving rise to $4n$ discrete values of $\varkappa_\nu$ ($\nu = 1, 2, ..., 4n$), and $\boldsymbol{k}_{0(\nu)} = \boldsymbol{K}_0 + \varkappa_\nu \hat{\boldsymbol{z}}$, respectively.

2.2.9 Eigenvectors and Wave Fields in the Crystal

Since $|\varkappa| \simeq |\chi_0|K \ll K$, the quadratic term $\varkappa^2$ in (2.40) can be very often omitted. The polynomial reduces in this case to a degree of $2n$ in $\varkappa$, giving rise to $2n$ discrete values of $\varkappa_\nu$ and $\boldsymbol{k}_{0(\nu)} = \boldsymbol{K}_0 + \varkappa_\nu \hat{\boldsymbol{z}}$, respectively. The quadratic terms still have to be retained if the incident or one of the reflected waves propagates almost parallel to the crystal surface (grazing incidence or grazing emergence), i.e., when one of γ_H is very small: $\gamma_H^2 \lesssim |\chi_0|$. Unless otherwise indicated, the quadratic terms will be neglected in the following[5].

Corresponding to each characteristic value $\varkappa_\nu$ (eigenvalue) the set of fundamental equations (2.40) has at least one non-trivial solution for the $2n$ amplitudes $\left\{ D^\sigma_{0(\nu)}, D^\pi_{0(\nu)}, D^\sigma_{H(\nu)}, D^\pi_{H(\nu)}, \ldots \right\}$ of the plane wave components, which is often referred to as the eigenvector. According to (2.9) there is a wave field

$$\boldsymbol{D}_\nu(\boldsymbol{r}) = e^{i\boldsymbol{k}_0(\nu)\boldsymbol{r}} \sum_H \boldsymbol{D}_{H(\nu)} e^{i\boldsymbol{H}\boldsymbol{r}}. \tag{2.43}$$

corresponding to each eigenvector. The eigenvectors are arbitrary to the extent of a constant multiplier. They can be normalized, e.g., in such a way that the eigenvector component corresponding to the forward-scattered plane wave with the polarization of the incident wave $\boldsymbol{s}_0$ equals unity:

$$D^{s_0}_{0(\nu)} = 1. \tag{2.44}$$

The net radiation field excited in the crystal is a sum of $2n$ wave fields

$$\boldsymbol{D}(\boldsymbol{r}) = \sum_\nu \Lambda_\nu \boldsymbol{D}_\nu(\boldsymbol{r}). \tag{2.45}$$

The quantities Λ_ν have to be determined. As follows from (2.45), they scale with the probability amplitudes of the excitation of the wave fields $\boldsymbol{D}_\nu(\boldsymbol{r})$.

Combining (2.43) and (2.45) one may obtain the following expression

$$\boldsymbol{D}(\boldsymbol{r}) = \sum_H e^{i(\boldsymbol{K}_0 + \boldsymbol{H})\boldsymbol{r}} \boldsymbol{D}_H(z),$$

$$\boldsymbol{D}_H(z) = \sum_\nu \Lambda_\nu \boldsymbol{D}_{H(\nu)} e^{i\varkappa_\nu z}, \tag{2.46}$$

for the radiation field in the crystal. The unknown coefficients Λ_ν can be determined from the boundary conditions for the components $\boldsymbol{D}_H(z)$ at the front (entrance) $z = z_1$ or rear $z = z_2 = z_1 + d$ crystal surface, respectively.

2.2.10 Boundary Conditions

By using (2.1), the boundary condition for the incident wave at the front surface can be written as

---

[5] We refer the reader to the paper of Colella (1974) for a general procedure of solving the problem, which takes the quadratic terms into account, i.e., includes the cases of grazing incidence and emergence. Modified algorithms were proposed more recently by Stepanov and Ulyanenkov (1994) and Stetsko and Chang (1997).

$$D_0(z_1) = \mathcal{E}_i. \tag{2.47}$$

As for the other components with $H \neq 0$, one has to distinguish between the Bragg-case and Laue-case waves. The Laue-case waves propagate towards the inside of the crystal. Therefore, they have to be zero at the front surface. The boundary condition for them then reads as

$$D_H(z_1) = 0: \qquad \text{Laue}-\text{case wave.} \tag{2.48}$$

The Bragg-case waves propagate towards the outside of the crystal. There is no wave with wave vector K_H outside the crystal rear surface. Therefore, for the in-crystal reflected wave the boundary condition at the rear surface $z = z_2$ is:

$$D_H(z_2) = 0: \qquad \text{Bragg}-\text{case wave.} \tag{2.49}$$

By using the boundary conditions (2.47)-(2.49), one obtains in general $2n$ linear equations in the $2n$ unknown coefficients Λ_ν. The resulting system of equations is not homogeneous. It depends on the incident vacuum wave amplitude - (2.47). The solution of the system provides a single set of the coefficients Λ_ν. With these quantities known and by using (2.46), one finally obtains the solution for the radiation field in the crystal.

2.2.11 Reflectivity and Transmissivity I

What is actually measured is the crystal reflectivity or transmissivity rather than the radiation fields D_H or their intensities $|D_H|^2$. The reflectivity is, by definition, the ratio of the total radiation flux received by the detector per unit time divided by the radiation flux of the incident beam in the same time interval. The radiation flux density is given by the Pointing vector, whose magnitude for plane waves is proportional to $|D_H|^2$. By multiplying this value by the cross-section of the beam one obtains the radiation flux.

Here one should note that the incident and reflected beams may differ considerably in cross-section. This happens, if the asymmetry angle $\eta_H \neq 0$, as is illustrated in Fig. 2.7. One may derive from (2.18),(2.29), and by using Fig. 2.7 that the cross-section of the reflected beam with wave vector K_H changes by the factor $|b_H|^{-1}$ as compared to the cross-section of the incident beam.

Taking this into account, the reflectivity for the reflected beam and the transmissivity for the forward transmitted beam are given by

$$R = |b_H|^{-1} |D_H(z_1)|^2 / |\mathcal{E}_i|^2 \qquad \text{Bragg}-\text{case wave,}$$

$$R = |b_H|^{-1} |D_H(z_2)|^2 / |\mathcal{E}_i|^2 \qquad \text{Laue}-\text{case wave,} \tag{2.50}$$

$$T = |D_0(z_2)|^2 / |\mathcal{E}_i|^2 \qquad \text{forward}-\text{transmitted wave.}$$

Now the problem is completely formulated. The way of obtaining its solution is described. Let us analyze the solutions in some particular cases.

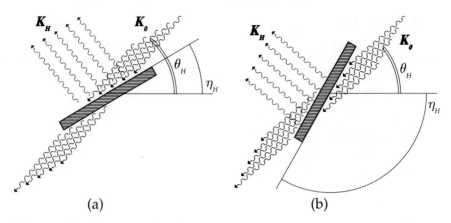

Fig. 2.7. Transformation of the beam cross-sections in asymmetric Bragg reflection (a) Bragg geometry, (b) Laue geometry.

2.3 Single-Beam Case

It can easily happen that only the component $\boldsymbol{D}_0$ with wave vector $\boldsymbol{k}_0$ is excited. In this case no other reciprocal lattice node except 0 is lying on the Ewald sphere - Fig. 2.1. The system of fundamental equations (2.40) reduces to two independent linear equations for each polarization component $\mathcal{D}_0^s$. The characteristic polynomial reduces to

$$\frac{\varkappa^2}{K^2} + 2\gamma_0 \frac{\varkappa}{K} - \chi_0 = 0. \tag{2.51}$$

In general, two values of $\varkappa$ are allowed as solutions of (2.51):

$$\frac{\varkappa}{K} = -\gamma_0 \pm \sqrt{\gamma_0^2 + \chi_0}. \tag{2.52}$$

However, if the incident wave is far from grazing incidence ($\gamma_0^2 \gg |\chi_0|$) then the quadratic term can be neglected in (2.51). As a result, a single solution exists:

$$\varkappa = K \frac{\chi_0}{2\gamma_0}. \tag{2.53}$$

By using (2.43)-(2.45) and the boundary condition (2.47) the following solution for the radiation field propagating in the crystal can be obtained

$$\boldsymbol{D}(\boldsymbol{r}) = \boldsymbol{\mathcal{E}}_i \, e^{i\boldsymbol{k}_0 \boldsymbol{r}}, \qquad \boldsymbol{k}_0 = \boldsymbol{K}_0 + K \frac{\chi_0}{2\gamma_0} \hat{\boldsymbol{z}}. \tag{2.54}$$

From (2.54) it follows that the magnitude of the crystal wave vector

$$k_0 = n'K, \qquad n = 1 + \chi_0/2 = 1 - r_e \frac{\lambda^2}{2\pi V} F_0, \tag{2.55}$$

where $n = n' + in''$ is the complex index of refraction. This is in agreement with (1.1). Since χ'_0 is a small and negative quantity typically $\approx -10^{-5}$, the real part of the index of refraction is slightly less than unity. Due to this fact the magnitude k_0 of the crystal wave vector is somewhat smaller than the magnitude K of the vacuum wave vector, or vice versa, the wavelength in the crystal λ_{crystal} is larger than the wavelength in vacuum λ and is given by

$$\lambda_{\text{crystal}} = \lambda/(1 + \chi'_0/2) \simeq \lambda \left(1 + r_e \frac{\lambda^2}{2\pi V} F'_0\right). \tag{2.56}$$

2.4 2-Beam Bragg Diffraction

We shall assume throughout the current section that the excitation condition is fulfilled for two particular waves with the wave vectors denoted by $\boldsymbol{k}_0$ and $\boldsymbol{k}_H$. Here we shall understand H as a specific rather than a running index. Two reciprocal lattice nodes 0 and H are lying on the Ewald sphere - Fig. 2.1. This is commonly referred to as two-beam case.

The system of fundamental equations (2.40) reduces in this case to a system of four linear equations for the amplitudes D_0^σ, D_0^π, D_H^σ, and D_H^π. The definition of the polarization vectors (2.39) stipulates that the polarization factors $P_{0H}^{ss'}$ in (2.41) are zero for $s \neq s'$. As a result, the system of four equations breaks up into two independent sets of two equations, one containing only the σ-components of the radiation field, and the other containing only the π-components:

$$\begin{aligned}(\chi_0 - \varepsilon) D_0^s + & \chi_{\overline{H}} P_{0H}^{ss} D_H^s = 0 \\ \chi_H \, b \, P_{H0}^{ss} D_0^s + & [(\chi_0 - \alpha) b - \varepsilon] D_H^s = 0,\end{aligned} \tag{2.57}$$

with

$$\varepsilon = 2\gamma_0 \frac{\varkappa}{K}, \tag{2.58}$$

and $s = \sigma$ or $s = \pi$. Since only one Bragg reflection is discussed, index H in α_H and b_H is omitted in (2.58)[6]. In the derivation of (2.57) the quadratic terms $\varkappa^2$ in (2.41) were neglected. As we know, physically this implies that both the incident wave is far from grazing incidence: $\gamma_0^2 \gg |\chi_0|$, and the reflected wave is far from grazing emergence: $\gamma_H^2 \gg |\chi_0|$. The notation $\chi_{\overline{H}}$ in (2.57) is equivalent to χ_{-H}.

The polarization factors are $P_{0H}^{\sigma\sigma} = P_{H0}^{\sigma\sigma} = 1$ for σ-polarization, and $P_{0H}^{\pi\pi} = P_{H0}^{\pi\pi} = \cos 2\theta$ for π-polarization. To reduce the number of indices

---
[6] Throughout this section index H will be also omitted for such quantities as η_H, ϕ_H, η_H, etc.

in the following equations, we shall omit also the polarization index $s(s')$. The polarization factors then become $P^{ss}_{0H} = P^{ss}_{H0} \equiv P$, where $P = 1$ for σ-polarization, and $P = \cos 2\theta$ for π-polarization.

The compatibility conditions for the set of linear homogeneous equations (2.57), i.e., determinant is zero, defines two possible values of ε (and respectively $\varkappa$), which will be distinguished by the index $\nu = 1, 2$:

$$\varepsilon_\nu - \chi_0 = -\tilde{\alpha} \pm \sqrt{\tilde{\alpha}^2 + P^2 b \chi_H \chi_{\overline{H}}}, \tag{2.59}$$

$$\tilde{\alpha} = \frac{1}{2}[\alpha b + \chi_0(1-b)]. \tag{2.60}$$

Hereafter, it is assumed that the imaginary part of the root in (2.59) is positive and the index $\nu = 1$ corresponds to the "+" sign in (2.59). Under these assumptions, $\text{Im}\{\varepsilon_1 - \varepsilon_2\} > 0$ and $\text{Im}\{\varkappa_1 - \varkappa_2\} > 0$.

Because of the two solutions for ε (and for $\varkappa$), the expression (2.45) for the radiation field in the crystal is given by:

$$\mathcal{D}(r) = e^{i\boldsymbol{K}_0 \boldsymbol{r}}\left[D_0(z) + D_H(z) e^{i\boldsymbol{H}\boldsymbol{r}}\right], \tag{2.61}$$

with

$$\begin{aligned} D_0(z) &= \sum_{\nu=1,2} A_\nu D_{0(\nu)} e^{i\varkappa_\nu z}, \\ D_H(z) &= \sum_{\nu=1,2} A_\nu D_{H(\nu)} e^{i\varkappa_\nu z}, \\ \varkappa_\nu &= \varepsilon_\nu K/(2\gamma_0). \end{aligned} \tag{2.62}$$

2.4.1 Radiation Fields in the Crystal and in Vacuum

In the subsequent discussion only the Bragg-case scattering geometry is considered, in which the incident and diffracted waves are on the same side of the vacuum-crystal interface, as in Fig. 2.8a. We shall distinguish between the front surface incidence, as shown in Fig. 2.8a, and the rear surface incidence - Fig. 2.8b, which is of importance for the problem of the x-ray Fabry-Pérot resonator addressed in Chap. 4.

Front Surface Incidence. As was previously mentioned, the amplitudes $D_{H(\nu)}$ and $D_{0(\nu)}$ are determined by the fundamental system of equations (2.57) to within a multiplicative constant. Following (2.44), one may use the normalization $D_{0(\nu)} = 1$ to determine them unequivocally. According to (2.57), the ratio of the amplitudes $D_{0(\nu)}$ and $D_{H(\nu)}$ is

$$\frac{D_{H(\nu)}}{D_{0(\nu)}} = \frac{\varepsilon_\nu - \chi_0}{P\chi_{\overline{H}}} = R_\nu. \tag{2.63}$$

This yields

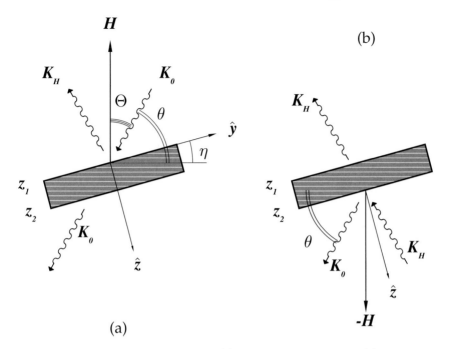

Fig. 2.8. Bragg scattering geometries: (a) front surface incidence, (b) rear surface incidence. The particular cases are shown, in which the wave vector of the incident, and reflected waves K_0, and K_H, respectively, are in the plane composed of the diffraction vector H and the inward normal $\hat{z}$ to the crystal surface. White lines depict the reflecting atomic planes normal to H (a), and $-H$ (b), respectively.

$$\{D_0(\nu), D_H(\nu)\} = \{1, R_\nu\} \tag{2.64}$$

for the eigenvector of the wave fields $\mathcal{D}_\nu(\mathbf{r})$ (2.43).

To determine Λ_ν in (2.62), the boundary condition (2.47) at the entrance surface $\mathbf{r} = \mathbf{r}_1$ ($z = z_1$) as well as the boundary condition (2.49) at the exit surface $\mathbf{r} = \mathbf{r}_2$ ($z = z_2$) is applied:

$$\begin{aligned}\Lambda_1\, e^{i\varkappa_1 z_1} &+ \Lambda_2\, e^{i\varkappa_2 z_1} = \mathcal{E}_i \\ \Lambda_1 R_1 e^{i\varkappa_1 z_2} &+ \Lambda_2 R_2 e^{i\varkappa_2 z_2} = 0.\end{aligned} \tag{2.65}$$

The solution of the system of equations (2.65) for Λ_ν together with the eigenvectors of the wave fields (2.64) being substituted into (2.62) yield for the forward $D_0(z)$ and Bragg scattered $D_H(z)$ components of the radiation field (2.61) inside the crystal at a depth z:

$$D_0(z) = \mathcal{E}_i \frac{R_1 \exp[i\varkappa_2(z-z_2)] - R_2 \exp[i\varkappa_1(z-z_2)]}{R_1 \exp[i\varkappa_2(z_1-z_2)] - R_2 \exp[i\varkappa_1(z_1-z_2)]},$$

$$D_H(z) = \mathcal{E}_i \frac{R_1 R_2 \{\exp[i\varkappa_2(z-z_2)] - \exp[i\varkappa_1(z-z_2)]\}}{R_1 \exp[i\varkappa_2(z_1-z_2)] - R_2 \exp[i\varkappa_1(z_1-z_2)]}. \quad (2.66)$$

Our next goal is to find the radiation fields transmitted through and reflected from the crystal, respectively.

By using (2.61), and (2.66), the radiation field component emerging from the rear crystal surface at $r = r_2$ ($z = z_2$) in the forward direction, and the reflected radiation field component emerging from the front crystal surface at $r = r_1$ ($z = z_1$), these being measured *in the crystal*, are given by

$$\mathcal{D}(r_2)_{\text{forward}} = \mathcal{E}_i \, t_{00} \, e^{i K_0 r_2},$$

$$\mathcal{D}(r_1)_{\text{reflected}} = \mathcal{E}_i \, r_{0H} \, e^{i(K_0+H)r_1}, \quad (2.67)$$

respectively. Here

$$t_{00} = e^{i\varkappa_1 d} \frac{R_2 - R_1}{R_2 - R_1 e^{i(\varkappa_1 - \varkappa_2)d}}, \quad (2.68)$$

$$r_{0H} = R_1 R_2 \frac{1 - e^{i(\varkappa_1 - \varkappa_2)d}}{R_2 - R_1 e^{i(\varkappa_1 - \varkappa_2)d}}. \quad (2.69)$$

are the transmission and reflection amplitudes at the rear and the front surfaces measured *in the crystal*. They are obtained from (2.66) and (2.67) and using $d = z_2 - z_1$ as the crystal thickness.

From (2.67) and the continuity of the radiation fields at the vacuum-crystal interfaces, we obtain for the transmitted and reflected radiation fields at an arbitrary point r in vacuum:

$$\mathcal{E}_0(r) = \mathcal{D}(r_2)_{\text{forward}} \, e^{i K_0(r-r_2)},$$

$$\mathcal{E}_H(r) = \mathcal{D}(r_1)_{\text{reflected}} \, e^{i K_H(r-r_1)}. \quad (2.70)$$

By using (2.70), (2.67), and (2.15) they can be written as:

$$\mathcal{E}_0(r) = \mathcal{E}_i \, T_{00} \, e^{i K_0 r},$$

$$\mathcal{E}_H(r) = \mathcal{E}_i \, \rho_{0H} \, e^{i K_H r}, \quad (2.71)$$

where

$$T_{00} = t_{00},$$

$$\rho_{0H} = r_{0H} \, e^{-i\Delta_H r_1}. \quad (2.72)$$

are the transmission and reflection amplitudes of the radiation field components at the rear and the front surfaces measured *in vacuum*. Keeping only the first term in the Taylor series expansion for the momentum transfer Δ_H (2.21) the reflection amplitude for the vacuum wave reads

$$\rho_{0H} = r_{0H} \exp\left(i\frac{Kz_1}{2\gamma_H}\alpha\right). \tag{2.73}$$

Therewith, the two-beam diffraction problem is basically solved, as the above equations allow evaluation of the scattered radiation fields at any space point: (2.61) and (2.66) allow calculations of the fields in the crystal, and (2.71), (2.72), (2.68), and (2.69) - of the fields in vacuum.

Rear Surface Incidence. In this case, the incident radiation wave

$$\mathcal{E}(r,t) = \tilde{\mathcal{E}}_i \exp[i(\boldsymbol{K}_H \boldsymbol{r} - \omega t)] \tag{2.74}$$

with wave vector $\boldsymbol{K}_H$ strikes the rear surface of the crystal at $r = r_2(z = z_2)$ - Fig. 2.8b.

Under these conditions, the field inside the crystal is

$$\mathcal{D}(\boldsymbol{r}) = e^{i\boldsymbol{K}_H \boldsymbol{r}}\left[D_H(z) + D_0(z)e^{-i\boldsymbol{H}\boldsymbol{r}}\right]. \tag{2.75}$$

For such a scattering geometry, one may use $D_{H(\nu)} = 1$ as the normalization condition for the eigenvectors of the wave fields. This, together with (2.63), yields for the eigenvectors:

$$\{D_{0(\nu)}, D_{H(\nu)}\} = \{1/R_\nu, 1\}. \tag{2.76}$$

The boundary conditions (2.47) and (2.49) then read:

$$\begin{aligned}(\Lambda_1/R_1)\,e^{i\varkappa_1 z_1} + (\Lambda_2/R_2)\,e^{i\varkappa_2 z_1} &= 0 \\ \Lambda_1\,e^{i\varkappa_1 z_2} \qquad\quad + \Lambda_2\,e^{i\varkappa_2 z_2} &= \tilde{\mathcal{E}}_i.\end{aligned} \tag{2.77}$$

All the variables are defined as previously. The field components inside the crystal become

$$D_0(z) = \tilde{\mathcal{E}}_i\,\frac{\exp[i\varkappa_2(z-z_1)] - \exp[i\varkappa_1(z-z_1)]}{R_2\exp[i\varkappa_2(z_2-z_1)] - R_1\exp[i\varkappa_1(z_2-z_1)]}, \tag{2.78}$$

$$D_H(z) = \tilde{\mathcal{E}}_i\,\frac{R_2\exp[i\varkappa_2(z-z_1)] - R_1\exp[i\varkappa_1(z-z_1)]}{R_2\exp[i\varkappa_2(z_2-z_1)] - R_1\exp[i\varkappa_1(z_2-z_1)]}. \tag{2.79}$$

From this we obtain for the transmission amplitude measured at the exit crystal surface $z = z_1$:

$$t_{HH} = e^{-i\varkappa_2 d}\,\frac{R_2 - R_1}{R_2 - R_1 e^{i(\varkappa_1 - \varkappa_2)d}}, \tag{2.80}$$

and for the reflection amplitude measured in the crystal at the crystal-vacuum interface at $z = z_2$:

$$r_{H0} = \frac{1 - e^{i(\varkappa_1 - \varkappa_2)d}}{R_2 - R_1 e^{i(\varkappa_1 - \varkappa_2)d}}, \tag{2.81}$$

respectively.

The vacuum radiation fields are obtained as follows:

$$\mathcal{E}_H(\mathbf{r}) = \tilde{\mathcal{E}}_i\, \tau_{HH}\, e^{i\mathbf{K}_H \mathbf{r}},$$
$$\mathcal{E}_0(\mathbf{r}) = \tilde{\mathcal{E}}_i\, \rho_{H0}\, e^{i\mathbf{K}_0 \mathbf{r}}, \tag{2.82}$$

where

$$\tau_{HH} = t_{HH},$$
$$\rho_{H0} = r_{H0}\, e^{+i\mathbf{\Delta}_H \mathbf{r}_1}, \tag{2.83}$$

are the transmission and reflection amplitudes of the radiation field components measured in vacuum at the vacuum-crystal interfaces. Again, using only the first term in the Taylor series expansion for the momentum transfer $\mathbf{\Delta}_H$ (2.21) the reflection amplitude for the vacuum wave reads

$$\rho_{H0} = r_{H0}\, \exp\left(-i\frac{Kz_2}{2\gamma_H}\alpha\right). \tag{2.84}$$

Effect of Crystal Translations. For the problem of the x-ray Fabry-Pérot resonator discussed in Chap. 4, it is also important to understand how the reflection and transmission amplitudes are transformed by crystal translations.

If the scattering crystal is shifted by the translation vector $\mathbf{U}$ the electric susceptibility, given (2.3), becomes $\chi(\mathbf{r} + \mathbf{U})$. As a result, the Fourier components of the susceptibility transform to

$$\chi_H \Rightarrow \chi_H\, e^{i\mathbf{H}\mathbf{U}}, \tag{2.85}$$

the ratios R_ν given by (2.63) transform to $R_\nu \exp(i\mathbf{H}\mathbf{U})$, and the reflection amplitudes become

$$\tilde{r}_{0H} = r_{0H}\, e^{i\mathbf{H}\mathbf{U}}, \qquad \tilde{r}_{H0} = r_{H0}\, e^{-i\mathbf{H}\mathbf{U}}. \tag{2.86}$$

In the particular case of the translation vector $\mathbf{U} = n_1 \mathbf{a}_1 + n_2 \mathbf{a}_2 + n_3 \mathbf{a}_3$, where $\mathbf{a}_i$ are the basis vectors for the direct crystal lattice, and n_i any integer, the scalar product becomes $\mathbf{H}\mathbf{U} = 2\pi m$, where $m = 0, \pm 1, \pm 2, \ldots$. Thus, in this particular case the reflection amplitudes (2.86) are invariant to such a translation. In contrast, transmission amplitudes t_{00} and t_{HH} are always invariant to crystal translations.

The reflection amplitudes ρ_{0H} and ρ_{H0} measured in vacuum contain explicitly the coordinates z_1 and z_2 of the front and rear surfaces, respectively. Therefore, they are transformed as follows:

$$\tilde{\rho}_{0H} = r_{0H} \exp\left(i\frac{K\tilde{z}_1}{2\gamma_H}\alpha\right) e^{i\boldsymbol{HU}} \tag{2.87}$$

$$\tilde{\rho}_{H0} = r_{H0} \exp\left(-i\frac{K\tilde{z}_2}{2\gamma_H}\alpha\right) e^{-i\boldsymbol{HU}}.$$

Here $\tilde{z}_1 = U_z + z_1$ and $\tilde{z}_2 = U_z + z_2$ are the new coordinates of the front and rear surfaces.

2.4.2 Reflectivity and Transmissivity II

The general expressions for the wave fields inside and outside the crystal in the two-beam diffraction case in Bragg scattering geometry were obtained in the previous paragraph. The reflectivity or transmissivity of the crystal - the quantities which can be measured - are calculated as the ratio of reflected or transmitted to the incident radiation flux, as we have discussed in Sect. 2.2.11. By using (2.50), and (2.67)-(2.69) we obtain

$$R = \frac{1}{|b|}|r_{0H}|^2, \qquad T = |t_{00}|^2, \tag{2.88}$$

in the case of front surface incidence. These expressions can be used to calculate numerically the reflectivity and transmissivity of crystals under conditions of two-beam Bragg diffraction. The expressions for the reflection amplitude r_{0H} and for the transmission amplitude t_{00} are unfortunately very complex. They do not offer a simple picture of the results of the theory.

To gain better insight, it is advantageous to proceed with the analysis of the crystal reflectivity and transmissivity in particular cases.

2.4.3 Reflectivity of Thick Non-Absorbing Crystals

Three assumptions will be used in the following part of this chapter through to Sect. 2.4.13.

First, the crystal is very thick, i.e., $d \gg d_e$, where $d_e = \text{Im}(\varkappa_1 - \varkappa_2)^{-1}$ is the effective penetration length of the radiation fields inside the crystal, which is obtained from (2.68)-(2.69). As follows from (2.69), the reflection amplitude r_{0H} can be closely approximated in this case by $r_{0H} = R_1$. As a result the crystal reflectivity becomes $R = |R_1|^2/|b|$ (2.88).

Secondly, there is no photo-absorption, i.e., the imaginary part f''_n of the atomic scattering amplitudes $f_n(\boldsymbol{H})$ in (2.6) is zero. In many cases it is a fairly good approximation, as typically $f''_n \ll f_n(\boldsymbol{H})$.

Thirdly, the crystal lattice is assumed for simplicity to be centrosymmetric so that the relation $\chi_H = \chi_{\bar{H}}$ holds. In this case all Fourier components of the electric susceptibility may be considered to be real.

With these assumptions the penetration length can be written as

$$d_e(y) = d_e(0) \operatorname{Im} \frac{1}{\sqrt{y^2 - 1}}, \qquad (2.89)$$

where

$$d_e(0) = \frac{\sqrt{\gamma_0 |\gamma_H|}}{K|P\chi_H|} \qquad (2.90)$$

is the penetration length at $y = 0$ usually termed the extinction length, and

$$y = \frac{\alpha b + \chi_0(1 - b)}{2|P\chi_H|\sqrt{|b|}}. \qquad (2.91)$$

is called reduced deviation parameter. The extinction length depends strongly on the photon energy, and is typically of the order of $10 - 100$ μm for photons with energies of about $10 - 20$ keV (see tables of Bragg reflections in Appendices A.3 and A.4).

By using (2.59), (2.63), and $R = |R_1|^2/|b|$, as above, we obtain finally the expression for the reflectivity in the approximation of a semi-infinite non-absorbing crystal:

$$R = \left| -y \pm \sqrt{y^2 - 1} \right|^2. \qquad (2.92)$$

The sign of the square root is chosen in such a way that $|R| \leq 1$. The reflectivity is a function of the parameter y alone. It is shown graphically in Fig. 2.9. A very important result, which will be used very often in the following, is that in the region $-1 \leq y \leq 1$ total reflection with $R = 1$ is achieved. The range of y with $-1 \leq y \leq 1$ is called the region of total reflection.

By using the notation $y = \cos\phi_r$, the reflection amplitude $r_{0H} = R_1$ in the region of total reflection can be presented as

$$r_{0H} = -\sqrt{|b|} \exp(-i\phi_r). \qquad (2.93)$$

The phase jump ϕ_r varies in the region of total reflection from $-\pi$ to 0, as y changes from -1 to 1.

The region of total reflection can be also expressed in terms of the deviation parameter α. Using (2.91) we obtain

$$\alpha_{+1} \leq \alpha \leq \alpha_{-1}, \qquad (2.94)$$

where

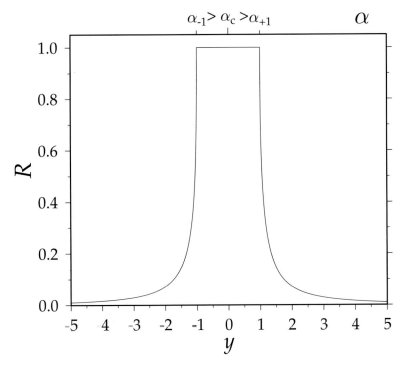

Fig. 2.9. Reflectivity R of a semi-infinite non-absorbing crystal in Bragg scattering geometry as a function of the reduced deviation parameter y and deviation parameter α.

$$\alpha_{\pm 1} = \alpha_c \mp 2|P\chi_H|/\sqrt{|b|}, \tag{2.95}$$

$$\alpha_c = \chi'_0\left(1 - \frac{1}{b}\right). \tag{2.96}$$

The center of the region of total reflection is at α_c, as shown in Fig. 2.9. It is a small negative value of the same order of magnitude as χ'_0 which is typically $\leq 10^{-5}$ for x rays of energy $E \simeq 15$ keV. The width of the region of total reflection is

$$\Delta\alpha = |\alpha_{-1} - \alpha_{+1}| = 4|P\chi_H|/\sqrt{|b|} \tag{2.97}$$

and has the same order of magnitude as $|\chi_H|$ which is typically $\leq 10^{-6}$ for x rays of energy $E \simeq 15$ keV. Since $|\chi_0| < |\chi_H|$, the width of the region is always smaller than $|\alpha_c|$: $\Delta\alpha < |\alpha_c|$.

2.4.4 Deviation Parameter

For the subsequent discussion it is important to establish the relationship between the formal parameter of the theory α, and the physical parameters of the problem such as: the glancing angle of incidence θ, x-ray wavelength λ (photon energy E), and crystal temperature T. By using (2.28), (2.7), and $K = 2\pi/\lambda$ we readily obtain (Shvyd'ko et al. 1998)

$$\alpha = \frac{2\lambda}{d_H(T)} \left[\frac{\lambda}{2d_H(T)} - \sin\theta \right]. \qquad (2.98)$$

Here we have taken explicitly into account that the interplanar distance $d_H(T)$ is temperature dependent. The above expression is valid for any glancing angle of incidence θ, including $\theta = \pi/2$.

Equivalently α can be expressed in terms of the x-ray photon energy $E = hc/\lambda$:

$$\alpha = 4 \frac{E_H(T)}{E} \left[\frac{E_H(T)}{E} - \sin\theta \right], \qquad (2.99)$$

where

$$E_H = \frac{H}{2} \hbar c \equiv \frac{hc}{2d_H} \qquad (2.100)$$

is termed the Bragg energy.

Since the width $\Delta\alpha$ (2.97) of the region of total reflection is typically $\leq 10^{-6}$, tiny variations of E (λ), θ or crystal temperature T may already produce, according to (2.98)-(2.99), sizable changes in the crystal reflectivity. Such variations leave the Fourier components of the electric susceptibility χ_0, χ_H, asymmetry factor b, polarization factor P, and thus α_c (2.96), and $\alpha_{\pm 1}$ (2.95) practically unchanged. From this, we conclude that the crystal reflectivity can be expressed as a function of the formal parameter α alone. If $R(\alpha)$ is known, then using equation (2.98) or (2.99), the reflectivity can be immediately determined also as a function of the glancing angle of incidence θ, the photon energy E, the wavelength λ, and the crystal temperature T.

Very often it is convenient to use approximations for the deviation parameter α, instead of the exact expressions (2.98) or (2.99). To proceed in this way, let us define for each wavelength λ, a so-called Bragg angle θ_B according to the rule:

$$\lambda = 2d_H \sin\theta_B. \qquad (2.101)$$

This expression is generally known as Bragg's law. We shall discuss its physical meaning in the next section.

For a constant wavelength λ, the deviation parameter α can be expressed in terms of the glancing angle θ and the Bragg angle θ_B as:

$$\alpha = 4\sin\theta_B \left[\sin\theta_B - \sin\theta \right]. \qquad (2.102)$$

In particular assuming that $|\theta_B - \theta| \ll 1$ one can easily derive from (2.102) an approximate expression for α:

$$\alpha = 2\,(\theta_B - \theta)\sin 2\theta_B, \tag{2.103}$$

which is used as a rule in the standard texts of the dynamical theory – see, e.g., Zachariasen (1945), Batterman and Cole (1964), Pinsker (1978), and others. This expression, however, fails for backscattering when $\theta \to \pi/2$.

The expression for the deviation parameter in backscattering can be approximated accurately by (Caticha and Caticha-Ellis 1982)

$$\alpha = 2\left[\Theta^2 - 2\epsilon\right], \tag{2.104}$$

with

$$\epsilon = -\frac{\lambda - 2d_H}{2d_H} \simeq \frac{E - E_H}{E_H}. \tag{2.105}$$

This expression is readily derived from the general expression (2.99), assuming that the incidence angle $\Theta = \pi/2 - \theta$, and the relative deviation ϵ of the x-ray energy E from Bragg's energy E_H are small: $|\Theta| \ll 1$, and $|\epsilon| \ll 1$.

As mentioned previously, the interplanar distance d_H and the Bragg energy are temperature dependent. Taking into account that the variation of the Bragg energy in a small temperature range can be expressed as

$$E_H(T + \delta T) = E_H(T)\,(1 - \beta_H\,\delta T), \tag{2.106}$$

with

$$\beta_H = \frac{1}{d_H(T)}\frac{\mathrm{d}d_H}{\mathrm{d}T} \tag{2.107}$$

as the linear thermal expansion coefficient in the direction of $\boldsymbol{H}$, one can generalize (2.104):

$$\alpha = 2\Theta^2 - 4\left[\frac{\delta E}{E_H(T)} + \beta_H(T)\,\delta T\right]. \tag{2.108}$$

Here $\delta E = E - E_H(T)$. This equation shows that in the backscattering region α varies linearly both with x-ray energy and crystal temperature and quadratically with deviation from normal incidence Θ.

Equation (2.108) can be used to establish the equivalence of a variation of x-ray energy and crystal temperature. Those values of E and T are equivalent which result in the same reflectivity. As long as the reflectivity of x rays is solely determined by α, the $E - T$ equivalence relation reads

$$\delta E = E_H(T)\,\beta_H(T)\,\delta T. \tag{2.109}$$

2.4.5 Bragg's Law

We have defined Bragg's law by (2.101).

For a wave with wavelength λ and glancing angle of incidence $\theta = \theta_B$ satisfying Bragg's law the deviation parameter (2.98) is zero: $\alpha = 0$ [7]. We know that the center of the region of total reflection is reached at $\alpha = \alpha_c$ (2.96). The magnitude of α_c is very small - of the order of $|\chi_0'|$, i.e., typically $\lesssim 10^{-5}$ or even less. In the very first approximation α_c may be neglected and taken as zero. Physically this would mean that the effect of refraction at the crystal-vacuum interface is neglected. This is generally known as the kinematic approximation. The theory of Bragg diffraction, which neglects refraction as well as multiple scattering effects, is called the kinematic theory, see, e.g., James (1950).

In the kinematic approximation, Bragg's law (2.101) relates the radiation wavelength λ to the glancing angle of incidence, which is termed the Bragg angle θ_B, at which the radiation wave is reflected from the atomic planes with maximum reflectivity.

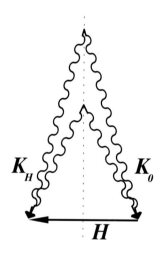

Fig. 2.10. Those wave vectors K_0 obey Bragg's condition (2.110), which start at the perpendicular bisector of the vector H, i.e., those wave vectors K_0 which start from a facet of the Brillouin zone normal to H.

It is often convenient to use Bragg's law in vector form. To derive it we apply the condition $\alpha = 0$ to the definition of α (2.19), and obtain [8]

$$2K_0 H + H^2 = 0. \tag{2.110}$$

Equation (2.110) implies, that those incident radiation waves fulfill Bragg's condition, whose wave vectors K_0 start at the perpendicular bisector of the

---

[7] Due to this α can be understood as a measure of deviation from Bragg's condition (2.101). Therefore, it is called the "deviation parameter".

[8] Alternatively, (2.110) can be derived from momentum conservation $K_H = K_0 + H$, and energy conservation $\hbar |K_H| c = \hbar |K_0| c$, with the additional momentum transfer Δ_H (2.15) due to refraction at the vacuum-crystal interface being neglected.

vector $\boldsymbol{H}$, i.e., whose wave vectors $\boldsymbol{K}_0$ start from a facet of the Brillouin zone normal to $\boldsymbol{H}$, as shown in Fig. 2.10.

Bragg's law (2.101) gives an approximate relation between the photon wavelength and the glancing angle of incidence of the wave, which is reflected with maximal reflectivity. It is a crude approximation, which does not account for a great many of subtle, and very important effects in Bragg diffraction from perfect crystals with far reaching consequences in many areas of high-resolution x-ray optics, as we shall see in the following chapters of this book. A more precise relation will be obtained in the following section. In the framework of the dynamical theory, the approximate relation (2.101) is often called the kinematic Bragg's law, to distinguish it from a more precise modified (or dynamical) Bragg's law.

2.4.6 Modified Bragg's Law for the Incident Wave

In Sect. 2.4.3 the region of total Bragg reflection was determined in terms of the deviation parameter α. Having established in Sect. 2.4.4 the relations between the deviation parameter and the physical parameters of the problem, the next step is to determine the region of total reflection in terms of the wavelength (photon energy), and the angles of incidence.

Using (2.4) we may write for the center of the region of total reflection α_c (2.96) in terms of the radiation wavelength λ_c:

$$\alpha_c = -2 w_H^{(s)} \left(\frac{\lambda_c}{2 d_H} \right)^2 \left(1 - \frac{1}{b} \right), \qquad (2.111)$$

where

$$w_H^{(s)} = -2 \chi_0' \frac{d_H^2}{\lambda^2} \equiv \frac{2 r_e d_H^2}{\pi V} F_0'. \qquad (2.112)$$

The glancing and azimuthal angles of incidence θ_c and ϕ_c, respectively, are indirectly included in (2.111) via the asymmetry factor b. The asymmetry factor is given by $b = \gamma_0 / \gamma_H$ (2.29), where the direction cosines γ_0, γ_H are defined according to (2.26)-(2.27)[9]. With these equations taken into account the asymmetry factor reads

$$b = \frac{\cos \theta_c \sin \eta \cos \phi_c + \sin \theta_c \cos \eta}{\cos \theta_c \sin \eta \cos \phi_c - \sin \theta_c \cos \eta}. \qquad (2.113)$$

Combining (2.111) with the expression (2.98) for the deviation parameter valid in the whole range of glancing angles we obtain

---

[9] In the framework of the theoretical model under discussion, the terms quadratic in $\varkappa$, and thus also in α, are ignored in the fundamental equations of the dynamical theory (2.40). For the same reason, we may also omit α in the definition of γ_H in (2.27).

$$2d_H \sin\theta_c = \lambda_c(1+w_H), \qquad (2.114)$$

where

$$w_H = w_H^{(s)}\frac{b-1}{2b}. \qquad (2.115)$$

The upper index (s) in $w_H^{(s)}$ stands for the symmetric scattering geometry.

Equivalently, the expression for the center of the region of total reflection in terms of the photon energy E_c reads

$$E_c \sin\theta_c = E_H(1+w_H). \qquad (2.116)$$

Equations (2.114)-(2.116) relate the radiation wavelength λ_c and photon energy E_c, respectively, to the glancing angle of incidence θ_c at the center of the region of total reflection. The other angles characterizing the scattering geometry: the azimuthal angle of incidence ϕ_c, and the asymmetry angle η, also enter equations (2.114)-(2.116), however, indirectly via the asymmetry factor b (2.113).

Equation (2.114) looks like Bragg's law (2.101); however, there is a small but significant correction w_H, which will be henceforth termed the "Bragg's law correction". Since the Bragg's law correction w_H is always positive in the Bragg scattering geometry, in which $b<0$, it shifts the reflectivity peak to larger angles θ_c or smaller wavelength λ_c (higher energies E_c), as compared to those given by Bragg's law. Henceforth, relation (2.114) or (2.116), is referred to as the modified (or dynamical) Bragg's law. The general expression for α (2.98) allows the dynamical Bragg law to be formulated in the whole range of glancing angles of incidence without division into "standard" and "backscattering" regimes.

The quantity $w_H^{(s)}$ (2.115) is the Bragg's law correction for the case of $b = -1$. As follows from (2.113), this occurs at least in three different cases. First, when the asymmetry angle $\eta = 0$, i.e., in the symmetric Bragg scattering geometry. This is the reason why $w_H^{(s)}$ is supplied with the upper index (s). Second, when the azimuthal angle of incidence $\phi_c = \pi/2$. Third, at normal incidence on the reflection atomic planes when $\theta_c = \pi/2$.

The quantity $w_H^{(s)}$ may be regarded as a constant for a given Bragg reflection, i.e., independent of incidence angle and wavelength (photon energy). This holds unless the wavelength (photon energy) dependence of the anomalous scattering corrections f' to F_0' (2.5)-(2.6) has to be taken into account[10].

---

[10] The anomalous scattering correction $f'(E)$ can be considered to be constant with a relatively high accuracy in a relatively large spectral range for a given Bragg reflection. For example, in the case of silicon the anomalous scattering correction f' varies at most by ± 0.003 around its average value of 0.119 in the spectral range $E_M - 200$ eV $\leq E_c \leq E_M + 200$ eV where $E_M = 14.4125$ keV, and thus w_H varies by less than 10^{-8} compared to the leading term 1 in (2.116). These estimates are based on the measurements of $f'(E)$ by Deutsch and Hart (1988).

If we neglect this weak energy (wavelength) dependence, the correction $w_H^{(s)}$ can be expressed using (2.112), and (2.55) as

$$w_H^{(s)} = 1 - n'(2d_H), \qquad (2.117)$$

i.e., it equals the deviation from unity of the real part of the refractive index taken at the *fixed* wavelength $\lambda = 2d_H$.

Typically $w_H^{(s)} \simeq 10^{-4} - 10^{-6}$, depending on the chosen crystal and on the interplanar distance of the reflecting atomic planes. The values of the quantity $w_H^{(s)}$ one can look up, e.g., in the tables of Bragg reflections in silicon and sapphire (α-Al$_2$O$_3$) single crystals presented in Appendices A.3 and A.4.

The radiation waves aimed at the crystal at the glancing angle of incidence θ_c are totally reflected not only if the wavelength equals λ_c (2.114). They can be also reflected if the wavelengths take values in the vicinity of λ_c in the range between λ_{+1} and λ_{-1}, determined by the edges α_{+1} and α_{+1} (2.95) of the region of total reflection. Using (2.95), (2.98), and (2.4) we find the following relation between $\lambda_{\pm 1}$ and the glancing angle of incidence θ_c:

$$2d_H \sin\theta_c = \lambda_{\pm 1}\left(1 + w_H \pm \frac{\epsilon_H}{2}\right), \qquad (2.118)$$

where

$$\epsilon_H = \frac{\epsilon_H^{(s)}}{\sqrt{|b|}}, \qquad \epsilon_H^{(s)} = \frac{4\,r_e\,d_H^2}{\pi\,V}\,|PF_H|. \qquad (2.119)$$

The quantity ϵ_H is the relative spectral width for a given Bragg reflection, assuming a plane wave at the glancing angle of incidence θ_c:

$$\epsilon_H = \frac{\lambda_{-1} - \lambda_{+1}}{\lambda} = \frac{\Delta\lambda}{\lambda} \equiv \frac{\Delta E}{E}. \qquad (2.120)$$

Since $\epsilon_H = \epsilon_H^{(s)}$ for $b = -1$, the parameter $\epsilon_H^{(s)}$ is the relative spectral width of the Bragg reflection in the symmetric scattering geometry. The upper index (s) in $\epsilon_H^{(s)}$ indicates the symmetric scattering geometry.

Applying similar argumentation to $\epsilon_H^{(s)}$ as to $w_H^{(s)}$ the quantity $\epsilon_H^{(s)}$ can be regarded as a constant in a broad spectral range for a given Bragg reflection. We shall discuss this point in more detail in Sect. 2.4.7.

Equations (2.114)-(2.115), and (2.118)-(2.119) describe the region of total Bragg reflection in terms of radiation wavelengths and angles of incidence. The glancing angle of incidence θ explicitly enters the aforementioned equations. The other angles characterizing the scattering geometry: the azimuthal angle of incidence ϕ, and the asymmetry angle η, enter these equations indirectly via the asymmetry factor b (2.113).

2.4.7 Reflection Region. Symmetric Scattering Geometry

We shall start the analysis of the equations obtained in the previous section with the simplest case of Bragg diffraction in symmetric geometry, when

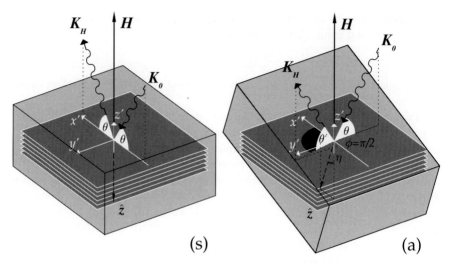

Fig. 2.11. Examples of Bragg scattering geometries characterized by the asymmetry factor $b = -1$: (s) asymmetry angle $\eta = 0$, and (a) $\eta \neq 0$, azimuthal angle of incidence $\phi = \pi/2$.

the asymmetry angle $\eta = 0$, as shown, e.g., in Fig. 2.11(s). In this case the asymmetry factor $b = -1$ for all glancing and azimuthal angles of incidence. The asymmetry factor may take the value of $b = -1$ also if $\eta \neq 0$. This happens, however, in the special scattering geometry with azimuthal angle of incidence $\phi = \pi/2$, as shown, e.g., in Fig. 2.11(a).

The trace of the region of total reflection in the whole (λ, θ) space calculated using the modified Bragg's law (2.114) and (2.118), is shown in Fig. 2.12[11]. All Bragg scattering geometries with $b = -1$ are represented by this graph. An equivalent presentation of the region of total reflection in the (E, θ) space is shown in Fig. 1.3 on page 5. For comparison, also the trace of the peak reflectivity as given by (kinematic) Bragg's law (2.101) is given.

Due to the Bragg's law correction w_H the region of total reflection predicted by the dynamical theory occurs for radiation wavelengths, which are shorter than the wavelength predicted by Bragg's law (2.101). It looks as if the distance between the reflecting planes is $d_H/(1+w_H)$ instead of d_H. This effect can be attributed to refraction. Indeed, according to (2.117), the Bragg's law correction is directly related to the real part of the refractive index taken, however, at a fixed wavelength $\lambda = 2d_H$.

The center of the region of total reflection on the wavelength scale λ_c changes almost linearly with the glancing angle of incidence θ - see Fig. 2.12 and (2.114). However, in the backscattering region, where the glancing angle of incidence $\theta \to \pi/2$, this dependence becomes practically quadratic.

---

[11] The graphical presentation of the Bragg reflection region in the (λ, θ) space is often referred to as a DuMond diagram.

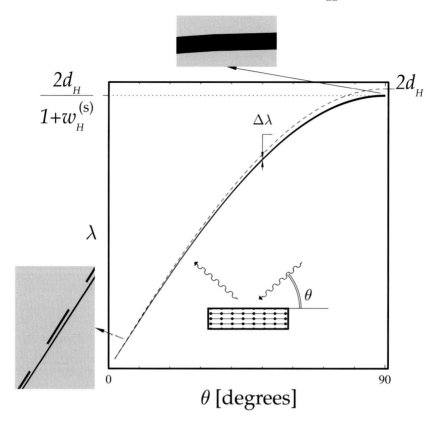

Fig. 2.12. Crystal reflectivity for x rays in the symmetric Bragg scattering geometry presented in the (λ, θ) space. Dashed line: trace of the Bragg peak reflectivity as given by the (kinematic) Bragg law (2.101). Solid line: the trace of the region of total reflection of x rays, as given by the dynamical theory of x-ray Bragg diffraction in perfect crystals (2.114), (2.118). The relative spectral width $\Delta\lambda/\lambda$ of the region is constant over the whole spectral and angular range. An equivalent presentation in the (E, θ) space is shown in Fig. 1.3 on page 5.

Modified Bragg's law (2.114)-(2.116) takes in backscattering the following form:

$$\lambda_c = \lambda_R^{(s)} \left(1 - \frac{\Theta_c^2}{2}\right), \quad E_c = E_R^{(s)} \left(1 + \frac{\Theta_c^2}{2}\right). \tag{2.121}$$

Here $\Theta_c = \pi/2 - \theta_c$, and

$$\lambda_R^{(s)} = \frac{2d_H}{1 + w_H^{(s)}}, \quad E_R^{(s)} = E_H(1 + w_H^{(s)}), \tag{2.122}$$

are the wavelength or photon energy, respectively, at the center of the region of total Bragg reflection at normal incidence, and $E_H = hc/2d_H$ (2.100) is the Bragg energy. The upper index (s) indicates the symmetric scattering geometry.

As follows from (2.121), because of the quadratic dependence on Θ_c, the energy (wavelength) of reflected photons changes extremely slowly with incidence angle in backscattering.

Spectral Width of the Reflection Region. The relative spectral width $\Delta\lambda/\lambda \equiv \Delta E/E$ of the region of total reflection in the symmetric scattering geometry ($b = -1$) is given by $\epsilon_H^{(s)}$ (2.119)-(2.120). Inserting (2.5)-(2.8) into (2.119), and neglecting the weak energy (wavelength) dependence of the atomic scattering amplitudes $f_n(\boldsymbol{H})$ (2.6), we obtain for $\epsilon_H^{(s)}$:

$$\epsilon_H^{(s)} = \frac{4|P|r_e d_H^2}{\pi V} \left| \sum_n f_n^{(0)}(\boldsymbol{H}) \exp\left(i\boldsymbol{H}\boldsymbol{r}_n - 2\pi^2 \frac{\langle v_H^2 \rangle_n}{d_H^2} \right) \right|. \qquad (2.123)$$

Equation (2.123) directly demonstrates that the relative spectral width of a given Bragg reflection depends only on the atomic and crystal parameters, such as the atomic form factor $f_n^{(0)}(\boldsymbol{H})$, the interplanar distance d_H, the scattering vector $\boldsymbol{H}$, and the mean square displacement of atoms from the equilibrium position $\langle v_H^2 \rangle_n$. The relative spectral width is, however, independent of the glancing angle of incidence or the energy of the incident radiation.

One may derive the following general rule from (2.123): the smaller the distance d_H between the reflecting atomic planes, or equivalently the greater the absolute value of the scattering vector $|\boldsymbol{H}| = 2\pi/d_H$, or the Bragg energy $E_H = hc/2d_H$ (2.100), the narrower is the relative spectral width of the Bragg reflection. The relative spectral width of the Bragg reflections decreases rapidly with the Bragg energy E_H.

The invariance of the relative spectral width implies that the *absolute* value of the spectral width for the given Bragg reflection in terms of energy units takes its smallest value for the smallest photon energy, i.e., at normal incidence ($\theta = \pi/2$). Figures 1.3 and 2.12 illustrate this statement graphically. The spectral width of Bragg reflections at normal incidence on different atomic planes in silicon and sapphire (α-Al$_2$O$_3$) single crystals is presented in Appendices A.3 and A.4[12].

The relative spectral width can be also related to the effective number of reflecting atomic planes participating in Bragg diffraction. For this, let us combine (2.119) with the expression (2.90) for the extinction length at the center $y = 0$ of the total reflection region. Assuming as before the symmetric diffraction geometry we obtain

---

[12] The values presented in Appendices A.3 and A.4 were calculated taking into account photo-absorption, whose influence is, however, relatively small.

$$\epsilon_H^{(s)} = \frac{d_H}{\pi d_e(0)} = \frac{1}{\pi N_e}. \tag{2.124}$$

Here $N_e = d_e(0)/d_H$ is the number of reflecting planes throughout the extinction length. Thus the relative spectral width of a Bragg reflection is the reciprocal of the effective number of the atomic planes contributing to Bragg reflection (times π). The general tendency is - the higher the x-ray energy (the smaller the wavelength) the larger is the extinction length and therefore the smaller is the relative spectral width.

For x rays with $E \simeq 15$ keV the extinction length is typically in the range of $d_e(0) \approx 10-100$ µm - see Appendices A.3 and A.4. Thus the relative spectral width is $\approx 10^{-6} - 10^{-7}$, and the energy widths of $\Delta E \simeq 10 - 1$ meV.

Invariance of the relative spectral width of Bragg reflection is valid in the symmetric scattering geometry with $b = -1$. The influence of the asymmetric scattering geometry on the spectral properties of the Bragg reflection will be discussed in Sect. 2.4.8.

Angular Width (1): Region of $\theta < \pi/2$. If the radiation wavelength (photon energy) is fixed, say at $\lambda_c = hc/E_c$, the incident wave can be totally reflected by aiming it at different angles θ to the atomic planes in the vicinity of θ_c (2.114). The glancing angles of incidence θ, for which total reflection takes place, are in the range between θ_{+1} and θ_{-1}, which are determined by the edges $\alpha_{\pm 1}$ (2.95) of the region of total reflection, and by (2.98). There is no simple analytical expression for the angular width $\Delta = \theta_{+1} - \theta_{-1}$ of the region for total reflection valid throughout the whole range of glancing angles of incidence.

We assume here that the glancing angle of incidence is *not* in the immediate proximity to normal incidence, i.e., $\theta_c < \pi/2$. Using (2.98), and (2.114) we obtain $\delta\alpha = -2\delta\theta \sin 2\theta_c$ for the variation of the deviation parameter with glancing angle of incidence. Combining this with the expression for the width $\Delta\alpha$ of the reflection region in terms of α (2.97) we readily obtain for the angular width

$$\Delta\theta = \Delta\theta^{(s)} \frac{1}{\sqrt{|b|}}, \qquad \Delta\theta^{(s)} = \frac{2|P\chi_H|}{\sin 2\theta_c}. \tag{2.125}$$

Here $\Delta\theta^{(s)}$ is the angular width of the region of total reflection in the symmetric scattering geometry ($b = -1$). The Fourier component of the electric susceptibility of the crystal χ_H contains an energy (wavelength) dependence, c.f., (2.4). Substituting E_c for $E_H/\sin\theta_c$ in χ_H (according to (2.116) with the small Bragg's law correction w_H being omitted) we obtain

$$\Delta\theta^{(s)} = \epsilon_H^{(s)} \tan\theta_c. \tag{2.126}$$

Thus, for a given Bragg reflection the angular width of the region of total reflection $\Delta\theta$ increases as $\tan\theta_c$ with increasing glancing angle of incidence θ_c.

The proportionality constant $\epsilon_H^{(s)}$ is the relative spectral width of the region of total reflection in the symmetric scattering geometry, as given by (2.119).

The angular width of Bragg reflections is small. Depending on the actual value of $\epsilon_H^{(s)}$ it is about $10^{-4} - 10^{-6}$, and even smaller. However, close to normal incidence, because of the $\tan\theta_c$-dependence, it increases. In the backscattering region where $\theta_c \simeq \pi/2$ (2.126) is no more valid.

Angular Width (2): Region of $\theta \to \pi/2$. In the backscattering geometry, another presentation of the deviation parameter α has to be used - namely $\alpha = 2\Theta^2 - 4\epsilon$ (2.104) - to derive the expression for the angular width of the reflection region. Here as previously, $\epsilon = (E - E_H)/E_H$ is a relative deviation of the photon energy E from the Bragg energy E_H. In the backscattering region $\alpha_c = -4w_H^{(s)}$, and $\alpha_{\pm 1} = \alpha_c \mp 2\epsilon_H^{(s)}$. The approximation $E \simeq E_H$ was used to derive these relations from (2.95)-(2.96). As a result, we obtain for the center, and for the boundaries of the region of total reflection in backscattering geometry:

$$\Theta_c^2 = 2\left(\epsilon - w_H^{(s)}\right), \qquad \Theta_{\pm 1}^2 = \Theta_c^2 \mp \epsilon_H^{(s)}, \tag{2.127}$$

and for the angular width $\Delta\Theta \equiv \Delta\theta = \Theta_{-1} - \Theta_{+1}$:

$$\Delta\Theta = \sqrt{\Theta_c^2 + \epsilon_H^{(s)}} - \sqrt{\Theta_c^2 - \epsilon_H^{(s)}}, \qquad \text{for } \epsilon - w_H^{(s)} > \frac{\epsilon_H^{(s)}}{2}. \tag{2.128}$$

If the photon energy is in the range $-\epsilon_H^{(s)}/2 \leq \epsilon - w_H^{(s)} \leq \epsilon_H^{(s)}/2$, then the angular region of total reflection for the azimuthal angle of incidence ϕ merges with the angular regions for the azimuthal angle $\phi + \pi$ to form a single region with an angular width $\Delta\Theta = 2\Theta_{-1}$:

$$\Delta\Theta = 2\sqrt{\Theta_c^2 + \epsilon_H^{(s)}}, \qquad \text{for } -\frac{\epsilon_H^{(s)}}{2} \leq \epsilon - w_H^{(s)} \leq \frac{\epsilon_H^{(s)}}{2}. \tag{2.129}$$

The relations (2.127)-(2.129) are valid at near-normal incidence on the reflecting planes. They demonstrate one of the most striking features of Bragg backscattering. The angular acceptance scales with the *square root* of the relative spectral width! In particular, if $\epsilon - w_H^{(s)} = 0$, the center of the region is at *exact* normal incidence with $\Theta_c = 0$, and from (2.129) we find that

$$\Delta\Theta = 2\sqrt{\epsilon_H^{(s)}}. \tag{2.130}$$

Figure 2.13 presents graphically the spectral-angular region of total reflection in backscattering from a semi-infinite non-absorbing crystal. In fact it represents the part of Fig. 2.12 at $\theta \simeq \pi/2$ drawn on different scales. The ordinate of the graph is $\epsilon = (E - E_H)/E_H$ - the deviation of the x-ray energy E from the Bragg energy E_H measured in units of E_H. The abscissa is the square of the angular deviation from normal incidence Θ^2. The spectral-angular region

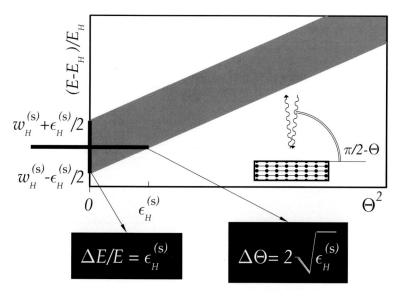

Fig. 2.13. Spectral-angular region of total reflection in backscattering, shown with the abscissa scaling as Θ^2.

of total reflection for a perfect non-absorbing crystal - a kind of DuMond diagram - is given in this representation by the almost straight stripe, according to (2.127)-(2.130). The width of the stripe along both axes scales with $\epsilon_H^{(s)}$ (2.119) - the relative spectral width of the given Bragg reflection in symmetric scattering geometry.

The spectral distribution of x rays reflected at normal incidence ($\Theta = 0$), is centered around $E = E_H(1 + w_H^{(s)})$. An important feature of the DuMond diagram for backscattering is that the angular scale - the abscissa - has a Θ^2 dependence, resulting in an unusually large width of the angular region of total reflection of $\Delta\Theta = 2\sqrt{\epsilon_H^{(s)}}$. This implies in particular that Bragg reflections with a relative spectral width as small as $\epsilon_H = 10^{-9}$ are still not blurred off if an x-ray beam has an angular divergence of $\lesssim 60$ μrad.

If the center of the angular region of total reflection deviates from normal incidence as much as $\Theta_c^2 \gg \epsilon_H^{(s)}$, then expression (2.128) can be transformed to $\Delta\Theta = \epsilon_H^{(s)}/\Theta_c$. In this form it agrees with (2.126) obtained for the case of glancing angles of incidence far from normal incidence. Thus, the value $2\sqrt{\epsilon_H^{(s)}}$, can be considered not only as a gauge for the angular width of a Bragg reflection in backscattering. The value

$$\Omega \sim 2\sqrt{\epsilon_H^{(s)}}$$

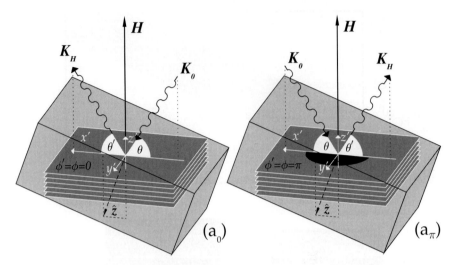

Fig. 2.14. Examples of asymmetric scattering geometries. (a_0) Scattering geometry characterized by an asymmetry factor $b < -1$, and azimuthal angles $\phi = \phi' = 0$. (a_π) Scattering geometry characterized by $-1 < b < 0$, and $\phi = \phi' = \pi$.

The configurations (a_0) and (a_π) represent particular cases with wave vectors K_0 and K_H in the dispersion plane (x', z'). The configuration (a_0) is transformed into (a_π) and vice versa by a 180° rotation of the wave vectors K_0 and K_H around the diffraction vector H. Note that $\theta(a_0) = \theta'(a_\pi)$ and $\theta(a_\pi) = \theta'(a_0)$.

can be also considered as a landmark on the scale of incidence angles, between the backscattering and non-backscattering regions.

2.4.8 Reflection Region. Asymmetric Scattering Geometry

The region of total reflection undergoes an essential change in the wavelength-angle space provided the asymmetry angle η becomes nonzero.

The effect of the asymmetry angle on the region of total reflection is formally described by the asymmetry factor b (2.113) which enters (2.114)-(2.115) and (2.118)-(2.119). The asymmetry factor is determined not only by the asymmetry angle η. It is also determined by the glancing angle of incidence θ_c, and the azimuthal angle of incidence ϕ_c. Therefore, both quantities $w_H = w_H^{(s)}(1-b)/2b$ and $\epsilon_H = \epsilon_H^{(s)}/\sqrt{|b|}$ defining the width and position of the reflection region are generally speaking functions of all these angles. Since $w_H^{(s)}$ and $\epsilon_H^{(s)}$ are small, and therefore, the angular width of the Bragg reflection is also small, one can consider the asymmetry factor to be a constant parameter in a small angular range in the vicinity of the reflection maximum. This is true, however, only if the incident wave is far from normal incidence on the reflecting atomic planes. If on the contrary it is close to normal incidence, i.e., $\theta \simeq \pi/2$, then the left hand part of (2.114) and (2.118) changes very slowly with θ and it is no longer possible to ignore the dependence of

b on θ and ϕ. As a consequence, the changes which occur to the region of total reflection due to a nonzero asymmetry angle η have to be discussed separately for the backscattering and off-backscattering geometries.

$\theta < \pi/2$. As a first step, we discuss the changes, which occur when the glancing angle of incidence is far from normal incidence.

Assuming the asymmetry factor b to be constant in a certain range of the angles of incidence, the changes in the region of total reflection due to a nonzero asymmetry angle η can be conveniently discussed in terms of a constant asymmetry factor b. Several (not equivalent) different scattering geometries may occur, which are characterized by the same value of the asymmetry factor b, but having different values of the angles η, θ, and ϕ. Despite this variety, one may subdivide the scattering geometries in two large groups representing two opposite tendencies in changes of spectral and angular properties of Bragg reflections. The first group is characterized by the asymmetry factor $b < -1$, and the second by $-1 < b < 0$. Figure 2.14(a_o) shows an example of the scattering geometry, in which the asymmetry factor is $b < -1$, while Fig. 2.14(a_π) shows an example of the scattering geometry with $-1 < b < 0$. Particular cases are presented, in which scattering takes place in the dispersion plane built up of the diffraction vector $\boldsymbol{H}$ and the inward normal $\hat{\boldsymbol{z}}$. That means, the azimuthal angle is either $\phi = 0$ or $\phi = \pi$. Similar scattering geometries in the two-dimensional presentation in the dispersion plane are shown in Fig. 2.5(a_o) and 2.5(a_π), respectively. The main difference between the scattering geometries shown in Figs. 2.14(a_o), 2.5(a_o), and those shown in Figs. 2.14(a_π), 2.5(a_π) is the value of the azimuthal angle of incidence. In the first case $\phi = 0$, and in the second case $\phi = \pi$.

The fragments of the region of total Bragg reflection in the (λ, θ) space corresponding to the asymmetric scattering geometries (a_o) and (a_π) are shown in Fig. 2.15. The reflection region labeled by (s) corresponds to the symmetric scattering geometry, as in Fig. 2.11(s). The dotted line labeled by (k) traces the peak reflectivity as given by the "kinematic" Bragg's law (2.101). Each region of total reflection is bounded by the borderlines given by (2.118)-(2.119) for a particular value of asymmetry factor b.

Figure 2.15 shows a small part of the (λ, θ) space with λ close to a selected wavelength λ_c, and θ close to $\theta_c^{(s)}$, the center of the reflection in the symmetric scattering geometry given by Bragg's law $\sin\theta_c^{(s)} = (\lambda_c/2d_H)(1 + w_H^{(s)})$. We assume that the glancing angles θ are far from normal incidence or grazing incidence.

Compared to the symmetric diffraction case $b = -1$ - Fig. 2.15(s) - two significant changes can be observed.

First, for a fixed glancing angle of incidence θ the reflection region shifts to larger wavelengths in the asymmetric case of $b < -1$ (a_o), and to smaller wavelengths in the case of $-1 < b < 0$ (a_π). The relative shift $\delta\lambda/\lambda_c$ is

$$\frac{\delta\lambda}{\lambda_c} = w_H^{(s)} - w_H = w_H^{(s)} \frac{b+1}{2b}$$

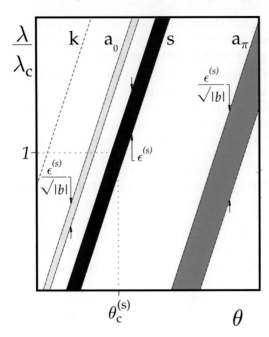

Fig. 2.15. Fragments of the regions of total Bragg reflection in the (λ, θ) space far from normal incidence, as given by the dynamical diffraction theory in the symmetric Bragg scattering geometry with $b = -1$ (s), in the asymmetric scattering geometry with $b < -1$ (a_o), and in the asymmetric scattering geometry with $-1 < b < 0$ (a_π). The schemes of the appropriate scattering geometries are shown in Figs. 2.11, and 2.14. The dashed line labeled by (k) traces the peak reflectivity in the kinematic approximation. See text for other notations.

as follows from (2.114)-(2.115). Similarly, for a fixed wavelength λ_c the center of the reflection region shifts by an angle

$$\delta\theta_c = -w_H^{(s)} \tan\theta_c^{(s)} \frac{b+1}{2b},$$

i.e., to a smaller angle in the case of $b < -1$ (a_o), and to a larger angles in the case of $-1 < b < 0$ (a_π).

Second, the relative spectral width of the reflection region, which is given by (2.119)

$$\epsilon_H = \frac{\epsilon_H^{(s)}}{\sqrt{|b|}},$$

becomes smaller in the asymmetric case of $b < -1$ (a_o), and vice versa, it becomes larger in the case of $-1 < b < 0$ (a_π). These changes are shown schematically in Fig. 2.15.

Similar changes occur with the angular width. It becomes smaller in the asymmetric case of $b < -1$ (a_o), and in the case of $-1 < b < 0$ (a_π) it becomes larger. The angular width is

$$\Delta\theta = \frac{\Delta\theta^{(s)}}{\sqrt{|b|}},$$

as given by (2.125).

These properties have great practical importance to x-ray optics. Due to these, the spectral and angular characteristics of Bragg reflections can be altered on demand by changing the asymmetry factor b. The application of these properties to monochromatization of x rays will be addressed in detail in Sect. 3.

Backscattering: $\theta \to \pi/2$. The fragments of the region of total Bragg reflection in the (λ, θ) space corresponding to the asymmetric scattering geometries (a_o) and (a_π) with $\phi = 0$ and $\phi = \pi$, respectively, are shown in Fig. 2.16. Each region of total reflection is shown as an area between the lines calculated by (2.118)-(2.119) for a particular value of the asymmetry angle $\eta \neq 0$. Figure 2.16 shows with magnification a small part of the (λ, θ) space for glancing angles θ adjacent to normal incidence ($\theta = \pi/2$). The angular range is from $\pi/2 - \Omega$ to $\pi/2$, where $\Omega \approx 5\sqrt{\epsilon_H^{(s)}}$. The reflection region labeled by (s) corresponds to the symmetric scattering geometry with $\eta = 0$, as in Fig. 2.11(s).

In the backscattering geometry the asymmetry factor b can no longer be considered as a constant. The explicit dependence of b (2.113) on the angles of incidence θ and ϕ has to be taken into account.

At normal incidence on the reflecting atomic planes, the glancing angle of incidence equals $\theta = \pi/2$, or equivalently, the angle of incidence equals $\Theta = 0$. Approaching normal incidence, the asymmetry factor tends to $b = -1$ for any asymmetry angle η (2.113). As a result, the widths of the reflection regions corresponding to scattering geometries with different asymmetry angle η tend to the same value, and the centers of the reflection regions themselves converge at normal incidence to $\lambda = \lambda_R^{(s)}$ (2.122) on the wavelength scale, as shown in Fig. 2.16.

Interestingly, the wavelength of the waves reflected from the crystal in the backscattering geometry (a_o) is no more a monotonically increasing function of the glancing angle of incidence θ. The wavelength achieves a maximum value at a glancing angle of incidence at $\theta \neq \pi/2$ and then drops at normal incidence to $\lambda = \lambda_R^{(s)}$.

As we know from the discussion in Sect. 2.2.6, exact backscattering (the incident wave reflects into itself) does not necessarily take place at normal incidence of the vacuum wave on the reflecting atomic planes. Exact backscattering takes place at normal incidence only if the asymmetry angle $\eta = 0$. For nonzero asymmetry angle η, exact backscattering takes place at glancing angles of incidence deviating from $\theta = \pi/2$. By energy and momentum conservation, exact backscattering is allowed only for those λ and θ, which are related by (2.37), and only for the azimuthal angle of incidence $\phi = 0$, i.e., in the scattering geometry (a_0). The locus of points in (λ, θ) space that obey the exact backscattering condition (2.37) is shown by the solid line in Fig. 2.16. The intersection of this line with the region of total Bragg reflection (a_o) presents the locus of points in (λ, θ) space for which total reflection in exact Bragg backscattering occurs. The intersection of this line with the

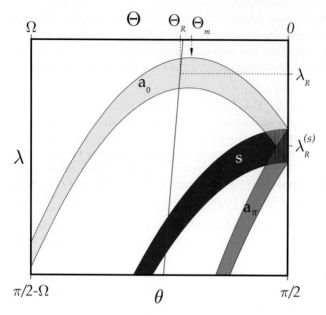

Fig. 2.16. Fragments of the regions of total Bragg reflection in (λ, θ) space in the vicinity of normal incidence $(\pi/2 - \Omega < \theta < \pi/2)$ to the reflecting atomic planes, as given by the dynamical diffraction theory in the symmetric Bragg scattering geometry with $\eta = 0$ (s), in asymmetric scattering geometry with $\eta \neq 0$ and $\phi = 0$ (a_o), and in asymmetric scattering geometry with $\eta \neq 0$ and $\phi = \pi$ (a_π). The schemes of the appropriate scattering geometries are shown in Figs. 2.11 and 2.14.

The solid line is the locus of points in (λ, θ) space that obeys the exact backscattering condition (2.37) in the asymmetric scattering geometry (a_o). It is not at $\theta = \pi/2$ due to refraction at the vacuum-crystal interface.

$\Omega \sim 4$ mrad, $\eta = 89.5°$, $w_H^{(s)} = 2 \times 10^{-5}$. See text for other notations.

region of total reflection (a_π) does not have the same physical meaning, since in the asymmetric scattering geometry (a_π) exact backscattering cannot be achieved at all.

Solving for λ and θ the modified Bragg's law (2.114) together with the condition for exact backscattering (2.37), we find the center of the region of *total exact Bragg back-reflection* in (λ, θ) space in the asymmetric scattering geometry. The solution denoted by θ_R and λ_R is given by

$$\theta_R = \pi/2 - \Theta_R$$

$$\tan \Theta_R = \frac{2 w_H^{(s)} \tan \eta}{1 + \sqrt{1 + 4 w_H^{(s)} \tan^2 \eta}}, \tag{2.131}$$

$$\frac{\lambda_R}{2 d_H} = \cos \Theta_R \left(1 - \frac{\tan \Theta_R}{\tan \eta}\right). \tag{2.132}$$

It is derived in Appendix A.8. If η is not too close to $\pi/2$, then the exact solutions (2.131)-(2.132) for the incidence angle Θ_R and the wavelength λ_R of exact backscattering can be accurately approximated by

$$\Theta_R \simeq w_H^{(s)} \tan\eta \left(1 - w_H^{(s)} \tan^2\eta\right), \tag{2.133}$$

$$\frac{\lambda_R}{2d_H} \simeq 1 - w_H^{(s)} \left(1 - \frac{w_H^{(s)} \tan^2\eta}{2}\right). \tag{2.134}$$

These expressions are obtained from the exact solutions (2.131)-(2.132) by performing Taylor expansions, and keeping terms up to second order in $w_H^{(s)}$.

It is important to note here that the point (λ_R, Θ_R) corresponding to exact backscattering does not coincide strictly speaking with the maximum of the reflection region (a_0), which is at $\Theta_m = \Theta_R - w_H^{(s)^2} \tan^3\eta$ (A.23). In most cases one can use $\Theta_m \simeq \Theta_R$. Yet, one has to respect that the difference $\Theta_R - \Theta_m$ increases with $\eta \to \pi/2$.

The difference between λ_R and the appropriate value $\lambda_R^{(s)}$ (2.122) valid in the symmetric scattering geometry is small because of the smallness of the Bragg's law correction $w_H^{(s)}$. For example, the relative difference

$$(\lambda_R - \lambda_R^{(s)})/\lambda_R^{(s)} = \left(w_H^{(s)} \tan\eta\right)^2/2.$$

In most cases this is a small shift, which can be neglected. Yet, for η close to $\pi/2$ the shift may become comparable or even greater than the width of the reflection region: $\left(w_H^{(s)} \tan\eta\right)^2 \gtrsim 2\epsilon_H^{(s)}$. Thus the effect of nonzero asymmetry becomes important if

$$\tan\eta \gtrsim \sqrt{2\epsilon_H^{(s)}/w_H^{(s)}}.$$

Such a case is shown in Fig. 2.16 where a large asymmetry angle of $\eta = 89.5°$ was chosen to draw examples of the reflection regions in the asymmetric scattering geometries.

The locus of points obeying the exact backscattering condition (2.37) is almost a straight line in Fig. 2.16. Assuming θ to be much closer to $\pi/2$ than η to $\pi/2$, we obtain by (2.37) for the inclination of the line

$$\frac{1}{2d_H} \frac{d\lambda}{d\theta} = \frac{\sin(\theta+\eta)}{\sin\eta} \simeq \frac{1}{\tan\eta}. \tag{2.135}$$

That means, the closer the asymmetry angle η to $\pi/2$, the smaller is the inclination of the line to the θ-axis.

The fact that the line is not vertical implies that at a given incidence angle the wave can be reflected into itself only at a single wavelength (photon energy) - that one obeying (2.37). If the wavelength deviates from this value

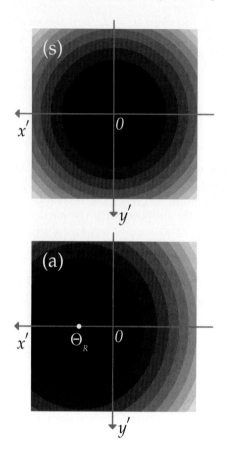

Fig. 2.17. Distribution of wavelengths determined by the modified Bragg's law (2.113)-(2.115) as a function of propagation direction of the incident wave with wave vector $\boldsymbol{K}_0$ (2.25).
The region of propagation directions corresponding to reflections of the incident waves with the greatest wavelengths is shown in black. The regions in white correspond to reflections with smallest wavelengths, which are only $\approx 10^{-5}$ smaller (in relative units) than the greatest wavelengths.
The propagation direction $\boldsymbol{K}_0/K$ is determined by the glancing angle of incidence θ and azimuthal angle ϕ, see Fig. 2.4, and, correspondingly, by the projection $(\cos\theta\cos\phi, \cos\theta\sin\phi)$ of $\boldsymbol{K}_0/K$ onto the (x',y') plane. The latter is parallel to the reflecting atomic planes.
(s) Symmetric Bragg scattering geometry with asymmetry angle $\eta = 0$.
(a) Asymmetric scattering geometry with asymmetry angle $\eta = 89°$.
$0 \leq \phi < 2\pi$,
$0 \leq \Theta < 2.6$ mrad $(\Theta = \pi/2 - \theta)$,
$w_H^{(s)} = 2 \times 10^{-5}$.

the wave can still be reflected, however not into itself. This is a manifestation of the angular dispersion of the reflected wave - the phenomenon which we have discussed briefly in general form in Sect. 2.2.4. We shall be concerned with angular dispersion in the next section in more detail.

Figure 2.16 illustrates the manifestation of the modified Bragg's law in symmetric and asymmetric backscattering geometries for the incident waves propagating parallel to the dispersion plane, i.e., when the azimuthal angle of incidence is either $\phi = 0$ or $\phi = \pi$. Figure 2.17 complements this graphical illustration and extends it to the general case of propagation directions with any azimuthal angle ϕ. The distribution of the wavelengths determined by the modified Bragg's law (2.113)-(2.115) is shown as a function of propagation direction of the incident wave, expressed in terms of the projection $(\cos\theta\cos\phi, \cos\theta\sin\phi)$ of the unit wave vector $\boldsymbol{K}_0/K$ onto the (x',y') plane (see Fig. 2.4).

The following main qualitative modification can be observed in the wavelength distribution in going from the case of symmetric Bragg scattering, presented in Fig. 2.17(s), to the case of asymmetric scattering, presented in Fig. 2.17(a): While the wavelength distribution remains axisymmetric[13] (an important circumstance!) the symmetry point shifts from $(0,0)$ to $(\Theta_R, 0)$.

2.4.9 Angular Dispersion

Angular dispersion is a phenomenon which accompanies Bragg diffraction in the asymmetric scattering geometry[14]. Angular dispersion can be described in the following way.

We start with the particular wavelength $\lambda_c = 2\pi/K_c$ of the incident wave, and fixed direction of its propagation $\boldsymbol{u}_0 = \boldsymbol{K}_0/K_c$ $(K_c = |\boldsymbol{K}_0|)$. The direction of the wave vector $\boldsymbol{K}_H$ of the reflected wave is given by the unit vector $\boldsymbol{u}_H = \boldsymbol{K}_H/K_c$. If the wavelength of the incident wave changes to $\lambda_c + \delta\lambda$ (without changing the direction of propagation) the direction of the reflected wave changes to $\boldsymbol{u}_H + \delta\boldsymbol{u}_H$, see Fig. 2.18. In Sect. 2.2.5 we have derived the general equation (2.22) for the variation $\delta\boldsymbol{u}_H$ of the wave vector direction with radiation wavelength. We continue the discussion of angular dispersion by analyzing (2.22).

If no grazing emergence of the reflected wave takes place, Δ_H can be closely approximated by the leading term $\Delta_H^{(1)}$ in (2.21). As a result the expression for angular dispersion (2.22) becomes

$$\delta\boldsymbol{u}_H = -\frac{\delta K}{K_c}\frac{\boldsymbol{H}}{K_c} - \frac{\delta\alpha}{2\gamma_H}\hat{\boldsymbol{z}}. \qquad (2.136)$$

It is important to stress here again that the variation $\delta\boldsymbol{u}_H$ of the direction of the wave vector $\boldsymbol{K}_H$ takes place only in the dispersion plane (see particular examples in Fig. 2.18 and Fig. 2.19).

The direction of the incident wave is assumed to be fixed with the angles of incidence equal to θ_c and ϕ_c. Taking this into account we can write for the variation of the deviation parameter α (2.98) with wavelength $\lambda_c \to \lambda_c + \delta\lambda$:

$$\delta\alpha = 4\frac{\delta\lambda}{\lambda_c}\sin^2\theta_c. \qquad (2.137)$$

Furthermore, inserting this expression into (2.136), using presentation (2.24) of the diffraction vector $\boldsymbol{H}$, and the surface normal $\hat{\boldsymbol{z}}$ in $\{x', y', z'\}$ coordinates, using the relation $\gamma_H \simeq \cos\theta_c \sin\eta \cos\phi_c - \sin\theta_c \cos\eta$ (2.27), and

---

[13] The distribution, in the case of asymmetric scattering, can be considered to be axisymmetric only in the very first approximation. This is valid for small angles of incidence Θ. In general, the distribution is *not* axisymmetric.

[14] It is worthwhile to recall here that the notion *dispersion* was introduced originally in optics in connection with the energy (wavelength) dependence of the index of refraction $n(E)$. An optical prism is a highly useful means for demonstrating dispersion and measuring $n(E)$. Interestingly, the situation with asymmetric diffraction of x rays, which we are discussing here, is similar to that of the optical prism.

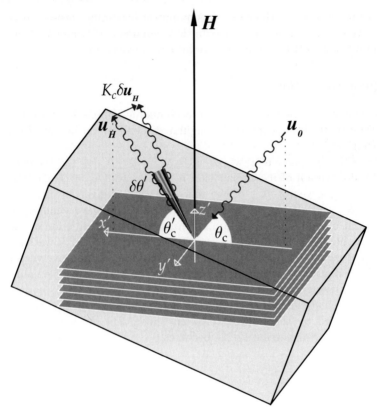

Fig. 2.18. Angular dispersion of the reflected radiation waves. The glancing angle of incidence is kept fixed at θ_c. For the radiation wavelength λ_c the glancing angle of reflection is θ_c'. Variation of the wavelength $\lambda_c \to \lambda_c + \delta\lambda$ results in variation of the reflection angle $\theta_c' \to \theta_c' + \delta\theta'$. The angular dispersion $\delta\theta'$ is greatly exaggerated compared to θ_c and θ_c' for clarity.
The particular case of the asymmetric scattering geometry of type a_0 ($\phi_c = \phi_c' = 0$, $b < -1$) is shown. In this case, the waves with larger wavelengths are reflected at smaller angles. The directions $\boldsymbol{u}_0$ and $\boldsymbol{u}_H$ of the vacuum wave vectors $\boldsymbol{K}_0$ and $\boldsymbol{K}_H$ of the incident and reflected waves, respectively, are shown with respect to the reflecting atomic planes (dark grey) in Bragg scattering geometry. The crystal contour is drawn with solid lines. For other notations see text.

Bragg's law $H/K_c \simeq 2\sin\theta_c$, and neglecting refraction correction terms, we obtain for the variation of the direction of the reflected wave with radiation wavelength

$$\delta\boldsymbol{u}_H = \frac{\delta\lambda}{\lambda_c} \frac{2}{\cot\theta_c \cos\phi_c - \cot\eta} \boldsymbol{v}_H. \qquad (2.138)$$

Here

$$\boldsymbol{v}_H = -\sin\theta_c\, \hat{\boldsymbol{x}}' + \cos\theta_c \cos\phi_c\, \hat{\boldsymbol{z}}'. \qquad (2.139)$$

is a vector perpendicular to $K(\cos\theta_c \cos\phi_c \hat{x}' + \sin\theta_c \hat{z}')$. The latter is the vector component of K_H (2.25) in the dispersion plane $(\hat{x}', \hat{z}')$, provided the difference between the angles of incidence and reflection are neglected.

As follows from (2.138), there is no angular dispersion, i.e., $\delta u_H = 0$, in the symmetric scattering geometry when the reflecting atomic planes are parallel to the crystal surface ($\eta = 0$).

The largest effect of angular dispersion is attained when the denominator in (2.138) approaches zero value, or equivalently when $\gamma_H \longrightarrow 0$. In other words, the largest effect of angular dispersion is attained while approaching the scattering geometry with grazing emergence of the reflected wave.

Let us discuss some particular cases.

$\theta < \pi/2$ and $\phi = 0$. First, we consider Bragg diffraction with wave vector K_0 of the incident wave lying in the dispersion plane ($\phi_c = 0$), as shown in Fig. 2.18. Equation (2.34′) requires automatically that the wave vector K_H of the reflected wave is also in the dispersion plane ($\phi'_c = 0$). We use $\gamma_H \simeq -\sin(\theta_c - \eta)$, $\gamma_0 = \sin(\theta_c + \eta)$, and $b = \gamma_0/\gamma_H$. Assuming also the incident wave to be far from normal incidence ($\theta_c < \pi/2$) we obtain from (2.138)

$$\delta u_H = \delta\theta'\, v_H, \qquad \delta\theta' = \frac{\delta\lambda}{\lambda_c}(1+b)\,\tan\theta_c. \qquad (2.140)$$

Apparently, the effect of angular dispersion increases with increasing the magnitude of the asymmetry factor $|b|$ and the glancing angle of incidence θ_c.

$\phi = \pi/2$. Another important Bragg scattering geometry of interest is with the wave vector K_0 of the incident wave lying in the plane (y', z') perpendicular to the dispersion plane (x', z'). In this case $\phi_c = \pi/2$, as shown in Fig. 2.19. The wave vector of the reflected wave $K_H = K_0 + H$ does not lie in the plane (y', z'), unless $\eta = 0$. This is because the total scattering vector $\tilde{H}$ is in the dispersion plane and not in the (y', z') plane. The reflection angles ϕ'_c and θ'_c of the wave with wavelength $\lambda = \lambda_c$ are given by:

$$\cot\phi'_c = -2w_H^{(s)}\tan\theta_c \tan\eta, \qquad (2.141)$$
$$\cos\theta'_c = \cos\theta_c\,(\sin\phi'_c)^{-1}.$$

These relations we have obtained using (2.34)-(2.34′), (2.27), (2.21), (2.111), (2.114), $b = -1$, and taking into account only the terms linear in $w_H^{(s)}$.

We would like to emphasize at this point, that the azimuthal angle of reflection ϕ'_c is larger than $\pi/2$, i.e., differs from the azimuthal angle of incidence $\phi_c = \pi/2$. In particular, the difference is greatest if either the glancing angle of incidence θ_c or asymmetry η approaches $\pi/2$.

If the wavelength deviates from λ_c by $\delta\lambda$ we find from (2.138) and (2.139) that the direction of the wave vector K_H changes by

90   2. Dynamical Theory of X-Ray Bragg Diffraction

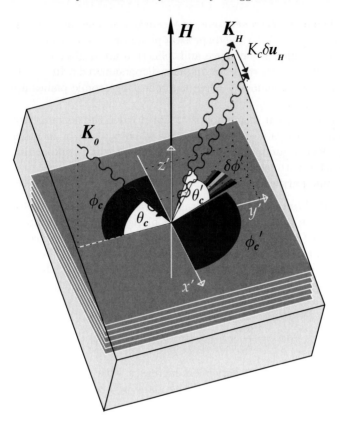

Fig. 2.19. Angular dispersion of the reflected x rays. The particular case of $\phi_c = \pi/2$ is shown. The glancing angle of incidence is fixed at θ_c. For the radiation wavelength λ_c the glancing angle of reflection is θ'_c and the azimuthal angle of reflection is ϕ'_c given by (2.141). Variation of the wavelength $\lambda_c \to \lambda_c + \delta\lambda$ results in variation of both the azimuthal reflection angle $\phi'_c \to \phi'_c + \delta\phi'$, and of the glancing angle of reflection $\theta'_c \to \theta'_c + \delta\theta'$ (not seen). The waves with larger wavelength are reflected with smaller ϕ'_c angle. The angular dispersion $\delta\phi'$ is shown greatly exaggerated compared to ϕ_c and ϕ'_c for clarity. Vacuum wave vectors $\boldsymbol{K}_0$ and $\boldsymbol{K}_H$ of the incident and reflected waves are shown with respect to the reflecting atomic planes (dark grey) in Bragg scattering geometry. For other notations see text.

$$\delta\boldsymbol{u}_H = 2\frac{\delta\lambda}{\lambda_c}\sin\theta_c\tan\eta\,\hat{\boldsymbol{x}}',\qquad(2.142)$$

as shown schematically in Fig. 2.19. Furthermore, using (2.25), (2.34′), (2.141), and taking $\theta \simeq \theta'$ we obtain that both the glancing and azimuthal angle of reflection are changing in this case as

$$\delta\phi' = -2\frac{\delta\lambda}{\lambda_c}\tan\theta_c\tan\eta\sin^2\phi'_c, \qquad (2.143)$$

$$\delta\theta' \simeq -2\frac{\delta\lambda}{\lambda_c}\tan\eta\cos\phi'_c.$$

In particular, if the wavelength variation is as large as the spectral width of the Bragg reflection, i.e., $\delta\lambda/\lambda_c = -\epsilon_H^{(s)}$, then by (2.126) we write $-\delta\lambda/\lambda_c \tan\theta_c = \Delta\theta^{(s)}$, where $\Delta\theta^{(s)}$ is the angular width of the Bragg reflection. As a result, we find (approximating $\sin\phi'_c$ by 1) that the waves are reflected with an angular spread of $\Delta\phi'$ in the azimuthal angle of reflection:

$$\Delta\phi' = 2\Delta\theta^{(s)}\tan\eta. \qquad (2.144)$$

The quantity $\Delta\phi'$ differs by a factor $2\tan\eta$ from the angular acceptance $\Delta\theta^{(s)}$ of the Bragg reflection, in terms of $\theta^{(s)}$. The angular spread $\Delta\phi'$ becomes especially large when the asymmetry angle η approaches $\pi/2$.

$\boldsymbol{\theta \simeq \pi/2}$. In backscattering geometry the glancing angle of incidence is in the range $\pi/2 - \Omega \lesssim \theta_c \leq \pi/2$, where $\Omega \sim 4\sqrt{\epsilon_H^{(s)}}$. The azimuthal angle of incidence ϕ_c may have any value in the range from 0 to 2π. We assume that $\theta_c > \eta$. We also assume as before that the radiation waves are not at grazing incidence or emergence. Under these conditions, and taking $\lambda_c \simeq 2d_H$ we obtain from (2.138)-(2.139) the expression for angular dispersion in backscattering:

$$\delta\boldsymbol{u}_H = 2\frac{\delta\lambda}{2d_H}\tan\eta\ \hat{\boldsymbol{x}}', \qquad \delta\theta' \simeq -2\frac{\delta\lambda}{2d_H}\tan\eta\ (\cos\phi'_c)^{-1}. \qquad (2.145)$$

As follows from all the discussed examples the greatest effect of angular dispersion is achieved in approaching backscattering geometry ($\theta_c \to \pi/2$), and with the asymmetry angle η approaching $\pi/2$. We shall see that angular dispersion is important for understanding the propagation of exit waves in asymmetric Bragg diffraction.

2.4.10 Modified Bragg's Law for the Exit Wave

We have ascertained in Sect. 2.2.6 that the angle of incidence is generally speaking not equal to the angle of reflection for the vacuum waves unless the asymmetry angle $\eta = 0$. This concerns both the glancing angles θ and θ', as well as the azimuthal angles ϕ and ϕ'. The effect of angular dispersion addressed in the previous section is closely related to this fact.

Having established in the previous sections the relation between the wavelength and the angles of incidence of the wave totally reflected off the crystal - modified Bragg's law - the next step is to derive the relation between the wavelength of the incident wave and the glancing angle of reflection.

Equations (2.34)-(2.34″) of Sect. 2.2.6 relate the angles of incidence and the reflection angles in the most general case. Substituting $H/K = 2\sin\theta +$

$\alpha K/H$ (2.28), as well as $\Delta_H = \Delta_H^{(1)} + \Delta_H^{(2)} + \ldots$ with $\Delta_H^{(1)} = -K\alpha/(2\gamma_H)$, etc., (2.21) into (2.34″), using $(H/K)\cos\eta = \gamma_0 - \gamma_H$ (2.27), and $b = \gamma_0/\gamma_H$ (2.29), we obtain the relation between the glancing angle of incidence θ and the glancing angle of reflection θ' in the following form:

$$\sin\theta' = \sin\theta - \xi, \tag{2.146}$$

where

$$\xi = \sum_{n=1,2,3,\ldots} \xi^{(n)},$$

$$\xi^{(1)} = \frac{\Delta_H^{(1)}}{K}\frac{b+1}{b-1}\cos\eta \equiv \frac{\Delta_H^{(1)}}{K}\frac{\sin\eta\cos\phi}{\tan\theta}, \tag{2.147}$$

$$\xi^{(n)} = \frac{\Delta_H^{(n)}}{K}\cos\eta, \quad n = 2, 3, \ldots.$$

If the asymmetry angle $\eta = 0$, one can show that $\xi = 0$, and the glancing angle of reflection equals the glancing angle of incidence: $\theta' = \theta$. As follows from (2.34′), simultaneously the azimuthal angles $\phi' = \phi$. If the asymmetry angle $\eta \neq 0$ then also the parameter $\xi \neq 0$, and the angles of incidence and reflection are not equal.

As follows from (2.21) $\Delta_H^{(1)} \gg \Delta_H^{(2)}$, $\Delta_H^{(2)} \gg \Delta_H^{(3)}$, etc. Therefore, $\xi^{(1)} \gg \xi^{(2)}$, $\xi^{(2)} \gg \xi^{(3)}$, etc., and the terms $\xi^{(n)}$ with $n \geq 2$ can be usually omitted in (2.147). However, this is not always the case: unless the glancing angle of incidence θ is close to $\pi/2$ (backscattering) or/and the azimuthal angle of incidence ϕ is close to $\pi/2$. If any of these cases, or both, take place, $\xi^{(1)}$ may become not only comparable or smaller than $\xi^{(2)}$, it may become zero. As a result, under these conditions $\xi^{(2)}$ is taking a leading role, and therefore may not be omitted.

Equations (2.146)-(2.147) are valid for any radiation wavelength. In particular, if the wave impinging on the crystal at a glancing angle of incidence θ_c has a wavelength of λ_c which relates to θ_c by the modified Bragg's law (2.114), then applying (2.146)-(2.147) to λ_c, θ_c, and θ'_c, using (2.21), (2.27), (2.111), and (2.114) we obtain

$$\xi^{(1)} = \frac{w_H^{(s)}}{2}\left(b - \frac{1}{b}\right)\frac{\lambda_c}{2d_H}$$

$$\xi^{(2)} = \frac{1}{2}\left(\frac{w_H^{(s)}}{2b\cos\eta}\right)^2 \left(\frac{1-b}{2}\right)^5 \frac{\lambda_c}{2d_H} \tag{2.148}$$

and for the glancing angle of reflection θ'_c:

$$2d_H \sin\theta'_c = \lambda_c \left(1 + w'_H\right), \tag{2.149}$$

$$w'_H = w_H^{(s)}\frac{1-b}{2} - \left(\frac{w_H^{(s)}}{2b\cos\eta}\right)^2 \left(\frac{1-b}{2}\right)^5 + \ldots.$$

The equation establishes a relation of the glancing angle of reflection θ'_c at the center of the total reflection region to the radiation wavelength λ_c, as well as to the glancing angle of incidence θ_c, and azimuthal angle of incidence ϕ_c via the asymmetry factor b given by (2.113). The equation is valid for any angle of incidence. Equation (2.149) looks similar to the modified Bragg's law (2.114). By analogy, it can be called "modified Bragg's law for the exit wave", and the quantity w'_H as the correction to Bragg's law for the exit wave. In general, $w'_H \neq w_H$, and therefore also $\theta'_c \neq \theta_c$. The term containing $(w_H^{(s)})^2$ originates from $\xi^{(2)}$ (2.148). As we have already mentioned, it can be very often neglected, however, it becomes important as one approaches the backscattering condition (provided also $\eta \neq 0$).

As we know, for a fixed glancing angle of incidence θ_c the wavelength λ_c marks the center of the spectral region of total Bragg reflection. The whole reflection region is in the range from λ_+ to λ_-, as given by (2.118)-(2.119). The glancing angles of reflection $\theta'_\pm$ corresponding to $\lambda_\pm$ can be determined by applying (2.146) to $\theta'_\pm$, θ_c, and $\lambda_\pm$. Further, by eliminating $\sin\theta_c$ from (2.146) with the help of (2.118) we obtain for the glancing angles of reflection $\theta'_\pm$:

$$2d_H \sin\theta'_\pm = \lambda_\pm \left(1 + w'_H \pm \frac{\epsilon'_H}{2}\right), \qquad \epsilon'_H = \epsilon_H^{(s)}\sqrt{|b|}. \qquad (2.150)$$

By analogy to ϵ_H, the quantity ϵ'_H will be referred to as the relative spectral width for the exit waves.

Equation (2.150) tells us that for a fixed glancing angle of incidence θ_c the glancing angle of reflection changes from θ'_+ to θ'_- with the radiation wavelength changing from λ_+ to λ_-. This is again a manifestation of angular dispersion of the reflected wave.

2.4.11 Reflection Region in the Space of Exit Waves

In Sect. 2.4.8, we discussed how the properties of the region of total reflection in Bragg diffraction change with the asymmetry angle η, and asymmetry factor b, respectively. The discussion was carried out in terms of the wave incident on the reflecting crystal, i.e., in (λ, θ) space. Having obtained in the previous sections the relationship between the wavelength of the incident wave and the glancing angle of reflection we can now study what does the reflection region look like in (λ, θ') space - in the space of the waves reflected from the crystal - for each particular case of asymmetric diffraction. The discussion of these matters will be pursued separately for the glancing angle of incidence far from normal incidence $(\theta < \pi/2)$, and for the glancing angle of incidence close to normal incidence, i.e., in backscattering $(\theta \to \pi/2)$.

$\theta < \pi/2$. Fragments of the regions of total Bragg reflection for incident waves in the (λ, θ) space, in scattering geometries with different asymmetry

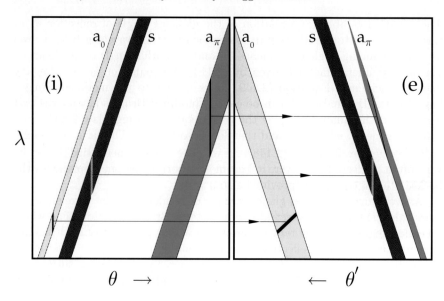

Fig. 2.20. Transformation of the Bragg reflection region from the (λ, θ) space of incident waves (i), into the (λ, θ') space of exit waves (e). Schematic presentation of fragments of the Bragg reflection regions according to (2.151), (2.152), and (2.153) in different scattering geometries: (s) symmetric Bragg scattering geometry with $b = -1$, (a_0) asymmetric scattering geometry with $b < -1$, and (a_π) asymmetric scattering geometry with $-1 < b < 0$. The schemes of the appropriate scattering geometries are shown in Fig. 2.11 and Fig. 2.14, respectively. The reflection regions in panel (e) are shown on the inverted angular scale. For more details, see text.

factors b, are shown in Fig. 2.15 and Fig. 2.20(i). Figure 2.20(e) shows the same reflection regions, however, now in the (λ, θ') space of the exit waves. As previously, the stripes denoted by a_π are fragments of the reflection region in asymmetric scattering geometry with $-1 < b < 0$, the stripes denoted by a_0 are fragments of the reflection region in asymmetric scattering geometry with $b < -1$, and s denotes the fragments of the reflection region in symmetric Bragg scattering geometry with $b = -1$. The reflection regions in Fig. 2.20(e) are shown on the inverted angular scale. For the transition from (λ, θ) to (λ, θ') space to be seen clearer, the reflection region, corresponding to the asymmetric scattering geometry with $-1 < b < 0$, is shown as an example in Fig. 2.21 in more detail.

Let λ_c be some specific wavelength, and θ_c the glancing angle of incidence, which is related to λ_c by the modified Bragg's law. A fragment of the total reflection region in the vicinity of (λ_c, θ_c) is shown in Fig. 2.21(i). The region of total reflection is bounded by the borderlines $\lambda_\pm(\theta)$, which are given by (2.118) with θ_c being substituted by the running angular variable θ. In a small angular range about θ_c far from normal incidence $\lambda_\pm(\theta)$ can be presented with

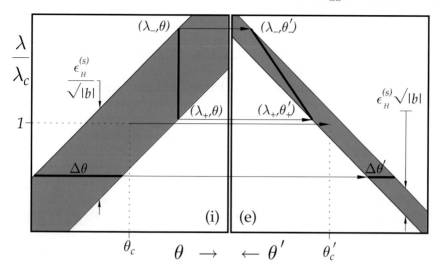

Fig. 2.21. Transformation of the Bragg reflection region from the (λ, θ) space of incident waves (i), into the (λ, θ') space of exit waves (e). The reflection region is shown as the green (dark grey) stripes. A particular case of Bragg reflection of type a_π is considered with an asymmetry factor $-1 < b < 0$. For other notations, see text.

a high degree of accuracy in linearized form as:

$$\frac{\lambda_\pm - \lambda_c}{\lambda_c} = \frac{\theta - \theta_c}{\tan\theta_c} \mp \frac{\epsilon_H^{(s)}}{2\sqrt{|b|}}. \tag{2.151}$$

The point (λ_c, θ_c) in (λ, θ) space becomes (λ_c, θ'_c) in the (λ, θ') space of the exit waves, see Fig. 2.21. The glancing angle of reflection θ'_c is given by (2.149). From (2.114) and (2.149) we obtain for the difference between θ_c and θ'_c to first approximation, i.e., with $(w_H^{(s)})^2$ terms being omitted[15]:

$$\theta_c - \theta'_c = \frac{w_H^{(s)}}{2}\left(b - \frac{1}{b}\right)\tan\theta_c. \tag{2.152}$$

The glancing angles of reflection and incidence are equal if $b = -1$, e.g., in the symmetric scattering geometry, as was already pointed out earlier. It should be emphasized here that (2.152) is not valid in backscattering ($\theta_c \to \pi/2$). Another relation has to be used. It will be discussed in the next subsection.

Recalling that $w_H^{(s)}$ is positive, it follows from (2.152) that the glancing angle of reflection θ'_c is greater than the glancing angle of incidence θ_c: $\theta'_c > \theta_c$, provided the asymmetry factor $b < -1$. On the other hand, if the asymmetry

---

[15] It should be noted that the quadratic terms can be of importance already for $\theta_c \simeq 70°$ and $b \simeq 10$.

factor b is in the range: $-1 < b < 0$, the glancing angle of reflection is smaller than the glancing angle of incidence: $\theta'_c < \theta_c$. Figure. 2.20 shows the transformation of the reflection regions in the (λ, θ) space to the (λ, θ') space in agreement with these rules.

Any boundary point $(\lambda_\pm, \theta)$ in Fig. 2.21(i) becomes $(\lambda_\pm, \theta'_\pm)$ in Fig. 2.21(e), with $\theta'_\pm$ being calculated by (2.150). The wave with wavelength λ_+ exits at an angle θ'_+, while the wave with wavelength λ_- exits at an angle $\theta'_- \neq \theta'_+$ (see also Fig. 2.18). Excitation points on the vertical solid line in Fig. 2.21(i) appear on the inclined solid line in Fig. 2.21(e).

The region of total reflection in the (λ, θ') space is bounded by the borderlines $\lambda_\pm(\theta')$, which are given by (2.150) with $\theta'_\pm$ being substituted by the running angular variable θ'. In a small angular range about θ_c far from normal reflection $\lambda_\pm(\theta')$ can be closely approximated in linearized form as:

$$\frac{\lambda_\pm - \lambda_c}{\lambda_c} = \frac{\theta' - \theta'_c}{\tan \theta'_c} \mp \frac{\epsilon_H^{(s)}}{2} \sqrt{|b|}. \tag{2.153}$$

As follows from (2.151), and is shown in Fig. 2.21(i), the incident waves with a fixed wavelength are reflected provided the glancing angle of incidence is within a range of $\Delta\theta = \tan\theta_c\, \epsilon_H^{(s)}/\sqrt{|b|}$. The glancing angles of reflection of the exit waves are in this case within a range of

$$\Delta\theta' = \epsilon_H^{(s)} \sqrt{|b|} \tan\theta'_c, \tag{2.154}$$

as follows from (2.153), and shown in Fig. 2.21(e). Combining these two expressions and taking to a good approximation $\tan\theta'_c = \tan\theta_c$, we obtain that the angular acceptance $\Delta\theta$ of the Bragg reflection for incident monochromatic waves, and the angular spread $\Delta\theta'$ of the exit waves are related by (Kohra and Kikuta 1967)

$$\Delta\theta' = |b|\, \Delta\theta. \tag{2.155}$$

Asymmetric Bragg diffraction thus offers a means of manipulating the angular divergence of x-ray beams. We recall at this point that asymmetric Bragg diffraction changes simultaneously the cross-section of x-ray beams, as shown in Fig. 2.7. The cross-section of the exit beam differs by a factor of $|b|^{-1}$ from the cross-section of the incident beam. This means that it has an inverse effect, which is consistent with the general requirement of phase volume conservation.

$\theta \to \pi/2$. Presentation of the borderlines $\lambda_\pm(\theta)$ and $\lambda_\pm(\theta')$ of the reflection region in the linearized form of (2.151), and (2.153), respectively, fails if θ_c approaches normal incidence ($\theta_c \to \pi/2$). The exact expressions (2.114)-(2.118) and (2.149)-(2.150) have to be used to graph the reflection regions in this case.

The reader is reminded that (2.34″) was used in Sect. 2.4.10 to derive (2.149)-(2.150). In the backscattering geometry, however, it is more instructive to apply another equation instead of (2.149)-(2.150). This is obtained from (2.34), which equivalently can be used to perform the transformation from (λ, θ) to (λ, θ') space. There are two advantages in applying this approach in backscattering. First, as we have seen in Sect. 2.4.10 for the transformation to be also correct in the backscattering geometry, it was necessary to keep in (2.34″) the second $\Delta_H^{(2)}$ and higher order terms in the Taylor expansion of the momentum transfer Δ_H (2.21). In contrast, in (2.34) it is sufficient to keep only the first oder term $\Delta_H \simeq \Delta_H^{(1)} = -K\alpha/(2\gamma_H)$ (2.21), as it does not vanish there even in backscattering geometry. Second, in the backscattering geometry the incidence and reflection angles are small: $\Theta = \pi/2 - \theta \ll 1$ and $\Theta' = \pi/2 - \theta' \ll 1$. Therefore, one can closely approximate $\cos\theta = \Theta$ and $\cos\theta' = \Theta'$ and use (2.34) and (2.34′) in the form linear in terms of Θ and Θ':

$$\Theta' \cos\phi' = \Theta \cos\phi - 2\epsilon \frac{\sin\eta}{\cos(\Theta + \eta)}, \qquad \epsilon = \frac{2d_H - \lambda}{2d_H}, \qquad (2.156)$$

$$\Theta' \sin\phi' = \Theta \sin\phi. \qquad (2.157)$$

To derive (2.156) we also used $\alpha = 2\Theta^2 - 4\epsilon$ (2.104), with Θ^2 being omitted for $\Theta \ll 1$, i.e., $\alpha = -4\epsilon$, and $\gamma_H \simeq -\sin(\Theta - \eta) = -\cos(\Theta + \eta)$.

With these equations any point in the (λ, θ) space in Fig. 2.22(i) can be conveniently transformed into the (λ, θ') space in Fig. 2.22(e). If the asymmetry angle $\eta \gg \Theta$ then (2.156)-(2.157) may be simplified to give

$$\Theta' \cos\phi' = \Theta \cos\phi - 2\epsilon \tan\eta, \qquad (2.158)$$

$$\Theta' \sin\phi' = \Theta \sin\phi. \qquad (2.159)$$

In the following, we shall consider this particular case of interest. In addition we shall assume that the vacuum wave vectors $\boldsymbol{K}_0$ and $\boldsymbol{K}_H$ of the incident and reflected waves are lying in the dispersion plane built up of the diffraction vector $\boldsymbol{H}$ and the inward normal $\hat{\boldsymbol{z}}$. That implies that the azimuthal angle of incidence ϕ and reflection ϕ' may be either 0 or π, see Fig. 2.4. Small deviations of $\boldsymbol{K}_0$ from the dispersion plane (i.e., small deviations of ϕ from 0 or π) would not lead to predictions significantly different from those presented in the following of this section - because of the $\cos\phi$ and $\cos\phi'$ dependences in (2.158).

Fragments of the regions of total Bragg reflection in the (λ, θ) space in the vicinity of normal incidence $\pi/2 - \Omega < \theta < \pi/2$ are shown in Figs. 2.16 and 2.22(i). The borderlines $\lambda_\pm(\theta)$ are given by (2.118). Figure 2.22(e) shows the appropriate reflection regions in the (λ, θ') space of exit waves. The borderlines of the reflection region in the (λ, θ') space are obtained by applying transformation (2.158) to the borderlines given by (2.118) in the (λ, θ) space.

The reflection region corresponding to the scattering geometry, in which the azimuthal angle of incidence is $\phi = 0$, is labeled by (a_0) and (a'_0), and drawn in the same color. The almost vertical solid borderline between these regions is the locus of points that obey the exact backscattering condition (2.37). Exact backscattering is thus attained in the transition from (a_0) to

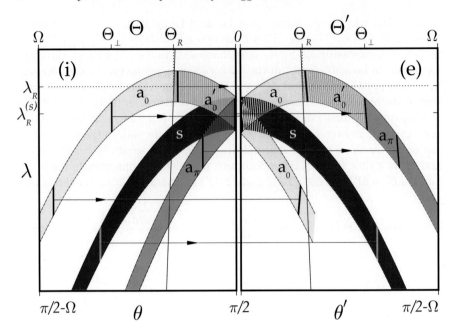

Fig. 2.22. Transformation of the Bragg reflection region from the (λ, θ) space of incident waves in the vicinity of normal incidence (i), into the (λ, θ') space of exit waves (e). Schematic presentation of fragments of the Bragg reflection regions according to (2.118) and (2.156) in different scattering geometries labeled by (a_0), (a'_0), and (a_π) with asymmetry angle $\eta \neq 0$. The schemes of the appropriate scattering geometries are shown in Fig. 2.23. The almost vertical solid line is a locus of points that obey the exact backscattering condition (2.37). It is a borderline between the regions marked by (a_0) and (a'_0). $\Omega \simeq 2.5$ mrad, $\eta = 89°$, $w_H^{(s)} = 2 \times 10^{-5}$. For other notations see text.

(a'_0), i.e., at the border of these regions. The schemes of the appropriate scattering geometries are shown in Figs. 2.23(a_0) and (a'_0).

It should be noted that while Θ keeps changing the azimuthal angle of incidence ϕ remains zero in the regions denoted by (a_0) and (a'_0). In contrast, the value of the azimuthal angle of reflection switches from $\phi' = 0$ to $\phi' = \pi$ as Θ attains $\Theta_\perp = 2\epsilon \tan \eta$. This follows from (2.158) and (2.159), the fact that $\epsilon > 0$, and the requirement $\Theta' \geq 0$. At the angle of incidence $\Theta_\perp$ the reflected wave emerges normal to the reflecting atomic planes: $\Theta' = 0$. Since, $\lambda \simeq 2d_H / \left(1 + w_H^{(s)}\right)$ (2.149) at $\Theta' = 0$, we obtain $\epsilon = w_H^{(s)}$ and

$$\Theta_\perp \simeq 2 w_H^{(s)} \tan \eta. \tag{2.160}$$

One can also derive a more precise relation $\Theta_\perp = 2\Theta_R$ with Θ_R being the incidence angle for exact back-reflection, which is given by (2.131) or (2.133). Thus, the incidence angle for normal reflection $\Theta_\perp$ is exactly twice as large as the incidence angle, at which exact back-reflection takes place.

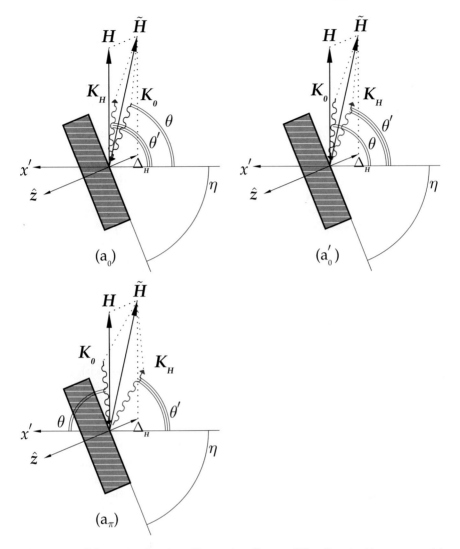

Fig. 2.23. Schematic drawing illustrating Bragg diffraction in the asymmetric backscattering geometry, with asymmetry angle $\eta \neq 0$. Particular cases of scattering are shown with the wave vectors $\boldsymbol{K}_0$ and $\boldsymbol{K}_H$ of the incident and reflected vacuum waves lying in the dispersion plane built up of the diffraction vector $\boldsymbol{H}$ and inward normal $\hat{\boldsymbol{z}}$. (a_0): $\phi = 0$, $\phi' = \pi$, and $\theta' \geq \theta$. Exact backscattering is attained in the transition from (a_0) to (a'_0) while increasing θ. (a'_0) differs from the previous case (a_0) by $\theta' \leq \theta$. (a_π) differs from the previous case (a'_0) in that now $\phi = \pi$. $\tilde{\boldsymbol{H}} = \boldsymbol{K}_H - \boldsymbol{K}_0$ is the total momentum transfer (2.15). The momentum transfer $\boldsymbol{\Delta}_H = \Delta_H \hat{\boldsymbol{z}}$ due to scattering from the vacuum-crystal interface is shown greatly exaggerated for clarity.

At normal incidence on the reflecting atomic planes when $\Theta = 0$, the reflected wave with wavelength $\lambda_R^{(s)}$ emerges at the angle of reflection Θ', which exactly equals $\Theta_\perp$. At this point, which is labeled appropriately in Fig. 2.22(e), the azimuthal angle of incidence switches from $\phi = 0$ to $\phi = \pi$ (the azimuthal angle of reflection remains $\phi' = \pi$). The reflection region corresponding to the scattering geometry, in which the azimuthal angle of incidence and reflection $\phi = \phi' = \pi$, is labeled by (a_π) in Figs. 2.22(i) and (e). The scheme of the appropriate scattering geometry is shown in Fig. 2.23(a_π).

Only small changes in the width of the reflection region can be observed in the transition from (λ, θ) to (λ, θ') space. This behavior is very different from the case addressed previously, as the glancing angle of incidence was far from normal incidence. This is attributed to the fact that in the vicinity of normal incidence the asymmetry factor b (2.113) takes values very close to $b = -1$ independent of how large is the asymmetry angle η.

Despite this, the reflection region in the space of incident waves is not mirrored into the space of reflected waves. Asymmetric diffraction in backscattering has the effect of, firstly, unequal glancing angle of incidence and reflection, and, secondly, angular dispersion. This is illustrated in Fig. 2.22 by the vertical excitation lines at the fixed angle of incidence in (λ, θ) space being transformed to inclined emission lines in (λ, θ') space arising at unequal glancing angles of reflection θ'. The inclination signifies the change of the angle of reflection with wavelength λ. The variation of Θ' (or equivalently θ') with λ is readily obtained by differentiating equation (2.156) with respect to λ for constant Θ:

$$\delta\Theta' = -\delta\theta' = \frac{\delta\lambda}{2d_H}\tau_D. \qquad (2.161)$$

Here $\tau_D = 2\sin\eta/(\cos(\Theta+\eta)\cos\phi')$. In particular, if $\Theta \ll 1$, then $\tau_D = 2\tan\eta/\cos\phi'$. Equation (2.161) is in agreement with (2.145) obtained earlier in Sect. 2.4.9 in a different manner to describe angular dispersion in a more general case of backscattering with arbitrary azimuthal angles. It is clear that the closer the asymmetry angle η to $\pi/2$, the greater is the effect of angular dispersion.

If the incident waves have an angular spread of width $\Delta\theta$ with the center at $\theta_c = \pi/2 - \Theta_c$, they can be graphically represented in phase space (λ, θ) by the vertical stripe of width $\Delta\theta$, as shown in Fig. 2.24. As one can easily show using (2.156) and (2.161), in the space of the reflected waves the stripe becomes inclined with the borderlines given by

$$\frac{\lambda_\pm - \lambda_c}{2d_H} \simeq -\frac{\theta' - \theta_c'}{\tau_D}\cos\phi' \mp \frac{\epsilon_D}{2}\cos\phi, \qquad (2.162)$$

$$\tau_D = \frac{2\sin\eta}{\cos(\theta_c - \eta)}, \qquad \epsilon_D = \frac{\Delta\theta}{\tau_D}.$$

This relation will be used later to address a special case of multiple crystal diffraction in backscattering.

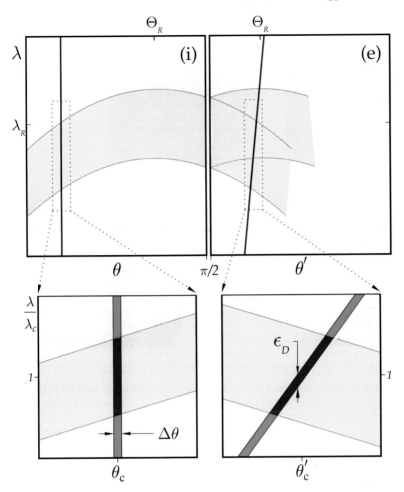

Fig. 2.24. Reflection region in asymmetric Bragg diffraction (a_0)-(a'_0) in backscattering geometry - marked in yellow (light grey) - is shown in the space of incident (i) and exit waves (e), c.f., Fig. 2.22. The fragments in the boxes marked-off by dotted lines are shown in the panels below on the enlarged angular scale. The vertical green (dark grey) stripe in (i) displays the phase volume of the incident waves with angular spread $\Delta\theta$. The intersection region marked in dark blue (black) depicts the phase volume of the waves reflected from the crystal. In the space of exit waves it is inclined due to angular dispersion, and has a relative spectral width of ϵ_D given by (2.162).

Figure 2.22 illustrates the transformation of the Bragg reflection region from the (λ, θ) space of incident waves (i) in the vicinity of normal incidence, into the (λ, θ') space of exit waves (e). Results are presented for the particular cases of the incident waves propagating parallel to the dispersion plane, i.e., when the azimuthal angle of incidence is either $\phi = 0$ or $\phi = \pi$ (both, in

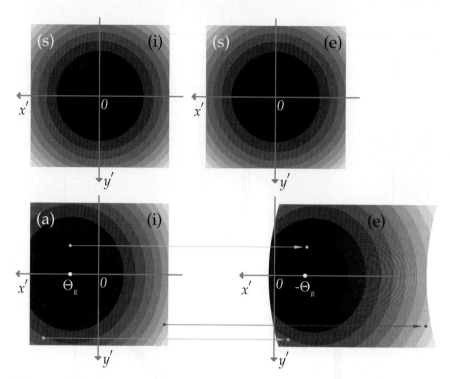

Fig. 2.25. Left (i): distribution of wavelengths determined by the modified Bragg's law (2.113)-(2.115) as a function of propagation direction of the incident wave with wave vector K_0 (2.25) - the same as in Fig. 2.17.
Right (e): distribution of wavelengths shown as a function of propagation direction of the reflected (exit) wave with wave vector K_H (2.25).
The region of propagation directions corresponding to reflections of the waves with greatest wavelengths is shown in black. The regions in white correspond to reflections with smallest wavelengths, which are only $\approx 10^{-5}$ smaller (in relative units) than the greatest wavelengths.
The propagation directions K_0/K and K_H/K are determined by the glancing angle of incidence θ, and reflection θ', as well as by the azimuthal angle of incidence ϕ, and reflection ϕ', respectively – see Fig. 2.4. Equivalently, they are determined by the projection $(\cos\theta\cos\phi, \cos\theta\sin\phi)$ of K_0/K, and $(\cos\theta'\cos\phi', \cos\theta'\sin\phi')$ of K_H/K onto the (x', y') plane. The latter is parallel to the reflecting atomic planes.
(s) Symmetric Bragg scattering geometry with asymmetry angle $\eta = 0$.
$0 \leq \phi^{(\prime)} < 2\pi$, $0 \leq \Theta^{(\prime)} < 2.5$ mrad ($\Theta^{(\prime)} = \pi/2 - \theta^{(\prime)}$), $w_H^{(s)} = 2 \times 10^{-5}$.
(a) Asymmetric scattering geometry with asymmetry angle $\eta = 89°$.

symmetric and asymmetric backscattering scattering geometries). Figure 2.25 complements this graphical illustration and extends it to the general case of incident and reflected wave propagation directions with any azimuthal angle ϕ, and ϕ', respectively.

In the case of symmetric Bragg scattering, presented by the graphs with labels (s) in Fig. 2.25, the distributions of wavelengths for the incident (i) and exit (e) waves are axisymmetric and equivalent.

In the case of asymmetric Bragg scattering, presented by the graphs with labels (a) in Fig. 2.25, the distributions of wavelength for the incident (i) and exit (e) waves remain in the first approximation axisymmetric[16]. However, the symmetry point of the distribution for the reflected (exit) waves is shifted from $(\Theta_R, 0)$ to $(-\Theta_R, 0)$. The symmetry points mark the propagation directions $\boldsymbol{K}_H = -\boldsymbol{K}_0$ in the case of *exact* backscattering.

Equations (2.158)-(2.159) are applied to transform the distribution of wavelengths from the space of incident into the space of exit waves. The horizontal arrows connect selected points in Fig. 2.25(a)(i) with their counterparts in Fig. 2.25(a)(e) according to (2.158)-(2.159). In the very first approximation, assuming $\epsilon = (2d_H - \lambda)/(2d_H) \approx w_H^{(s)}$ to be constant, the distribution in Fig. 2.25(a)(i) is shifted as a whole antiparallel to the x'-axis by $-2 w_H^{(s)} \tan \eta \simeq -2\Theta_R$ to result in the distribution shown in Fig. 2.25(a)(e). More accurately, every point is shifted antiparallel to the x'-axis proportional to the relative wavelength deviation ϵ, resulting in a greater shift for the points corresponding to smaller wavelengths (lighter areas).

2.4.12 Reflectivity of Thin Crystals

In the previous sections we have analyzed the Bragg reflectivity of thick non-absorbing crystals. Those crystals have been defined in Sect. 2.4.3 as having thickness larger than the penetration depth: $d \gg d_e$. In fact, this condition is rather indefinite. Indeed, according to (2.89), d_e is a function of the reduced deviation parameter y, and therefore of the wavelength λ, and the glancing angle of incidence θ. The penetration depth $d_e(y)$ takes its minimal value - the extinction length $d_e(0)$ (2.90) - at the center of the total reflection region when $y = 0$. It increases rapidly with deviation of y, and respectively of θ, and λ from the center of the reflection region. Therefore, in general the assertion of whether the crystal is thin or thick depends not only on the physical thickness of the crystal d, but also on the proximity to the center of the total reflection region.

Since the extinction length $d_e(0)$ is the smallest penetration depth, the thin-crystal limit can be defined unequivocally, i.e., independent of the parameter y, in contrast to the thick crystal limit. Those crystals can be defined as thin whose thickness

$$d \lesssim d_e(0). \tag{2.163}$$

---

[16] The distributions, in the case of asymmetric scattering, can be considered to be axisymmetric only in the very first approximation. This is valid for small angles of incidence Θ and reflection Θ'. In general, strictly speaking, the distributions are *not* axisymmetric.

In the present section we shall derive expressions for the Bragg reflectivity in the thin-crystal limit, given by (2.163). We assume still that the crystal is centrosymmetric, i.e., the relation $\chi_H = \chi_{\overline{H}}$ holds, and that the photoabsorption is negligible.

Using this condition and (2.58)-(2.60), we find that the phase of the exponents $\exp[\mathrm{i}(\varkappa_1 - \varkappa_2)d]$ in (2.68) and (2.69) changes significantly only if the deviation parameter $|\tilde{\alpha}| \gg |P\chi_H|\sqrt{|b|}$. We shall use this condition to obtain simplified analytical expressions for the reflectivity in the thin crystal limit. At this point, it may be noted that the same condition can be also used to obtain analytical expressions for asymptotics (wings) of the Bragg reflectivity for parallel-sided crystals of any thickness.

To calculate the reflectivity, we use as previously (2.88) and (2.69). In the limit $|\tilde{\alpha}| \gg |P\chi_H|\sqrt{|b|}$, and assuming $\tilde{\alpha} \geq 0$, the quantities R_1, R_2, and $(\varkappa_1 - \varkappa_2)$ read:

$$R_1 \simeq \frac{bP\chi_H}{|\tilde{\alpha}|}, \tag{2.164}$$

$$R_2 \simeq -\frac{2\tilde{\alpha}}{P\chi_H}, \tag{2.165}$$

$$\varkappa_1 - \varkappa_2 \simeq \frac{K|\tilde{\alpha}|}{2\gamma_0}. \tag{2.166}$$

As a result, $|R_2| \gg |R_1|$, and we have for the reflection amplitude (2.69):

$$r_{0H} \simeq R_1 \left\{1 - \exp[\mathrm{i}(\varkappa_1 - \varkappa_2)d]\right\}. \tag{2.167}$$

Inserting this expression into (2.88), and using (2.166), we obtain for the reflectivity in the thin-crystal limit:

$$R \simeq \frac{4|R_1|^2}{|b|} \sin^2 \frac{(\varkappa_1 - \varkappa_2)d}{2}$$

$$= \frac{|\chi_H|^2 P^2 K^2 d^2}{4|\gamma_H|\gamma_0} \frac{\sin^2 \xi}{\xi^2}, \tag{2.168}$$

where

$$\xi = \frac{K|\tilde{\alpha}|d}{4\gamma_0}. \tag{2.169}$$

The same expression for the reflectivity (2.168) is also obtained for negative $\tilde{\alpha}$, i.e., it is valid for any $\tilde{\alpha}$.

Using (2.60), the representation (2.98) of the deviation parameter α, and the definition (2.115) of the parameter w_H, we find

$$\xi = \pi N_d \left| (1 + w_H) \frac{\lambda}{2d_H} - \sin\theta \right|, \tag{2.170}$$

where

$$N_d = \frac{d}{d_H |\gamma_H|} \qquad (2.171)$$

is the number of reflecting planes throughout the crystal thickness.

The reflectivity, given by (2.168), has the form of the diffraction function well known in optics. Peak reflectivity is achieved at $\xi = 0$. Inserting $\xi = 0$ into (2.170) we obtain the modified Bragg law (2.114). Thus, the peak reflectivity is achieved at the same angle (wavelength) independent of whether the crystal is thin or thick. Assuming symmetric Bragg diffraction, and thus $|\gamma_H|\gamma_0 = \sin^2\theta$, using Bragg's law, and definition (2.4) of χ_H, we obtain from (2.168) for the peak reflectivity $R(\xi = 0)$ in the thin-crystal limit:

$$R(0) \simeq 4 \left(\frac{r_e d_H d}{V} \right)^2 |P F_H|^2. \qquad (2.172)$$

It is proportional to the square of the crystal thickness d^2, as is usual for a coherent scattering process.

The width of the reflection curve (2.168) is $\Delta\xi \simeq \pi$, or equivalently in terms of α:

$$\Delta\alpha \simeq |\gamma_H| \frac{2\lambda}{d}. \qquad (2.173)$$

The wings of the reflectivity curve decay as $\propto 1/\xi^2$ ($\propto 1/|\alpha|^2$) and oscillate with a period of $\delta\xi = \pi$, or equivalently with a period of

$$\delta\alpha = |\gamma_H| \frac{2\lambda}{d}. \qquad (2.174)$$

Thus, the width of the reflection curve $\Delta\alpha$ and the period of oscillation $\delta\alpha$ are equal. Using (2.170), it follows that for a given glancing angle of incidence θ the period of oscillations $\delta\lambda$ (and equivalently the width $\Delta\lambda$) measured in units of $2d_H$ is a constant independent of the wavelength λ:

$$\frac{\delta\lambda}{2d_H} = \frac{d_H |\gamma_H|}{d(1 + w_H)} = \frac{1}{N_d(1 + w_H)}. \qquad (2.175)$$

It is determined only by the number of reflecting planes N_d (if we neglect the small value w_H). On the energy scale, the period of oscillations, and also the reflection width is

$$\delta E = \Delta E = \frac{hc}{2d(1 + w_H)} \frac{|\gamma_H|}{\sin^2\theta}. \qquad (2.176)$$

At exact backscattering it is simply

$$\delta E = \Delta E = \frac{hc \cos\eta}{2d(1 + w_H)}. \qquad (2.177)$$

The thinner the crystal the larger the energy width, and vice versa.

Interestingly, both in the thin-crystal and thick-crystal limit, the relative spectral width of the Bragg reflection is the reciprocal of the effective number of reflecting atomic planes contributing to Bragg diffraction c.f., (2.124) and (2.175). The difference is in the meaning of the "effective number of reflecting planes". In the thick-crystal limit it is the number of reflecting planes $N_e = d_e(0)/d_H$ throughout the extinction length, while in the thin-crystal limit it is the number of reflecting planes N_d throughout the crystal thickness.

Similarly, (2.168) and (2.170) describe the angular dependence of the reflectivity. However, due to the nonlinear dependence of ξ on θ in (2.170) - the intensity oscillations of the reflected beam are aperiodic on the angular scale. The aperiodicity is especially pronounced at glancing angles of incidence close to $\pi/2$, i.e., in the backscattering geometry.

The intensity oscillation is an effect of crystal thickness. It occurs due to interference of the waves created at the entrance and rear surfaces, see, e.g. (Azaroff et al. 1974) for some more details. They can be referred to as "thickness" oscillations. This effect could be especially well recognized in the time dependence of Bragg diffraction, as will be discussed in Sect. 2.4.15.

Figure 2.26 shows the reflectivity of a parallel-sided α-Al$_2$O$_3$ crystal of thickness (a) $d = d_e(0)$, and (b) $d = 3d_e(0)$, as a function of x-ray photon energy E (left panels) and angle of incidence Θ (right panels) close to normal incidence $\Theta = 0$. The (0 0 0 30) symmetric Bragg reflection with extinction length $d_e(0) = 16$ μm is used in the calculations. Calculations performed with exact expressions of the dynamical theory are given by the solid lines. The dotted lines represent results of calculations in the framework of the thin-crystal approximation given by (2.168)-(2.170).

The comparison shows that the approximate expressions describe perfectly the reflectivity of the thiner crystal with $d = d_e(0)$. This confirms, that in the thin-crystal limit defined by (2.163) the approximate expressions can be applied in the whole energy (angular) range.

The approximate expressions do not describe the region of high reflectivity of the thicker crystal with $d = 3d_e(0)$. However, they describe the wings of the Bragg reflection curves. This is due to the fact, that in the wings the penetration depth d_e gets larger than the crystal thickness d, and therefore the thin-crystal approximation becomes valid again. It should be noted, however, that since no photo-absorption is taken into account in (2.168)-(2.170) the amplitude of oscillation appears to be overestimated at large deviations from the peak reflectivity.

2.4.13 Reflectivity of Thick Absorbing Crystals

We drop now the assumption that the photo-absorption is negligibly small. Non-zero photo-absorption makes the analysis of the Bragg reflectivity more complicated. Peak reflectivity, its position in terms of the deviation parameter α_c, as well as the width of the reflection curve $\Delta\alpha = \alpha_{+1} - \alpha_{-1}$ can

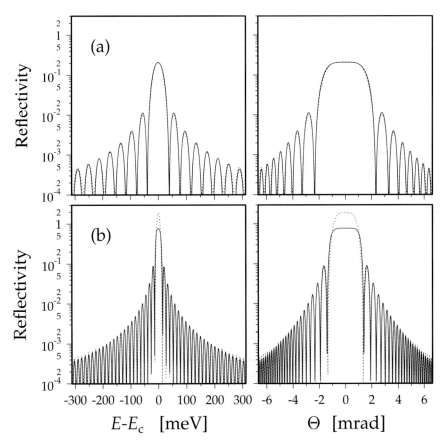

Fig. 2.26. Bragg reflectivity of a parallel-sided α-Al$_2$O$_3$ crystal of thickness (a) $d = d_e(0)$, and (b) $d = 3d_e(0)$, as a function of the x-ray photon energy E (left panels) and the glancing angle of incidence Θ (right panels). Negative Θ signifies the change of the azimuthal angle of incidence ϕ to $\phi + \pi$. The solid lines represent calculations with exact expressions of the dynamical theory, the dotted lines - calculations in the thin-crystal approximation (2.168)-(2.170). The (0 0 0 30) Bragg reflection is used with extinction length $d_e(0) = 16$ μm. At normal incidence on the (0 0 0 30) atomic planes ($\Theta = 0$) x rays with energy $E_c = 14.315$ keV are reflected. The calculations are performed in the angular and energy ranges corresponding to the same range of the deviation parameter α.

be determined numerically using the general expressions derived in previous sections 2.4.1 and 2.4.2. $\Delta\alpha$ differs from (2.97), but is still of the order of $|\chi_H| \lesssim 10^{-5}$. The conclusion that the reflectivity is solely a function of α remains valid.

From numerical calculations one determines $R(\alpha)$, the peak reflectivity $R(\alpha_c)$ for α_c, and the half-width $\Delta\alpha$. The dependence on θ, E, or T and

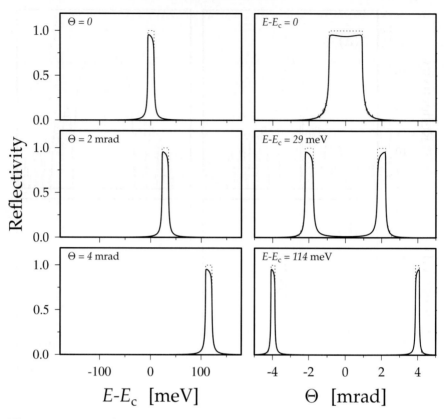

Fig. 2.27. Bragg reflectivity of a semi-infinite crystal vs x-ray energy E for a fixed glancing angle of incidence Θ (left panels), and vs angle of incidence Θ for a fixed x-ray energy E (right panels). Negative Θ signifies the change of the azimuthal angle of incidence ϕ to $\phi + \pi$. The (0 0 0 30) Bragg reflection in α-Al$_2$O$_3$ is used in the dynamical theory calculations with photo-absorption taken into account (solid lines). Dashed lines show the results of the same calculations with zero photo-absorption. The center of the region of total reflection at normal incidence on the (0 0 0 30) atomic planes ($\Theta = 0$) is at $E_c = 14.315$ keV.

corresponding widths and positions can be obtained from the relation $\alpha = f(\theta, E, T)$ as given by (2.98) in Sect. 2.4.4.

The energy and angular dependence of Bragg reflectivity for an absorbing crystal calculated using the dynamical theory are shown in Fig. 2.27 as solid lines. The (0 0 0 30) Bragg reflection in an α-Al$_2$O$_3$ crystal is used as an example with the following values for the coefficients of the dynamical equations: $\chi_0 = (-7.96 + \mathrm{i}\, 0.0297) \times 10^{-6}$, $\chi_H = (-8.63 + \mathrm{i}\, 0.209) \times 10^{-7}$, $\chi_{\overline{H}} = \chi_H$. The coefficients are calculated using the crystallographic data given in Appendix A.2. The crystals for which $\chi'_0 \gg \chi''_0$ and $\chi'_H \gg \chi''_H$ are

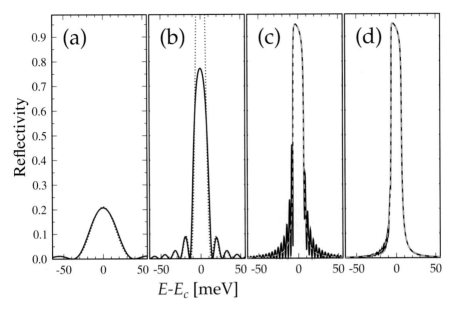

Fig. 2.28. Energy dependence of the Bragg reflectivity from α-Al$_2$O$_3$ crystals of different thickness: (a) $d = d_e(0)$, (b) $d = 3d_e(0)$, (c) $d = 10d_e(0)$, (c) $d - 100d_e(0)$. X rays are at normal incidence on the (0 0 0 30) atomic planes of sapphire at $T = 300$ K. Extinction length $d_e(0) = 16$ μm. $E_c \simeq 14.315$ keV. The solid lines are dynamical theory calculations in the two-beam approximation. The dotted lines in (a-b) are calculations within the thin-crystal approximation. The light dashed lines in (c-d) are calculations within the thick-crystal approximation.

termed weakly-absorbing. Silicon, sapphire (α-Al$_2$O$_3$), and diamond belong to this class of crystals, provided hard x rays (≥ 5 keV) are considered.

The dashed lines in Fig. 2.27 show the result of the same calculations but with zero photo-absorption, i.e., with the imaginary parts dropped in χ_0, χ_H, and $\chi_{\overline{H}}$. As is clearly seen, photo-absorption reduces the peak reflectivity and makes the reflectivity curve asymmetric in the region of "total" reflection. The peak reflectivity is achieved not at the center of the region E_c, but at an energy, which will be denoted in the following as E_P. In general $E_P \neq E_c$. However, the reflectivity curves are practically un-shifted, and have almost the same width. This implies that the simple equation (2.114) determining the center of the region of total reflection for non-absorbing crystals as well as the expressions (2.119), (2.125) for the angular and energy widths can also be used with high precision for weakly-absorbing crystals.

2.4.14 Reflectivity of Crystals with Intermediate Thickness

The transition from the thin-crystal to the thick-crystal limit is illustrated in Fig. 2.28 by the energy dependence of the Bragg reflectivity calculated for crystals of different thickness: (a) $d = d_e(0)$, (b) $d = 3d_e(0)$, (c) $d = 10d_e(0)$, (c) $d = 1000d_e(0)$. The solid lines in Fig. 2.28 are calculated using the exact expressions of the dynamical theory in the two-beam approximation taking into account photo-absorption. The dotted lines show the calculations of the reflectivity in the framework of the thin-crystal approximation using (2.168)-(2.170). As we have just discussed in the previous section the thin-crystal approximation works quite well for crystals with a thickness not exceeding the extinction length, in agreement with the condition $d \lesssim d_e(0)$ of (2.163). It also describes quite well the wings of the reflection curves of thicker crystals, as can be seen in Fig. 2.28(b). However, the peak reflectivity is calculated wrong.

In contrast to this, the reflection dependence calculated in the thick-crystal approximation (dashed lines Fig.2.28(c-d)) describes well the region of high reflectivity for crystals with thickness $d \gtrsim 10d_e(0)$, but not the wings where the oscillations are still present. The oscillations disappear only if $d \gg d_{\rm ph}$, where $d_{\rm ph}$ is the photo-absorption length, which is given by $d_{\rm ph} = \left(K\chi_0''\right)^{-1}$ as can be derived from (2.54). In the particular case presented in Fig. 2.28 the photo-absorption length $d_{\rm ph} = 464$ μm$\simeq 30d_e(0)$. Only for crystals with $d \gtrsim 3d_{\rm ph}$ does the thick-crystal approximation reproduce, as shown in Fig. 2.28(d), the results of calculations using the exact expressions of the dynamical theory.

2.4.15 Time Dependence of Bragg Diffraction

We have discussed the energy and angular dependence of Bragg diffraction in some detail. Now we would like to tackle the subject of the time dependence. An interesting question is, how long does this special scattering process, known as Bragg diffraction, last?

The answer to this question is of interest not only on its own. A similar question arises in conjunction with scattering in x-ray Fabry-Pérot resonators, see Sect. 4.3. However, since the components of a resonator are single crystals, one should at the outset be able to answer the former question. There exists another reason why it is important to address this question. The upcoming fourth generation synchrotron radiation sources based on free-electron lasers will generate radiation pulses as short as 100 fs (LCLS 1998, Materlik and Tschentscher 2001, Ayvazyan et al. 2002b,a). How will Bragg crystal optics affect such short radiation pulses? This problem was addressed in publications by Chukhovskii and Förster (1995), Missalla et al. (1999), Shastri et al. (2001b), Shastri et al. (2001a), Graeff (2002), and Shvyd'ko (2002).

The elementary scattering process from a single atom is very short. Typical widths of atomic resonances are in the eV-range. Thus, from the energy-time uncertainty relation, one should expect that the typical scattering times are less than 1 fs, provided the photon energy is close to the K, L, or M, etc.

x-ray absorption edge. Under "off-resonance" conditions, which are typical for Bragg diffraction, the scattering time should be even shorter. The dependence of the collision time on the proximity to resonance was experimentally investigated by Smirnov and Shvyd'ko (1989) using long-lived nuclear resonances.

At this point it is important to note that the distance that light covers in 1 fs is about 0.3 μm, which is much smaller than typical extinction lengths. Thus, the collision time with a single scatterer is *not* the main contribution to the duration of x-ray Bragg diffraction. This is, by the way, very different from the case of nuclear resonant Bragg diffraction from Mössbauer nuclei, where the collision time with a single scatterer determines the time scale of nuclear Bragg diffraction (Kagan et al. 1979, Smirnov and Shvyd'ko 1986, Rüffer et al. 1987, van Bürck et al. 1987, Shvyd'ko and Smirnov 1990).

The basic physical phenomenon underlying Bragg diffraction in thick perfect crystals is multiple scattering. Due to multiple scattering and because of time delays for the radiation to propagate from one scatterer to another, Bragg diffraction occupies a finite duration independent of the duration of the elementary scattering process. One can roughly estimate the typical duration of Bragg diffraction as the time lag $\Delta t \sim d_e(0)/c$ for the radiation to travel the penetration length which is determined by the extinction length $d_e(0)$, see (2.90). By virtue of the relation $\Delta E = \hbar c/d_e(0)\sin\theta$ (cf. (2.120) and (2.124)) between the energy width of a Bragg reflection and its extinction length, it follows that $\Delta t \sim \hbar/\Delta E \sin\theta$. The latter expression (up to the constant factor $\sin\theta$) could have been derived directly from the uncertainty relation $\Delta t \Delta E = \hbar$. This fact shows that indeed multiple scattering, whose spatial extent is given by the extinction length, determines the duration of Bragg scattering. All these suggest also a technique for the rigorous calculation of the time dependence of Bragg scattering – the Fourier transform of the energy dependence of Bragg diffraction.

The time dependence of the reflection amplitude $r_{0H}(t)$ is calculated as the Fourier transform of the energy dependence of the reflection amplitude $r_{0H}(E)$:

$$r_{0H}(t) = \int_{-\infty}^{\infty} \frac{dE}{2\pi\hbar} e^{-iEt/\hbar} r_{0H}(E). \tag{2.178}$$

$r_{0H}(t)$ represents the response amplitude of the system to the instantaneous excitation with a δ-function type radiation pulse. The time dependence of the Bragg reflectivity, similar to (2.88), is given by $R(t) \propto |r_{0H}(t)|^2$. Similar relations can be used to calculate the time dependence of transmission through a single crystal[17].

---

[17] We consider here only the case of *symmetric* Bragg scattering geometry. In all other cases, angular dispersion changes the wavefront of the reflected wave. It is no more one and the same for all the frequency components. To allow for angular dispersion, this would require a theory more complicated than the one we are

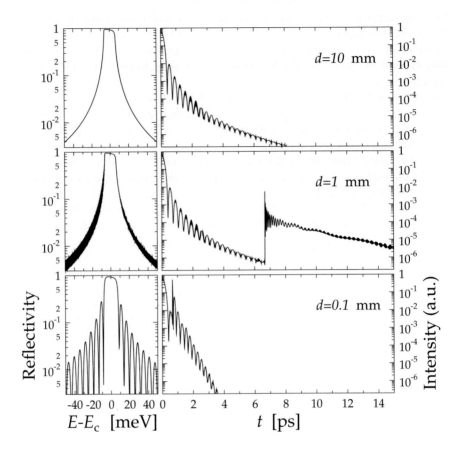

Fig. 2.29. Left: Bragg reflectivity vs x-ray energy E for crystals of different thickness d. Right: time dependence of Bragg diffraction. It represents the time response to the instantaneous excitation with a δ-function type x-ray radiation pulse. The (0 0 0 30) Bragg reflection in α-Al$_2$O$_3$ is used in the dynamical theory calculations. X rays are at normal incidence ($\theta = \pi/2$) to the (0 0 0 30) atomic planes, $E_c = 14.315$ keV.

Figure 2.29 shows an example of the numerical dynamical theory calculations for the (0 0 0 30) Bragg reflection in α-Al$_2$O$_3$ crystals of different thicknesses. The extinction length for this reflection is $d_e(0) = 16.0$ μm. For clarity, x rays are assumed to be incident normal ($\theta = \pi/2$) to the (0 0 0 30) atomic planes. The left panels in Fig. 2.29 represent the energy dependence of Bragg reflectivity with peaks at $E_c = 14.315$ keV. Examples are given

---

considering here. Time dependence of Bragg diffraction in the Laue scattering geometry with angular dispersion taken into account was considered by Graeff (2002).

for three different crystal thicknesses d: a thick one $d \gg d_e(0)$ on the top, $d \sim 10 d_e(0)$ in the middle, and with $d \sim d_e(0)$ on the bottom.

From the aforementioned value of the extinction length one expects an energy width of $\Delta E = \hbar c/d_e(0) = 12.5$ meV, which agrees with 13.0 meV ascertained from the numerical calculations in Fig. 2.29 ($d = 10$ mm). Similarly, one can estimate that the response time $\simeq c/d_e(0) = 0.055$ ps. One obtains practically the same value from the uncertainty relation $\simeq \hbar/\Delta E = 0.051$ ps. However, the duration – full width at half maximum (FWHM) – of the time response ascertained from the numerical calculations in Fig. 2.29 ($d = 10$ mm) turns out to be $\simeq 0.17$ ps, which is a factor π larger.

This discrepancy can be attributed to the fact, that the "classical" uncertainty energy-time relation $\Delta E \Delta t \simeq \hbar$ is valid only for the wave packets with Lorentzian energy and exponential time distributions. The energy dependence of the Bragg reflectivity is nearly rectangular. The Fourier transform of the rectangular energy dependence of a width ΔE is proportional to $\sin(\Delta E t/2\hbar)/\Delta E t$. As a result, the time dependence of the response is proportional to $[\sin(\Delta E t/2\hbar)/\Delta E t]^2$ with a response time Δt (FWHM) given by the "corrected" uncertainty energy-time relation for Bragg reflections

$$\Delta t\, \Delta E = \pi \hbar. \tag{2.179}$$

The oscillations accompanying the subsequent decay has a period of $T = 2\pi\hbar/\Delta E$. Using these relations one obtains estimates for the response time of Bragg diffraction, $\Delta t = 0.17$ ps, and for the period of subsequent oscillations $T = 0.34$ ps, which agree with the values obtained from the rigorous numerical calculations in Fig. 2.29 ($d = 10$ mm).

The following expressions can be derived to estimate the energy width and the response time of Bragg diffraction by using the extinction length with arbitrary glancing angle of incidence θ:

$$\Delta E = \frac{\hbar c}{d_e(0)} \frac{1}{\sin\theta} \qquad \Delta t = \pi \frac{d_e(0)}{c} \sin\theta. \tag{2.180}$$

According to the definition (2.90) the extinction length $d_e(0) \propto \gamma_0$. In the case of symmetric Bragg diffraction the latter equals $\sin\theta$. Therefore, the response time Δt scales as $\sin^2\theta$ with glancing angle of incidence θ. Thus, the smaller the glancing angle of incidence (Bragg angle) the smaller is the response time for a given Bragg reflection.

The next graph in Fig. 2.29 is calculated for a thinner crystal with $d = 1$ mm. A remarkable feature appears in this case - an intensity jump at $t = 6.7$ ps. This is an effect due to the wave field being reflected from the rear surface of the crystal. It appears at a time, which equals the back and forth time-of-flight $2d/c$ through the crystal. A jump of the same origin is observed at $t = 0.67$ ps in Fig. 2.29 ($d = 0.1$ mm) with a more pronounced amplitude. The thinner the crystal - the larger the amplitude of the field reaching the rear surface, because of the weaker influence of both extinction and photo-absorption.

2.5 Multiple-Beam Diffraction in Backscattering: Kinematic Treatment

The discussion of x-ray Bragg diffraction was restricted in the previous section to the two-beam diffraction case, a regime where the incident wave excites only two waves inside the crystal: the forward-transmitted and the Bragg-reflected waves. As a result, the system of fundamental equations (2.40) of the dynamical theory was reduced to a system of two linear equations (2.57) for two vector field amplitudes $\boldsymbol{D}_0$ and $\boldsymbol{D}_H$, for which the excitation condition (2.13) was fulfilled. It was ascertained in the previous section that the reflected wave in the two-beam case can be very strong, which is of great practical importance.

In general, the excitation condition (2.13) can be fulfilled for more than two waves. As a result, multiple-beam Bragg diffraction takes place with more than two strong waves excited in the crystal. Multiple-beam diffraction in crystals was observed for the first time by Renninger (1937). Different aspects of this diffraction regime are discussed, e.g., in the texts on the dynamical theory of Pinsker (1978) and Authier (2001). Also many references to the original publications are given there.

To fulfill the Bragg excitation condition for one set of atomic planes is much easier than for several sets simultaneously. In this sense, multiple-beam Bragg diffraction is rather an exception to the rule than a rule. However, under some special conditions multiple-beam Bragg diffraction occurs systematically. For example, Bragg reflection *in backscattering* is very often accompanied simultaneously by other Bragg reflections. As discussed in this section, multiple-beam diffraction becomes rather a rule than an exception in backscattering. The accompanying reflections can remove intensity from the backscattered beam and thus reduce the crystal reflectivity for the backscattering channel. It is a very unfavorable effect for many applications of Bragg backscattering in x-ray crystal optics. On the other hand, interference of several coherent waves, which takes place in multiple-beam diffraction, may provide valuable information on the phases of the crystal structure factors, and can be used in structure determination. One should note, that extracting the phase information was the main application of multiple-beam diffraction so far. The recent advances in this research area have been extensively reviewed, e.g., by Colella (1995), Weckert and Hümmer (1997), and Chang (1998).

The discussion of the multiple-beam effect in Bragg backscattering we commence in this section with simplified considerations based on the kinematic Bragg's law. The effects of refraction and multiple scattering will be ignored at this stage. Afterwards, in Sect. 2.6 the dynamical treatment will be applied to gain more insight into this complicated scattering process. The main emphasis in this discussion will be on how to avoid the influence of accompanying reflections on the backscattering channel.

2.5 Multiple-Beam Diffraction in Backscattering: Kinematic Treatment 115

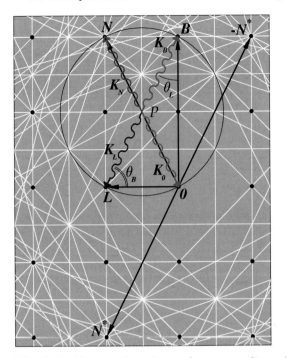

Fig. 2.30. Bragg backscattering from a two-dimensional square lattice in the presence of multiple-beam diffraction. N, B, L are reciprocal lattice vectors. The white lines are perpendicular bisectors of the reciprocal lattice vectors, which compose the Brillouin zones. K_0 is the wave vector of the incident radiation fulfilling Bragg's condition for backscattering with diffraction vector N: $K_N = -K_0 = N/2$. Scattering channels are open not only for the waves with wave vector K_N but also for the waves with wave vectors K_B and K_L. P is the crossing point of the facets of the Brillouin zone and simultaneously the tie point. Reciprocal vectors $N^* = L - B$, and $\overline{N}^* = B - L$ are symmetric for back-diffraction vector N with respect to a reflection in mirror planes perpendicular to B, and L, respectively.

2.5.1 Accompanying Reflections

To start with, the multiple-beam effects in backscattering will be discussed for some particular crystal lattices.

For crystals with a cubic lattice the following rule can be observed. The wave vector K_0 of the incident radiation fulfilling Bragg's condition $K_0 = -N/2$ for backscattering with the diffraction vector N always originates at an intersection of *many* facets of one of the Brillouin zones of the reciprocal lattice. Figure 2.30 illustrates this statement using the example of a two-dimensional square lattice. The intersection point is marked by P. In Sect. 2.4.4, it was pointed out that any wave vector K_0 that starts from the Brillouin zone facet, which is normal to the reciprocal vector, say H as in Fig. 2.10, fulfills Bragg's condition with the diffraction vector H. Since many facets are crossing at point P in Fig. 2.30, this implies that Bragg's condition is simultaneously fulfilled for many diffraction vectors $\boldsymbol{B}$, $\boldsymbol{L}$, etc., which are normal to the facets of the Brillouin zone crossing at P. In crystals with a cubic lattice exact backscattering thus co-exists with multiple-beam Bragg scattering. Scattering channels become open not only for the back-reflected radiation with wave vector

$$\boldsymbol{K}_N = -\boldsymbol{K}_0 = \boldsymbol{N}/2, \tag{2.181}$$

but also for the radiation with wave vectors

$$\boldsymbol{K}_B = \boldsymbol{K}_0 + \boldsymbol{B}, \qquad \boldsymbol{K}_L = \boldsymbol{K}_0 + \boldsymbol{L}, \qquad \text{etc.} \tag{2.182}$$

These additional Bragg reflections will be termed in the following as "accompanying" reflections. Point P is not only the crossing point of the facets of the Brillouin zone. It is simultaneously the tie point - the common origin of the wave vectors, c.f., Fig. 2.1.

The particular case of a cubic lattice illustrated here graphically can be generalized to any reciprocal crystal lattice. To discuss this point, we still use the kinematic approximation. By using the kinematic Bragg's law in vector form $2\boldsymbol{K}_0 \boldsymbol{B} + \boldsymbol{B}^2 = 0$, as given by (2.110), and the Bragg reflection condition in backscattering $\boldsymbol{K}_0 = -\boldsymbol{N}/2$, one can find the following relation between the diffraction vector $\boldsymbol{N}$ for the back-reflection and the diffraction vector $\boldsymbol{B}$ for the accompanying reflection:

$$\boldsymbol{B}\boldsymbol{N} = \boldsymbol{B}^2. \tag{2.183}$$

The relation is valid in the kinematic approximation for the reciprocal lattice of any kind, not only cubic. If a solution of (2.183) exists, it is not necessarily unique. All possible solutions, i.e., the vectors $\boldsymbol{B}$ satisfying (2.183) for the given $\boldsymbol{N}$, can be found by exhaustive computer search.

2.5.2 Examples of Si and α-Al$_2$O$_3$ Crystals

The general graphical and analytical considerations can be supported by numerical analysis of Bragg back-reflections and their accompanying reflections for some particular crystal lattices.

The results of the numerical analysis for silicon single crystals are summarized in Appendix A.3. Silicon crystallizes in a cubic structure with the face-centered cubic Bravais lattice. C (diamond), Ge, GaAs, and many other crystals have the same structure and belong to the space group $Fdm3(O_h^7)$ (No. 227). All allowed Bragg reflections (hkl) in backscattering are listed in the table of Appendix A.3 for which the energy E of backscattered photons does not exceed 40 keV. Altogether, there are about 260 reflections. Almost all Bragg reflections in backscattering are accompanied by other Bragg reflections. The accompanying reflections are ascertained numerically by using

2.5 Multiple-Beam Diffraction in Backscattering: Kinematic Treatment   117

(2.183). The number of simultaneously excited waves in the crystal is always even and can be as large as 96. There are only two exceptions. Multiple-beam diffraction is not excited if Bragg's condition for back-diffraction is fulfilled for $\boldsymbol{N} = (1\ 1\ 1)$, or $\boldsymbol{N} = (2\ 2\ 0)$, and equivalent reciprocal vectors.

The situation changes drastically in crystal lattices with lower symmetry. Bragg reflections in backscattering are still very frequently accompanied by other Bragg reflections. However, in lattices with lower symmetry there exist many more multiple-beam-free Bragg reflections in backscattering.

This statement is confirmed by the numerical analysis of Bragg reflections in sapphire (α-Al$_2$O$_3$) single crystals. α-Al$_2$O$_3$ has a rhombohedral crystal lattice and belongs to the space group $R\bar{3}c$ (D_{3d}^6) (No. 167). For more information on α-Al$_2$O$_3$, see Appendix A.2. The results of the numerical analysis are given in Appendix A.4. All allowed *two-beam-case* Bragg back-reflections ($hkil$) in α-Al$_2$O$_3$ for photon energies $E \leq 40$ keV are listed there whose peak reflectivity exceeds $R = 0.1$. Altogether, about 150 reflections of this kind exist. That means, that the number of the multiple-beam-free Bragg reflections in backscattering in α-Al$_2$O$_3$ and the number of all allowed reflections in silicon is about the same.

2.5.3 Conjugate Pairs of Accompanying Reflections

It is still an open question, whether for a certain vector $\boldsymbol{N}$ there exists at least one vector $\boldsymbol{B}$ fulfilling (2.183). However, if a such vector exists, the relation (2.183) implies that the vectors $\boldsymbol{N}$, $\boldsymbol{B}$, and $\boldsymbol{L} = \boldsymbol{N} - \boldsymbol{B}$ compose a right-angled triangle ($\boldsymbol{N}^2 = \boldsymbol{B}^2 + \boldsymbol{L}^2$) with a right angle between the vectors $\boldsymbol{B}$, and $\boldsymbol{L} = \boldsymbol{N} - \boldsymbol{B}$ as illustrated in Fig. 2.30. Moreover, if the relation (2.183) is valid for the vectors $\boldsymbol{N}$ and $\boldsymbol{B}$, a similar relation

$$\boldsymbol{L}\boldsymbol{N} = \boldsymbol{L}^2 \tag{2.184}$$

is valid for the reciprocal vector $\boldsymbol{L}$. Thus, if a back-reflection with diffraction vector $\boldsymbol{N}$ is simultaneously accompanied by a reflection with diffraction vector $\boldsymbol{B}$, it is also simultaneously accompanied by the reflection with diffraction vector $\boldsymbol{L}$, where

$$\boldsymbol{B} + \boldsymbol{L} = \boldsymbol{N} \quad \text{with} \quad \boldsymbol{L} \perp \boldsymbol{B}. \tag{2.185}$$

The plane built up of the vectors $\boldsymbol{B}$, $\boldsymbol{L}$, and, which certainly contains $\boldsymbol{N}$, we shall call as "basal plane of the reflection pair $\boldsymbol{B}$, $\boldsymbol{L}$", or simply basal plane.

The reflections with diffraction vector $\boldsymbol{L}$ and $\boldsymbol{B}$ are termed "conjugate pairs" of the accompanying reflections, and the pair of the related waves are called "conjugate waves". Such conjugate pairs were discussed by Sutter (2000), and Sutter et al. (2001) for a particular case of 24-beam diffraction at normal incidence on the (0 4 12) planes in the cubic lattice of silicon. The present analysis shows that the appearance of the accompanying reflections in pairs is a general property independent of the reflection and crystal symmetry.

118   2. Dynamical Theory of X-Ray Bragg Diffraction

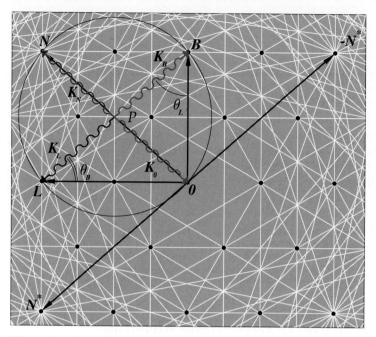

Fig. 2.31. Bragg backscattering from a two-dimensional hexagonal lattice in the presence of multiple-beam diffraction. $\boldsymbol{K}_0$ is the wave vector of the incident radiation fulfilling Bragg's condition for backscattering with diffraction vector $\boldsymbol{N}$: $\boldsymbol{K}_N = -\boldsymbol{K}_0 = \boldsymbol{N}/2$. Other notations as in Fig. 2.30.

Accompanying reflections appear always in conjugate pairs, whereas the vectors $\boldsymbol{N}$, $\boldsymbol{B}$, and $\boldsymbol{L}$ compose a right-angled triangle. The latter implies that the accompanying reflections appear systematically in rectangular reciprocal lattices. However, they can also appear in non-rectangular reciprocal lattices, if a rectangular lattice occurs as a sub-lattice. Such a case is shown in Fig. 2.31 using the example of a two-dimensional hexagonal lattice.

By using (2.181), (2.182), and (2.185) one readily obtains for the wave vectors of the waves reflected from atomic planes belonging to the conjugate pair

$$\boldsymbol{K}_B = \frac{1}{2}(\boldsymbol{B}-\boldsymbol{L}), \qquad \boldsymbol{K}_L = \frac{1}{2}(\boldsymbol{L}-\boldsymbol{B}), \tag{2.186}$$

and

$$\boldsymbol{K}_B = -\boldsymbol{K}_L. \tag{2.187}$$

The wave vectors $\boldsymbol{K}_B$ and $\boldsymbol{K}_L$ are exactly opposite to each other. In other words, conjugate waves propagate in opposite directions. This is also clearly seen in Figs. 2.30 and 2.31. This property could be used to generate coherent waves propagating in opposite directions, i.e., standing waves.

2.5 Multiple-Beam Diffraction in Backscattering: Kinematic Treatment

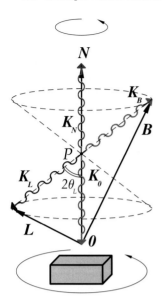

Fig. 2.32. Rotation of the crystal around the back-diffraction vector N does not violate the Bragg condition for the accompanying reflections with diffraction vectors B and L. By rotating the crystal, the accompanying reflected waves with wave vectors K_B and K_L together with B and L change their direction in space and travel along the generating lines of the cones with apices at the tie point P with opening angle $2\theta_L$ or $2\theta_B = \pi - 2\theta_L$, respectively.

By using the preceding relations, it may also be readily shown that the sum of the Bragg angles of the reflections belonging to the same conjugate pair compose the right angle:

$$\theta_L + \theta_B = \pi/2, \qquad (2.188)$$

as shown in Figs. 2.30 and 2.31. The angles θ_L and θ_B are defined as $\sin\theta_L = L/N$ and $\sin\theta_B = B/N$, respectively. These are the Bragg angles for x-ray photon energy $E = N\hbar c/2$ and for the reflections with the diffraction vectors B and L, respectively.

It is of interest to note that rotation of the crystal around the back-diffraction vector N does not violate the Bragg condition for the accompanying reflections. By rotating the crystal the accompanying reflected waves do not disappear. They only change their direction in space, as is shown in Fig. 2.32. The locus of the endpoints of each wave vector K_B and K_L is obviously the arc of a circle. The reflected rays will travel along the generating lines of a cone whose apex is at the tie point P and with opening angle $2\theta_L$ or $2\theta_B = \pi - 2\theta_L$, respectively. This property could be used as a tool for changing directions of reflected x-ray beams over a very wide angular range (up to 2π) with the direction and wavelength of the incident beam fixed. In the two-beam diffraction case such angular changes are impossible. Also, if N is *not* a back-diffraction vector, this effect does not take place in multiple-beam diffraction.

From (2.187), (2.182) and (2.18) one can find the following relation for the direction cosines $\gamma_B = (K_0 + B)\hat{z}/K$ and $\gamma_L = (K_0 + L)\hat{z}/K$ (2.18) of the angles between the surface normal and the wave vectors of the conjugate

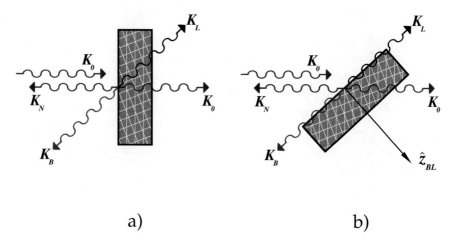

Fig. 2.33. Bragg backscattering in the presence of multiple-beam diffraction. $\boldsymbol{K}_0$ is the wave vector of the incident and of the forward-transmitted wave. $\boldsymbol{K}_N$ is the wave vector of the Bragg backscattered wave. $\boldsymbol{K}_L$ and $\boldsymbol{K}_B$ are the wave vectors of a pair of conjugate waves. Scattering geometries in direct space of the crystal: a) the conjugate waves are propagating towards the inside and outside of the crystal; b) the conjugate waves are propagating parallel to the crystal surface. $\hat{\boldsymbol{z}}_{BL}$ shows the direction of the critical surface normal, for which $\hat{\boldsymbol{z}}_{BL} \perp \boldsymbol{K}_B$ and $\hat{\boldsymbol{z}}_{BL} \perp \boldsymbol{K}_L$. White lines indicate the reflecting atomic planes.

pairs:

$$\gamma_B = -\gamma_L. \tag{2.189}$$

By using the definition (2.29), the asymmetry factors of the conjugate waves are given by

$$b_B = -b_L. \tag{2.190}$$

These relations imply that one wave of the conjugate pair with $\gamma_B < 0$ is always directed towards the outside of the crystal (Bragg-case wave) and the other wave with $\gamma_L > 0$ is directed towards the inside of the crystal (Laue-case wave) as seen relative to the crystal front surface - Fig. 2.33a.

The only exception is when the conjugate waves are propagating parallel to the crystal surface, i.e., $\gamma_B = -\gamma_L = 0$, or $b_B = -b_L = \infty$, as is shown in Fig. 2.33b. The direction of the inward normal to the crystal surface, which may be called in this case the "critical surface normal" for the pair of accompanying reflections B, and L, may be given by $\hat{\boldsymbol{z}}_{BL} = \boldsymbol{K}_B \times (\boldsymbol{K}_0 \times \boldsymbol{K}_B)/K^3$. By using (2.186), (2.181), and the rule $\boldsymbol{X} \times (\boldsymbol{Y} \times \boldsymbol{Z}) = \boldsymbol{Y}(\boldsymbol{X} \cdot \boldsymbol{Z}) - \boldsymbol{Z}(\boldsymbol{X} \cdot \boldsymbol{Y})$ for the triple vector product, this expression for the vector of the critical surface normal can be transformed to

$$\hat{z}_{BL} = -\frac{2}{N^3}\left(\boldsymbol{L}B^2 + \boldsymbol{B}L^2\right) \equiv -\frac{2}{N}\left(\boldsymbol{L}\sin^2\theta_B + \boldsymbol{B}\sin^2\theta_L\right). \quad (2.191)$$

Such cases, when the waves of the conjugate pair are propagating parallel to the crystal surface, are encountered quite often. For example, the Bragg reflection $(0\ 0\ n)$ in backscattering in silicon crystals is accompanied by the following two conjugate pairs of Bragg reflections $(n/2\ 0\ n/2)$, $(\bar{n}/2\ 0\ n/2)$ and $(0\ n/2\ n/2)$, $(0\ \bar{n}/2\ n/2)$, respectively. Here n takes the values: $n = 2^m, m = 2, 3, \ldots$ It can be readily shown, that the wave vectors of the conjugate pairs are perpendicular to $\boldsymbol{K}_0$. Therefore, if $\boldsymbol{K}_0$ is normal to the entrance surface the wave vectors of the conjugate pairs are parallel to the entrance surface.

Since the backscattered wave is always directed towards the outside of the crystal and the incident wave is directed towards the inside - there are always the same number of waves directed towards the inside and the outside of the crystal.

2.5.4 Umweganregung - Detour Excitation

The wave with wave vector $\boldsymbol{K}_B$ can be scattered directly into the wave with wave vector $\boldsymbol{K}_L$ with momentum transfer $\boldsymbol{N}^* = \boldsymbol{K}_L - \boldsymbol{K}_D = \boldsymbol{L} - \boldsymbol{B}$ - see Figs. 2.30, and 2.31. Since, $\boldsymbol{K}_L = -\boldsymbol{K}_B$, the vector $\boldsymbol{N}^*$ is the diffraction vector, which is responsible for backscattering of the wave with wave vector $\boldsymbol{K}_L$ into the wave with wave vector $\boldsymbol{K}_B$. One may readily check, that $|\boldsymbol{N}^*| = |\boldsymbol{N}|$. Vector $\boldsymbol{N}^*$ may be considered as the mirror image of the back-diffraction vector $\boldsymbol{N}$, i.e., they are symmetric with respect to a mirror reflection across the plane perpendicular to $\boldsymbol{B}$.

Similarly, $-\boldsymbol{N}^*$ is the diffraction vector, which is responsible for backscattering of the wave with wave vector $\boldsymbol{K}_B$ into the wave with wave vector $\boldsymbol{K}_L$. It is symmetric to $\boldsymbol{N}$ with respect to a mirror reflection across the plane perpendicular to $\boldsymbol{L}$. The existence of the scattering channels with momentum transfer $\boldsymbol{N}^*$ and $-\boldsymbol{N}^*$ has the following important consequences.

It may happen that the structure factor for one of the two conjugate reflections, say with diffraction vector $\boldsymbol{L}$, is zero, i.e., this reflection is forbidden[18]. One may then think that one wave less will be generated in the scattering process. However, in fact, all four waves are present. Although the wave with $\boldsymbol{K}_L$ cannot be excited directly, still it can be excited indirectly: due to backscattering with momentum transfer $\boldsymbol{N}^*$ of the wave with wave vector $\boldsymbol{K}_B$. This is the so-called Umweganregung (detour excitation) well known in the field of multiple-beam diffraction. We may call such a case "degenerate" n-beam diffraction. The intensity of the Umweganregung remains an open question, which will be addressed in some more detail in Sect. 2.6.2.

---

[18] For example, the $(11\bar{2}3)$ back-reflection in α-Al$_2$O$_3$ is accompanied by the $(11\bar{2}0)$ and (0003) conjugate pair of reflections, among which the latter is forbidden.

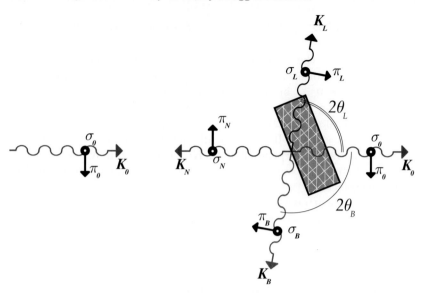

Fig. 2.34. Schematic drawing illustrating the definition of the polarization vectors $\boldsymbol{\sigma}_H$ and $\boldsymbol{\pi}_H$ ($H = 0, N, L, B$) in a four-beam coplanar Bragg diffraction case with backscattering channel N. $\boldsymbol{K}_0$ and $\boldsymbol{K}_N$ are the wave vectors of the incident (or forward-transmitted) and Bragg backscattered waves, respectively. $\boldsymbol{K}_L$ and $\boldsymbol{K}_B$ are the wave vectors of a conjugate pair of the accompanying Bragg reflections.

In this case, only one accompanying reflection of the conjugate pair – the reflection with diffraction vector $\boldsymbol{B}$ influences *directly* the intensity of the Bragg back-reflection.

2.5.5 Polarization Effects

Two-beam Bragg diffraction in backscattering is practically independent of the polarization state of the incident wave. As distinct from this, the polarization state of the incident wave may be a very important factor in multiple-beam Bragg diffraction. In particular, the result of the excitation of a conjugate pair of the accompanying reflections may be very different for the incident wave in different polarization states. Figure 2.34 illustrates the definition of the polarization vectors $\boldsymbol{\sigma}_H$ and $\boldsymbol{\pi}_H$ ($H = 0, N, L, B$) in a four-beam coplanar Bragg diffraction case with backscattering channel N.

All polarization vectors $\boldsymbol{\sigma}_H$ are parallel to each other and perpendicular to the basal plane going through the wave vectors $\boldsymbol{K}_0$, $\boldsymbol{K}_N$, $\boldsymbol{K}_B$, and $\boldsymbol{K}_L$. All polarization vectors $\boldsymbol{\pi}_H$ are lying in the basal plane. They are defined according to (2.39). Each polarization vector $\boldsymbol{\sigma}_H$ is perpendicular to all polarization vectors $\boldsymbol{\pi}_H$ and vice versa. This means that the waves in the linear polarization states σ are not coupled to the waves in the linear polarization states π. This situation is similar to the two-beam diffraction case.

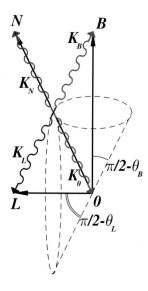

Fig. 2.35. Reflection cones for the conjugate pair of Bragg reflections with diffraction vectors B and L accompanying the Bragg back-reflection with diffraction vector N. The wave vectors K_0, K_N, K_L, and K_B are those of the incident and Bragg-reflected waves, respectively.

Furthermore, since $\sigma_H \sigma_{H'} = \pm 1$, the coupling between the waves in the polarization σ-states is strong. By contrast, the coupling between the waves in the π-polarization states can be very weak. For example, in the special case when $K_0 \perp K_B$ and respectively $K_0 \perp K_L$, the polarization factor $P^{\pi\pi}_{0B} = \pi_0 \pi_B = 0$, and also $P^{\pi\pi}_{0L} = \pi_0 \pi_L = 0$. In this particular case, the pair of conjugate waves in the linear polarization state π is not coupled to the incident wave and thus cannot be excited. Thus, by choosing a proper polarization of the incident wave, one can suppress or enhance the excitation of the accompanying Bragg reflections.

A more complicated situation arises if more than one conjugate pair of accompanying Bragg reflections is excited. The Bragg diffraction in this case is most probably non-coplanar. Such situations are not simple and should in general be analyzed by using computer simulations.

2.5.6 Region of Multiple-Beam Excitation

A practical question, which often arises, is how to avoid multiple-beam excitation? To give an answer, it is worth mapping the region of the reciprocal space in which the condition of multiple-beam excitation is fulfilled. One may understand the principles by considering only one conjugate pair of the accompanying reflections. Still, even in this case, a rigorous treatment requires the numerical solution of the diffraction problem. Instead, one may use simple geometrical constructions suggested by Sutter (2000) to solve the problem qualitatively.

Take any Bragg reflection with diffraction vector $\boldsymbol{B}$. The kinematic Bragg's law $2\boldsymbol{K}_0\boldsymbol{B} + \boldsymbol{B}^2 = 0$ (2.110) tells us that all wave vectors $\boldsymbol{K}_0$, which compose an angle $\pi/2 - \theta_B$ with the diffraction vector $\boldsymbol{B}$, where $\sin\theta_B = B/2K_0$, satisfy this law. All these vectors are lying on the cone, termed the "reflection cone", with opening angle $\pi/2-\theta_B$, as in Fig. 2.35. The reflection cone for back-reflection with diffraction vector $\boldsymbol{N}$ has an opening angle of zero ($\pi/2 - \theta_N = 0$), and therefore degenerates to a line antiparallel to $\boldsymbol{N}$. The reflection cone for another Bragg reflection of the conjugate pair with diffraction vector $\boldsymbol{L}$ has an opening angle of $\pi/2 - \theta_L$. For any conjugate pair $\theta_L + \theta_B = \pi/2$ and $\boldsymbol{L} \perp \boldsymbol{B}$. Therefore, the reflection cones of the conjugate pair intersect along a line which coincides with $\boldsymbol{N}$, see Fig. 2.35. This line gives us the direction of $\boldsymbol{K}_0$, for which all three Bragg reflections are excited simultaneously. At this point, one may remember that in the two-beam diffraction case high reflectivity is achieved not only when $\boldsymbol{K}_0$ fulfills Bragg's law exactly, but also in some small region in its vicinity. As a result, the wall of the cone should have a finite thickness, which is dependent on the angular width of the appropriate two-beam-case Bragg reflection.

Figure 2.36 shows the cross-section of the reflection cones with the plane normal to $\boldsymbol{N}$. Only a small part of the traces of the cones are seen shown in light blue (grey). The circle around the point N indicates the angular region of the back-reflection belonging to the diffraction vector $\boldsymbol{N}$. In fact, it is a cross-section of the aforementioned degenerated cone; however, now it is shown with a finite opening because of the finite angular width of the back-reflection. The radius of the circle is shown much larger than the width of the traces of the walls of the reflection cones, which is in agreement with the 100 to 1000 times larger angular widths of the back-reflections. The radius of the circle and the thicknesses of the cones' walls are greatly exaggerated compared to the cones' opening for clarity. The area in white, including the point N, shows the intersection region of the three objects. Thus, this area maps the region of the reciprocal space in which the condition of multiple-beam excitation is fulfilled.

The widths of the reflection cones' walls are in fact very small compared to the width of the back-reflection region. Therefore, the range over which the three Bragg reflections are excited simultaneously is greatly stretched-out in the direction normal to the basal plane.

This gives us a clear prescription of how multiple-beam Bragg diffraction in backscattering can be avoided. One leaves the region of multiple-beam diffraction by changing the direction of the incident wave vector from $\boldsymbol{K}_0 = -\boldsymbol{N}/2$ valid for exact backscattering to

$$\boldsymbol{K}_0 = -\frac{\boldsymbol{N}}{2} + \boldsymbol{q}, \qquad |\boldsymbol{q}| \ll |\boldsymbol{N}|. \tag{2.192}$$

If $\boldsymbol{q}$ is directed in the basal plane, as in Fig. 2.36, already a very small q ($q/K_0 \approx 10^{-6} - 10^{-5}$) will move $\boldsymbol{K}_0$ outside the region. The direction normal to the basal plane is much less favorable. One needs much larger q's to move

2.5 Multiple-Beam Diffraction in Backscattering: Kinematic Treatment 125

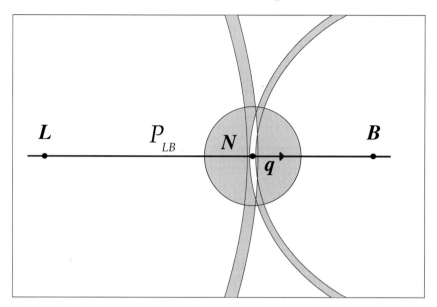

Fig. 2.36. Cross-section in the reciprocal space of the Bragg reflection cones. The incident wave excites the Bragg back-reflection with diffraction vector N, and the conjugate pair of Bragg reflections B and L and (see also Fig. 2.35). The intersection points of the cross-section plane with N, B and L are shown with small black circles. The cross-section plane is normal to N, and to the basal plane P_{BL} of the reflection pair B, L, seen as a line connecting the points B and L. The walls of the reflection cones in light blue (grey) have a finite thickness due to the finite angular width of the Bragg reflections. The large circular area in light blue (grey) shows the region of Bragg back-reflection. The area in white maps the region in which the condition of multiple-beam excitation is fulfilled. Vector q points into the direction, in which one can escape the region in the shortest possible distance.

K_0 out of the region. From the properties of two-beam Bragg backscattering one may expect that $q/K_0 \approx 2\sqrt{|\chi_H|} \approx 10^{-3}$.

Generalization of the considerations presented above to the case of more than one pair of the conjugate accompanying Bragg reflections is shown in Fig. 2.37. Projection in the reciprocal space onto the plane perpendicular to diffraction vector N is presented. All the basal planes $P_{L_iB_i}$, $i = 1, 2, ...$ of the reflection pairs B_i, L_i contain N and therefore intersect in N. Projection of the basal planes are seen as dark blue (black) solid and dotted lines going through N. The area where multiple beam diffraction takes place is shown as white stripes perpendicular to $P_{L_iB_i}$. The most favorable strategy, of how to avoid multiple-beam diffraction, is clear: one has to move in the reciprocal space off exact backscattering (off point N) in the direction, in which one escapes the multiple-beam diffraction area in the shortest distance.

One should be cautious, since these predictions have a more qualitative character, as only the properties of the two-beam-case Bragg diffraction were

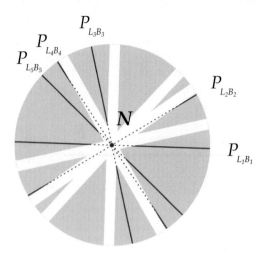

Fig. 2.37. Projection in the reciprocal space onto the plane perpendicular to diffraction vector N and basal planes $P_{L_i B_i}$, $i = 1, 2, ...$ The projections of the basal planes are seen as dark blue (black) solid and dotted lines going through N. The area where multiple beam diffraction takes place is shown as white stripes perpendicular to $P_{L_i B_i}$.

used for their derivation without taking into account the interaction between the conjugate reflection pairs.

2.6 4-Beam Diffraction in Backscattering: Dynamical Treatment

For a deeper understanding of the influence of multiple-beam effects on Bragg reflection in backscattering, we shall examine this scattering mode by using the system of fundamental equations of the dynamical theory (2.40). This will also allow us to calculate the crystal reflectivity into different diffraction channels.

We start with the assumption that the excitation condition (2.13) is fulfilled for n plane wave components with vector amplitudes D_H and wave vectors k_H, where the index H takes n different values. This implies that n equations have to be retained in the fundamental system with n unknown vector amplitudes D_H and unknown k_0. As in the two-beam diffraction case the problem reduces to the eigenvalue problem with boundary conditions. However, a large number of unknowns greatly increases the difficulties in solving the problem. Analytical solutions usually cannot be obtained, except for the regions where the multiple-beam case transforms to the two-beam case. Such kind of an asymptotic analysis in the four-beam case was carried out, e.g., by Afanas'ev and Kohn (1976). In general, the problem must be solved numerically. Colella (1974) has developed a general formalism for the solution of the n-beam x-ray Bragg diffraction problem and proposed a numerical procedure by which the amplitudes of the diffracted waves can be determined. Modified algorithms were proposed more recently by Stepanov and Ulyanenkov (1994) and Stetsko and Chang (1997). Stepanov et al. (1991) have also studied a particular four-beam diffraction with the (0 2 6) Bragg back-reflection in germanium under grazing incidence conditions. Kohn et al. (1999) have considered a particular case of the six-beam Bragg diffraction with the (1 9 9) Bragg back-reflection in silicon. Sutter et al. (2001) have discussed the 24-beam case with the (12 4 0) Bragg back-reflection in the same crystal (see also Sutter (2000)).

As was shown in the preceding section, if the Bragg back-reflection with the diffraction vector $\boldsymbol{N}$ is accompanied by the reflection with the diffraction vector $\boldsymbol{B}$ then it is accompanied also by one more reflection with diffraction vector $\boldsymbol{L} = \boldsymbol{N} - \boldsymbol{B}$, i.e., the accompanying reflections appear in pairs - the conjugate pairs. In turn, this means that the minimum number of waves excited in the crystal is four, with wave vectors $\boldsymbol{k}_0$, $\boldsymbol{k}_N$, $\boldsymbol{k}_B$, and $\boldsymbol{k}_L$. The four-beam case is thus the simplest multiple-beam case, which regularly occurs in backscattering. If the structure factor for one of the two conjugate reflections, however, is zero, then only one accompanying reflection is allowed. Thus, in this case only one accompanying reflection can *directly* influence the backscattering channel. Such regime we have termed "degenerate" four-beam diffraction.

Because of their relative simplicity, and because we encounter these cases in the applications discussed in this book, the "degenerate" four-beam diffraction and four-beam diffraction regimes will be addressed in more detail. The first focus is to find the particular form of the fundamental equations (2.40) in the four-beam diffraction regime including one Bragg reflection in backscattering.

2.6.1 Fundamental Equations

The diffraction vector $\boldsymbol{N}$, which relates to the back-reflection, and the diffraction vectors $\boldsymbol{B}$, $\boldsymbol{L}$, which relate to the conjugate pair of accompanying reflections, are coplanar and compose a right-angled triangle: $\boldsymbol{N} = \boldsymbol{B} + \boldsymbol{L}$ with $\boldsymbol{B} \perp \boldsymbol{L}$ (2.185), as in Fig. 2.35.

The wave vector $\boldsymbol{K}_0$ of the incident wave does not necessarily lie in the basal plane. The same is valid for the in-crystal wave vectors $\boldsymbol{k}_0$, $\boldsymbol{k}_N$, $\boldsymbol{k}_B$, and $\boldsymbol{k}_L$ as well as for the corresponding vacuum wave vectors $\boldsymbol{K}_N$, $\boldsymbol{K}_B$, and $\boldsymbol{K}_L$. However, as follows from our discussion in Sect. 2.5.6, if $\boldsymbol{K}_0$ deviates from the plane by more than $\simeq 1$ mrad, no significant excitations of the diffracted waves can be achieved. Therefore, in all cases of interest the wave vectors of the transmitted and diffracted waves make a small angle with the basal plane. To a good approximation, one may regard this regime as four-beam *coplanar* Bragg diffraction.

We shall be concerned with the problem of how the crystal scatters the radiation wave propagating close to normal incidence on the set of atomic planes associated with the reciprocal vector $\boldsymbol{N}$. The parameters of the prob-

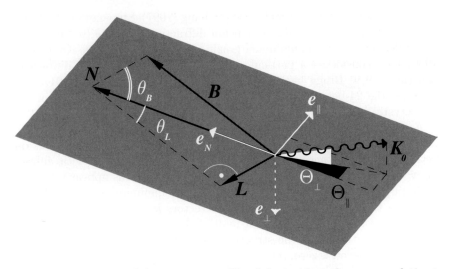

Fig. 2.38. Direction of the wave vector $\boldsymbol{K}_0$ of the incident plane wave relative to coplanar diffraction vectors $\boldsymbol{N}$, $\boldsymbol{B}$, and $\boldsymbol{L}$ ($\boldsymbol{B} \perp \boldsymbol{L}$, $\boldsymbol{N} = \boldsymbol{B} + \boldsymbol{L}$). The vectors $\boldsymbol{N}$, $\boldsymbol{B}$, and $\boldsymbol{L}$ compose the basal plane (shown in grey). The reference unit vectors $\boldsymbol{e}_\parallel$ and $\boldsymbol{e}_N$ are lying in the same plane, while $\boldsymbol{e}_\perp$ is normal to it. $\boldsymbol{K}_0$ is at near normal incidence on the atomic planes (not shown) perpendicular to $\boldsymbol{N}$. The angles $\Theta_\parallel$, and $\Theta_\perp$ specify the deviation from normal incidence in the basal plane, and in the direction perpendicular to it.

lem are the magnitude K of the wave vector $\boldsymbol{K}_0$ of the incident wave (or equivalently the photon energy $E = \hbar c K$), and its direction relative to $\boldsymbol{N}$.

According to the kinematic Bragg's law (2.110), the radiation wave is reflected backwards if its wave vector $\boldsymbol{K}_0 \simeq -\boldsymbol{N}/2$. The photon energy $E = \hbar c K$ is equal in this case to the Bragg energy $E_N = \hbar c N/2$. Therefore, it is convenient to express the wave vector $\boldsymbol{K}_0$ in terms of small deviations from $-\boldsymbol{N}/2$ as

$$\boldsymbol{K}_0 = -\frac{\boldsymbol{N}}{2} + \boldsymbol{q}, \qquad (2.193)$$

where $|\boldsymbol{q}| \ll N$. In the reference system with the unit basis vectors $\boldsymbol{e}_N = \boldsymbol{N}/N$, $\boldsymbol{e}_\perp = \boldsymbol{L} \times \boldsymbol{B}/|\boldsymbol{L} \times \boldsymbol{B}|$, and $\boldsymbol{e}_\parallel = \boldsymbol{e}_\perp \times \boldsymbol{e}_N$, as is shown in Fig. 2.38, we obtain for $\boldsymbol{q}$:

$$\boldsymbol{q}/K = \Theta_\parallel \boldsymbol{e}_\parallel + \Theta_\perp \boldsymbol{e}_\perp - \left[\epsilon - \frac{1}{2}\left(\Theta_\perp^2 + \Theta_\parallel^2\right)\right] \boldsymbol{e}_N, \qquad (2.194)$$

$$E = \hbar c K, \qquad \epsilon = (E - E_N)/E_N.$$

Here ϵ is the relative deviation of the photon energy E from the Bragg energy E_N. The angles $\Theta_\parallel$ and $\Theta_\perp$ describe small deviations from normal incidence

2.6 4-Beam Diffraction in Backscattering: Dynamical Treatment

in the basal plane, and in the direction perpendicular to it, respectively. They can take both positive and negative values.

Further, one may show with the help of the definition (2.19), equations (2.183)-(2.184), and (2.194) that the parameters of deviation from Bragg's condition α_H entering the fundamental equations (2.40) are equal to

$$\alpha_H = \frac{2qH}{K^2}, \qquad (2.195)$$

where $H = 0, N, B$ or L. By using (2.195) and $N = L + B$ (2.185) the following dependency can be established between deviation parameters associated with different diffraction vectors:

$$\alpha_N = \alpha_B + \alpha_L. \qquad (2.196)$$

With the help of (2.194) and (2.195), one may also derive the expressions for the deviation parameters α_H expressed in terms of angular deviations $\Theta_\parallel$, $\Theta_\perp$, and relative energy deviation $\epsilon = (E - E_N)/E_N$ from the Bragg energy E_N of the back-reflection:

$$\alpha_N = 2\left(\Theta_\perp^2 + \Theta_\parallel^2\right) - 4\epsilon$$

$$\alpha_B = 2\Theta_\parallel \sin 2\theta_B + \left[2\left(\Theta_\perp^2 + \Theta_\parallel^2\right) - 4\epsilon\right]\sin^2\theta_B, \qquad (2.197)$$

$$\alpha_L = -2\Theta_\parallel \sin 2\theta_L + \left[2\left(\Theta_\perp^2 + \Theta_\parallel^2\right) - 4\epsilon\right]\sin^2\theta_L.$$

Here θ_L and θ_L are the angles, which are defined as $\sin\theta_L = L/N$ and $\sin\theta_B = B/N$, respectively. The angles are shown in Fig. 2.38.

One may note that in all the three cases given by (2.197) there exists only a quadratic dependence on the angular deviation $\Theta_\perp$ perpendicular to the basal plane. Recalling that $\Theta_\perp \ll 1$, we can immediately predict a weak dependence of both the crystal reflectivity and the crystal transmissivity on the parameter $\Theta_\perp$.

Another physical parameter of the problem is the polarization state of the incident wave. By a judicious choice of the polarization components of the crystal wave fields, the system of fundamental equations (2.40) can be simplified in our particular case of four-beam diffraction. We use the definition (2.39) and neglect possible slight deviations of the wave vectors from the basal plane. With this approximation, the σ-polarization vectors are directed perpendicular to the basal plane. Let us order them all in the same direction. The π-polarization vectors lie in the basal plane, as in Fig. 2.34. Due to this, the mixed polarization factors $P_{HH'}^{\pi\sigma} = (\pi_H \sigma_{H'}) = 0$ (indices H and H' take any of the four values 0, N, B, and L). In other words, there is no polarization mixing. As a result, the elements $G_{HH'}^{ss'}$ of the scattering matrix (2.41) with $s \neq s'$ are zero, and the system of eight fundamental equations (2.40) breaks up into two independent sets of four equations, one containing only the σ-components of the radiation field, and the other containing only the π-components. Later on it will be assumed that the incident radiation wave is either in the σ- or π-polarization state.

One more physical parameter of the problem is the direction of the crystal surface normal $\hat{z}$ (2.24). At the moment we shall impose no constraints on its possible direction. To simplify the problem, we shall only assume that the crystal is cut in such a way that none of the four waves propagates parallel to the crystal slab surface, i.e., $|\gamma_H| \gg \sqrt{|\chi_0|} \simeq 10^{-3}$ for all the waves. For example, cases like the one illustrated in Fig. 2.33b are not considered. As a result, the system (2.40) can by linearized by omitting the quadratic terms $\varkappa^2$ in (2.40). The asymmetry factors $b_H = \gamma_0/\gamma_H$ are given by

$$b_0 = 1, \quad b_N \simeq -1, \quad b_L = -b_B = b > 0. \tag{2.198}$$

This follows from the definition (2.29) and equation (2.190).

With these restrictions taken into account the system of fundamental equations can be written as:

$$\sum_{H'} \tilde{G}^s_{HH'} D^s_{H'} - \varepsilon D^s_H = 0, \quad \varepsilon = 2\gamma_0 \frac{\varkappa}{K}, \tag{2.199}$$

$$\tilde{G}^s_{HH'} = \left(\chi_{H-H'} P^{ss}_{HH'} - \alpha_H \delta_{HH'} \right) b_H, \quad b_H = \frac{\gamma_0}{\gamma_H}, \tag{2.200}$$

with $H(H') = 0, N, B,$ and L; $s = \sigma$ or π. Here $\tilde{G}^s_{HH'}$ are the s-polarization components of the elements of the scattering matrix, whose rank is reduced to four.

Finally, in the four-beam diffraction case with one Bragg-reflection in backscattering, one may introduce modified polarization factors P^s_F with a reduced number of indices according to the rule: $P^s_F = P^{ss}_{H, H \pm F}$, where F, H, and $F \pm H$ take the values $0, N, B, L$ or N^*. In particular, since $P^{\sigma\sigma}_{HH'} = 1$ for all possible H and H', the modified polarization factors with $s = \sigma$ are equal to unity: $P^\sigma_F = 1$. The modified polarization factors with $s = \pi$ are defined as $P^\pi_N = P^\pi_{N^*} = -1$, $P^\pi_L = \cos 2\theta_L$, and $P^\pi_B = \cos 2\theta_B = -\cos 2\theta_L = -P^\pi_L$, respectively. To derive these relations, the above defined rule, the definition of the polarization factors (2.42), and Fig. 2.34 were used.

For the four-beam Bragg diffraction case with one Bragg-reflection in backscattering, $\tilde{G}^s$ can be presented in the following final form

$$\tilde{G}^s = \begin{pmatrix} \chi_0 & \chi_{\overline{N}} P^s_N & \chi_{\overline{B}} P^s_B & \chi_{\overline{L}} P^s_L \\ -\chi_N P^s_N & -(\chi_0 - \alpha_N) & -\chi_{\overline{L}} P^s_L & -\chi_B P^s_B \\ -\chi_B P^s_B b & -\chi_{\overline{L}} P^s_L b & -(\chi_0 - \alpha_B) b & -\chi_{\overline{N}^*} P^s_N b \\ \chi_L P^s_L b & \chi_{\overline{B}} P^s_B b & \chi_{N^*} P^s_N b & (\chi_0 - \alpha_L) b \end{pmatrix} \tag{2.201}$$

with $N^* = L - B$.

By solving the eigenvalue problem for the system of fundamental equations (2.199) with scattering matrix (2.201), and applying the appropriate boundary conditions, as described in detail in Sect. 2.2, the crystal reflectivity and transmissivity (2.50), can be obtained numerically with an automated computer code that implements the theory.

To gain more insight into how the accompanying reflections influence Bragg backscattering, we shall consider some particular cases. Clearly, they will not give a complete understanding of the problem. Still, they will enrich our comprehension, and additionally will provide recommendations on how the accompanying reflections could be suppressed, or their excitation avoided.

2.6.2 4-Beam Diffraction - "Degenerate" Case

As a first step, special cases of four-beam Bragg diffraction are considered, in which one of the reflections is forbidden.

Bragg-case Reflection is Forbidden. First, it is assumed that the Bragg-case reflection with diffraction vector B is forbidden, i.e., $\chi_B = 0$. The Laue-case reflection with diffraction vector L is allowed: $\chi_L \neq 0$. At first glance, one may think that due to this the problem is reduced to a pure three-beam diffraction case. However, the Fourier component of the electric susceptibility χ_{N^*} may be nonzero, and it is the Fourier component, which is responsible for backscattering of the plane wave with wave vector K_L into the wave with wave vector $K_B = K_L - N^*$. Therefore, one can still expect nonzero amplitude for the wave with wave vector K_B. This is an Umweganregung (detour excitation) briefly discussed in Sect. 2.5.4. All four equations of the system of fundamental equations (2.199) have therefore to be preserved.

To be specific, the reflections with $N = (1\ 1\ \bar{2}\ 45)$, $B = (0\ 0\ 0\ 45)$, and $L = (1\ 1\ \bar{2}\ 0)$ are considered in α-Al$_2$O$_3$. The Bragg energy for the back-reflection is $E_N \simeq 21.630$ keV. The crystal thickness $d = 1$ mm.

The following nonzero Fourier components of the electric susceptibility of α-Al$_2$O$_3$ are used in the scattering matrix (2.201): $\chi_0 = (-34.8 + i\ 0.0546) \times 10^{-7}$, $\chi_N = \chi_{\bar{N}} = (0.635 - i\ 0.00292) \times 10^{-7}$, $\chi_{N^*} = \chi_{\bar{N}^*} = -\chi_N$, $\chi_L = \chi_{\bar{L}} = (-7.08 + i\ 0.0423) \times 10^{-7}$, and $\chi_B = \chi_{\bar{B}} = 0$.

Figure 2.39 shows results of numerical calculations for the energy dependence of reflection and transmission at normal incidence on the atomic planes normal to reciprocal lattice vector N. The dependence for each wave with wave vector K_0, $K_N \simeq K_0 + N$, or $K_L \simeq K_0 + L$ is shown in panel 0, N, or L, respectively. The results for the wave with $K_B \simeq K_0 + B$ are not shown, as the reflectivity into this channel is about 10^{-6}. The right-hand panels in Fig. 2.39 present the results of the calculations for a smaller energy range, so that the details of the energy dependence of the reflectivity into the backscattering channel can be clearly seen. The center of the region of high reflectivity into the backscattering channel is at E_c, which is shifted from the Bragg energy E_N by $E_c - E_N = 37.6$ meV.

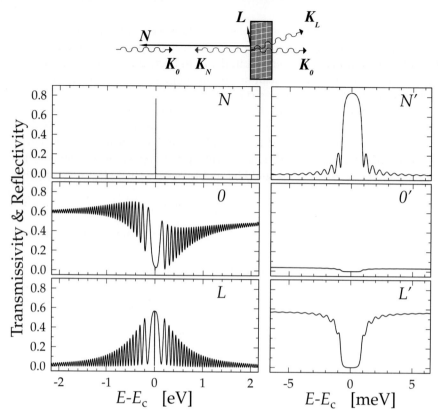

Fig. 2.39. "Degenerate" four-beam diffraction with diffraction vectors $\boldsymbol{N}$ (back-reflection), $\boldsymbol{L}$ (Laue-case reflection), and $\boldsymbol{B} = \boldsymbol{N} - \boldsymbol{L}$ (Bragg-case reflection, not shown). Bragg reflection with $\boldsymbol{B}$ is forbidden. Energy dependence of reflection or transmission is shown at normal incidence on the atomic planes perpendicular to $\boldsymbol{N}$. Dynamical theory calculations were performed for an α-Al_2O_3 crystal with $\boldsymbol{N} = (1\ 1\ \bar{2}\ 45)$, $\boldsymbol{L} = (1\ 1\ \bar{2}\ 0)$, and $\boldsymbol{B} = (0\ 0\ 0\ 45)$. The dependence for each wave with wave vector $\boldsymbol{K}_0$, $\boldsymbol{K}_N \simeq \boldsymbol{K}_0 + \boldsymbol{N}$, or $\boldsymbol{K}_L \simeq \boldsymbol{K}_0 + \boldsymbol{L}$ is shown in panel 0, N, or L, respectively. The right-hand panels $0'$, N', and L' present the dependence calculated for a smaller energy range. Crystal thickness $d = 1$ mm, surface normal $\hat{z} = -\boldsymbol{N}/N$, temperature $T = 300$ K. $E_c = 21.630$ keV. The radiation waves are in the σ-polarization state.

The results of the calculations in the two-beam approximation with only one diffraction vector $\boldsymbol{N}$ are also presented in Fig. 2.39(N'). However, they cannot be distinguished from the results of calculations for the backscattering channel in the "degenerate" four-beam case, as they are practically identical. This fact demonstrates that in the "degenerate" four-beam case the accompanying scattering channel with diffraction vector $\boldsymbol{L}$ does not essentially influence backscattering! The accompanying channel L receives significant

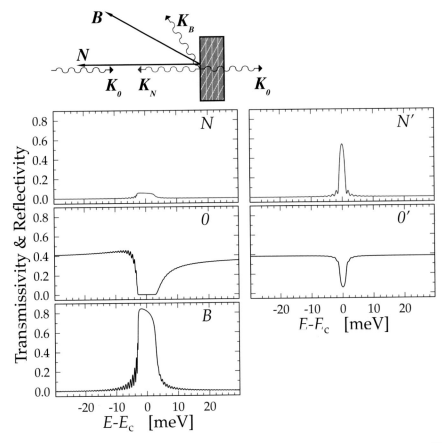

Fig. 2.40. Left: "Degenerate" four-beam diffraction with diffraction vectors N (back-reflection), B (Bragg-case reflection), $L = N - B$ (Laue-case reflection, not shown). The Bragg reflection with L is forbidden. The energy dependence is shown of reflection and transmission at normal incidence to the atomic planes perpendicular to N. Dynamical theory calculations were performed for an α-Al$_2$O$_3$ crystal with $N = (5\ 5\ \overline{10}\ 15)$, $B = (5\ 5\ \overline{10}\ 0)$, and $L = (0\ 0\ 0\ 15)$. The dependence for each wave with wave vector K_0, $K_N \simeq K_0 + N$, or $K_B \simeq K_0 + B$ is shown in panel 0, N, or B, respectively. Crystal thickness $d = 0.5$ mm, surface normal $\hat{z} = -N/N$, temperature $T = 300$ K. $E_c = 14.8624$ keV. The radiation waves are in the σ-polarization state.
Right: the energy dependence calculated in the two-beam approximation with a single diffraction vector N.

intensity only outside the region of high reflection into the backscattering channel. In other words, the Laue-case diffraction channel does not compete with the back-reflection channel. The comparison of Fig. 2.39(L) and Fig. 2.39(0) shows that the Laue-case diffraction wave competes only with the forward-transmitted wave.

134   2. Dynamical Theory of X-Ray Bragg Diffraction

Laue-case Reflection is Forbidden. A quite different situation occurs if, conversely, the Bragg-case accompanying reflection is allowed: $\chi_B \neq 0$, and the Laue-case reflection is forbidden: $\chi_L = 0$. The reflections with $\boldsymbol{N} = (5\ 5\ \overline{10}\ 15)$, $\boldsymbol{B} = (5\ 5\ \overline{10}\ 0)$, and $\boldsymbol{L} = (0\ 0\ 0\ 15)$ are considered in an α-Al$_2$O$_3$ crystal as an example. The Bragg energy for the back-reflection is $E_N = 14.8624$ keV. The results of calculations of the crystal reflectivity into different channels as a function of the photon energy is shown in the left panels of Fig. 2.40. Exact normal incidence on the atomic planes perpendicular to $\boldsymbol{N}$ is assumed. For comparison, the results of calculations in the two-beam approximation with a single diffraction vector $\boldsymbol{N}$ are also shown in the right-hand panels of Fig. 2.40.

What is in common with the former case, in which the Bragg-case reflection was forbidden? The reflectivity into the channel associated with the forbidden reflection, although not zero, is again very weak. It is not shown in Fig. 2.40. Unlike the previous case, however, the back-reflection is strongly influenced by the accompanying reflection.

A dramatic reduction of the peak reflectivity into the backscattering channel is observed by going over from a pure two-beam case with single diffraction vector $\boldsymbol{N}$ - Fig. 2.40(N'), to the four-beam case - Fig. 2.40(N). The accompanying Bragg-case wave receives the main intensity - Fig. 2.40(B). As a consequence, the backscattered wave is suppressed.

Back-reflection is Forbidden. The third situation is also possible when the back-reflection is forbidden: $\chi_N = 0$, while both accompanying reflections are allowed: $\chi_B \neq 0$, and $\chi_L \neq 0$. The backscattering channel can nevertheless receive some photons due to Umweganregung according to the scattering sequence: $\boldsymbol{K}_0 \Rightarrow \boldsymbol{K}_B \Rightarrow \boldsymbol{K}_N$ or $\boldsymbol{K}_0 \Rightarrow \boldsymbol{K}_L \Rightarrow \boldsymbol{K}_N$. One may expect the scattering amplitude for the first path to be proportional to $\chi_{B-0}\chi_{N-B} = \chi_B\chi_L$, and for the second path $- \chi_{L-0}\chi_{N-L} = \chi_L\chi_B$, respectively. One cannot expect high reflectivity into the backscattering channel in this case. Therefore, this scattering regime is hardly of importance for x-ray crystal optics. Still, this regime could be of interest for extracting the phases of the crystal structure factors in applications to crystal structure determination.

2.6.3 4-Beam Diffraction - General Case

We now drop the assumption that one of the reflections is forbidden. Back-reflection and both accompanying reflections are allowed. Each of the four waves can receive an appreciable fraction of the energy of the incident wave.

To be specific, the back-reflection with diffraction vector $\boldsymbol{N} = (1\ 3\ \overline{4}\ 28)$, and accompanying reflections with diffraction vectors $\boldsymbol{B} = (\overline{1}\ 1\ 0\ 28)$, and $\boldsymbol{L} = (2\ 2\ \overline{4}\ 0)$ are considered in an α-Al$_2$O$_3$ crystal. This particular back-reflection is of interest for many applications, since at a crystal temperature of $T = 373$ K the Bragg energy of the $(1\ 3\ \overline{4}\ 28)$ reflection matches the 14.4125 keV nuclear transition energy in ^{57}Fe.

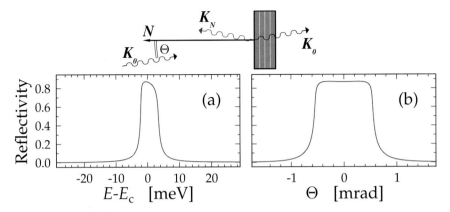

Fig. 2.41. Two-beam Bragg diffraction in backscattering. Dynamical theory calculations for an α-Al$_2$O$_3$ crystal with back-diffraction vector $\bm{N} = (1\ 3\ \bar{4}\ 28)$. (a) Energy dependence of reflection at *exact* normal incidence ($\Theta = 0$) to the atomic planes perpendicular to $\bm{N}$. (b) Angular dependence of reflection for x-ray photon energy $E = E_c$. Crystal thickness $d = 1$ mm, surface normal $\hat{\bm{z}} = -\bm{N}/N$, temperature $T = 373$ K. $E_c = 14.4125$ keV.

The following nonzero Fourier components of the electric susceptibility of α-Al$_2$O$_3$ are considered in the scattering matrix (2.201): $\chi_0 = (-78.4 + i\,0.289) \times 10^{-7}$, $\chi_N = \chi_{\overline{N}} = (-3.81 + i\,0.122) \times 10^{-7}$, $\chi_L = \chi_{\overline{L}} = (-14.1 + i\,0.214) \times 10^{-7}$, $\chi_B = \chi_{\overline{B}} = (-3.78 + i\,0.123) \times 10^{-7}$, and $\chi_{N^*} = \chi_{\overline{N}^*} = \chi_N$.

Before addressing this four-beam case, it is important to ascertain what the dynamical theory predicts for the reflectivity into the backscattering channel in the two-beam approximation with a single diffraction vector $\bm{N} = (1\ 3\ \bar{4}\ 28)$. Figure 2.41 shows the results of the calculations. The left panel presents the reflectivity at *exact* normal incidence on the atomic planes perpendicular to $\bm{N}$ as a function of the incident photon energy. A peak reflectivity of 0.87 is achieved. The center of the region of high reflectivity into the backscattering channel is at E_c, which is shifted from the Bragg energy E_N by: $E_c - E_N = 55.9$ meV.

The crystal reflectivity in the right panel of Fig. 2.41 is calculated as a function of incidence angle Θ for the fixed x-ray energy E_c. We do not distinguish here between the angles $\Theta_\perp$ and $\Theta_\parallel$, as in the two-beam case the results of the calculations are equal. The angular dependence shows a broad angular width of high reflectivity of about about 1 mrad, which is typical for Bragg backscattering.

An account of the accompanying reflections results in dramatic changes in the reflectivity at *exact* normal incidence. Figure 2.42 shows the results of the numerical calculations for the energy dependence of reflection and transmission at normal incidence on the atomic planes perpendicular to $\bm{N}$. The dependence for each wave with wave vector $\bm{K}_0$, $\bm{K}_N \simeq \bm{K}_0 + \bm{N}$, $\bm{K}_B \simeq$

$K_0 + B$, or $K_L \simeq K_0 + L$ is shown in panel 0, N, B, or L, respectively. The right-hand panels $0'$, N', B', or L' present the results of calculations for a smaller energy range, so that the details of the energy dependence can be clearly seen. The energy E_c is the same as in the two-beam case - Fig. 2.41(a).

Unlike the two-beam case - see Fig. 2.41(a) - the energy dependence of the reflectivity into the backscattering channel in the four-beam case becomes a complicated structure with two peaks of very different widths - see Figs. 2.42(N), (N'). The broader peak is shifted by about 10 meV from E_c. The peak reflectivity in both cases is a factor of two smaller compared to the two-beam case reflectivity.

Figure 2.43 shows the results of calculations for the angular dependence of reflection and transmission at fixed x-ray energy $E = E_c$ - the same energy as indicated in Figs. 2.41 and 2.42. The right-hand panels present the results of calculations for a smaller angular range, presenting more details in the angular dependence. The angular deviation $\Theta_\parallel$ from exact normal incidence *in the basal plane* is considered. The solid lines show the results of calculations for the incident radiation in the σ-polarization state, while the dashed lines show the results of calculations for the radiation in the π-polarization state. Comparison with the angular dependence of the reflectivity in the two-beam case - Fig. 2.41(b) - shows a very strong influence of the accompanying reflections on the crystal reflectivity into the backscattering channel in the immediate vicinity of exact normal incidence ($\Theta_\parallel = 0$). The effect of accompanying reflections upon the backscattering channel, however, vanishes beyond a minuscule angular deviation $\Theta_\parallel$ of about ± 5 μrad.

At this point, we note that our qualitative picture for the region of multiple-beam excitation discussed in Sect. 2.5.6, and graphically illustrated in Figs. 2.36 and 2.37, are in agreement with the results of the numerical dynamical theory calculations, presented here.

Importantly, the angular range, in which the accompanying reflections have an effect upon the backscattering channel, is even smaller for the incident wave in the π-polarization state (Fig. 2.43). This is due to the fact that the polarization factors $P_L^\pi = \cos 2\theta_L = 0.74$, and $P_B^\pi = \cos 2\theta_B = -0.74$ reduce the elements of the scattering matrix $\tilde{G}^s$ (2.201) responsible for the scattering into the accompanying channels. A more drastic effect would take place if $2\theta_L = 2\theta_B = \pi/2$. Under this condition the accompanying reflections would vanish completely.

This example shows that by a judicious choice of the polarization state of the incident wave one can diminish the influence of accompanying reflections.

The effect of angular deviation $\Theta_\perp$ in the direction normal to the basal plane is very different from the effect of the just considered $\Theta_\parallel$-angular-deviation. Departure from normal incidence in the direction perpendicular to the basal plane causes very weak changes in the reflectivity and transmissivity. The effect of the accompanying reflections is present for a much broader angular range of about 1 mrad. Such behaviour was already qualita-

2.6 4-Beam Diffraction in Backscattering: Dynamical Treatment

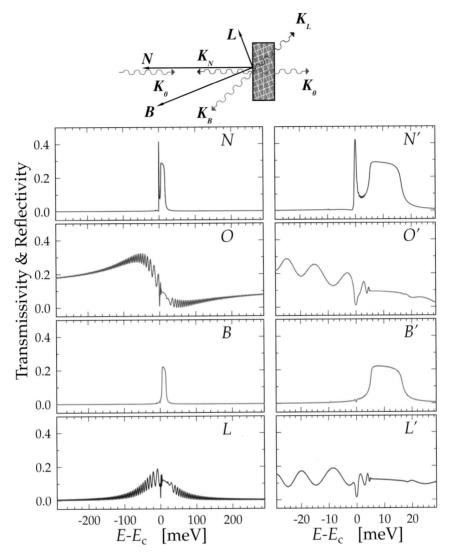

Fig. 2.42. 4-beam diffraction with diffraction vectors N (back-reflection), L (Laue-case reflection), and B (Bragg-case reflection). Energy dependence of reflection and transmission at *exact* normal incidence on the atomic planes perpendicular to N. Dynamical theory calculations for an α-Al$_2$O$_3$ crystal with $N = (1\ 3\ \bar{2}\ 28)$, $B = (\bar{1}\ 1\ 0\ 28)$, and $L = (2\ 2\ \bar{4}\ 0)$. The dependence for each wave with wave vector K_0, $K_N \simeq K_0 + N$, or $K_L \simeq K_0 + L$ is shown in panel 0, N, B, or L, respectively. The right-hand panels present the dependence calculated for a smaller energy range. Crystal thickness $d = 1$ mm, surface normal $\hat{z} = -N/N$, temperature $T = 373$ K. $E_c = 14.4125$ keV. The radiation waves are in σ-polarization state.

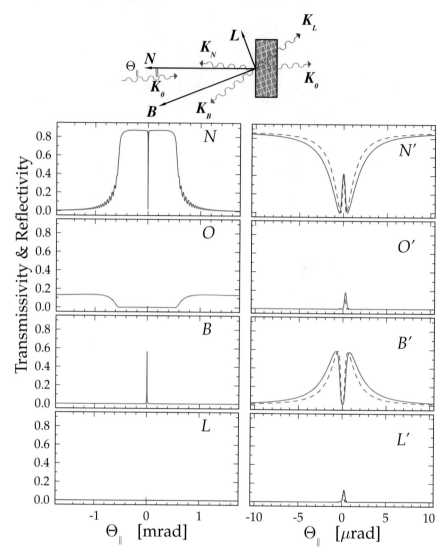

Fig. 2.43. 4-beam diffraction with diffraction vectors N, L, and B, as in Fig. 2.42. Angular dependence of reflection and transmission at near-normal incidence on the atomic planes perpendicular to N. $\Theta_{\|}$ is the angular deviation from normal incidence *in the basal plane*. Dynamical theory calculations with photon energy $E = E_c$. The dependence for each wave with wave vector K_0, $K_N \simeq K_0 + N$, or $K_L \simeq K_0 + L$ is shown in panel 0, N, B, or L, respectively. Right-hand panels show the dependence calculated for a smaller angular range. The solid lines show the calculations for the incident beam in the σ-polarization state, and the dashed lines - in the π-polarization state, respectively.

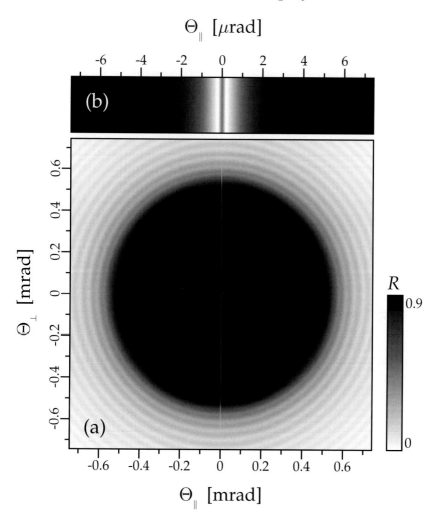

Fig. 2.44. 4-beam diffraction with diffraction vectors N, L, and B, as in Figs. 2.42 and 2.43. (a) Two-dimensional plot of the angular dependence of back-reflectivity R at near-normal incidence on the atomic planes perpendicular to N. $\Theta_\parallel$ is the angular deviation from normal incidence *in the basal plane*. $\Theta_\perp$ is the angular deviation in the direction *perpendicular to the basal plane*. Dynamical theory calculations with photon energy $E = E_c$. (b) The central part around exact normal incidence ($\Theta_\parallel = 0$, $\Theta_\perp = 0$) is shown with magnification 100x.

tively predicted in Sect. 2.5.6, Fig. 2.37. It was also suggested in Sect. 2.6.1, by the reason of the quadratic dependence of the deviation parameters α_H (2.197) on $\Theta_\perp$. The numerical simulations, which are shown in Fig. 2.44, confirm this prediction.

2.7 n-Beam Diffraction: Suppression of Accompanying Reflections

We know that one pair of accompanying reflections can be suppressed provided the belong to them reflected waves propagate perpendicular (or almost perpendicular) to the direction of the incident wave. This situation is shown in Fig. 2.34 on page 122, assuming $2\theta_B = 2\theta_L = \pi/2$. The suppression is effected, as discussed in Sect. 2.5.5, by choosing the incident wave in the π-polarization state, i.e., the polarization vector lies in the basal plane of this reflection pair.

The accompanying reflections can also be easily suppressed in the four-beam diffraction regime, as we have ascertained in Sect. 2.6.3. For this, the direction of propagation of the incident wave, given by wave vector $\boldsymbol{K}_0$ has to be turned *in the basal plane* slightly off normal incidence. A few microradians is sufficient in the four-beam case, as is shown in the example of Fig. 2.43(N').

However, if *exact* normal incidence is required, or the number of the conjugate pairs of accompanying reflections is more than one, and therefore a great many accompanying waves exist, these procedures may be not very efficient or sufficient.

Suppression of accompanying reflections can also be nevertheless achieved in the general case of n-beam diffraction. To understand how this suppression technique works, let us examine again four-beam diffraction with one conjugate pair of accompanying reflections, as an example. The results can be then easily generalized to the n-beam case.

Imagine a scattering geometry, in which $\boldsymbol{K}_0$ the wave vector of the incident radiation makes a small angle with the crystal surface, as shown in Fig. 2.45. Under this condition the direction cosine $\gamma_0 = \boldsymbol{K}_0 \hat{\boldsymbol{z}}/K$ (2.18) of the angle between the surface normal $\hat{\boldsymbol{z}}$ and $\boldsymbol{K}_0$ is also small[19]: $\gamma_0 \ll 1$.

The back-reflected wave propagates in this case almost at the same small angle to the crystal surface. The asymmetry angle η_N - the angle between the crystal surface and the reflecting atomic planes normal to $\boldsymbol{N}$ is close to 90°. As a result, the direction cosine $\gamma_N = (\boldsymbol{K}_0 + \boldsymbol{N})\hat{\boldsymbol{z}}/K$ associated with the back-reflected wave is also small.

In contrast to this, the direction cosines $\gamma_B = (\boldsymbol{K}_0 + \boldsymbol{B})\hat{\boldsymbol{z}}/K$ and $\gamma_L = (\boldsymbol{K}_0 + \boldsymbol{L})\hat{\boldsymbol{z}}/K = -\gamma_B$ associated with the accompanying waves are not small, unless the surface normal $\hat{\boldsymbol{z}}$ lies in the basal plane.

However small γ_0, and $|\gamma_N|$ may be, the asymmetry factors for the forward- and back-reflected waves are not small and equal to $b_0 = 1$, and

---

[19] As previously, we shall still require that the angle is not too small, i.e., it is larger than the critical angle of total reflection, so that $\gamma_0 \gtrsim \sqrt{2w_H^{(s)}} \simeq 10^{-3}$ (see Appendices A.3 and A.4). Otherwise, diffraction at grazing incidence takes place with the additional strong wave specularly reflected from the crystal surface, as was discussed by Stepanov et al. (1991). To obtaining high reflectivity into the backscattering channel, one has to avoid specular reflections.

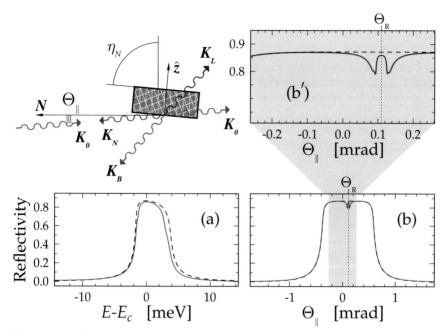

Fig. 2.45. 4-beam coplanar Bragg diffraction in α-Al_2O_3. Dynamical theory calculations under the same conditions as in Figs. 2.42(N), and 2.43(N), respectively, with the exception of the surface normal $\hat{z}$ composing now an angle of $\eta_N = 88°$ (asymmetry angle) with $-N$.
(a) Solid line: energy dependence of *exact Bragg back-reflection* (reflection of the wave with wave vector K_0 into K_N) at the incidence angle $\Theta_\| = \Theta_R$ to the atomic planes normal to N. Θ_R is the incidence angle of exact backscattering - (2.133).
(b) Solid line: angular dependence of back-reflection at near-normal incidence on the atomic planes perpendicular to N, for the photon energy $E = E_c$. $\Theta_\|$ is the angular deviation from normal incidence *in the basal plane*.
(b') Central part of the angular dependence (b) is shown on the expanded scale. The dashed lines in panels (a),(b), and (b') display the energy and angular dependence calculated in the two-beam approximation.

$b_N \simeq -1$. On the contrary, the magnitudes of the asymmetry factors for the accompanying reflections are very small $|b_B| = |b_L| = |\gamma_0/\gamma_L| \ll 1$. This also makes the elements $G^s_{HH'}$ of the scattering matrix - (2.200) and (2.201) - of the system of fundamental equations very small (note that in (2.201) $b = b_L = -b_B$ according to (2.198)), except those related to the forward- and back-reflected channels, i.e., for $H = 0$ and $H = N$. As a result the accompanying reflections become suppressed.

The solid lines in Fig. 2.45 show the results of the dynamical theory calculations of the energy and angular dependence of reflectivity for the backscattering channel under these conditions. The same parameters are used as for the calculations presented in Figs. 2.42(N), (N'), and 2.43(N), (N'), with the exception of the direction of the surface normal $\hat{z}$, which now makes an angle of $\eta_N = 88°$ with $-\boldsymbol{N}$. It is also assumed that $\hat{z}$ lies in the basal plane. In this particular geometry, $\gamma_0 \simeq 0.035$, and $|b_B| = |b_L| = b = 0.05$.

Figure 2.45(a) shows the energy dependence of *exact Bragg back-reflection* (reflection of the wave with wave vector $\boldsymbol{K}_0$ into $\boldsymbol{K}_N$) at the incidence angle $\Theta_\parallel = \Theta_R$ to the atomic planes normal to $\boldsymbol{N}$. $\Theta_\parallel$ is the angular deviation from normal incidence *in the basal plane*. Θ_R is the incidence angle of exact backscattering (2.133).

Figures 2.45(b),(b') displays the angular dependence of back-reflection at near-normal incidence on the atomic planes perpendicular to $\boldsymbol{N}$, for x rays with photon energy $E = E_c$.

The dashed lines in Figs. 2.41(a),(b), and (b') show the results of similar calculations in the two-beam approximation. A small difference can be observed. For example: a small additional modulation on the top of the angular dependence in the four-beam case - Figs. 2.45(b),(b') - in the vicinity of $\Theta_\parallel = \Theta_R$, the incidence angle of exact backscattering. Otherwise, the dependence calculated in the four-beam and in the two-beam approximations look similar, both, with almost equal maximum reflectivity at exact backscattering.

The results of the numerical simulations confirm our expectations, that the accompanying reflections can be suppressed by using asymmetric Bragg diffraction in backscattering with the incident wave propagating at a small angle to the entrance crystal surface (close to grazing incidence).[20]

It is evident that this technique of suppression of the accompanying reflections is applicable also in the general n-beam diffraction case.

---

[20] We have discussed the case in which the crystal surface is cut perpendicular to the basal plane. If it is on the contrary cut almost parallel to the basal plane, then B and L waves also propagate close to grazing emergence and the desired suppression cannot be achieved. Thus, to achieve suppression of the accompanying waves one has to take care, that the crystal surface is cut *not* parallel to any basal plane of the accompanying reflections.

3. Principles of Multiple-Crystal X-Ray Diffraction

3.1 Introduction

Having examined Bragg diffraction from individual crystals in the previous chapter, one may now proceed to study diffraction involving an arbitrary number of crystals. As we shall see, x rays beams with desired spectral and angular characteristics - monochromatic beams, beams with extremely narrow angular divergence, etc. - can be obtained by Bragg reflections off several crystals.

The theory of multiple-crystal diffraction has been discussed extensively in the literature. One can refer here to the popular texts of Compton and Allison (1935), Zachariasen (1945), Pinsker (1978), and Authier (2001). DuMond (1937) has introduced a simple method of graphical analysis of transformations in the (λ, θ) space of x rays in multiple-crystal diffraction. DuMond diagram analysis provides a graphical method for defining and mapping the boundary x rays in the angular-wavelength space, which propagate from one crystal to another. Nakayama et al. (1973), Kohra et al. (1978), and Matsushita and Hashizume (1983) have extended the DuMond diagram analysis to asymmetric diffraction. Xu Shunsheng and Li Runsheng (1988) have introduced the three-dimensional (λ, θ, ϕ) DuMond diagram analysis. Davis (1990) has derived a set of transformation equations to map DuMond diagrams between crystals in a multi-crystals system. Phase-space analysis is another graphical technique for predicting the transformation of x-ray beams by x-ray optical elements. It was discussed, e.g., by Matsushita and Hashizume (1983), and Suorti and Freund (1989).

Two different approaches will be considered in this book to treat Bragg diffraction from multiple-crystal systems. In the first approach the crystals are considered to be independent. This means, it is assumed that the radiation scattered from one of the crystals is incident on the next one, and does not act back on the previous one. In other words, there is a successive but no multiple scattering of the waves within the multiple-crystal system. The second approach takes these multiple scattering effects into account. In the present chapter, the first approach of successive scattering from independent crystals will be considered. The second approach will be applied in Chap. 4 to develop the theory of x-ray Fabry-Pérot resonators.

As long as the solution for Bragg diffraction from an individual single crystal is known, the solution of the diffraction problem involving an arbitrary number of crystals can be obtained by applying it successively to all constituent crystals according to the following rule: the wave reflected from the previous crystal is the incident wave for the next one. For example, for a two-crystal system the total reflectivity R_{12} can be written as:

$$R_{12} = R_2\left(\bm{H}_2, \bm{K}_0 + \tilde{\bm{H}}_1\right) R_1\left(\bm{H}_1, \bm{K}_0\right). \tag{3.1}$$

Here $R_1\left(\bm{H}_1, \bm{K}_0\right)$ is the Bragg reflectivity of the first crystal with $\bm{H}_1$ being the diffraction vector, and $\bm{K}_0$ the wave vector of the incident wave. $R_2\left(\bm{H}_2, \bm{K}_0 + \tilde{\bm{H}}_1\right)$ is the Bragg reflectivity of the next crystal with diffraction vector $\bm{H}_2$, and the wave vector of the incident wave being the wave vector $\bm{K}_0 + \tilde{\bm{H}}_1$ of the wave reflected from the first crystal (2.15). This rule can be further generalized to a sequence of reflections from any number of crystals and applied to reflectivity calculations in multiple-crystal systems.

Computers afford the most definite and straightforward method of solving the problem numerically for an arbitrary number of crystals by using (3.1). Many computer codes exist, which can be applied for this purpose. Being very convenient and precise, they can be used to predict the result of multiple-crystal Bragg diffraction exactly, in the framework of the dynamical theory. However, more instructive (although possibly less precise because of approximations) analytical solutions of the multiple-crystal diffraction problem can be also obtained.

The approximation of a non-absorbing crystal has proved instructive in Sect. 2.4 for the analysis of the main features of Bragg diffraction from individual crystals. It predicts rather precisely the angular and spectral widths of Bragg reflections in terms of the parameter $\epsilon_H^{(s)}$, the relative spectral width of the symmetric Bragg reflection, and of the asymmetry parameter b, or asymmetry angle η. The same approximation could be applied to obtain analytical expressions for the angular and spectral widths of multiple Bragg reflections from crystal ensembles in terms of $\epsilon_H^{(s)}$ and b of the individual Bragg reflections of the crystals participating in the scattering process.

In the present chapter we shall discuss multiple crystal diffraction along these lines. We shall start with the classification of two-crystal configurations, and shall define the so-called $(+,+)$ and $(+,-)$ configurations. Then we shall obtain analytical expressions for the spectral and angular spreads of the waves successively reflected from two crystals in the $(+,+)$ and $(+,-)$ configurations. A generalization to more-than-two-crystal configurations will be discussed as well. Examples of four-crystal and three-crystal configurations of particular interest to x-ray monochromator and analyzer applications will be addressed. Among scattering geometries the exact back-reflection geometry

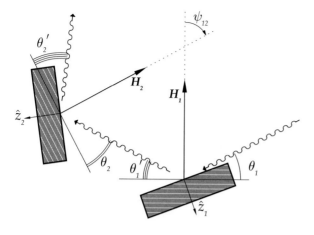

Fig. 3.1. Schematic view of two successive Bragg reflections from crystals in the $(+,+)$ configuration. $\theta_2 = \psi_{12} - \theta'_1(\theta_1, \lambda)$.

will also be considered. The analytical solutions will be graphically illustrated in the angular-wavelength space with the help of DuMond diagrams.

3.2 Two-Crystal Configurations

Consider two successive Bragg reflections from two plane crystal plates, as shown in Figures 3.1, and 3.2. The incident wave is reflected from the lattice planes normal to the reciprocal lattice vector $\boldsymbol{H}_1$ in the first crystal, and from the lattice planes normal to the reciprocal lattice vector $\boldsymbol{H}_2$ in the second crystal. A practically important special case is considered, in which the vectors $\boldsymbol{H}_1$, and $\boldsymbol{H}_2$ are coplanar. It is also assumed that the dispersion planes $(\hat{\boldsymbol{z}}_1, \boldsymbol{H}_1)$, and $(\hat{\boldsymbol{z}}_2, \boldsymbol{H}_2)$ for the first, and for the second reflection, respectively, are parallel to each other. The angle ψ_{12} between the vectors $\boldsymbol{H}_1$ and $\boldsymbol{H}_2$ is measured clockwise from $\boldsymbol{H}_1$ to $\boldsymbol{H}_2$.

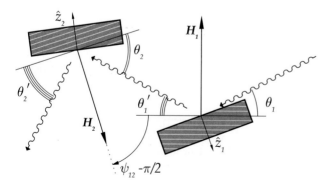

Fig. 3.2. Schematic view of two successive Bragg reflections from crystals in the $(+,-)$ configuration. $\theta_2 = (\pi - \psi_{12}) + \theta'_1(\theta_1, \lambda)$.

3. Principles of Multiple-Crystal X-Ray Diffraction

Two types of scattering geometries for the system of two crystals can be distinguished. In the first case, shown in Fig. 3.1, reflection from the second crystal deviates the beam in the same sense as from the first one, e.g., always clockwise, as shown in Fig. 3.1. This is called $(+,+)$ configuration of the crystals. In the second case, shown in Fig. 3.2, the beam is deviated by the second crystal in the opposite sense. This is called $(+,-)$ configuration[1].

The reflectivity R_{12} of a system of two crystals is a product of the reflectivities R_1 and R_2 of the constituent crystals as given by (3.1). It should be averaged over the distribution of the angles of incidence θ_1, ϕ_1 on the first crystal, and the distribution of radiation wavelengths λ. The angles of incidence on the second crystal θ_2, ϕ_2 are determined by the reflection angles θ'_1, ϕ'_1 from the first crystal, and by ψ_{12}. In the coplanar scattering geometry discussed here the azimuthal angles are either zero or π [2]. Therefore, with the exception of the extreme backscattering case discussed at the end of the chapter, only the glancing angles of incidence and reflection are considered.

As seen in Fig. 3.1, the glancing angle of incidence θ_2 to the second crystal is determined by the angle ψ_{12} between the diffraction vectors $\boldsymbol{H}_1$ and $\boldsymbol{H}_2$, and the glancing angle of reflection θ'_1 of the first crystal. In the $(+,+)$ configuration the relation

$$\theta_2 = \psi_{12} - \theta'_1(\theta_1, \lambda), \tag{3.2}$$

is valid, while in the $(+,-)$ configuration another relation has to be used:

$$\theta_2 = (\pi - \psi_{12}) + \theta'_1(\theta_1, \lambda). \tag{3.3}$$

We remind the reader that θ'_1 is a function of θ_1 and λ, as discussed in detail in Sect. 2.4.10 and Sect. 2.4.11.

The reflectivity of the two-crystal system under these assumptions is a function of ψ_{12}, θ_1, and λ and is given by

$$R_{12}(\psi_{12}, \theta_1, \lambda) = R_2\left[\psi_{12} - \theta'_1(\theta_1, \lambda), \lambda\right] R_1(\theta_1, \lambda). \tag{3.4}$$

in the $(+,+)$ configuration, and by

$$R_{12}(\psi_{12}, \theta_1, \lambda) = R_2\left[\pi - \psi_{12} + \theta'_1(\theta_1, \lambda), \lambda\right] R_1(\theta_1, \lambda) \tag{3.5}$$

in the $(+,-)$ configuration. The reflectivity, being averaged over the distribution of wavelengths and glancing angles of incidence on the first crystal reads

---

[1] If the sign for the first crystal is chosen as "$-$", then the configuration of two crystals is labeled as $(-,-)$, or $(-,+)$, respectively.

[2] Small deviations of the direction of the incident wave from the dispersion plane, i.e., small deviations of ϕ from 0 or π, do not lead to predictions significantly different from those presented in the remainder of this chapter - because of the $\cos\phi$ and $\cos\phi'$ dependence in (2.34).

$$R_{12}(\psi_{12}) = \langle R_2(\theta_2, \lambda) R_1(\theta_1, \lambda) \rangle_{\theta_1, \lambda}, \tag{3.6}$$

where θ_2 is given by either (3.2) or (3.3) in the $(+,+)$ or $(+,-)$ configuration, respectively.

3.3 Diffraction from Crystals in $(+,+)$ Configuration

We commence with the discussion of diffraction from crystals in the $(+,+)$ configuration, as shown schematically in Fig. 3.1. Unless otherwise indicated, the glancing angles in the remainder of this chapter are assumed to be not in the immediate proximity to $\pi/2$.

Figure 3.3(1) shows a fragment of the region of total reflection from the first crystal for entrance waves in the (λ, θ_1) space. The region marked in green is bounded between the borderlines $\lambda(\theta_1)$ given by

$$\frac{\lambda - \lambda_c}{\lambda_c} = \frac{\theta_n - \theta_{c_n}}{\tan \theta_{c_n}} \mp \frac{\epsilon_n^{(s)}}{2\sqrt{|b_n|}}, \tag{3.7}$$

with index $n = 1$, an obvious generalization of (2.151) for this multi-crystal case. Here λ_c is some specific wavelength of interest. It is related to the glancing angle of incidence θ_{c_1} at the center of the region of total reflection by the dynamical Bragg's law (2.114) in the case of diffraction from the lattice planes with interplanar distance d_1.

The reflection region of the second crystal for the entrance waves is shown in the (λ, θ_2) space by the stripe marked in yellow in Fig. 3.3(1'+2). It is bounded between the borderlines, given by (3.7) with $n = 2$. The glancing angle of incidence θ_{c_2} is given by (2.114) with the wavelength equal to λ_c, and the interplanar distance d_2.

The wave reflected from the first crystal emerges at the glancing angle of reflection θ_1', which is in the angular range centered at θ_{c_1}', as given by the modified Bragg's law for the exit wave (2.149). It is further reflected from the second crystal if the glancing angle of reflection θ_1' from the first crystal is related to the glancing angle of incidence on the second crystal by $\theta_2 = \psi_{12} - \theta_1'$ (3.2). In particular, the wave with wavelength λ_c is reflected successively from both crystals at the center of their reflection region if

$$\theta_{c_2} = \psi_{12} - \theta_{c_1}'. \tag{3.8}$$

Combining (3.8) with $\theta_2 = \psi_{12} - \theta_1'$ (3.2) we may write the condition of reflection from both crystals in the following form:

$$\theta_2 - \theta_{c_2} = -(\theta_1' - \theta_{c_1}'). \tag{3.9}$$

The green stripe in Fig. 3.3(1'+2) shows a fragment of the region of total reflection from the first crystal for exit waves in the (λ, θ_1') space. The reflection region is bounded between the borderlines $\lambda(\theta_1')$ given by

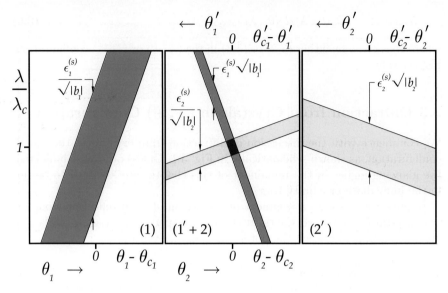

Fig. 3.3. DuMond diagrams showing reflection regions for a sequence of two asymmetric Bragg reflections from crystals in the $(+,+)$ configuration.
(1): green stripe - reflection region of the first crystal for the entrance waves.
($1'+2$): green stripe - reflection region of the first crystal for the exit waves, yellow stripe - reflection region of the second crystal for the entrance waves.
($2'$): yellow stripe - reflection region of the second crystal for the exit waves.

$$\frac{\lambda - \lambda_c}{\lambda_c} = \frac{\theta'_n - \theta'_{c_n}}{\tan \theta_{c_n}} \mp \frac{\epsilon_n^{(s)}}{2} \sqrt{|b_n|} \tag{3.10}$$

with index $n = 1$, according to (2.153) applied to this multi-crystal case. In (3.10), we have used the approximation $\tan \theta'_{c_n} \simeq \tan \theta_{c_n}$. The reflection regions of both crystals are drawn in Fig. 3.3($1'+2$) in such a way that (3.9) is fulfilled. For this, the reflection region of the first crystal is shown on an inverted scale.

The dark blue (black) intersection region $ABDE$ in Figs. 3.3($1'+2$) and 3.4($1'+2$) determines the range of the wavelengths and angles of incidence (reflection) of the waves, which can be reflected successively from both crystals. The wavelengths λ_X corresponding to the intersection points, with $X = A, B, E,$ or D, can be determined as solutions of four systems of two linear equations (3.10) with $n = 1$ and of (3.7) with $n = 2$. The solutions are:

$$\frac{\lambda_X - \lambda_c}{\lambda_c} = \pm \epsilon_1^{(s)} \tau_1 \frac{\sqrt{|b_1|}}{2} \pm \epsilon_2^{(s)} \tau_2 \frac{1}{2\sqrt{|b_2|}}, \tag{3.11}$$

where

$$\tau_1 = \frac{\tan \theta_{c_1}}{\tau}, \quad \tau_2 = \frac{\tan \theta_{c_2}}{\tau}, \quad \tau = \tan \theta_{c_1} + \tan \theta_{c_2}. \tag{3.12}$$

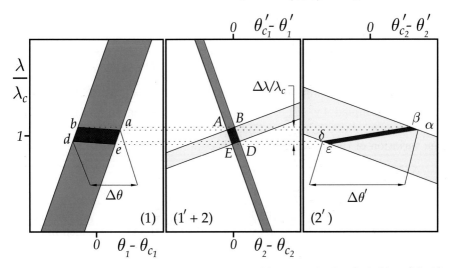

Fig. 3.4. The same as Fig. 3.3, showing in addition with the dark blue (black) parallelograms the phase space of the waves reflected from both crystals.

The sequence of signs in (3.11) is $-+$ if $X = A$; $++$ if $X = B$; $--$ if $X = E$; $+-$ and if $X = D$.

By using (3.11) we can obtain for the relative spectral width $\Delta\lambda/\lambda_c = \lambda_B - \lambda_E/\lambda_c$ of the two successive Bragg reflections[3]

$$\frac{\Delta\lambda}{\lambda_c} = \epsilon_1^{(s)} \tau_1 \sqrt{|b_1|} + \epsilon_2^{(s)} \frac{\tau_2}{\sqrt{|b_2|}}. \qquad (3.13)$$

The center λ_c of the spectral region of reflection from the crystals in the $(+,+)$ configuration is a fixed quantity for the given angle ψ_{12} between the vectors $\boldsymbol{H}_1$ and $\boldsymbol{H}_2$ (Fig. 3.1), and is determined by the condition (3.8) of reflection from both crystals. Combining (3.8) with the modified Bragg's law (2.114)-(2.115) for the incident wave, and the modified Bragg's law for the exit wave (2.149), and assuming the asymmetry factors to be constant, we obtain for the center λ_c of the spectral region of reflection from the crystals in the $(+,+)$ configuration:

$$\frac{1}{\lambda_c} = \frac{1}{\sin\psi_{12}} \sqrt{\frac{1}{(2d_1^*)^2} + \frac{1}{(2d_2^*)^2} + \frac{\cos\psi_{12}}{2d_1^* d_2^*}}, \qquad (3.14)$$

---

[3] Defined in this simple graphical way, $\Delta\lambda/\lambda_c$ gives an estimate of the upper bound for the spectral bandwidth of two reflections in the $(+,+)$ configuration. The exact value of $\Delta\lambda/\lambda_c$ depends on the actual form of the reflection function of the single Bragg reflections, and has to be evaluated numerically by using exact expressions of the dynamical theory. The exact value is usually a factor 2 to 1.5 smaller than $\Delta\lambda/\lambda_c$ derived with the help of the DuMond graphical analysis and given by (3.13).

where

$$d_1^* = \frac{d_1}{1 + w_1^{(s)}(1 - b_1)/2}, \qquad d_2^* = \frac{d_2}{1 + w_2^{(s)}(1 - 1/b_2)/2}. \qquad (3.15)$$

The variation of ψ_{12} results in a change of λ_c, as well as of θ'_{c_1} and θ'_{c_2}, which are coupled by $\delta\theta_{c_2} = \delta\psi_{12} - \delta\theta'_{c_1}$ according to (3.8). Applying $\delta\theta'_{c_1} \simeq \tan\theta_{c_1} \delta\lambda/\lambda$ (2.149) and $\delta\theta_{c_2} \simeq \tan\theta_{c_2} \delta\lambda/\lambda$ (2.114), we obtain for the variation of the wavelength λ_c with ψ_{12}:

$$\frac{\delta\lambda_c}{\lambda_c} = \frac{\delta\psi_{12}}{\tau}. \qquad (3.16)$$

The region of intersection $ABDE$ in Fig. 3.4(1'+2) can be projected onto the reflection region in Fig. 3.4(1). The projection is $abde$. Its width in terms of θ_1 yields the angular acceptance of the two successive reflections. The angular width, denoted as $\Delta\theta$, is the quantity, which is the largest of $\Delta\theta = \theta_{1a} - \theta_{1d}$ (when $\tau_1 |b_1| \geq 1$) and $\Delta\theta = \theta_{1e} - \theta_{1b}$ (when $\tau_1 |b_1| < 1$). The quantities θ_{1x} ($x = a, b, c$ or d) are the angle coordinates of the points a, b, c, and d. They can be calculated by inserting λ_x of (3.11) into (3.7). We obtain for the angular width of the two successive reflections

$$\Delta\theta = \left(\epsilon_1^{(s)} \frac{|1 - \tau_1 |b_1||}{\sqrt{|b_1|}} + \epsilon_2^{(s)} \frac{\tau_2}{\sqrt{|b_2|}} \right) \tan\theta_{c_1}. \qquad (3.17)$$

The reflection region of the second crystal for the exit waves is shown in Fig. 3.3(2') as yellow stripe in the (λ, θ'_2) space. It is bounded between the borderlines, given by (3.10) with $n = 2$. The glancing angle of reflection θ'_{c_2} is given by (2.149) with the wavelength equal to λ_c and interplanar distance d_2. The projection of the intersection region $ABDE$ in Fig. 3.4(1'+2) onto the reflection region in Fig. 3.4(2') is $\alpha\beta\varepsilon\delta$. The quantity, which is the largest of $\Delta\theta' = \theta'_{2\alpha} - \theta'_{2\delta}$ (when $\tau_2 \leq |b_2|$) and $\Delta\theta' = \theta'_{2\varepsilon} - \theta'_{2\beta}$ (when $\tau_2 > |b_2|$), where θ_{2x} ($x = \alpha, \beta, \delta$ or ε) are the angle coordinates of the points α, β, δ or ε, gives the angular divergence of the waves emerging after the two successive reflections from the crystals in the $(+,+)$ configuration. It is calculated in a similar way as $\Delta\theta$, and is equal to

$$\Delta\theta' = \left(\epsilon_1^{(s)} \sqrt{|b_1|}\, \tau_1 + \epsilon_2^{(s)} \sqrt{|b_2|}\, \left|1 - \frac{\tau_2}{|b_2|}\right| \right) \tan\theta_{c_2}. \qquad (3.18)$$

Expressions equivalent to (3.13), (3.17), and (3.18) were derived by Nakayama et al. (1973) and Matsushita and Hashizume (1983). They were given in terms of angular widths of Bragg reflections $\Delta\theta = \epsilon^{(s)} \tan\theta_c$. We write (3.13), (3.17), and (3.18) instead in terms of the relative spectral width $\epsilon^{(s)}$, since it is a universal (wavelength, and angle independent) parameter of any given Bragg reflection.

If both reflections are equivalent, i.e., $\epsilon_1^{(s)} = \epsilon_2^{(s)} = \epsilon^{(s)}$, $\tan\theta_{c_1} = \tan\theta_{c_2} = \tan\theta_c$, and additionally $b_1 = 1/b_2$, then the expression for the relative spectral bandwidth of the radiation after two successive *equivalent* reflections from the crystals in the $(+, +)$ configuration simplifies to

$$\frac{\Delta\lambda}{\lambda_c} = \epsilon^{(s)} \sqrt{|b_1|}. \tag{3.19}$$

Thus, the relative spectral bandwidth of the radiation, which is spectrally filtered by two equivalent Bragg reflections in the $(+, +)$ configuration, is determined in this case solely by the relative spectral bandwidth $\epsilon^{(s)}$ of the single reflection and the asymmetry factor b_1. It is independent of the selected wavelength λ_c or angle of incidence θ_c.

If additionally $|b_1| \ll 1$ then the expressions for the angular acceptance $\Delta\theta$ - (3.17) - and for the angular spread $\Delta\theta'$ of the radiation transmitted through the system - (3.18) - can be closely approximated by:

$$\Delta\theta = \Delta\theta' \simeq \frac{\epsilon^{(s)}}{\sqrt{|b_1|}} \tan\theta_c \equiv \frac{\Delta\theta^{(s)}}{\sqrt{|b_1|}}. \tag{3.20}$$

Thus, the application of two identical Bragg reflections in the $(+, +)$ configuration with asymmetry parameters $|b_1| = 1/|b_2| \ll 1$ increases the angular acceptance, and simultaneously decreases the spectral band of transmission. This property is very important for monochromator design.

Inspection of (3.13), (3.17), and (3.18) shows that by a judicious choice of Bragg reflections (i.e., by a judicious choice of $\epsilon^{(s)}$), and asymmetry factors, it is possible to select x rays with the desired spectral and angular spread. By proper choice of these parameters it is also possible to control the angular acceptance of the incident, and angular spread of the emerging x rays. The use of the $(+, +)$ configuration is therefore an important practical method for building x-ray monochromators and achieving parallel x-ray beams.

3.4 Diffraction from Crystals in $(+, -)$ Configuration

In the next step, we consider diffraction from crystals in the $(+, -)$ configuration shown schematically in Fig. 3.2. We shall consider here a particular case of two identical Bragg reflections, which one encounters frequently in practice. The interplanar distance, and the relative spectral width for both reflections are equal to d, and $\epsilon^{(s)}$, respectively. Although the Bragg reflections are identical, the two reflecting crystals may be cut asymmetrically and may differ considerably in asymmetry angle ($\eta_1 \neq \eta_2$), and thus also in asymmetry factor ($b_1 \neq b_2$). To be specific, we shall assume that $|b_1| < 1$ and $|b_2| < 1$.

Figure 3.5 shows reflection regions for a sequence of two asymmetric Bragg reflections from the crystals in the $(+, -)$ configuration. In particular, the

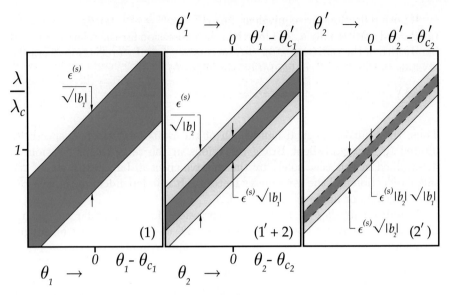

Fig. 3.5. DuMond diagrams showing reflection regions for a sequence of two asymmetric Bragg reflections from crystals in the $(+,-)$ configuration.
(1): green stripe - reflection region of the first crystal for entrance waves. $(1'+2)$: green stripe - reflection region of the first crystal for exit waves, yellow stripe - reflection region of the second crystal for entrance waves.
$(2')$ yellow stripe - reflection region of the second crystal for exit waves. Green stripe shows the phase space of the waves reflected from both crystals.

green stripe in panel (1) shows the reflection region of the first crystal for entrance waves. The same region for exit waves is shown by the green stripe in panel $(1'+2)$, while the yellow stripe in the same panel shows the reflection region of the second crystal for entrance waves. The reflection regions are bounded between the borderlines given by (3.7) and (3.10).

As distinct from the case of the $(+,+)$ configuration, discussed in the previous section and shown in Figs. 3.3 and 3.4, the reflection regions of the two crystals do not intersect - they are parallel to each other. They may, however, overlap, as in Fig. 3.5$(1'+2)$, and thus two successive reflections may take place. This difference is due to the fact, that in the $(+,-)$ configuration the condition for reflection from both crystals is given by (3.3), and the condition for reflection at the center of the reflection regions of both crystals reads:

$$\theta_2 - \theta_{c_2} = +(\theta'_1 - \theta'_{c_1}). \tag{3.21}$$

It is obtained from (3.3) in the same manner as the condition (3.9) is obtained from (3.2). Unlike (3.9), the present condition contains the positive multiplier before the parentheses. As a consequence, the scales $\theta_2 - \theta_{c_2}$ and $\theta'_1 - \theta'_{c_1}$ are

3.4 Diffraction from Crystals in (+, −) Configuration

drawn in Fig. 3.5(1' + 2) in the same sense, as opposed to Fig. 3.3(1' + 2) where they are drawn in the opposite sense.

The relative angle ψ_{12} between the crystals, for which the reflection from both crystals takes place at the center of their reflection region, as in Fig. 3.5(1' + 2), can be calculated in the following way. From (2.114)-(2.115) and (2.149) we may find in the first approximation:

$$\theta'_{c_1} - \theta_{c_2} = \frac{w^{(s)}}{2}\left(\frac{1}{b_2} - b_1\right)\tan\theta_{c_1}. \qquad (3.22)$$

Since the Bragg reflections are exactly alike, we have used the relations $w_2^{(s)} = w_1^{(s)} \equiv w^{(s)}$. Inserting (3.22) into (3.3) we obtain that the successive reflections from both crystals take place, provided

$$\psi_{12} = \pi + \frac{w^{(s)}}{2}\left(\frac{1}{b_2} - b_1\right)\tan\theta_{c_1}. \qquad (3.23)$$

As follows from this equation, the atomic planes in both crystals must be set antiparallel ($\psi_{12} = \pi$) if $b_2 = 1/b_1$. In all other cases a small canting angle $\psi_{12} - \pi$ must be introduced. Bragg's law correction $w^{(s)}$ is typically much less than 10^{-4}. Therefore, the canting angle $|\psi_{12} - \pi|$ is usually small. Nevertheless it is significant, as the angular width of the Bragg reflections as a rule are even smaller.

The angle ψ_{12} given by (3.23) is independent of the selected wavelength λ_c. This situation is quite different from that of diffraction in the (+, +) configuration, where the angle ψ_{12} defines unequivocally (in the sense of (3.14)) the wavelength λ_c of the wave reflected from both crystals. It is due to these facts that the (+, +) configuration is also called dispersive configuration, and the (+, −) configuration with two equivalent Bragg reflections as nondispersive configuration. If the Bragg reflections are different the (+, −) configuration becomes dispersive (or slightly dispersive).

If ψ_{12} differs from the value given by (3.23), the reflection regions may not overlap at all, and therefore successive reflections from both crystals may not take place. Varying ψ_{12} and measuring the intensity of the radiation reflected from both crystals, as well as the width of the intensity dependence on ψ_{12}, provides a means of measuring precisely the angular widths of Bragg reflections (Kohra and Kikuta 1967).

An important question is: what is the region of reflection from both crystals? The reflection region of the second crystal for exit waves is shown as the yellow stripe in panel (2') in Fig. 3.5. The green stripe shows the region excited by the radiation from the first crystal, which is simultaneously the reflection region in question - the region of wavelengths and glancing angles of reflection for the waves reflected from both crystals. It is a stripe of breadth $\epsilon^{(s)}|b_2|\sqrt{|b_1|}$ bounded by the borderlines:

$$\frac{\lambda - \lambda_c}{\lambda_c} = \frac{\theta'_2 - \theta'_{c_2}}{\tan\theta_{c_2}} \pm \frac{\epsilon^{(s)}}{2}|b_2|\sqrt{|b_1|}. \qquad (3.24)$$

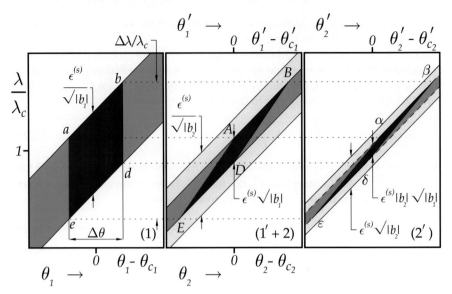

Fig. 3.6. The same as in Fig. 3.5. In addition, the dark blue (black) parallelogram $abde$ in panel (1) displays the part of the reflection region of the first crystal excited by the radiation waves with an angular spread $\Delta\theta$ of the glancing angles of incidence. The parallelograms $ABDE$, and $\alpha\beta\gamma\delta$ show the transform into the space of the waves reflected from the first and the second crystal, respectively.

We conclude that the first reflection changes the spectral and angular spread of the reflected radiation by a factor of $\sqrt{|b_1|}$, as compared to the spread, which one would obtain after the symmetric Bragg reflection. The application of the second asymmetric reflection in the $(+, -)$ configuration enhances this effect by a factor of $|b_2|$. In other words, the application of the second reflection changes the spread much more effectively, as the enhancement factor scales linearly with the asymmetry parameter rather than as a square root. In particular, the angular spread of the waves *with a fixed wavelength* after the two reflections becomes (Kohra and Kikuta 1967)

$$\Delta\theta' = \Delta\theta^{(s)} |b_2| \sqrt{|b_1|}, \tag{3.25}$$

where $\Delta\theta^{(s)} = \epsilon^{(s)} \tan\theta_c$ is the width of the appropriate symmetric Bragg reflection.

A generalization of the present results to the case of Bragg diffraction from n crystals in the nondispersive configuration $(+, -, +, -, ...)$ is straightforward. Applying similar arguments, one arrives at the following result. The angular spread of the waves *with a fixed wavelength* is concentrated within

$$\Delta\theta' = \Delta\theta^{(s)} |b_n ... b_3 b_2| \sqrt{|b_1|} \tag{3.26}$$

after n successive reflections from crystals in the $(+, -, +, -, ...)$ nondispersive configuration.

Although the width of the reflection range after two asymmetric reflections changes dramatically, by a factor of $|b_2|/\sqrt{|b_1|}$, this does not change the spectral range of x rays reflected from both crystals. This statement can be elucidated by Fig. 3.6. Assume incident radiation waves with a broad wavelength spectrum, and with an angular spread $\Delta\theta$ of the incidence angles. The glancing angles of incidence fall into a broad angular range $\Delta\theta$ around θ_{c_1}. The reflection region for these waves is bounded by the parallelogram $abde$. The radiation waves are reflected from the first crystal in the spectral band with a relative width of

$$\frac{\Delta\lambda}{\lambda_c} = \frac{\Delta\theta}{\tan\theta_{c_1}} + \frac{\epsilon^{(s)}}{\sqrt{|b_1|}}. \tag{3.27}$$

The reflection region $abde$ for the incident waves transforms to $ABDE$ for the waves exiting the first crystal, and finally to the reflection region $\alpha\beta\delta\gamma$ for the waves exiting the second crystal. Despite the drastic transformations of the width of the reflection region, the spectral range, in which the two successive reflections take place, remains constant, i.e., it is not changed by the second reflection.

3.5 Diffraction in $(+,-,-,+)$ Configuration

To summarize the discussion of the previous sections, the application of two successive Bragg reflections from two asymmetrically cut crystals in the $(+,-)$ configuration may drastically squeeze the common reflection region in the angular-wavelength space of the reflected waves; however, this does not yet restrict the spectral range of the waves reflected from both crystals. In contrast to this, the common region of reflection from two crystals in the $(+,+)$ configuration is restricted both on the angular and the wavelength scales, which offers a possibility of filtering radiation waves with a small spectral or angular spread. As we shall see, a combination of these two configurations may lead to enhanced possibilities in spectral filtering of x rays. Let us analyze diffraction from crystals in such a mixed configuration.

We consider four crystals labeled as $n = 1, 2, 3$, and 4. The first pair of crystals ($n = 1, 2$) is in the $(+,-)$ nondispersive configuration, as in Fig. 3.2. Also the second pair ($n = 3, 4$) is in the nondispersive configuration. The second crystal of the first pair ($n = 2$) and the first crystal of the second pair ($n = 3$) are, however, in the dispersive configuration, as in Fig. 3.1, which has to be labeled as $(-,-)$. Altogether the whole configuration of the four crystals can be described as $(+,-,-,+)$. It is shown schematically in Fig. 3.7.

As in the previous sections, we shall use the quantities $\boldsymbol{H}_n$, d_n, $\epsilon_n^{(s)}$, b_n, and θ_{c_n} to characterize the corresponding Bragg reflections of the radiation wave with the wavelength of interest λ_c. We shall consider here a particular case

156   3. Principles of Multiple-Crystal X-Ray Diffraction

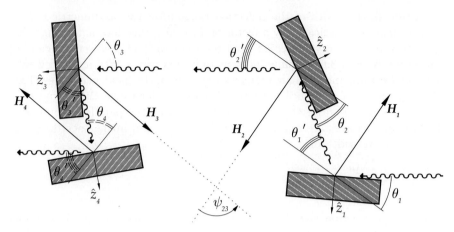

Fig. 3.7. Schematic view of four successive asymmetric Bragg reflections from crystals in the mixed $(+,-,-,+)$ configuration.

of identical Bragg reflections in each pair. That is, the interplanar distances, and the relative spectral widths for each pair of reflections are considered to be equal: $d_1 = d_2$, $d_3 = d_4$, and $\epsilon_1^{(s)} = \epsilon_2^{(s)}$, $\epsilon_3^{(s)} = \epsilon_4^{(s)}$. The asymmetry factors may take any values. However, to be concrete, we shall consider that $|b_n| < 1$ for $n = 1, 2$, and $|b_n| > 1$ for $n = 3, 4$, as it is shown in Fig. 3.7.

A fragment - in the vicinity of λ_c - of the region of reflection from the first two crystals ($n = 1, 2$) presented in the angular-wavelength space of the waves emerging from the second crystal is shown in Fig. 3.8($2' + 3$) by the green stripe of width $\epsilon_1^{(s)} |b_2| / \sqrt{|b_1|}$. Its origin was discussed in the previous section and is illustrated by the DuMond diagrams in Fig. 3.5. The turquoise stripe of width $\epsilon_3^{(s)} / \sqrt{|b_3|}$ is the reflection region of the third crystal for the incident waves. The second and the third crystals are in the dispersive configuration. Therefore, the reflection regions are drawn in such a way that the appropriate angular scales are oppositely directed. The region of reflection from the third and the fourth crystals for the subsequent incident and exit waves are shown successively in panels ($3' + 4$), and ($4'$) in Fig. 3.8.

The common reflection region for the waves incident on the third crystal and reflected successively from *both* the third and the fourth crystals is given by the dark turquoise stripe of width $\epsilon_4^{(s)} / |b_3| \sqrt{|b_4|}$ in Fig. 3.8($2' + 3$).

The intersection region $ABDE$ shown in white in Figs. 3.8($2' + 3$) and 3.9($2' + 3$) is the common region of reflection for the waves reflected successively from *all four* crystals, presented in the space of waves emerging from the second, and waves incident on the third crystal. It can be projected onto any other space. For example, the region $\alpha\beta\varepsilon\delta$ in Fig. 3.9($4'$) is the same reflection region, however, it is shown in the angular-wavelength space of waves emerging from fourth crystal.

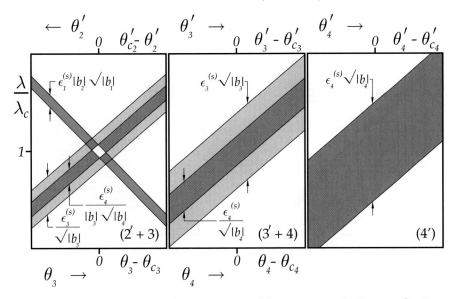

Fig. 3.8. DuMond diagrams for a sequence of four asymmetric Bragg reflections from crystals in the $(+,-,-,+)$ configuration.
$(2'+3)$: Green stripe is the region of reflection from the first two crystals in the space of the waves leaving the second crystal - see Fig. 3.5. Turquoise stripe - reflection region of the third crystal for the incident waves. Dark turquoise stripe - region of reflection from the third *and* the fourth crystal in the space of the waves incident on the third crystal.
$(3'+4)$: Turquoise stripe is the reflection region of the third crystal for the exit waves. Dark turquoise stripe - reflection region of the fourth crystal for the incident waves.
$(4')$: Dark turquoise stripe is the reflection region of the fourth crystal for the exit waves.

Following similar steps to those used to derive (3.13), (3.17), and (3.18) in Sect. 3.3, we obtain for the relative spectral width of the waves reflected from *all the four* crystals[4]

$$\frac{\Delta\lambda}{\lambda_c} = \epsilon_2^{(s)} \tau_2 |b_2| \sqrt{|b_1|} + \epsilon_3^{(s)} \frac{\tau_3}{|b_3|\sqrt{|b_4|}}, \qquad (3.28)$$

where

$$\tau_2 = \frac{\tan\theta_{c_2}}{\tau_{23}}, \qquad \tau_3 = \frac{\tan\theta_{c_3}}{\tau_{23}}, \qquad \tau_{23} = \tan\theta_{c_2} + \tan\theta_{c_3}.$$

The angular width (acceptance) of the four successive reflections is given by

$$\Delta\theta = \left(\epsilon_2^{(s)} \frac{|1 - \tau_2 |b_1 b_2||}{\sqrt{|b_1|}} + \epsilon_3^{(s)} \frac{\tau_3}{|b_3|\sqrt{|b_4|}} \right) \tan\theta_{c_2}. \qquad (3.29)$$

---

[4] See the remark in footnote 3 on page 149

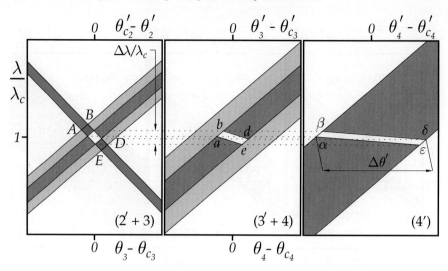

Fig. 3.9. The same as in Fig. 3.8. In addition, the white parallelograms in all the panels show the phase space of the waves reflected from *all four* crystals.

The angular spread of the waves after the four reflections is

$$\Delta \theta' = \left(\epsilon_2^{(s)} \sqrt{|b_1|} |b_2| \tau_2 + \epsilon_3^{(s)} \sqrt{|b_4|} \left| 1 - \frac{\tau_3}{|b_3 b_4|} \right| \right) \tan \theta_{c_3}. \quad (3.30)$$

Comparing these results with those of (3.13), (3.17), and (3.18) describing the $(+,+)$ dispersive configuration case, we may notice their apparent similarity in form. The form is similar because of the double reflections from the crystal pairs in the nondispersive configuration, but the energy and angular widths are greatly changed by additional asymmetry factors to the first power.

The center λ_c of the spectral region of reflection from the four crystals in the $(+,-,-,+)$ configuration is fixed for any given angle ψ_{23} between the vectors $\boldsymbol{H}_2$ and $\boldsymbol{H}_3$ (Fig. 3.7). Similar to the method used in Sect. 3.3, we obtain that λ_c is determined by (3.14)-(3.15), however, with index "1" being substituted by "2", and index "2" by "3". The variation of ψ_{23} results in a change of the central wavelength λ_c, which is given by

$$\frac{\delta \lambda_c}{\lambda_c} = \frac{\delta \psi_{23}}{\tau_{23}}. \quad (3.31)$$

To simplify the analysis of (3.28)-(3.30), let us assume that all the four reflections are equivalent, i.e., $\epsilon_2^{(s)} = \epsilon_3^{(s)} = \epsilon^{(s)}$, $\tan \theta_{c_2} = \tan \theta_{c_3} = \tan \theta_c$. Additionally, let $b_1 = 1/b_4$, and $b_2 = 1/b_3$. Under these assumptions, the relative spectral width of the radiation reflected from the system of four crystals in the $(+,-,-,+)$ configuration becomes (Yabashi et al. 2001)

$$\frac{\Delta\lambda}{\lambda_c} = \epsilon^{(s)} |b_2| \sqrt{|b_1|}. \tag{3.32}$$

If additionally $|b_1| \ll 1$, we obtain from (3.29) and (3.30) the following simplified expressions for the angular acceptance $\Delta\theta$ and for the angular spread $\Delta\theta'$ of the radiation transmitted through the system:

$$\Delta\theta = \Delta\theta' \simeq \frac{\epsilon^{(s)}}{\sqrt{|b_1|}} \tan\theta_c \equiv \frac{\Delta\theta^{(s)}}{\sqrt{|b_1|}}. \tag{3.33}$$

Let us compare these results with (3.19), and (3.20) describing a similar situation in the two-crystal $(+,+)$ configuration. A very important factor is the additional asymmetry factor $|b_2|$ appearing in (3.32) to the first power. By choosing this factor to be $|b_2| \ll 1$, one may narrow drastically the spectral spread of the radiation waves reflected from the four crystals in the $(+,-,-,+)$ configuration, as compared to the two-crystal $(+,+)$ configuration. In other words, by a judicious choice of the asymmetry parameters one may drastically enhance the effect of spectral filtering in the $(+,-,-,+)$ four-crystal configuration.

As for the angular acceptance of the incident radiation, and the angular spread of the reflected radiation, they remain almost unchanged, as compared to the two-crystal $(+,+)$ case.

3.6 Diffraction in $(+,+,-)$ Configuration

By looking closely at Fig. 3.4(2'), the idea suggests itself that the spectral spread of the radiation waves leaving the second crystal, as determined by the height of the dark blue (black) parallelogram, could be further narrowed by selecting the waves within a reduced angular range, i.e., smaller than $\Delta\theta'$ (the length of the larger diagonal of the parallelogram projected on to the θ'_2 scale). To realize this idea, a third crystal has to be applied performing this selection. As Fig. 3.4(2') also suggests, it is most favorable to place the third crystal in the $(+,-)$ configuration with respect to the second one. In this case the reflection region of the third crystal - the dark green stripe in Fig. 3.10(2' + 3) - cuts the dark blue parallelogram at the greatest angle and thus ensures the smallest spectral spread of the waves leaving the third crystal.

The schematic view of three successive Bragg reflections from the three crystals is shown in Fig. 3.11. The first two crystals are arranged in the dispersive $(+,+)$, while the second and the third are in the $(+,-)$ configuration. Altogether the three-crystal configuration may be labeled as $(+,+,-)$. We shall examine here in this section x-ray diffraction in this $(+,+,-)$ three-crystal configuration.

By using Fig. 3.11, we find the following relations between the glancing angles of incidence and reflection:

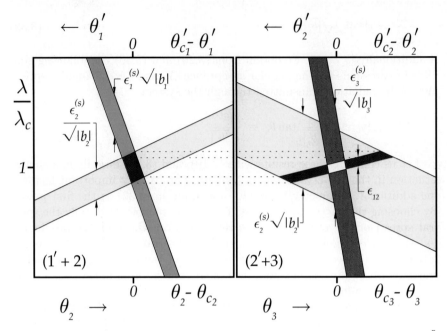

Fig. 3.10. DuMond diagrams presenting reflection regions for a sequence of three asymmetric Bragg reflections from crystals in the $(+, +, -)$ configuration, as shown schematically in Fig. 3.11.
$(1' + 2)$: green stripe is the reflection region of the first crystal for the exit waves, yellow stripe - reflection region of the second crystal for the entrance waves, as in Fig. 3.4($1' + 2$).
$(2'+3)$: yellow stripe - reflection region of the second crystal for the exit waves, dark green stripe - reflection region from the third crystal for the incident waves.
Dark blue (black) parallelograms in both panels show the phase space of the waves reflected from the first two crystals. The white parallelogram shows the phase space of the waves reflected from the three crystals.

$$\theta_2 = \psi_{12} - \theta'_1,$$
$$\theta_3 = \psi_{23} - \pi + \theta'_2. \tag{3.34}$$

The first one is of the same type as (3.2) relating the glancing angles of incidence and reflection in the case of the $(+, +)$ configuration. As discussed in Sect. 3.3, it defines the wavelength λ_c (3.14) at the center of intersection of the reflection regions of the first two crystals. The second relation in (3.34) is of the same type as (3.3) relating the glancing angles of incidence and reflection in the case of the $(+, -)$ configuration. It can be used (together with the modified Bragg's laws) to determine the relative angle ψ_{23} between the diffraction vectors $\boldsymbol{H}_2$ and $\boldsymbol{H}_3$ - Fig. 3.11 - which has to be set for the radiation wave of wavelength λ_c to be reflected from the third crystal at the center θ_{c_3} of its reflection region, as shown in Fig. 3.10($2' + 3$).

3.6 Diffraction in $(+,+,-)$ Configuration   161

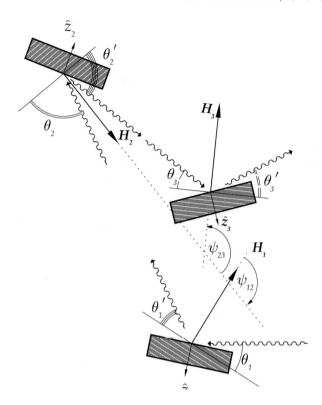

Fig. 3.11. Schematic view of three successive Bragg reflections from crystals in $(+,+,-)$ configuration. Bragg reflection from the second crystal is set close to backscattering.

Diffraction from the first two crystals in the $(+,+)$ configuration was described in Sect. 3.3 and graphically illustrated with the help of the DuMond diagrams in Figs. 3.3 and 3.4. Figure 3.10($1'+2$) is a copy of Fig. 3.3($1'+2$). Figure 3.10($2'+3$) is supplemented, as compared to Fig. 3.3($2'$), with the region of total reflection from the third crystal. It is a dark green stripe bounded between the borderlines $\lambda(\theta_3 - \theta_{c_3})$ given by (3.7) with $n = 3$. Our aim is to calculate the relative spectral bandwidth $\Delta\lambda/\lambda_{\mathrm{c}}$ for the waves reflected from all the three crystals. It is shown in Fig. 3.12($2'+3$) as the height of the white parallelogram - the intersection of the region of total reflection from the first two crystals (the dark blue parallelogram) and the region of total reflection from the third crystal (dark green stripe).

The dark blue (black) region in Figs. 3.10($2'+3$) and 3.12($2'+3$) is bounded from "top" by the line $\delta\beta$, and from "bottom" - by $\varepsilon\alpha$. The wavelength coordinates of the points α, β, ε, and δ are given by λ_X ($X = A, B, E, D$) of (3.11). The angle coordinates of the points are calculated by using λ_X and (3.10) with $n = 2$. Having calculated the coordinates of the points α, β, ε, and δ in (λ, θ'_2) space, we derive in the next step equations for the borderlines $\delta\beta$, and $\varepsilon\alpha$. The result is:

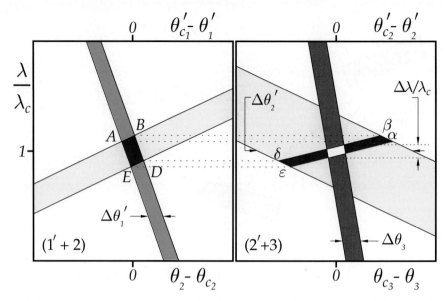

Fig. 3.12. The same as in Fig. 3.10. In addition, dark blue (black) parallelograms in both panels show the phase space of the waves reflected from the first two crystals. The white parallelogram shows the phase space of the waves reflected from the three crystals.

$$\frac{\lambda - \lambda_c}{\lambda_c} = \frac{\theta'_2 - \theta_{c_2}}{t_{12}} \pm \frac{\epsilon_{12}}{2}, \qquad (3.35)$$

where

$$t_{12} = \left(1 + \frac{b_2}{\tau_2}\right) \tan \theta_{c_2}, \qquad (3.36)$$

$$\epsilon_{12} = \epsilon_1^{(s)} \tau_1 \sqrt{|b_1|} \frac{|b_2|}{|\tau_2 + b_2|}, \qquad (3.37)$$

and τ_1 and τ_2 are defined in (3.12). The intersection points of the borderlines $\varepsilon\alpha$ and $\delta\beta$ with the borderlines of the region of total reflection from the third crystal - the vertices of the white parallelogram - can be easily calculated (by a procedure similar to the one used in Sect. 3.3 to derive (3.11)-(3.12)). This allows us to determine finally the relative spectral spread of the waves successively reflected from the three crystals in the $(+, +, -)$ configuration:

$$\frac{\Delta \lambda}{\lambda_c} = \epsilon_{12} \tilde{\tau}_1 + \epsilon_3^{(s)} \frac{\tilde{\tau}_2}{\sqrt{|b_3|}}, \qquad (3.38)$$

$$\tilde{\tau}_1 = \frac{|t_{12}|}{\tilde{\tau}}, \quad \tilde{\tau}_2 = \frac{\tan \theta_{c_3}}{\tilde{\tau}}, \quad \tilde{\tau} = |t_{12}| + \tan \theta_{c_3}. \qquad (3.39)$$

3.6 Diffraction in (+, +, −) Configuration

Two conditions have to be fulfilled for the $(+,+,-)$ configuration to be of interest from the viewpoint of filtering x rays with a spectral spread smaller than that obtainable by diffraction in the two-crystal $(+,+)$ configuration. First, the relative spectral width ϵ_{12} associated with the dark blue parallelogram has to be smaller than the relative spectral width $\epsilon_2^{(s)}\sqrt{|b_2|}$ of the second Bragg reflection: $\epsilon_{12} \ll \epsilon_2^{(s)}\sqrt{|b_2|}$ - see Fig. 3.10(2' + 3). Second, the angular width of the reflection region of the third crystal $\Delta\theta_3 = (\epsilon_3^{(s)}/\sqrt{|b_3|})\tan\theta_{c_3}$ has to be chosen much smaller than the angular spread $\Delta\theta_2' = (\epsilon_2^{(s)}\sqrt{|b_2|})\tan\theta_{c_2}$ of the waves leaving the second crystal: $\Delta\theta_3 \ll \Delta\theta_2'$.

To fulfill these conditions, the Bragg reflections have to be chosen with proper values of $\epsilon_n^{(s)}$, and of the asymmetry parameters b_n. Further, the glancing angles of incidence have to match the relations: $\tan\theta_{c_2} \gg \tan\theta_{c_1}$, $\tan\theta_{c_2} \gg \tan\theta_{c_3}$. If the latter is fulfilled, we may write to good accuracy: $\tau_1 \simeq \tan\theta_{c_1}/\tan\theta_{c_2}$, $\tau_2 \simeq 1$, $\tilde{\tau}_1 \simeq 1$, and $\tilde{\tau}_2 \simeq \tan\theta_{c_3}/|t_{12}|$. With these relations and (3.36)-(3.39) we obtain the following expression for the spectral bandwidth of the waves after three reflections from the crystals in the $(+,+,-)$ configuration:

$$\frac{\Delta\lambda}{\lambda_c} = \left(|b_2|\Delta\theta_1' + \Delta\theta_3\right)\frac{1}{|1 + b_2|\tan\theta_{c_2}}, \qquad (3.40)$$

where

$$\Delta\theta_1' = \epsilon_1^{(s)}\tan\theta_{c_1}\sqrt{|b_1|}, \qquad \Delta\theta_3 = \frac{\epsilon_3^{(s)}\tan\theta_{c_3}}{\sqrt{|b_3|}}$$

are the angular spread of the waves leaving the first crystal, and the angular acceptance of the Bragg reflection from the third crystal, respectively[5] - see Fig. 3.12.

In the case of extreme backscattering when θ_{c_2} approaches $\pi/2$, the asymmetry factor $b_2 \simeq -1$, and $(1+b_2)\tan\theta_{c_2} = -2\sin\eta_2\sin\theta_{c_2}/\sin(\theta_{c_2}-\eta_2) \simeq -2\tan\eta_2$. In this case (3.40) becomes

$$\frac{\Delta\lambda}{\lambda_c} = \left(\Delta\theta_1' + \Delta\theta_3\right)\frac{1}{2\tan\eta_2}. \qquad (3.41)$$

We emphasize here a remarkable property of x-ray diffraction in the $(+,+,-)$ configuration, which follows from (3.40) and (3.41), namely, the relative spectral spread $\Delta\lambda/\lambda_c$ of the waves successively reflected from the three crystals is *independent* of the relative spectral widths $\epsilon_n^{(s)}$ of the chosen Bragg reflections[6] in general, and of the second Bragg reflection in particular.

---

[5] $\Delta\theta_1'$ and $\Delta\theta_3$ can be viewed from another point: $\Delta\theta_1'$ is the angular spread of the waves incident on the second crystal, and $\Delta\theta_3$ is the angular spread of the waves leaving the second crystal, but not of all, only of those which are selected by the third crystal.

[6] With the only exception that we have required $\epsilon_{12} \ll \epsilon_2^{(s)}\sqrt{|b_2|}$, and $(\epsilon_2^{(s)}\sqrt{|b_2|})\tan\theta_{c_2} \gg (\epsilon_3^{(s)}/\sqrt{|b_3|})\tan\theta_{c_3}$.

Other factors are determining the relative spectral spread. First, the angular spread $\Delta\theta'_1$ of the waves incident on the second crystal, and the angular spread $\Delta\theta_3$ of those waves leaving the second crystal, which are selected by the third crystal. Second, the relative spectral width, e.g., in (3.40), is determined by the factor $1/(|1+b_2|\tan\theta_{c_2})$, which is a manifestation of angular dispersion in Bragg reflection from the second crystal.

Indeed, assuming the waves with a relative wavelength spread $\Delta\lambda/\lambda$ incident at a fixed angle (i.e., no angular spread: $\Delta\theta'_1 = 0$) on the second crystal, angular dispersion in Bragg reflection results according to (2.140) in an angular spread of $\Delta\theta' = (\Delta\lambda/\lambda_c)(1+b_2)\tan\theta_c$ of the exit waves. The same expression we would obtain for $\Delta\theta_3$ from (3.40), having assumed $\Delta\theta'_1 = 0$. This proves a key role of angular dispersion in x-ray diffraction in the $(+,+,-)$ configuration.

A similar interrelationship one can find between the factor $1/2\tan\eta_2$ in (3.41) and the expression for angular dispersion in the extreme backscattering geometry (2.145).

The mechanism of obtaining the spectral spread given by (3.40) and (3.41) is now clear: the first crystal collimates the beam, the second crystal acts as a dispersing element that transforms the wavelength into angular spread, and the third crystal acts as a wavelength selector, selecting the waves with a desired angular and thus spectral spread.

Regarding the angular acceptance $\Delta\theta$ of the whole three-crystal system for the incident waves, it is easy to show that under the conditions used to derive (3.40) and (3.41) it is determined primarily by the angular acceptance of the Bragg reflection from the first crystal:

$$\Delta\theta \simeq \frac{\epsilon_1^{(s)}}{\sqrt{|b_1|}}\tan\theta_{c_1} \equiv \Delta\theta_1^{(s)}\frac{1}{\sqrt{|b_1|}}. \qquad (3.42)$$

In conclusion, diffraction in the three-crystal $(+,+,-)$ configuration offers a new way of filtering x rays with a small spectral bandwidth. Remarkably, and as distinct from the previous cases discussed in this chapter, angular dispersion plays a leading role, resulting in reflected waves with relative spectral spreads, which are independent of the relative intrinsic spectral widths of the participating Bragg reflections, and determined solely by geometrical parameters, such as angular spread of the incident waves, asymmetry angle of the crystal acting as the dispersing element, and angular acceptance of the wavelength selector crystal.

This property offers a radically different means of x rays monochromatization, which does not rely on intrinsic spectral properties of Bragg reflections.

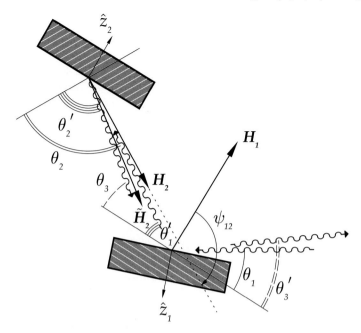

Fig. 3.13. Schematic view of three successive Bragg reflections from two crystals. Bragg reflection from the second crystal is set almost to exact backscattering. X-ray diffraction in such a two-crystal configuration is equivalent to diffraction in the $(+,+,+)$ three-crystal configuration. Notations are explained in the text.

3.7 Exact Backscattering in $(+,+,+)$ Configuration

Figure 3.13 displays a special case of Bragg diffraction in the multiple-crystal configuration discussed in the previous section. There are three distinctions between the scheme presented in Fig. 3.13 and that of Fig. 3.11.

First, the second crystal in Fig. 3.13 is set almost to exact back-reflection.

Second, the third crystal is removed, as now the first crystal is playing the role of both the collimator and the wavelength selector. There are two crystals and three successive Bragg reflections with glancing angles of incidence θ_n and reflection θ'_n ($n = 1, 2, 3$).

Third, in approaching exact backscattering a qualitatively different configuration of the Bragg reflections arises. Indeed, we know that in asymmetric diffraction, exact backscattering takes place for the wave with the glancing angle of incidence $\theta_2 = \theta_R$ (2.131), which is smaller than $\pi/2$, with the azimuthal angle of incidence $\phi_2 = 0$, and the azimuthal angle of reflection $\phi'_2 = \pi$ (see Sect. 2.2.7). The deviation of θ_R from $\pi/2$ is due to the fact that the actual momentum transfer in Bragg diffraction is given by the total momentum transfer vector $\tilde{\boldsymbol{H}}_2$ (2.15) rather than by the diffraction vector $\boldsymbol{H}_2$, as shown in Fig. 3.13. We remind the reader, that the near to normal incidence scattering geometry for one crystal is shown in some more detail in Fig. 2.6, and in Fig. 2.23(a_0)-(a'_0). It is essential that now the azimuthal angle of reflection is $\phi'_2 = \pi$ (due to this the incident and reflected wave are shown in Fig. 3.13 on the same side of $\boldsymbol{H}_2$, as distinct from Fig. 3.11, where they are shown on different sides). The azimuthal angle of reflection switches from $\phi'_2 = 0$ to $\phi'_2 = \pi$ (i.e., the wave is reflected normal to the atomic planes) as θ_2 attains $\pi/2 - \Theta_\perp$ (2.160). At this point the $(+,+,-)$ configuration of Bragg reflections becomes a $(+,+,+)$ configuration. Indeed, by using Fig. 3.13, we find the following relations between the glancing angles of incidence and reflection

$$\begin{aligned}\theta_2 &= \psi_{12} - \theta'_1, \\ \theta_3 &= \psi_{12} - \theta'_2,\end{aligned} \quad (3.43)$$

instead of those in (3.34), which are valid in the $(+,+,-)$ configuration. Both relations in (3.43) are of the same type as the relation (3.2) valid for diffraction from two crystals in the $(+,+)$ configuration. X-ray diffraction from the two-crystal configuration with three Bragg reflections is, therefore, equivalent to diffraction from three crystals in the $(+,+,+)$ configuration. We shall term this crystal configuration as double-dispersive configuration.

Presentation of the borderlines of the region of total Bragg reflection in the linearized form (3.7) and (3.10) fails in the vicinity of exact backscattering. Therefore, it is not evident that the expression (3.41), derived in the previous section, for the relative spectral spread $\Delta\lambda/\lambda_c$ of the waves reflected successively from the three crystals also remains valid when θ_2, the glancing angle of incidence on the second crystal, is very close to $\pi/2$. Besides, it is also not clear whether switching from the $(+,+,-)$ to the $(+,+,+)$ configuration causes any changes in the three-crystal spectral bandwidth.

Therefore, in the remainder of this section, we shall initially reconsider the three-crystal diffraction case on the condition that the second crystal is set almost to exact back-reflection, and then study the transition from the $(+,+,-)$ - Fig. 3.11 - to the $(+,+,+)$ - Fig. 3.13 - configuration. Finally, we shall derive an expression for the relative spectral spread $\Delta\lambda/\lambda_c$ of the waves after the three successive reflections by using equations valid in the extreme backscattering geometry.

The region of total reflection from the second crystal set to extreme backscattering ($\theta_2 \to \pi/2$) is shown highlighted in yellow in Figures 3.14($1'+2$) and 3.15($1'+2$) in the space of incident waves. The area highlighted in yellow in Figs. 3.14($2'+3$) and 3.15($2'+3$) shows the same reflection region, however, in the space of the reflected waves. The reflection region in question is equivalent to that shown in Fig. 2.22(i) and 2.22(e) on page 98, and labeled by (a_0) and (a'_0). Its origin and details we have discussed in Sect. 2.4.11.

The narrow vertical green stripe in Figs. 3.14($1'+2$) and 3.15($1'+2$) represents the phase space of the waves incident on the second crystal. In our particular case this is determined by the region of total reflection from the

3.7 Exact Backscattering in (+, +, +) Configuration 167

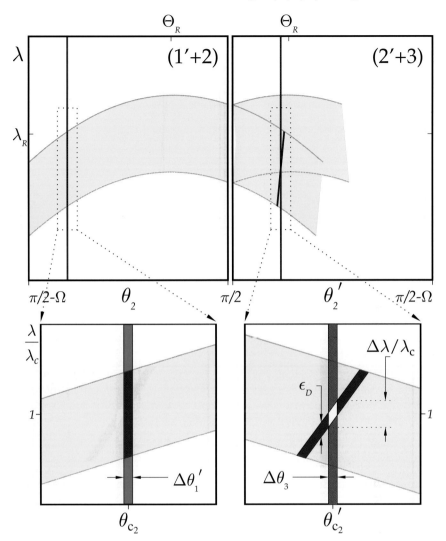

Fig. 3.14. DuMond diagrams for a sequence of three asymmetric Bragg reflections in (+, +, −) configuration, as in Fig. 3.11. The second crystal is set to extreme backscattering ($\theta_2 \to \pi/2$). The area in the boxes marked-off by dotted lines are shown below on the enlarged angular scale. $(1' + 2)$: Green stripe is the region of reflection from the first crystal in the space of exit waves. Yellow area is the region of reflection from the second crystal in the space of incident waves. $(2' + 3)$: Yellow area is the reflection region of the second crystal in the space of exit waves. Dark green stripe is the reflection region of the third crystal in the space of incident waves. Dark blue (black) areas display the common reflection region from the first and second crystals. The white parallelogram displays the common reflection region from all the three crystals.

168   3. Principles of Multiple-Crystal X-Ray Diffraction

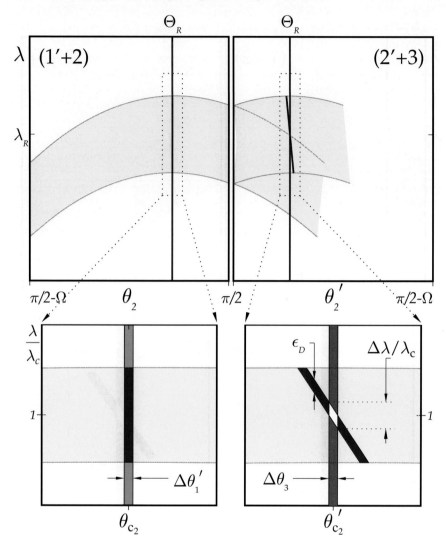

Fig. 3.15. DuMond diagrams for a sequence of three asymmetric Bragg reflections in $(+,+,+)$ configuration from two crystals, as in Fig. 3.13. The second crystal is set to exact backscattering ($\theta_{c_2} = \pi/2 - \Theta_R$). Other notations as in Fig. 3.14.

first crystal. Since, we assume $\tan\theta_2 \gg \tan\theta_1$ (as in the previous section), the vertical stripe with an angular width of $\Delta\theta'_1$ centered at θ_{c_2} closely approximates the phase space of the waves reflected from the first crystal and incident on the second crystal. The parts of the DuMond diagrams in the boxes marked-off by dotted lines are shown below in Figs. 3.14 and 3.15 for clarity on an expanded angular scale. The difference between Fig. 3.14 and Fig. 3.15 is in the position of the center θ_{c_2} of the distribution of glancing angles of incidence of the waves impinging on the second crystal. In the case shown in Fig. 3.15 the angle $\theta_{c_2} = \pi/2 - \Theta_R$, i.e., the waves incident on the second crystal at θ_{c_2} are reflected exactly backwards. The azimuthal angle of reflection is $\phi'_2 = \pi$. The two crystal arrangement in Fig. 3.13 represents this situation. As distinct from this, in the case shown in Fig. 3.14 the angle θ_{c_2} is smaller than $\pi/2 - \Theta_R$. It is chosen such that the azimuthal angle of reflection is $\phi'_2 = 0$. The three crystal arrangement in Fig. 3.11 represents this situation.

Dark blue (black) parallelograms in both figures display the common reflection region from the first and the second crystals. Due to angular dispersion, the vertical stripe in Figs. 3.14(1' + 2) and 3.15(1' + 2) becomes inclined in the space of the waves reflected from the second crystal, as shown in Figs. 3.14(2' + 3) and 3.15(2' + 3), respectively. As we have discussed in Sect. 2.4.11, it is bounded between the borderlines $\lambda(\theta'_2)$ given by (2.162), which in our particular case reads

$$\frac{\lambda_\pm - \lambda_c}{2 d_H} \simeq -\frac{\theta'_2 - \theta'_{c_2}}{\tau_D} \cos\phi'_2 \mp \frac{\epsilon_D}{2} \cos\phi_2, \tag{3.44}$$

$$\tau_D = \frac{2\sin\eta_2}{\sin(\theta_{c_2} - \eta_2)}, \qquad \epsilon_D = \frac{\Delta\theta'_1}{\tau_D}. \tag{3.45}$$

The intersection area marked as a white parallelogram is the common reflection region for the waves three times successively reflected either from three crystals - Fig. 3.14(2'+3), or from *two crystals* - Fig. 3.15(2'+3). It is shown in the space of waves leaving the second, and the waves incident on the third - in the $(+, +, -)$ arrangement - or again on the first crystal - in the $(+, +, +)$ arrangement. The spectral width of the intersection region is in both cases obviously $\Delta\lambda/\lambda_c = \epsilon_D + \Delta\theta_3/\tau_D$, which by using (3.45) can be written as

$$\frac{\Delta\lambda}{\lambda_c} = (\Delta\theta'_1 + \Delta\theta_3) \frac{\sin(\theta_{c_2} - \eta_2)}{2\sin\eta_2}. \tag{3.46}$$

If we take $\theta_{c_2} \simeq \pi/2$, which is adequate for our case, then $\sin\eta_2/\sin(\theta_{c_2} - \eta_2) \simeq \tan\eta_2$, and (3.46) becomes

$$\frac{\Delta\lambda}{\lambda_c} = \frac{\Delta\theta'_1 + \Delta\theta_3}{2\tan\eta_2}, \tag{3.47}$$

i.e., equivalent to (3.41) obtained in the previous section. The first thing to note at this point is that the present analysis confirms the validity of (3.41) when the second crystal is set to extreme backscattering.

One may easily show, that the angular acceptance $\Delta\theta$ for the incident waves in the case of the triple Bragg reflection from crystals either in the $(+, +, -)$ or in the $(+, +, +)$ configuration, is determined primarily by the angular acceptance of the Bragg reflection from the first crystal, and is given by (3.42).

3. Principles of Multiple-Crystal X-Ray Diffraction

Equation (3.46) can be applied to calculate the relative spectral width both for a sequence of three asymmetric Bragg reflections in the $(+,+,-)$ configuration with three crystals, as shown schematically in Fig. 3.11, as well as in the $(+,+,+)$ configuration with two crystals, as shown in Fig. 3.11.

In the $(+,+,+)$ configuration with two crystals $\Delta\theta'_1 = \Delta\theta_3 = \Delta\theta_1^{(s)} \sqrt{|b_1|}$, where $\Delta\theta_1^{(s)} = \epsilon_1^{(s)} \tan\theta_{c_1}$ is the angular width of the symmetric Bragg reflection from the first crystal. By using this relations equation (3.46) can be transformed to

$$\frac{\Delta\lambda}{\lambda_c} = \frac{\Delta\theta'_1}{\tan\eta_2} = \frac{\Delta\theta_1^{(s)} \sqrt{|b_1|}}{\tan\eta_2}. \qquad (3.48)$$

It tells us, that the relative spectral width of the triple Bragg reflection (with the second reflection in almost exact backscattering) is equal to the angular spread $\Delta\theta'_1$ of the radiation waves incident on the back-reflecting crystal, decreased by a factor of $\tan\eta_2$. It is independent of the chosen Bragg reflection. It is determined solely by the angular spread of the radiation waves incident on the back-reflecting crystal and by the proximity of the asymmetry angle η_2 to $\pi/2$.

The analysis of the results obtained here may bring us to the same conclusions as in the previous section. Namely, the spectral spread given by (3.46)-(3.48) is a result of the following sequence of events: the first crystal collimates the beam, the second crystal acts as a dispersing element that transforms the wavelength into angular spread, and the first crystal executing the third reflection acts as a wavelength selector, selecting the waves with a desired angular and thus spectral spread.

4. Theory of X-Ray Fabry-Pérot Resonators

4.1 Introduction

Fabry-Pérot resonators (or interferometers) are standard instruments in visible-light optics[1]. Invented by Fabry and Pérot (1899), they have been used for more than a century to explore the visible spectral region of the electromagnetic radiation. They are used for high-precision wavelength measurements in atomic spectroscopy, in astrophysics, and many other physical as well as life sciences (Born and Wolf 1999, Vaughan 1989). They are used as interference filters with very high spectral resolution. They are widely used as resonators in laser physics.

The main components of the simplest Fabry-Pérot interferometer are two highly reflecting low absorbing parallel mirrors. The system becomes transparent despite the high reflectivity of each mirror when the gap d_g between the mirrors is an integer multiple of half of the radiation wavelength as then the resonance condition for standing wave formation in the gap is fulfilled. The energy separation between two successive transmission resonances - the free spectral range $E_f = hc/2d_g$ - is a constant independent of the incident photon energy. The spectral width of the transmission resonances $\Gamma = E_f/F$ is smaller the higher the finesse $F = \pi\sqrt{R}/(1-R)$, or equivalently, the higher the mirror reflectivity R (Born and Wolf 1999). With a reflectivity of $R = 0.85$ ($F = 19.3$) and a gap of 5 cm, the width of the transmission resonances is in the sub-micro-eV range. Physically, such a small spectral width is due to interference of a large number $\simeq F$ of coherent waves arising from multiple reflections.

Beam divergence blurs the resonances and degrades the interferometer's performance. To avoid this, the back-reflection geometry is essential. Unlike the visible spectral region where back-reflection mirrors of high reflectivity are available, this is a problem for x rays.

Steyerl and Steinhauser (1979) pioneered the idea of a Fabry-Pérot-type interference filter for x rays and performed the first theoretical analysis. They

---

[1] According to the established nomenclature in optics, Fabry-Pérot interferometers with a fixed gap are called Fabry-Pérot étalons. As the formation of resonance modes is the main and general feature we shall use the word "resonator" as a generic name for such devices.

have applied the dynamical theory of Bragg diffraction. A matrix approach of matching boundary conditions was employed. A particular case of *exact* normal incidence to the reflecting atomic planes was considered. The transmission energy spectrum with a single sharp transmission peak was calculated at 17.4 keV for the (14 6 0) reflection in silicon. The calculated relative spectral bandwidth of the transmitted peak was $\Delta E/E \approx 2 \times 10^{-8}$. Interestingly, the first theory of the x-ray Fabry-Pérot resonator by Steyerl and Steinhauser (1979) was developed at a time when there was no consistent theory of Bragg backscattering.

Caticha and Caticha-Ellis (1990b) extended the theoretical treatment to an arbitrary angle of incidence. They also introduced an additional parameter allowing for a possible relative shift of the crystal plates. A similar matrix approach was used to match the boundary conditions. The matrix technique for a system of N thin parallel crystals diffracting at near-normal incidence was formulated. Numerical simulations of the energy dependence of the transmissivity of the x-ray Fabry-Pérot resonator were performed. It was shown that for crystal plates with low absorption the transmission function has sharp resonances of defined energy that can be controlled by tuning the gap of the x-ray resonator. The theory of a thermal neutron resonator of Fabry-Pérot type was presented by the same authors (Caticha and Caticha-Ellis 1996). The matrix technique is convenient for numerical calculations and it can be easily applied to systems containing many diffracting crystals. However, it does not allow simple analytical analysis.

To overcome this drawback Caticha et al. (1996) applied a geometrical approach, which is often used in theories of the optical resonator (Born and Wolf 1999, Vaughan 1989). A summation of the multiple reflection amplitudes from the crystal plates was used to calculate the reflectivity and transmissivity of the system. The reflection and transmission amplitudes of the resonator elements - the crystal plates - were calculated using the dynamical theory of Bragg diffraction (Caticha and Caticha-Ellis 1990a). The transmission spectra of the x-ray resonator were also calculated.

Shvyd'ko and Gerdau (1999) have revised the theory of Caticha et al. (1996). It was necessary to introduce an important phase factor that was lacking in the previously mentioned theories. Numerical calculations for an x-ray Fabry-Pérot resonator built of sapphire single crystals were performed. Both transmission and reflection intensities were calculated as a function of x-ray energy and incidence angle. The validity of the newly incorporated phase factor was rigorously substantiated by Kohn et al. (2000).

Kohn et al. (2000) have developed the theory of the x-ray Fabry-Pérot resonator as a particular case of the dynamical theory of x-ray Bragg diffraction in layered crystalline systems. It has been shown that the performance of the x-ray resonator is similar to that of the optical Fabry-Pérot resonator. Both show a fine resonance structure in the transmission and reflection dependence. However, for the x-ray Fabry-Pérot resonator this occurs only inside the region of the Bragg back-reflection peak. The influence of possible imperfections, such as the roughness of the crystal plate surfaces and the error in the parallelism of the atomic planes has been discussed. Numerical estimates for the sapphire x-ray resonator have been given.

In the present chapter the theory of the x-ray Fabry-Pérot resonator is developed following the results of Shvyd'ko and Gerdau (1999), Kohn et al. (2000), and Shvyd'ko (2002). At the beginning, a perfect x-ray resonator, is considered: the x-ray Fabry-Pérot resonator built as a system of two parallel perfect crystal plates, playing the role of resonator mirrors. The derivation of the mathematical expressions for the transmissivity and reflectivity of the perfect x-ray resonator uses the geometrical approach based on the summation of the probability amplitudes of multiple reflections from the crystal plates. The single reflection and transmission amplitudes of the crystal plates are calculated using the dynamical theory of x-ray Bragg diffraction, as described in Sect. 2.4.1. The expressions obtained are analyzed in detail. In particular, the position and width of the Fabry-Pérot transmission resonances are estimated analytically as a function of the gap between the crystal mirrors, of the x-ray energy, of the angle of incidence, and of the crystal temperature. Numerous results of computer simulations of the energy, time and angular dependence are presented to illustrate the performance of the x-ray Fabry-Pérot resonator.

Special attention is payed to addressing the time response of the x-ray Fabry-Pérot resonator. This is of particular interest as the time response plays an important role in the experiments on x-ray Fabry-Pérot resonators presented in Chap. 7. An analytical expression for the time response of an x-ray Fabry-Pérot resonator to an excitation with a short radiation pulse is derived. Examples of numerical calculations of the time response are also presented.

The influence of possible imperfections, such as roughness of the crystal surfaces, error in the parallelism of the atomic planes, and unequal crystal temperatures is discussed in Sect. 4.4. It is shown that these factors may significantly deteriorate the performance of the x-ray resonator. Analytical estimates are given.

In conclusion to this chapter, some model applications of x-ray Fabry-Pérot resonators are discussed.

Before going into detail, it is worthwhile to recall the theory of the optical Fabry-Pérot resonator and to discuss in particular the effect of multiple-wave interference underlying this theory. This will allow us to reveal similarities and differences between the optical Fabry-Pérot resonator and its x-ray counterpart and thus to gain deeper insight into the physical nature of the theoretical results.

4.2 Multiple-Beam Interference in Optics

Multiple-wave interference is the basic physical phenomenon underlying Fabry-Pérot resonators, and is its main distinguishing feature.

The purpose of this section is to give a short account of the theory of multiple-wave interference in visible-light optics and, based upon that, high resolution interference filters. A more detailed treatment can be found, e.g., in the standard optics texts such as Born and Wolf (1999), Vaughan (1989), or Hecht and Zajac (1974). Plane parallel (Fabry-Pérot) and wedge (Fizeau) multiple-wave resonators are discussed to provide a link between optical and x-ray multiple-wave interference devices.

4.2.1 Plane Resonators

Consider a plate of thickness d_g with parallel surfaces coated with thin layers of high reflectivity, which play the role of mirrors, see Fig. 4.1. These could be thin metal films or multi-layer dielectric films. It is assumed here for simplicity, that the mirrors are infinitely thin and are not absorbing. The refractive index of the medium filling the gap between the mirrors is n_g, and of the surrounding medium is n. Let the amplitude-reflection coefficient of the mirrors be r and the amplitude-transmission coefficient be t, for a wave traveling from the surrounding medium to the reflecting plates. Similarly, let $\tilde{r}$ and $\tilde{t}$ be the corresponding coefficients for a wave traveling from the gap to the surrounding medium.

The transmission of a plane monochromatic wave through the system of two mirrors and filling medium is considered in terms of the wave frequency ω, amplitude E_0, glancing angle θ, gap width d_g, and some other parameters, which will be discussed below.

The radiation passing through the first mirror may experience multiple reflections between the mirrors before it escapes through one of them, as is shown schematically in Fig. 4.1. It should be noted that the rays are actually lines drawn perpendicular to the wavefronts and thus to the optical fields.

The amplitude of the transmitted wave after having experienced in total $2p$ successive reflections is $E_{t_p} = E_0 \, t \, \tilde{r}^{2p} \, \tilde{t} \, \exp(i\, p\, \delta\varphi)$. Here $\delta\varphi = K\Lambda$ is the phase difference between adjacent rays, with Λ being the difference in optical path lengths between adjacent rays, $K = 2\pi/\lambda$ is the wavenumber, and $\lambda = c/2\pi\omega$ is the wavelength. It is readily shown (see Fig. 4.1) that $\Lambda = n_g(AB + BC) - n AD = 2d_g n_g \sin\theta_g$. Expressed in terms of the glancing angle of incidence θ, the optical path difference reads

$$\Lambda = 2d_g \sqrt{n_g^2 - n^2 \cos^2\theta}. \tag{4.1}$$

The total amplitude of the radiation transmitted through the system of two reflecting mirrors is the sum of all possible multiply reflected waves:

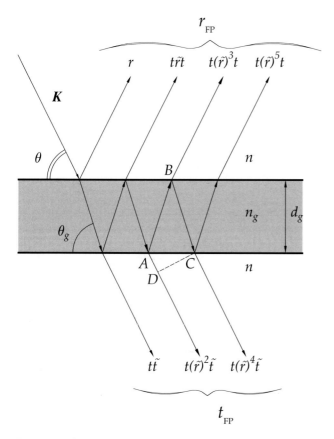

Fig. 4.1. Scheme of formation of multiple-waves in an optical Fabry-Pérot resonator.

$$E_{\rm t} = \sum_{p=0}^{\infty} E_{{\rm t}_p} = E_0 \sum_{p=0}^{\infty} t\,\tilde{r}^{2p}\,\tilde{t}\exp(\mathrm{i}\,p\,\delta\varphi). \tag{4.2}$$

The summation of the geometrical series results in

$$E_{\rm t} = E_0 \frac{t\tilde{t}}{1 - \tilde{r}^2\,\exp(\mathrm{i}\,\delta\varphi)}. \tag{4.3}$$

Let us term the reflectivity of the mirrors $\mathcal{R} = |\tilde{r}^2|$, transmissivity $\mathcal{T} = |t\tilde{t}|$, the phase jump on reflection $\phi_{\mathcal{R}} = \arg[\tilde{r}^2]$, and the phase jump on transmission $\phi_{\mathcal{T}} = \arg[t\tilde{t}]$. Using these notations one arrives at the following expression for the amplitude-transmission coefficient of the system

$$t_{\rm FP} = \frac{E_{\rm t}}{E_0} = \frac{\mathcal{T}\,\exp(\mathrm{i}\,\phi_{\mathcal{T}})}{1 - \mathcal{R}\,\exp(\mathrm{i}\,\varphi_{\rm A})}, \tag{4.4}$$

where
$$\varphi_A = 2K d_g \sqrt{n_g^2 - n^2 \cos^2 \theta} + \phi_\mathcal{R} \tag{4.5}$$

will be termed henceforth the Airy phase.

The transmissivity of the system of two parallel reflecting mirrors $T_{FP} = |t_{FP}|^2$ is the ratio of the transmitted flux density $I_t = |E_t|^2/2$ to incident flux density $I_0 = |E_0|^2/2$. Using (4.4) one obtains

$$T_{FP} = \frac{\mathcal{T}^2}{1 + \mathcal{R}^2 - 2\mathcal{R}\cos(\varphi_A)} = \frac{\mathcal{T}^2}{(1-\mathcal{R})^2 + 4\mathcal{R}\sin^2(\varphi_A/2)}. \tag{4.6}$$

Defining the parameters

$$F = \frac{4\mathcal{R}}{(1-\mathcal{R})^2}, \qquad H = \frac{\mathcal{T}^2}{(1-\mathcal{R})^2} \tag{4.7}$$

one arrives at the expression

$$T_{FP} = \frac{H}{1 + F\sin^2(\varphi_A/2)}. \tag{4.8}$$

One should note that in the absence of absorption, which was assumed previously, the parameter $H = 1$. Equation (4.8) is widely known in optics as the Airy formula, which describes the intensity distribution of the transmitted pattern due to multiple reflections in a system of two parallel plane reflecting surfaces.

By using similar considerations and summations of the reflected wave trains one can obtain the equation for the reflectivity of the system of the two mirrors. In the absence of absorption, however, it can be easily calculated using $R_{FP} = 1 - T_{FP}$.

The behavior of the transmissivity (4.8) as a function of the Airy phase is shown in Fig. 4.2, calculated for different values of $\mathcal{R}$ and F, respectively. The transmitted intensity is a periodic function of φ_A with a period of 2π. The fringes of the order of interference $m-1$, m, and $m+1$ are shown. The transmission maxima (interference fringes) are sharper the higher the reflectivity of the mirrors $\mathcal{R}$. Alternatively, the higher the reflectivity, the more multiply reflected waves contribute to the net signal, and the sharper are the fringes. This demonstrates in particular the great significance of multiple-wave interference and its difference from the more usual two-wave interference. If the reflectivity approaches unity, the number of interfering beams increases tremendously and the transmission becomes very small except in the immediate vicinity of the maxima of total transmission. The interference pattern consists of narrow bright fringes on an almost completely dark background.

The ratio of the separation of adjacent fringes (2π) to their width (full-width at half-maximum) $\Delta\varphi_A$ is called the *finesse*:

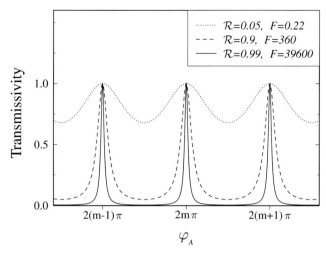

Fig. 4.2. multiple-wave fringes in transmitted light as a function of the Airy phase φ_A shown for a few orders of interference m. The curves are calculated with (4.8) for different values of the mirrors' reflectivity $\mathcal{R}$ and $F = 4\mathcal{R}/(1-\mathcal{R})^2$.

$$\mathcal{F} = 2\pi/\Delta\varphi_A. \tag{4.9}$$

For large F it is readily shown to be given by

$$\mathcal{F} = \frac{\pi\sqrt{F}}{2} = \frac{\pi\sqrt{\mathcal{R}}}{1-\mathcal{R}}. \tag{4.10}$$

Fabry and Pérot (1899) have proposed the use of two parallel mirrors of high reflectivity as a high-resolution spectrometer based on multiple-wave interference. Because of the sharpness of the multiple-wave interference fringes, such devices have an enhanced resolving power.

One can observe interference fringes by changing the spacing d_g between the mirrors. If the spacing can be scanned, such devices are usually called Fabry-Pérot interferometers. Since the Airy phase φ_A is proportional to d_g (4.5) the interference pattern on the d_g-scale is similar to that of Fig. 4.2. Let us assume for simplicity that $n_g = 1$ and $n = 1$, i.e., the mirrors are in vacuum. Then the interference fringes appear every time if the variation of the spacing δd_g obeys the following rule

$$2\,\delta d_g \sin\theta = m\,\lambda, \quad m = 0, \pm 1, \pm 2, \ldots \tag{4.11}$$

Thus the highest transmittance is achieved if the condition of standing wave formation between the mirrors is fulfilled. This is the usual resonator condition in optics. It is equivalent to Bragg's law[2], see (2.101). Due to this

---

[2] One could think of the parallel atomic planes reflecting x rays in a crystal in the Bragg diffraction mode as a kind of Fabry-Pérot resonator. A significant

property and due to the enhanced resolving power, Fabry-Pérot resonators have been applied since their advent to the precise comparison of optical wavelengths and to the measurement of the wavelengths in terms of the standard meter. For more details see, e.g., (Vaughan 1989).

Alternatively, one can fix the spacing and change the frequency of the incident radiation. If the spacing is fixed, such devices are usually called Fabry-Pérot étalons [3]. In this case the interference pattern for transmitted light on the frequency (energy) scale will also be similar to that of Fig. 4.2. The separation between the adjacent fringes on the energy scale is given by

$$E_{\mathrm{f}} = \frac{hc}{2 d_{\mathrm{g}} \sin \theta}, \qquad (4.12)$$

which is called the free spectral range. The width of the fringe on the energy scale is then

$$\Gamma = \frac{E_{\mathrm{f}}}{\mathcal{F}}. \qquad (4.13)$$

One can estimate that at normal incidence with mirrors spaced by 1 cm the free spectral range is $E_{\mathrm{f}} = 62 \ \mu\mathrm{eV}$. With a reflectivity of $\mathcal{R} = 0.99$ the finesse $\mathcal{F} = 314$ and therefore the spectral width of the transmission lines is in the sub-micro-electronvolt range. This demonstrates the high resolving power of optical Fabry-Pérot resonators. It is for this reason, that they were intensively and extensively applied over the 20th century to study fine and hyperfine structure of atomic spectra, among other applications.

The fringes can be also observed on the angular scale. Assuming monochromatic light and taking again for simplicity $n_{\mathrm{g}} = 1$, one obtains for the Airy phase $\varphi_{\mathrm{A}} = 2 K d_{\mathrm{g}} \sin \theta + \phi_{\mathcal{R}}$. Thus the fringes of equal inclination are observed at glancing angles, for which the order of interference is given by

$$m = \frac{\varphi_{\mathrm{A}}}{2\pi} = \frac{2 d_{\mathrm{g}}}{\lambda} \sin \theta + \frac{\phi_{\mathcal{R}}}{2\pi}. \qquad (4.14)$$

Figure 4.3 shows simulations of the multiple-wave fringes of equal inclination using the Airy formula (4.8) with $F = 1520$ and $\mathcal{R} = 0.95$. The centers of the graphs correspond to normal incidence: $\Theta = \pi/2 - \theta = 0$. Figure 4.3a represents the case for which the interference order $m = 100$ appears nearest to the center ring. In Fig. 4.3b it is seen exactly at the center ($\Theta = 0$), while in Fig. 4.3c it does not appear at all. Physically these situations happen when the wavelength of the light increases gradually from (a) to (c), or equivalently, the spacing between the mirrors decreases from (a) to (c).

---

difference is the very low reflectivity of the individual atomic planes which is however compensated by their great number. In fact, Bragg diffraction from a crystal could be compared with the performance of a multiple (consisting of many reflecting planes) Fabry-Pérot resonator.

[3] Unless otherwise indicated, the name Fabry-Pérot resonator or simply resonator is used for both options.

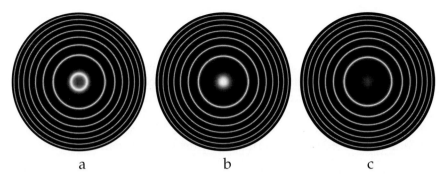

Fig. 4.3. Simulations of the multiple-wave fringes of equal inclination in transmission.

At near-normal incidence, where the approximation $\sin\theta \simeq 1 - (\Theta)^2/2$ is valid, the Airy phase has a quadratic dependence on the incidence angle Θ. As a result the fringes of equal inclination are not equidistant, as is clearly seen in Fig. 4.3. Also, the angular widths of the fringes are larger the closer they are to normal incidence. For the fringe appearing at exact normal incidence the angular width is

$$\Delta\theta = 2\sqrt{\frac{\lambda}{2d_g \mathcal{F}}}. \tag{4.15}$$

Equation (4.9) and the quadratic dependence of the Airy phase was used to derive this expression. For high interference orders $m \simeq 2d_g/\lambda \gg 1$ the width of the central fringe can be given as $\Delta\theta = 2/\sqrt{m\mathcal{F}}$.

These formulae demonstrate that the use of the Fabry-Pérot resonator as a high resolution interference filter is most favorable under near-normal incidence conditions where the angular width of the transmission resonances is largest.

4.2.2 Wedge Resonators

The discussion of multiple-wave interference in the preceding section was confined to the condition of parallel mirrors. Naturally the question arises, what happens to multiple-wave interference if the mirrors are not parallel? It is of practical importance to obtain an answer to the question: what inclination of the mirrors does not destroy the very sharp Fabry-Pérot fringes?

For a fixed wavelength and a fixed incidence angle, the light is either transmitted, or not, through a Fabry-Pérot resonator, depending on how well it satisfies the resonator condition (4.11). If one of the plates is tilted slightly, the spacing varies across the plate, and transmission takes place if the condition (4.11) is fulfilled. Thus the transmission pattern observed behind the resonator is a contour map of the resonator spacing. If the plates are flat, the contours are straight lines. This arrangement of a Fabry-Pérot resonator with tilted plates, also called in the literature a multiple-wave Fizeau resonator, is most useful for testing surfaces (Tolansky 1960).

The first detailed study of the multiple-wave wedge fringes, was made by Brossel (1947a,b). As outlined in these papers, phase lag due to the wedge angle can create asymmetry and subsidiary fringes. The first detailed numerical evaluation of a multiple-wave fringe profile was carried out by Kinosita (1953). Rogers (1982) has performed more detailed calculations of the shape and location of multiple-wave transmission Fizeau fringes. Moos et al. (1963) have demonstrated sharp wedge fringes for plate separations as large as 20 cm. A useful general account of multiple-wave interferometry can be found in the texts of Born and Wolf (1999), Polster (1969) and Vaughan (1989).

4.3 Perfect X-Ray Fabry-Pérot Resonator

In the x-ray spectral region high reflectivity may be achieved either by employing scattering at grazing incidence or using Bragg diffraction from single crystals. In Bragg diffraction the backscattering geometry is feasible. Therefore, crystals haven take over the role of high reflectivity mirrors, in the x-ray spectral region.

4.3.1 General Equations

The perfect x-ray Fabry-Pérot resonator is a device which consists of two perfectly parallel crystal plates of thicknesses d and $\tilde{d}$, separated by a gap of width d_g filled with a non-diffracting medium - Fig. 4.4a. It is assumed in the following that the crystals are made of the same material. The reflecting atomic planes in the second crystal plate are assumed to be perfectly parallel to the planes in the first crystal plate and have the same temperature.

The relative position of the crystal plates in space can be described by a translation vector $\boldsymbol{u}$. The translation is defined in such a way that $\boldsymbol{u} = 0$ corresponds to a configuration where both crystals compose a single crystal of thicknesses $d + \tilde{d}$ - Fig. 4.4b. With this choice the gap d_g is related to $\boldsymbol{u}$ by $d_g = \boldsymbol{u}\hat{\boldsymbol{z}}$.

Alternatively, another translation vector $\boldsymbol{U}$ can be introduced defined in such a way that $\boldsymbol{U} = 0$ corresponds to a configuration where the entrance surfaces (for the incoming x rays) of both crystals coincide. Additionally, it is required that the positions of the atoms coincide in the overlapping volumes of both crystals. The latter implies that $\boldsymbol{U} - \boldsymbol{u} = \boldsymbol{A}$, where $\boldsymbol{A} = \sum_{i=1}^{3} n_i \boldsymbol{a}_i$ is a translation period of the crystal lattice. Here, $\boldsymbol{a}_i$ are the basis vectors of the crystal unit cell, and n_i are any integers. As vector $\boldsymbol{A}$ connects points on the front and the rear surface of the first crystal, one can write that $d = \boldsymbol{A}\hat{\boldsymbol{z}}$ or $d = U_z - d_g$.

4.3 Perfect X-Ray Fabry-Pérot Resonator

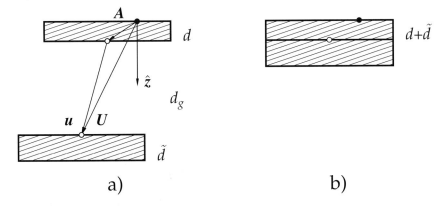

Fig. 4.4. (a) Arrangement of crystals in an x-ray Fabry-Pérot resonator. Two parallel crystal plates of thicknesses d and $\tilde{d}$ are separated by a gap d_g; $\hat{z}$ is an internal normal to the crystal surface. Reflecting atomic planes are shown by the set of parallel lines. The relative position of the crystals can be described in two different ways, by using translation vectors $\boldsymbol{u}$ or $\boldsymbol{U}$, respectively. Translation vector $\boldsymbol{U} = 0$ corresponds to a configuration where the front surfaces of both crystals and the reflecting atomic planes coincide in the overlapping volumes. Translation vector $\boldsymbol{u} = 0$ corresponds to a configuration where both crystals compose a single crystal of thickness $d + \tilde{d}$, as shown in panel (b). $\boldsymbol{A} = \boldsymbol{U} - \boldsymbol{u}$ is a translation period of the crystal lattice.

The incident wave is reflected either directly off the first crystal or after multiple reflections off the resonator crystals, as shown schematically in Fig. 4.5. Similarly, the incident wave is either transmitted directly or after multiple reflections off the crystals. The amplitude r_{FP} of the radiation field reflected off the x-ray Fabry-Pérot resonator and the amplitude t_{FP} of the radiation transmitted through the x-ray Fabry-Pérot resonator is a superposition of the probability amplitudes of all possible multiple scattering paths. With each path, a probability amplitude is associated. It is a product of successive reflection and transmission amplitudes related to both crystal plates and the gap. The scheme of multiple scattering paths and the amplitudes of the individual scattering events r_{0H}, $\tilde{r}_{0H}$, $e^{i\phi_g}$, etc., are shown in Fig. 4.5.

The reflection amplitude $r_{0H}(d)$ and transmission amplitude $t_{00}(d)$ for radiation incident on the front surface of the first crystal are given by (2.69), and (2.68) of the dynamical theory, respectively. The reflection amplitude $r_{H0}(d)$ and transmission amplitude $t_{HH}(d)$ for radiation incident on the rear surface of the first crystal are given by (2.81), and (2.80), respectively.

The second crystal is shifted by the translation vector $\boldsymbol{U}$ from the position where the front surfaces and reflecting atomic planes coincide for both crystals. From (2.86), the reflection amplitude for the radiation incident on the front surface of the second crystal is $\tilde{r}_{0H}(\tilde{d}) = r_{0H}(\tilde{d}) \exp(\mathrm{i}\boldsymbol{H}\boldsymbol{U})$. As it

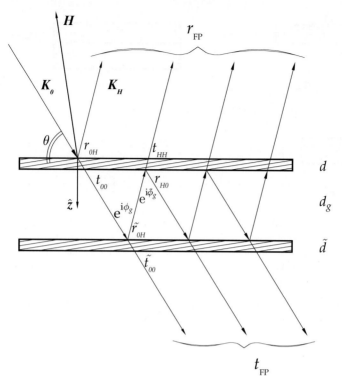

Fig. 4.5. An x-ray Fabry-Pérot resonator with two parallel single crystal plates used as reflecting mirrors. The crystals are of a thickness d and $\tilde{d}$ separated by a gap d_g filled with non-diffracting medium. The scheme of multiple scattering and the amplitudes of the individual scattering events r_{0H}, r_{H0}, etc. are shown.

was argued in Sect. 2.4.1, the transmission amplitudes for the second are the same as for the first crystal, i.e., $\tilde{t}_{00}(\tilde{d}) = t_{00}(\tilde{d})$.

The transmission amplitudes for radiation propagating through the gap in the direction $\boldsymbol{K}_0$ of the incident and in the direction $\boldsymbol{K}_H$ of the reflected wave are given by $e^{i\phi_g}$ and $e^{i\tilde{\phi}_g}$, respectively. The complex phases ϕ_g and $\tilde{\phi}_g$ are:

$$\phi_g = \frac{Kd_g}{2\gamma_0}\chi_g, \qquad \tilde{\phi}_g = \frac{Kd_g}{2\gamma_H}(\alpha - \chi_g). \tag{4.16}$$

They are derived from (2.68) and (2.80), respectively, by replacing χ_0 with χ_g and taking $\chi_H = 0$. Here χ_g is the Fourier component of zero order of the electric susceptibility of the medium filling the gap, cf. (2.3), and (2.4).

By using the notation of Fig. 4.5, and assuming that all reflected and respectively transmitted waves are superimposed, we obtain for the reflection and transmission amplitude coefficients the infinite series

4.3 Perfect X-Ray Fabry-Pérot Resonator

$$r_{\rm FP} = r_{0H} + t_{00} e^{{\rm i}\phi_g} \tilde{r}_{0H} e^{{\rm i}\tilde{\phi}_g} \left(t_{HH} + r_{H0} e^{{\rm i}\phi_g} \tilde{r}_{0H} e^{{\rm i}\tilde{\phi}_g} \left(t_{HH} + \ldots \right. \right., \quad (4.17)$$

$$t_{\rm FP} = t_{00} e^{{\rm i}\phi_g} \left(\tilde{t}_{00} + \tilde{r}_{0H} e^{{\rm i}\tilde{\phi}_g} r_{H0} e^{{\rm i}\phi_g} \left(\tilde{t}_{00} + \ldots \right. \right.. \quad (4.18)$$

The series can be readily transformed to the following form

$$r_{\rm FP} = r_{0H} + t_{00} \tilde{r}_{0H} t_{HH} e^{{\rm i}(\phi_g + \tilde{\phi}_g)} \left[1 + \sum_{m=1}^{\infty} \left(r_{H0} \tilde{r}_{0H} e^{{\rm i}(\phi_g + \tilde{\phi}_g)} \right)^m \right], \quad (4.19)$$

$$t_{\rm FP} = t_{00} \tilde{t}_{00} e^{{\rm i}\phi_g} \left[1 + \sum_{m=1}^{\infty} \left(r_{H0} \tilde{r}_{0H} e^{{\rm i}(\phi_g + \tilde{\phi}_g)} \right)^m \right]. \quad (4.20)$$

The summation of the series results in

$$r_{\rm FP} = r_{0H} + \frac{t_{00} t_{HH} \tilde{r}_{0H} e^{{\rm i}(\phi_g + \tilde{\phi}_g)}}{1 - r_{H0} \tilde{r}_{0H} e^{{\rm i}(\phi_g + \tilde{\phi}_g)}},$$

$$t_{\rm FP} = \frac{t_{00} \tilde{t}_{00} e^{{\rm i}\phi_g}}{1 - r_{H0} \tilde{r}_{0H} e^{{\rm i}(\phi_g + \tilde{\phi}_g)}}. \quad (4.21)$$

Combining (4.21) with (2.68)-(2.69), (2.80)-(2.81), (2.86), and (4.16) we obtain for the reflection and transmission amplitudes of the x-ray Fabry-Pérot resonator

$$t_{\rm FP} = \frac{t_{00}(d)\, \tilde{t}_{00}(\tilde{d})\, \exp({\rm i}\phi_g)}{1 - r_{H0}(d)\, r_{0H}(\tilde{d})\, \exp({\rm i}\phi)}, \quad (4.22)$$

$$r_{\rm FP} = r_{0H}(d) + \frac{t_{00}(d)\, t_{HH}(d)\, r_{0H}(\tilde{d})\, \exp({\rm i}\phi)}{1 - r_{H0}(d)\, r_{0H}(\tilde{d})\, \exp({\rm i}\phi)} \quad (4.23)$$

with the complex phase

$$\phi = \boldsymbol{H}\boldsymbol{u} + \phi_g (1-b) + \frac{K d_g}{2\gamma_H} \alpha, \qquad d_g = u\hat{z}. \quad (4.24)$$

In the derivation of (4.24) use was made of the fact that $\exp({\rm i}\boldsymbol{H}\boldsymbol{U}) = \exp({\rm i}\boldsymbol{H}\boldsymbol{u})$ by virtue of $\boldsymbol{U} = \boldsymbol{u} + \boldsymbol{A}$ and $\exp({\rm i}\boldsymbol{H}\boldsymbol{A}) = 1$.

One may verify that for zero gap ($d_g = 0$) and in the absence of the translation of the second crystal relative to the first one ($\boldsymbol{u} = 0$), the reflection and transmission amplitudes satisfy the obvious relation

$$t_{\rm FP} = t_{00}(d + \tilde{d}), \qquad r_{\rm FP} = r_{0H}(d + \tilde{d}), \quad (4.25)$$

i.e. t_{FP} and r_{FP} are equal to the reflection and transmission amplitudes of a crystal with thickness $d + \tilde{d}$. The same result is obtained for a nonzero gap, provided $\phi = 2\pi n$ ($n = 0, \pm 1, \ldots$) and $\chi_g = 0$.

Let us introduce effective transmission and reflection coefficients of the crystal mirrors $\mathcal{T}$, $\mathcal{R}$

$$\begin{aligned} \mathcal{T} &= |t_{00}(d)\, t_{00}(\tilde{d})|\, \exp(-\phi''_g), \\ \mathcal{R} &= |r_{H0}(d)\, r_{0H}(\tilde{d})|\, \exp\left[-\phi''_g (1-b)\right], \end{aligned} \qquad (4.26)$$

and relevant phases $\phi_{\mathcal{T}}$, $\phi_{\mathcal{R}}$:

$$\begin{aligned} \phi_{\mathcal{T}} &= \arg\left[t_{00}(d)\, \tilde{t}_{00}(\tilde{d})\right] + \phi'_g, \\ \phi_{\mathcal{R}} &= \arg\left[r_{H0}(d)\, r_{0H}(\tilde{d})\right]. \end{aligned} \qquad (4.27)$$

By using these notations and $\phi = \phi' + \mathrm{i}\phi''$, the transmission amplitude (4.22) can be written as

$$t_{FP} = \frac{\mathcal{T} \exp(\mathrm{i}\phi_{\mathcal{T}})}{1 - \mathcal{R} \exp(\mathrm{i}\varphi_{A})}, \qquad (4.28)$$

where $\varphi_{A} = \phi' + \phi_{\mathcal{R}}$. From (4.24) the phase φ_{A} can be presented in the form

$$\varphi_{A} = \boldsymbol{H}\boldsymbol{u} + \frac{Kd_g}{2\gamma_0}\chi'_g(1-b) + \phi_{\mathcal{R}} + \frac{Kd_g}{2\gamma_H}\alpha. \qquad (4.29)$$

In a manner similar to the optical case discussed in Sect. 4.2, the phase φ_{A} is termed as the x-ray Airy phase, or simply Airy phase unless this produces ambiguity.

Remarkably, the formal expression (4.28) for the transmission amplitude of the x-ray Fabry-Pérot resonator is identical to that of the optical Fabry-Pérot resonator, cf. (4.4). Despite this, there are also differences, which are in the physical content of the parameters entering these formulae and their dependence on the photon energy, glancing angle of incidence, mirror characteristics, etc.

The reflectivity and transmissivity of the x-ray Fabry-Pérot resonator are calculated by

$$R_{FP} = \frac{1}{|b|}|r_{FP}|^2, \quad T_{FP} = |t_{FP}|^2. \qquad (4.30)$$

With (4.22), (4.23) and (4.28), we write the expressions for the transmissivity and the reflectivity in the form

$$T_{FP} = \frac{H}{1 + F \sin^2(\varphi_{A}/2)}, \quad R_{FP} = 1 - T_{FP} - A_{FP} \qquad (4.31)$$

where $A_{\rm FP}$ is the absorption factor of the resonator,

$$H = \frac{\mathcal{T}^2}{(1-\mathcal{R})^2}, \quad \text{and} \quad F = \frac{4\mathcal{R}}{(1-\mathcal{R})^2}. \tag{4.32}$$

Within the Bragg diffraction region, $\mathcal{R}$ is close to one and the parameter F has a large value. In this case, the transmissivity has sharp maxima when the condition

$$\varphi_{\rm A} = 2\pi n \equiv \varphi_{{\rm A}_n}, \qquad n = 0, \pm 1, \pm 2, \ldots. \tag{4.33}$$

is fulfilled. Hereafter, equation (4.33) is called the "resonance condition".

For negligible photo-absorption one can put $A_{\rm FP} = 0$ in (4.31). As a result $R_{\rm FP} = 1 - T_{\rm FP}$. In this case the analysis of the transmissivity and reflectivity of the x-ray resonator is rather straightforward. In the following discussion we shall use this approximation to obtain analytical expressions, which will allow us to gain deeper insight into the physics of x-ray Fabry-Pérot resonators. Rigorous equations are used in the following to perform computer simulations of the transmissivity and reflectivity of the x-ray Fabry-Pérot resonator. In most of the following example, the resonator composed of two α-Al$_2$O$_3$ single crystal plates and the (0 0 0 30) Bragg reflection is considered.

4.3.2 Comparison with Optical Resonators

Equation (4.31) is similar to the Airy formula (4.8) which describes multiple-wave interference phenomena with two identical parallel non-absorbing mirrors of high reflectivity, and describes the performance of optical Fabry-Pérot resonators. This implies that, x-ray and optical Fabry-Pérot resonators are described by similar or even identical formal expressions. However, one has also to observe significant differences, which are in the physical content of the parameters entering these formulae.

First of all, the effective reflection coefficient $\mathcal{R}$ takes significant values only within the range of Bragg reflection. Secondly, and this is even more important, the expression for the x-ray Airy phase $\varphi_{\rm A}$ given by (4.29) differs from the expression (4.5) known from the theory of the optical Fabry-Pérot resonator.

The difference starts with the first term $\boldsymbol{Hu}$ in (4.29) which is unique in the case of x-ray resonators. This term appears due to the internal structure of the x-ray mirrors. It describes whether the reflecting atomic planes of the second crystal mirror are in perfect registry with the atomic planes in the first one. It describes how coherent is the internal structure of the first mirror with respect to the second one.

The fourth term in (4.29) is also unique. The most pronounced dependence of the Airy phase on photon energy, angle of incidence, and crystal temperature is included in the deviation parameter α. A detailed analysis of the influence of different physical parameters on the Airy phase is given in the subsequent discussion.

The third term $\phi_\mathcal{R}$ in (4.29) is a phase jump upon Bragg reflection, which is similar to ϕ_R in (4.5) - the phase jump on reflection off optical mirrors. The latter is a constant, which takes the values 0 or π depending on the optical properties of the mirrors. Unlike this, $\phi_\mathcal{R}$ in Bragg reflection may change significantly within the range of Bragg reflection. Indeed, for non-absorbing crystals we can approximate $\phi_\mathcal{R} = 2\phi_r$ where ϕ_r is given by (2.93). As was discussed in Sect. 2.4.3, for thick non-absorbing crystals the phase ϕ_r changes in the range from $-\pi$ to 0 within the Bragg reflection peak. Thus, $\phi_\mathcal{R}$ may change from -2π to 0 in the same range.

In particular cases, one can find similarities between the optical and x-ray Airy phases. For this we have to note that the magnitude of the scattering vector can be represented as $H = 2K\sin\theta$. Assuming that $\boldsymbol{H} \parallel \hat{\boldsymbol{z}}$ we can write for the first term in (4.29): $\phi_1' = \boldsymbol{H}\boldsymbol{u} = 2Kd_\text{g}\sin\theta$. If we then assume symmetric diffraction, i.e., $b = -1$ and thus $\gamma_0 = \sin\theta$, the second term in (4.29) can be written as $\phi_2' = \phi_\text{g}(1-b) = Kd_\text{g}\chi_\text{g}/\sin\theta$. By using $n_\text{g} = 1 + \chi_\text{g}/2$ for the refractive index in the gap and taking $|\chi_\text{g}| \ll 1$ into account we can ascertain that under these conditions $\phi_1' + \phi_2' \simeq 2Kd_\text{g}\sqrt{n_\text{g}^2 - \cos^2\theta}$. Thus, in this particular case, the sum of the first and the second terms in (4.29) can be reduced to the first term of the Airy phase given by in the theory of the optical resonator - (4.5).

4.3.3 Enhanced Phase Sensitivity

If we assume (to a first approximation) that the effective reflection coefficient $\mathcal{R}$ is constant, then the transmissivity and reflectivity of the x-ray Fabry-Pérot resonator given by (4.31)-(4.32) varies only with the Airy phase φ_A (4.29) as is shown in Fig. 4.2. The relative position of the mirrors, given by the translation vector $\boldsymbol{u}$, the x-ray energy E (or wavelength λ), the glancing angle of incidence θ, the crystal temperature T, the refractive index of the material filling the gap $n_\text{g} = 1 + \chi_\text{g}/2$, and some other quantities, have an effect on the phase φ_A.

The influence of each of these parameters on the phase φ_A and thus on the performance of the x-ray Fabry-Pérot resonator will be discussed. As a first step, however, the transmissivity of the x-ray Fabry-Pérot resonator is analyzed as a function of the Airy phase φ_A without specifying its dependence on any physical value.

As was already pointed out the transmissivity of the x-ray resonator has sharp transmission peaks - Fabry-Pérot resonances - when the Airy phase φ_A takes the values $\varphi_{\text{A}_n} = 2\pi n$ defined by the resonance condition (4.33). The resonance values of the phase are separated by 2π.

To analyze the shape of the transmission peak as a function of φ_A let us consider a small deviation of the phase from the resonance value φ_{A_n}, so that

4.3 Perfect X-Ray Fabry-Pérot Resonator

$|\varphi_A - \varphi_{A_n}| \ll 1$. By using this assumption we obtain from (4.31) and (4.32)

$$T_{\text{FP}}(\varphi_A) \simeq \left(\frac{\Delta\varphi_A}{2}\right)^2 \frac{H}{(\varphi_A - \varphi_{A_n})^2 + (\Delta\varphi_A/2)^2}, \tag{4.34}$$

where

$$\Delta\varphi_A = \frac{4}{\sqrt{F}} = \frac{2(1-\mathcal{R})}{\sqrt{\mathcal{R}}} = \frac{2\pi}{\mathcal{F}}. \tag{4.35}$$

Thus the Fabry-Pérot resonances have a Lorentzian shape with a maximum value of H (4.32) and phase width (FWHM) of the Fabry-Pérot resonance $\Delta\varphi_A$. The quantity $\mathcal{F}$ is the so-called finesse. As in the theory of the optical resonator (4.9)-(4.10) the finesse $\mathcal{F}$ of the x-ray resonator is defined as a ratio of the separation between the resonances to the resonance width:

$$\mathcal{F} = \frac{2\pi}{\Delta\varphi_A} = \frac{\pi\sqrt{\mathcal{R}}}{1-\mathcal{R}}. \tag{4.36}$$

The higher the reflectivity of the crystal mirrors and the smaller the absorption in the gap, the larger is the effective reflection coefficient $\mathcal{R}$ (4.26), and as a consequence the larger is the finesse. For example, for the mirrors with $\mathcal{R} = 0.9$ the finesse $\mathcal{F} = 29.8$. As is discussed in the following, the finesse scales with the effective number $\mathcal{N}_s$ of multiple reflections within the gap: $\mathcal{F} = 2\mathcal{N}_s$, see Sect. 4.3.9.

Equation (4.35) demonstrates the enhanced phase sensitivity of the x-ray Fabry-Pérot resonator. The enhancement factor is $\mathcal{F}$, as compared to the two-wave type interferometers, such as Michelson, Mach-Zehnder, etc., interferometers (Born and Wolf 1999, Hecht and Zajac 1974), for which $\mathcal{F} = 1$. The enhanced phase sensitivity is a distinguishing feature of resonators of the Fabry-Pérot-type, which is due to *multiple*-wave interference

The absolute position of the Fabry-Pérot resonances on the phase scale is fixed, and is given by the resonance condition (4.33). However, the Airy phase is a function of practically all physical parameters of the problem: the x-ray photon energy, the glancing angle of incidence, the gap width, the polarizability of the crystal, the polarizability of the medium in the gap, the crystal temperature, etc. As a consequence, the position of the Fabry-Pérot resonances, for instance, on the x-ray energy scale, is a function of all the physical parameters, with the exception of the photon energy. By the same arguments, the position of the resonances on the angular scale is a function of all the physical parameters (including the x-ray photon energy), with the exception of the angle of incidence, etc. The practical significance of this peculiarity is that all the parameters have to be precisely controlled, in order to observe the resonance behavior of the x-ray Fabry-Pérot resonator.

At this point, let us proceed with a detailed discussion of the influence of these physical parameters on the resonance behavior of x-ray Fabry-Pérot resonators.

4.3.4 Influence of the Gap Width

Three terms of the four in the expression for the Airy phase (4.29) depend on the relative position of the crystal mirrors, which is described by the translation vector $\boldsymbol{u}$ and the gap width $d_g = \boldsymbol{u}\hat{\boldsymbol{z}}$. Therefore, the transmissivity of the x-ray resonator can be changed by shifting the mirrors relative to each other. Before going into further detail, let us specify the expression for the Airy phase. It is assumed that the glancing angle of incidence θ and the x-ray wavelength λ are fixed at the center of the region of total reflection ($\lambda = \lambda_c$, $\theta = \theta_c$, $\alpha = \alpha_c$), as given by the modified Bragg law (2.114). Therefore, (4.29) can be written as

$$\varphi_A = \boldsymbol{H}\boldsymbol{u} + \frac{Kd_g}{2\gamma_H}\alpha_c + \frac{Kd_g}{2\gamma_0}\chi'_g(1-b). \tag{4.37}$$

The phase jump $\phi_\mathcal{R}$ is omitted in (4.37) as it is not affected by crystal translations. By also using $\alpha_c = \chi'_0(1 - 1/b)$ - (2.96) - and $d_g = \boldsymbol{u}\hat{\boldsymbol{z}}$ we arrive at the following expression for the Airy phase

$$\varphi_A = \boldsymbol{H}\boldsymbol{u} - \frac{K\boldsymbol{u}\hat{\boldsymbol{z}}}{2|\gamma_H|}\left(\chi'_0 - \chi'_g\right)\left(1 - \frac{1}{b}\right). \tag{4.38}$$

The first term in (4.38) is henceforth called the "diffractive" phase and the second one - the "refractive" phase. The reasons for this are obvious. The diffractive phase depends on the diffraction vector $\boldsymbol{H}$, while the refractive phase depends on the difference of the refractive index $n = 1 + \chi_0/2$ of the crystals and the refractive index $n_g = 1 + \chi_g/2$ of the medium filling the gap. Both terms in (4.38) depend on the translation vector $\boldsymbol{u}$. If the refractive indices in both media coincide, i.e., $\chi'_0 = \chi'_g$, the refractive phase becomes zero.

The diffractive phase changes rapidly - a shift of one of the crystals by only $u = d_H$ along $\boldsymbol{H}$ alters the phase by 2π. This effect is seen clearly in the calculated transmission and reflection spectra in Fig. 4.7 (p. 194). The diffractive phase, which can also be represented as $\boldsymbol{H}\boldsymbol{u} = 2\pi s$, describes the spatial coherence of the atomic planes in both crystal plates. If $s = n$, where $n = 0, \pm 1, \pm 2, \ldots$, the atomic planes in both crystals are in perfect registry relative to each other. The atomic planes, for instance, are always in perfect registry if the mirrors are monolithically cut out of a single crystal piece. It should be noted that a shift $\boldsymbol{u} \perp \boldsymbol{H}$ does not change the diffractive phase.

The refractive phase changes more slowly than the diffractive phase. To change the refractive phase by 2π the gap $d_g = \boldsymbol{u}\hat{\boldsymbol{z}}$ should be changed by $\delta d_g \simeq \lambda|\gamma_H|/|\chi'_0 - \chi'_g|$. For $\lambda = 1$Å, $\chi'_0 \approx 10^{-5}$ and $\chi'_g = 0$ we obtain $\delta d_g \approx 10$ μm. The refractive phase is $\approx 10^5$ less sensitive to translations than the diffractive phase.

A unique property of the x-ray Fabry-Pérot resonator shows up when the reflecting atomic planes and the crystal plate surface (or equivalently their normals $\hat{z}$ and $\boldsymbol{H}$) are not parallel, as is shown in Fig. 4.5. In this case, if one adds to $\boldsymbol{u}$ an additional shift $\delta\boldsymbol{u}$ with $\delta\boldsymbol{u}\hat{z} = 0$ (i.e., parallel to the crystals surfaces) but $\delta\boldsymbol{u}\boldsymbol{H} \neq 0$, then obviously, according to (4.37), the transmissivity of the x-ray resonator can be varied by a transverse translation of the mirrors without varying the gap. This property is due to the intrinsic structure of the mirrors of the x-ray resonator.

In the particular case of $\hat{z} \parallel \boldsymbol{H}$, i.e., the crystal surface and atomic planes are parallel to each other, $b = -1$ and $|\gamma_H| = \sin\theta_c$. If one additionally assumes that there is vacuum in the gap, i.e., $\chi_g = 0$, the diffractive phase can then be presented as $\boldsymbol{Hu} = Hd_g = 2\pi d_g/d_H = 4\pi d_g \sin\theta_c/\lambda_c(1+w_H)$, and the refractive phase as $-Kd_g\chi'_0(1-1/b)/2|\gamma_H| = 4\pi d_g \sin\theta_c w_H/\lambda_c(1+w_H)^2$. Substituting both terms into (4.38) we obtain for the Airy phase:

$$\varphi_A = 2\pi \frac{2d_g \sin\theta_c}{\lambda_c}. \tag{4.39}$$

This expression for φ_A is accurate to within $w_H^2 \approx 10^{-10}$.

If one varies the gap width, it follows from (4.39), that Fabry-Pérot resonances emerge at a distance δd_g given by

$$2\,\delta d_g \sin\theta_c = m\,\lambda_c, \qquad m = \pm 1, \pm 2, \pm 3, \ldots, \tag{4.40}$$

if at d_g a resonance was found.

The result expressed by (4.39) and (4.40) is remarkable in two ways. First, the Airy phase is a function of only the gap width and the x-ray wavelength in vacuum (for a fixed glancing angle). Second, the same resonance condition as (4.40) is valid for the visible-light Fabry-Pérot resonator with the optical wavelength substituted for the x-ray wavelength λ_c; see (4.11). This fact opens up the possibility of direct precise comparison of the radiation wavelengths in visible and hard x-ray spectral regions. Practical aspects of this problem were addressed in Sect. 1.2.3.

4.3.5 Spectral Dependence

Next, we consider the wavelength (energy) dependence of the transmissivity of the x-ray resonator. In the subsequent discussion it is assumed for simplicity that there is a vacuum gap between the mirrors, i.e., $\chi_g = 0$.

By using (2.98) the Airy phase (4.29) reads

$$\varphi_A = \boldsymbol{Hu} - 2\pi N_{go}\left(\frac{\lambda}{2d_H} - \sin\theta\right) + \phi_\mathcal{R}. \tag{4.41}$$

Here

$$N_{go} = \frac{d_g}{d_H |\gamma_H|} \tag{4.42}$$

is the effective number of interplanar distances d_H fitting into the gap of the x-ray Fabry-Pérot resonator. For a fixed glancing angle of incidence θ the center of the region of total reflection λ_c is given by the modified Bragg law (2.114). As a result we arrive at the expression for the Airy phase convenient for the current analysis:

$$\varphi_A(\lambda) = \boldsymbol{H}\boldsymbol{u} - 2\pi\frac{\lambda - \lambda_c}{\lambda_{fo}} + \phi_H + \phi_\mathcal{R}(\lambda), \qquad (4.43)$$

$$\lambda_{fo} = \frac{2d_H}{N_{go}} = \frac{2d_H^2|\gamma_H|}{d_g}, \qquad (4.44)$$

$$\phi_H = 2\pi N_{go}\, w_H\, \sin\theta.$$

The correction w_H, as well as the phase ϕ_H, can be considered constant within the broad spectral range, cf. discussion in Sect. 2.4.6.

Positions of Resonances. The positions of the Fabry-Pérot resonances on the wavelength (energy) scale are defined by (4.43) and the resonance condition (4.33). The resonances can be shifted easily. They are most sensitive to the translation of mirrors, as this is expressed formally by the first term in (4.43), the diffractive phase. The positions can also be easily shifted by changing the glancing angle of incidence. Though the resonance positions can be easily changed, the separation between them is rather constant. Indeed, using the resonance condition (4.33) and neglecting for a moment the energy (wavelength) dependence of the phase $\phi_\mathcal{R}$ we find that the separation between two neighboring points of high transparency on the wavelength scale is λ_{fo}.

In analogy with the theory of optical resonators, λ_{fo} can termed the free spectral range (cf. with (4.12) defining the free spectral on the energy scale). It is worth noting that the expression for λ_{fo} (4.44) does not depend on the x-ray wavelength λ. Equally noteworthy is the fact that the expression for the free spectral range (4.44) is identical to the expression for the period of thickness oscillation in the wings of the Bragg reflection curves (2.175) for parallel-sided crystals. The only difference is that the gap width d_g is substituted in (4.44) for the crystal thickness d. The underlying physics is the same: interference of the radiation fields reflected off two surfaces at a distance d_g or d, respectively. The difference is in the number of reflections from the surfaces. In the case of the x-ray Fabry-Pérot resonator the number of reflections is large, which makes the transmission bandwidth narrow.

By replacing the wavelength λ in (4.43) with the photon energy $E = hc/\lambda$, the expression for the Airy phase φ_A reads

$$\varphi_A(E) = \boldsymbol{H}\boldsymbol{u} + 2\pi\frac{E - E_c}{E_{fo}} + \phi_H + \phi_\mathcal{R}(E). \qquad (4.45)$$

Here

$$E_{fo} = \frac{E_H}{N_{go}}\frac{1}{\sin^2\theta} \equiv \frac{hc}{2L_{go}} \simeq \frac{0.6199\ \text{meV mm}}{L_{go}} \qquad (4.46)$$

is the free spectral range on the energy scale. The notation

$$L_{g_o} = d_g \frac{\sin^2 \theta}{|\gamma_H|} \qquad (4.47)$$

is used to denote *half* of the optical path length traveled by x rays between successive reflections from one mirror. At normal incidence and with symmetrically cut crystals ($|\gamma_H| = \sin\theta$) L_{g_o} simply equals the gap width d_g. Unlike (4.44), the expression for the energy dependence of the Airy phase (4.45) is valid only for relatively small energy deviations $|E - E_c|/E_c \ll 1$.

Influence of $\phi_\mathcal{R}$. The weak energy dependence of the phase $\phi_\mathcal{R}$ was so far ignored in the analysis. To understand whether this could be of importance, we consider the simplified case of crystal mirrors with a large thickness $d > d_e(0)$. Here $d_e(0) = \sqrt{\gamma_0 |\gamma_H|}/(K|P\chi_H|)$ is the extinction length at the center $y = 0$ of the region of total reflection – (2.90). In this case $\phi_\mathcal{R} = 2\phi_r$ where ϕ_r is given by (2.93). As pointed out in Sect. 2.4.3 the phase ϕ_r changes from $-\pi$ to 0 within this region. Because of this, an additional transmission resonance should appear within the Bragg peak. As a result, the actual free spectral range should be slightly less than that given by (4.46)

The phase $\phi_r(E)$ can be closely approximated at the center of the Bragg peak, i.e. at $E = E_c$, by a first-order Taylor expansion $\phi_r(E) \simeq \phi_r(E_c) + (E - E_c)\partial\phi_r/\partial E$. By using $\cos\phi_r = y$, $\phi_r(E_c) = -\pi/2$, and the definition of $d_e(0)$, we arrive at the following expression for the derivative: $\partial\phi_r/\partial E \simeq 4\pi d_e(0)\sin^2\theta/|\gamma_H|hc = 2\pi E_{fo}(d_e(0)/d_g)$ at the center of the Bragg peak. Combining all these expressions and substituting them into (4.45) we find for the Airy phase the following approximation

$$\varphi_A(E) = 2\pi \frac{E - E_o}{E_f}, \qquad E_o = E_c - E_f \frac{\mathbf{H}\mathbf{u} + \phi_H - \pi}{2\pi} \qquad (4.48)$$

where

$$E_f = \frac{hc}{2L_g}, \qquad L_g = L_{g_o}\left(1 + \frac{2d_e(0)}{d_g}\right) \qquad (4.49)$$

is the improved approximation of the free spectral range. For x rays with $E \simeq 15$ keV the extinction lengths are typically $10 - 100$ μm. Therefore, the modified free spectral range may differ considerably from (4.46), especially for a small gap width $d_g \approx d_e(0)$. Equation (4.49) is valid only within the central part of the Bragg reflection maximum. At the boundaries $\partial\phi_r/\partial E$ becomes strongly dependent on photo-absorption, crystal thickness, etc., and differs from the expression used. The approximation of a thick transparent mirror is no longer applicable.

It should be noted that the values of $\mathcal{T}, \mathcal{R}, \mathcal{F}$ vary slightly inside the Bragg reflection region which leads, for example, to different height of the peaks. This is best seen from the results of the subsequent numerical calculations.

4. Theory of X-Ray Fabry-Pérot Resonators

Numerical Analysis. The results of the computer simulations of the energy dependence of transmission and reflection at normal incidence to the (0 0 0 30) reflecting planes in α-Al_2O_3 are shown in Fig. 4.6 (left and right panels respectively). The calculations for different values of the gap d_g between the sapphire mirrors are shown. The mirrors have identical thickness $d = \tilde{d} = 50\mu$m. The reflecting atomic planes in both crystals are assumed to be in perfect registry, i.e., $\boldsymbol{Hu} = 2\pi n$, where n is an integer. For example, this is the case where the two thin parallel crystal plates are monolithically cut from a crystal block such that spatial lattice coherence between the two slabs is preserved. It is also assumed $\chi_g = 0$, i.e., vacuum between the crystal plates. Under these conditions only the third and the fourth terms of the Airy phase (4.29) are significant.

The dependence in Fig. 4.6(a) corresponds to the configuration with zero gap ($d_g = 0$), i.e., to a single crystal plate of a thickness $d + \tilde{d}$. The dependence reveals the region of high reflectivity and, respectively, very low transmissivity, which is typical for Bragg diffraction as discussed, e.g., in Sect. 2.4.1. For a zero gap the Bragg curve for backscattering is reproduced. The superimposed oscillation is a result of the finite crystal thickness $d + \tilde{d}$, cf. Sect. 2.4.12.

The transmission spectra with non-zero gaps marked by (b),(c), or (d) in Fig. 4.6 show sharp maxima (resonances) in the region of initially low transmissivity as in (a). The spacing between the resonances, the free spectral range, decreases inversely proportional to the gap width. The reflection spectra show sharp minima at the same positions. The sum of both curves is not one due to photo-absorption. For the resonator with a gap of 27.2 μm there - Fig. 4.6(b) - is only one transmission resonance with an energy width of $\simeq 800$ μeV. The gap is chosen in such a way that the transmission resonance is exactly at the center of the region.

By increasing the gap the number of resonances increases. The resonator gaps in the subsequent examples of Fig. 4.6(c) and Fig. 4.6(d) are chosen as odd integer multiples of the gap used in the example of Fig. 4.6(b): $d_g(c) = 3d_g(b)$; $d_g(d) = 17d_g(b)$. Under these conditions the transmission resonances in the subsequent graphs appear at the same positions and additionally in between.

The calculations demonstrate that in the region of high reflectivity the x-ray Fabry-Pérot resonator behaves similarly to the optical Fabry-Pérot resonator. At well defined x-ray energies a system of two strongly reflecting parallel Bragg mirrors becomes transparent.

Single crystal transmission and reflection curves are shown for comparison with dashed lines in the panels (d). In the very first approximation, the curves of the x-ray Fabry-Pérot resonator look like the reflection and transmission curves of individual crystal mirrors with additional periodic modulation having a period equal to the free spectral range. This resembles the situation for

4.3 Perfect X-Ray Fabry-Pérot Resonator

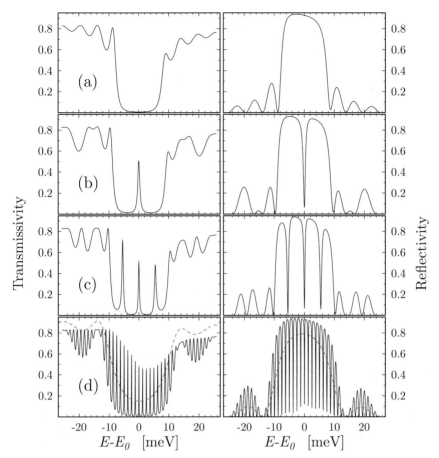

Fig. 4.6. Energy dependence of transmission (left panels) and reflection (right panels) for an x-ray Fabry-Pérot resonator made of two single crystal plates of α-Al_2O_3. $E_0 = 14.3148$ keV. The incident radiation is a plane monochromatic wave at normal incidence ($\Theta = 0$) to the reflecting (0 0 0 30) atomic planes. The reflecting atomic planes in both crystal plates are in perfect registry relative to each other, i.e., $\boldsymbol{Hu} = 2\pi n$ (n is an integer). The thicknesses of the crystal plates are $d = \tilde{d} = 50$ μm, and their temperature $T = 300$ K. The vacuum gap between the plates is taken to be (a) $d_g = 0$, (b) $d_g = 27.2$ μm, (c) $d_g = 81.6$ μm, (d) $d_g = 462.4$ μm. The appropriate individual crystal curves are shown for comparison with dashed lines in panels (d).

the diffraction patterns of a slit and a system of two slits, respectively, see e.g., Born and Wolf (1999) or Hecht and Zajac (1974).

The condition of perfect coherence $\boldsymbol{Hu} = 2\pi n$ between the atomic planes in both crystals was assumed in the simulations shown in Fig. 4.6. Figure 4.7 shows what happens if the perfect coherence is violated. This occurs when

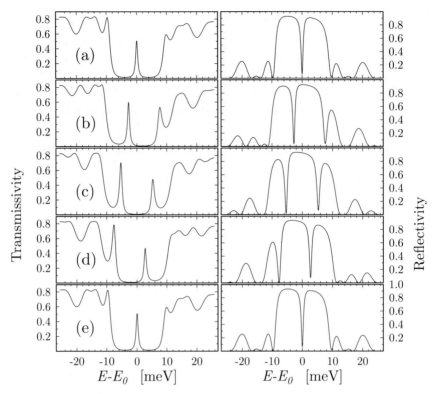

Fig. 4.7. Energy dependence of transmission (left panels) and reflection (right panels) for an x-ray Fabry-Pérot resonator made of two single crystal plates of α-Al$_2$O$_3$. $E_0 = 14.3148$ keV. (a) The same as Fig. 4.6(b). (b) The crystals plates are shifted additionally relative to each other along the scattering vector $\boldsymbol{H}$ by a translation vector $\boldsymbol{u}$ of a magnitude $u = 0.25 d_H$, (c) $u = 0.5 d_H$, (d) $u = 0.75 d_H$, (e) $u = d_H$, where $d_H = 0.433$ Å.

one of the mirrors is translated by a fraction of the interplanar distance. An additional shift of the crystal lattice of the second crystal leads to a shift of the diffractive phase and the transmission pattern on the energy scale. A shift of $u = d_H$ along the diffraction vector $\boldsymbol{H}$ produces no change in the energy spectrum[4].

Resonance Width. The energy width of the resonances decreases with the gap width, as shown in Fig. 4.6. The resonances in Fig. 4.6(c) have a width of $\simeq 400$ μeV. The width of the resonances in Fig. 4.6(d) is $\simeq 100$ μeV. The spectra in Fig. 4.6 show that the x-ray Fabry-Pérot resonator acts as a high-resolution energy filter. The energy resolution can be improved by increasing

---

[4] Except for a small change of the free spectral range.

the gap width, however, this increases the number of transmission resonances per energy interval.

Such behavior is readily explained. The phase width of the Fabry-Pérot resonances $\Delta\varphi_A$ (4.35) is a constant determined only by the finesse $\mathcal{F}$. From this, the analytical expression for the energy width Γ can therefore be obtained by using the linear dependence (4.48) of φ_A on E:

$$\Gamma = E_f/\mathcal{F}. \tag{4.50}$$

From (4.49) and (4.46) the energy width can also be presented in the form

$$\Gamma = \frac{E_H}{\mathcal{F} N_g} \frac{1}{\sin^2\theta} \equiv \frac{hc}{2L_g \mathcal{F}}. \tag{4.51}$$

Here $N_g = N_{go}(1 + 2d_e(0)/d_g)$. The relative spectral width of the Fabry-Pérot resonances is then given by

$$\epsilon_{FP} = \frac{\Gamma}{E} \simeq \frac{1}{N_{go} \mathcal{F}} \frac{|\gamma_H|}{\sin\theta}. \tag{4.52}$$

To obtain this result, we have neglected small extinction and refraction corrections.

The energy width is thus inversely proportional to the gap width, or to the effective number N_{go} of interplanar distances fitting into the resonator gap. This explains the results of the numerical simulations in Fig. 4.6.

Additionally, the energy width is inversely proportional to the finesse $\mathcal{F}$, which in turn is determined by the reflectivity of the mirrors. The finesse increases with the reflectivity. For a fixed Bragg reflection this can be achieved either by increasing the thickness of the crystal mirrors, or by cooling the crystal mirrors. The latter results in increased an Debye-Waller factor and thus larger atomic scattering amplitudes.

Making the Fabry-Pérot resonances sharper by changing the crystal thickness is illustrated by the results of numerical simulations shown in Fig. 4.8. The gap between the mirrors is now fixed at $d_g = 200$ μm. The thickness of the mirrors changes from $d = 50$ μm in panel (a) to $d = 100$ μm in panel (d), respectively. The right panels display parts of the same spectra with enhanced resolution around the resonance at $E - E_0 \simeq -5$ meV. The smallest resonance width is obtained for the thickest crystals. Indeed, for a crystal thickness $d = 50$ μm, the resonance width is $\Delta E = 250$ μeV – panel (a); for $d = 75$ μm, $\Delta E = 115$ μeV – panel (b); for $d = 85$ μm, $\Delta E = 88$ μeV – panel (c); and for $d = 100$ μm, $\Delta E = 72$ μeV – panel (d). However, the energy resolution improves at a cost of decreasing peak transmissivity, because of nonzero photo-absorption in the mirrors. The energy resolution can be reduced further, e.g., to $\Delta E = 48$ μeV at $d = 125$ μm with a maximum transmissivity of 5%.

196   4. Theory of X-Ray Fabry-Pérot Resonators

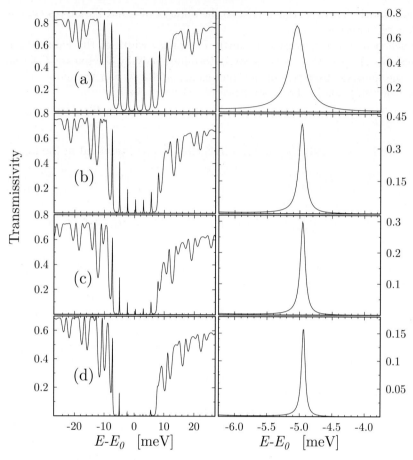

Fig. 4.8. Energy dependence of transmission for an x-ray Fabry-Pérot resonator made of two α-Al_2O_3 single crystal plates of thickness (a) $d = 50$ μm, (b) $d = 75$ μm, (c) $d = 85$ μm, and (d) $d = 100$ μm. $E_0 = 14.3148$ keV. The vacuum gap between the plates is $d_g = 200$ μm. The incident radiation is a plane monochromatic wave at normal incidence ($\Theta = 0$) to the reflecting (0 0 0 30) atomic planes. The reflecting atomic planes in both crystal plates are in perfect registry relative to each other. The right panels display fragments of the same spectra with enhanced resolution.

Notably, different Fabry-Pérot resonances do not behave in the same way. With increasing crystal thickness some resonances disappear. Those resonances which reside in the vicinity of the highest Bragg reflectivity tend to survive. In the example of Fig. 4.8 this is the region around $E - E_0 \simeq -5$ meV. In the dynamical theory of x-ray diffraction it is known that the highest reflectivity within the Bragg peak is at the point where the photo-absorption of the electromagnetic field inside the crystal reaches its minimum. This anomalously low photo-absorption is achieved when the field is in the form of a standing wave with nodes at the atomic planes, see, e.g., (Batterman and Cole 1964, Pinsker 1978).

Even though such behavior is very special, one can also find here similarities between the x-ray and optical resonator: The higher the reflectivity and the lower the absorption, the sharper are the Fabry-Pérot resonances. From this point of view, Be single crystals would be the best material for the x-ray resonator. Unfortunately, the crystal lattice perfection of current Be crystals is not sufficient. The next best choice is diamond. For example, Fig. 5.45 on page. 285 shows the calculated transmission spectrum of a diamond x-ray resonator. Due to low photo-absorption and very high Bragg reflectivity strong and narrow Fabry-Pérot resonances can be expected. Because of the cubic lattice, and therefore problems with multiple-beam diffraction, diamond resonators can be used only under near-normal incidence condition.

Thus, the capabilities of x-ray Fabry-Pérot resonators as high energy resolution filters surpass by far the capabilities of single Bragg reflections. Indeed, as mentioned previously in Sect. 2.4.3, (2.124), the relative energy width of a single Bragg reflection is inversely proportional to the number of reflecting planes N_e throughout the extinction length. The relative energy width of Bragg reflections for x rays of $\simeq 15$ keV is therefore $\simeq 10^{-7}$, see Appendices A.3 and A.4). In contrast, the performance of the x-ray Fabry-Pérot resonator as an interference filter is characterized by two enhancement factors; see (4.52). First, the energy width is inversely proportional to the effective number N_g of interplanar distances fitting into the gap of an x-ray Fabry-Pérot resonator, which can be chosen much larger than the extinction length and thus $N_g \gg N_e$. Second, due to multiple Bragg reflections within the gap, the energy width is further reduced by a factor of the finesse $\mathcal{F} \gg 1$.

One more important feature should be pointed out. The positions of the resonances shift slightly in the direction of increasing energy with increasing thickness of the mirrors; see the right-hand panels in Fig. 4.8. This can be attributed to the fact that the phase jump $\phi_\mathcal{R}$ upon the Bragg reflection, which enters the Airy phase, depends on the thickness of the reflecting mirrors.

Multiple Resonator as Single-Line Interference Filter. By looking closely at Fig. 4.6, the idea suggests itself that several resonators with different but matched gaps placed successively one after another, can be used to obtain a single transmission maximum. The energy width of such a multiple resonator will be as small as the energy width of the resonator with the largest gap – Fig. 4.6(d). The idea of the multiple resonator in visible-light optics was introduced and realized by Houston (1927). Such devices are also called multi-pass resonators (Vaughan 1989).

4.3.6 Time Dependence

Time and energy are complementary. Therefore, the sharp energy resonances of the Fabry-Pérot resonator should be observable in the time domain. This is in complete analogy with Mössbauer (or nuclear resonant) spectroscopy, which was briefly discussed in Sect. 1.2. Figure 1.17 shows an example of time spectra of resonant nuclear forward scattering (NFS). Very sharp nuclear resonances (about neV to μeV broad) are observed either by measuring nuclear resonant energy or time spectra. More details on the modern capabilities of nuclear resonance spectroscopy in the energy and time domains can be found in the collection of review papers (Gerdau and de Waard 1999/2000).

According to the previous considerations, the transmission spectrum of the x-ray resonators can be approximated by a set of Lorentzians of energy width Γ, separated by the free spectral range E_f. By analogy to nuclear resonance spectroscopy in the time domain, one could then expect that the time response of the x-ray Fabry-Pérot resonator to the excitation with a δ-function-like radiation pulse should be a decaying exponential function with superimposed oscillations. The decay constant is evidently $\tau = \hbar/\Gamma$, or by using $\Gamma = E_f/\mathcal{F}$ we find

$$\tau = \frac{\hbar}{E_f} \mathcal{F}. \tag{4.53}$$

The oscillations are a kind of quantum beats, generally known to be due to interference of the path amplitudes of scattering via neighboring resonances. In applications to nuclear resonant scattering they were first discussed by Trammell and Hannon (1978). The quantum beat period is

$$t_f = \frac{2\pi\hbar}{E_f}. \tag{4.54}$$

By using (4.49) we obtain

$$t_f = \frac{2L_{go}}{c}\left(1 + \frac{2d_e(0)}{d_g}\right). \tag{4.55}$$

Thus the quantum beat period t_f is equal to the time lag between successive scattering events. It should be noted here that the time lag is determined not only by the optical path in the gap L_{go} but also includes the part related to the extinction length $d_e(0)$ in both mirrors; see (4.49).

These results can be obtained in a more rigorous fashion. The time response $T_{FP}(t)$ of an x-ray Fabry-Pérot resonator in the forward direction to the excitation with a radiation pulse is given by $T_{FP}(t) = |t_{FP}(t)|^2$. Here $t_{FP}(t)$ is the amplitude of the time response. The time dependence of the amplitude is calculated as a Fourier transform of the product of the energy dependence of the transmission amplitude $t_{FP}(E)$ and the spectral amplitude of the incident radiation pulse $a_{\text{inc}}(E)$:

$$t_{FP}(t) = \int_{-\infty}^{\infty} \frac{dE}{2\pi\hbar}\, e^{-iEt/\hbar}\, t_{FP}(E)\, a_{\text{inc}}(E). \tag{4.56}$$

4.3 Perfect X-Ray Fabry-Pérot Resonator

By putting $t_{\rm FP}(E) = 1$ we find from (4.56) the amplitude of the time dependence $a_{\rm inc}(t)$ of the incident radiation pulse, as a particular case.

The transmission spectrum amplitude of the x-ray Fabry-Pérot resonator, given by (4.28), can be written as a power series of $\mathcal{R}{\rm e}^{{\rm i}\varphi_{\rm A}(E)}$:

$$t_{\rm FP}(E) = \mathcal{T}{\rm e}^{{\rm i}\phi_{\mathcal{T}}} \sum_{n=0}^{\infty} \left(\mathcal{R}{\rm e}^{{\rm i}\varphi_{\rm A}(E)}\right)^n. \tag{4.57}$$

To obtain an analytical expression for $t_{\rm FP}(t)$, we assume for simplicity that $\mathcal{R}$, $\mathcal{T}$, and $\phi_{\mathcal{T}}$ are energy independent in the Bragg reflection region, which can be defined as $E_{\rm c} - \Delta E/2 \leq E \leq E_{\rm c} + \Delta E/2$, with ΔE approximately given by (2.120) and (2.119). Outside the region we assume $\mathcal{R} = 0$. By using (4.48) and (4.54) the Airy phase can be written: $\varphi_{\rm A}(E) = (E - E_0)t_{\rm f}/\hbar$. It is assumed further that very many Fabry-Pérot resonances fit into the width ΔE, i.e., $\Delta E \gg E_{\rm f}$. Furthermore, $a_{\rm inc}(E) \simeq$ constant within the region of total reflection. The latter means that the spectrum of the incident radiation pulse is assumed to be much broader than ΔE. Alternatively, the duration of the incident pulse is much shorter than the duration of Bragg diffraction from an individual crystal mirror, cf. Sect. 2.4.15. As a result, to good accuracy, we find for the response amplitude

$$t_{\rm FP}(t) = \mathcal{T}{\rm e}^{{\rm i}\phi_{\mathcal{T}}} \sum_{n=0}^{\infty} \mathcal{R}^n {\rm e}^{-{\rm i}nE_0 t_{\rm f}/\hbar} \delta(t - n\,t_{\rm f}). \tag{4.58}$$

Thus the time response amplitude is a sequence of very sharp pulses reappearing every time interval $t_{\rm f}$ with the next one attenuated by $\mathcal{R}$. This structure is clearly due to multiple reflections between the mirrors of the resonator.

Since the time response amplitude is only significant at $t = n t_{\rm f}$ we substitute $t/t_{\rm f}$ for n, and arrive at

$$t_{\rm FP}(t) \simeq \mathcal{T}{\rm e}^{{\rm i}\phi_{\mathcal{T}}} \exp\left(-{\rm i}\frac{E_0 t}{\hbar} - \frac{t}{2\tau^*}\right) \sum_{n=0}^{\infty} \delta(t - n\,t_{\rm f}), \tag{4.59}$$

with

$$\tau^* = \frac{t_{\rm f}}{2\ln(1/\mathcal{R})}. \tag{4.60}$$

By using (4.30), we find finally for the time response of the resonator

$$T_{\rm FP}(t) \simeq |\mathcal{T}|^2 \exp(-t/\tau^*) \sum_{n=0}^{\infty} \delta(t - n\,t_{\rm f}). \tag{4.61}$$

The characteristic response time of the x-ray Fabry-Pérot resonator is τ^*. If $\mathcal{R} \simeq 1$ then $\ln(1/\mathcal{R}) \simeq 1 - \mathcal{R} \simeq \pi/\mathcal{F}$. As a result $\tau^* \simeq t_{\rm f}\mathcal{F}/2\pi$, or $\tau^* = \tau$.

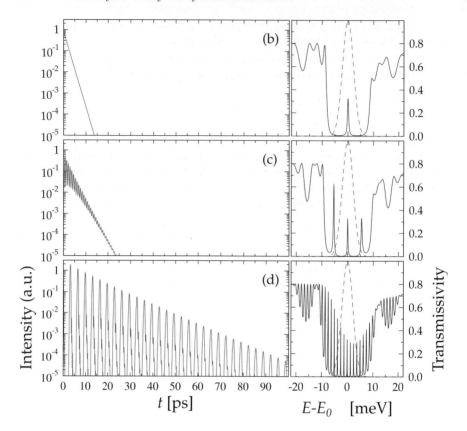

Fig. 4.9. Left: the time dependence of transmission of a short radiation pulse through an x-ray Fabry-Pérot resonator. Right: the energy spectrum of transmission through the resonator (the same as in Fig. 4.6). The dotted lines show the spectrum of the incident radiation pulse, 5 meV spectral width, corresponding to a 0.32 ps pulse duration (of the Gaussian wave-packet). The x-ray Fabry-Pérot resonator is made of two α-Al$_2$O$_3$ single crystal plates: $d = \tilde{d} = 50$ μm at $T = 300$ K, with a gap (b) $d_\mathrm{g} = 27.2$ μm, (c) $d_\mathrm{g} = 81.6$ μm, and (d) $d_\mathrm{g} = 462.4$ μm. The incident radiation is a plane wave at normal incidence ($\Theta = 0$) to the reflecting (0 0 0 30) atomic planes. $E_0 = 14.3148$ keV.

Thus τ^* is equivalent to τ obtained in (4.53) from the energy width of the Fabry-Pérot resonances.

The separation t_f between the successive pulses is equivalent to the quantum beat period, derived in (4.54) from the distance between the Fabry-Pérot resonances on the energy scale. The width of the periodic δ-pulses in (4.59) and (4.61) is zero, because an infinite number of Fabry-Pérot resonances was assumed to fall into the Bragg reflection region ($\Delta E/E_\mathrm{r} \gg 1$). In fact the number of resonances is finite. A better approximation for the terms with

$n > 0$ (instead of the δ-function) would be the $\sin^2(\Delta Et/2)/t^2$ function, with a width of $\hbar/\Delta E$.

By using the above derived formulae, one can estimate that for an x-ray Fabry-Pérot resonator with a gap of 5 cm the time lag between successive multiply scattered waves is $t_f = 333$ ps. Thus, in this case, an extinction length of $\simeq 15$ μm gives only a 0.06% correction to t_f. If the finesse $\mathcal{F} = 15$, then the characteristic response time of the resonator is $\tau^* \simeq 0.8$ ns, which corresponds to an energy width of the Fabry-Pérot resonances $\Gamma \simeq 0.8$ μeV.

We can use (4.56) to perform non-approximated numerical calculations. Figure 4.9 shows an example of such calculations for an x-ray Fabry-Pérot resonator with the same parameters as used for the evaluation of the energy spectra in Fig. 4.6. The time dependence labeled with (b) correspond to a gap $d_g = 27.2$ μm, (c) $d_g = 81.6$ μm, and (d) $d_g = 462.4$ μm, respectively. They represent time responses to the excitation with 0.32 ps radiation pulses (5 meV spectral width of the Gaussian wave-packet). The corresponding energy transmission spectra of the x-ray Fabry-Pérot resonator are shown in the right panels (the same as in Fig. 4.6). The dotted lines show the spectrum of the incident radiation pulse.

As is clearly seen, in all three cases the decay is given by an exponential function. This is attributed to the Lorentzian form of the Fabry-Pérot resonances on the energy scale. The decay time increases with the gap (with decreasing free spectral range E_f), in agreement with (4.53). This is also in agreement with the corresponding behavior of the width of the resonances, cf. Sect. 4.3.5. In the case presented in panel (d), where the incident radiation pulse excites many resonances simultaneously, the decay is modulated by quantum beats. The period of quantum beats is a reciprocal value of the energy difference between the resonances E_f - (4.54), or equivalently it scales with $L_g = L_{g_o}\left(1 + 2d_e(0)/d_g\right)$. With the extinction length $d_e(0) = 15$ μm for the Bragg reflection in question, the quantum beat period is 6.7% longer than the time of flight between the mirrors. This agrees perfectly with the calculated time dependence in Fig. 4.9(d).

4.3.7 Angular Dependence

The preceding discussion was general, in the sense that it was not restricted to any specific glancing angle of incidence θ. By changing the glancing angle of incidence θ the Fabry-Pérot resonances are shifted on the energy scale. The divergent incident beam can thus blur the resonances. In practice it is important to know the admissible angular divergence of the primary beam which does not yet smear out the resonances. In other words, knowledge of the angular widths of the Fabry-Pérot resonances is of practical importance.

For a fixed x-ray energy E (wavelength λ), the center of the angular region of total reflection is given by θ_c (2.114). The Airy phase φ_A (4.41) as a function of θ then reads

$$\varphi_{\rm A} = 2\pi N_{\rm go}\left(\sin\theta - \sin\theta_{\rm c}\right) + \boldsymbol{Hu} + \phi_H + \phi_{\mathcal{R}}. \tag{4.62}$$

The angular width of a Fabry-Pérot resonance can be calculated from the phase width $\Delta\varphi_{\rm A} = 2\pi/\mathcal{F}$ (4.35) of the Fabry-Pérot resonance.

Let us first assume that $\theta_{\rm c}$ is far enough from $\pi/2$ so that the approximation $\sin\theta - \sin\theta_{\rm c} \simeq (\theta - \theta_{\rm c})\cos\theta_{\rm c}$ can be applied. Then the angular width is

$$\Delta\theta = \frac{1}{N_{\rm go}\mathcal{F}\cos\theta_{\rm c}}. \tag{4.63}$$

By using this relation, one can estimate that for $\cos\theta_{\rm c} > 0.1$, i.e., for $\theta_{\rm c} < 85°$, the angular width of the Fabry-Pérot resonances is very small. Indeed, for a Fabry-Pérot resonator with a gap of $d_{\rm g} = 1$ cm, $d_H = 0.5$ Å, and $\mathcal{F} = 30$, the angular width of the Fabry-Pérot resonances is typically $\Delta\theta \approx 10^{-9}/\cos\theta_{\rm c}$. For a gap of $d_{\rm g} = 0.1$ mm, the width $\Delta\theta \approx 10^{-7}/\cos\theta_{\rm c}$ is still very small. In any event it is much smaller than the angular divergences of available or planned x-ray sources. Thus, it is practically impossible to excite selectively separate Fabry-Pérot resonances at glancing angles of incidence far from $\pi/2$ ($\cos\theta_{\rm c} > 0.1$).

From (4.63) it follows that the angular width of the resonances can be enhanced drastically by choosing $\theta_{\rm c}$ close to $\pi/2$. For example, for $\cos\theta_{\rm c} \approx 10^{-4}$ (the angular deviation from normal incidence is $\Theta = \pi/2 - \theta_{\rm c} \approx 1.4 \times 10^{-2}$) the corresponding widths are $\approx 10^3$ larger. It is therefore favourable to operate the x-ray resonator at glancing angles of incidence very close to $\pi/2$.

The effect is most pronounced when $\theta_{\rm c} = \pi/2$. Under this condition the approximation $\sin\theta - \sin\theta_{\rm c} \simeq -\Theta^2/2, (\Theta = \pi/2 - \theta)$ can be applied. As suggested by (4.62) the phase $\varphi_{\rm A}$ is proportional to the square of the angular deviation from normal incidence Θ:

$$\varphi_{\rm A} = -\pi N_{\rm go}\Theta^2 + \boldsymbol{Hu} + \phi_H + \phi_{\mathcal{R}}. \tag{4.64}$$

As a result the angular positions of the resonances are not equidistant and their widths are not equal.

By using (4.35) and the definition of $N_{\rm go}$ we find for the width of the transmission peak at near-normal incidence

$$\Delta\theta = 2\sqrt{\frac{d_H|\gamma_H|}{d_{\rm g}\mathcal{F}}}. \tag{4.65}$$

We note that $\Delta\theta = \Delta\Theta$. By using this equation and (4.52), we can obtain a very simple relation between the angular acceptance and the relative spectral width $\epsilon_{\rm FP}$ of the Fabry-Pérot resonances

$$\Delta\theta = 2\sqrt{\epsilon_{\rm FP}}. \tag{4.66}$$

The relation is valid at near-normal incidence to the reflecting planes. It demonstrates one of the most striking features of these devices. The angular acceptance scales with the *square root* of the relative spectral resolution! Thus, resonances with a relative spectral width as small as $\epsilon_{\rm FP} = 10^{-10}$ could still be selectively excited by an x-ray beam of a 20 µrad divergence. Such beams are routinely available at modern synchrotron radiation facilities.

This square root dependence is due to the backscattering geometry. A similar situation is also observed for single Bragg reflections, see (2.130). The crucial difference is in the enhanced spectral resolution of the resonators owing to multiple-wave interference effects.

Numerical Analysis. Figure 4.10 shows an example of numerical calculations. The angular dependence of transmission (left panels) and reflection (right panels) are calculated with the same parameters for the x-ray Fabry-Pérot resonator used in the calculations of the energy dependence in Fig. 4.6. The difference is that the x-ray energy is now fixed at $E = E_0$, i.e., at the energy at which maximum reflectivity occurs at normal incidence. The curves labeled with (a) correspond to a zero gap ($d_{\rm g} = 0$), i.e., a single crystal plate of a thickness $d + \tilde{d}$. The angular dependence shows a typical broad Bragg back-diffraction region (width $\approx$ 2 mrad) of high reflectivity and very low transmissivity, as discussed, e.g., in Sect. 2.4.1.

The angular dependence of transmission calculated for non-zero gaps are marked by (b),(c), or (d) in Fig. 4.10. They show maxima in the region of initially low transmissivity. The reflection curves show sharp minima at the same positions. The angular widths of the transmission maxima are not constant. The largest width is observed at near-normal incidence. It decreases rapidly with increasing incidence angle Θ.

The angular width of the region of high transmissivity at near-normal incidence decreases with increasing gap width. For a gap of $d_{\rm g} = 27.2$ µm the angular width is 500 µrad - Fig. 4.10(b), while for a gap of $d_{\rm g} = 462.4$ µm the angular width is 160 µrad - Fig. 4.10(d). However, it is still much larger than the typical angular divergence of $\simeq$ 20 µrad for synchrotron radiation at third generation facilities. This fact demonstrates that it is favorable to operate the x-ray resonator as a high energy resolution filter at near-normal incidence, where the high transmissivity is preserved over a large angular range.

The unfavorable influence of multiple-beam diffraction in crystals with the diamond structure (Si, Ge, diamond, etc) on crystal reflectivity at normal incidence was discussed in Sections 2.5, and 2.6. Because of the low reflectivity the performance of x-ray resonators built of such mirrors will be poor. To avoid this problem, Steyerl and Steinhauser (1979) have proposed operating the resonators *off* normal incidence. As we know, from the discussion on page 125, see also Fig. 2.37, the more accompanying reflections can be excited the larger offset angle is required. The studies of Shvyd'ko and Gerdau (1999) and Sutter et al. (2001) (see Figs. 5.16 and 5.3.6) show that in some cases one has to deviate the incident wave as far as 200 µrad off normal incidence to avoid completely the multiple-beam diffraction effects. It is therefore of

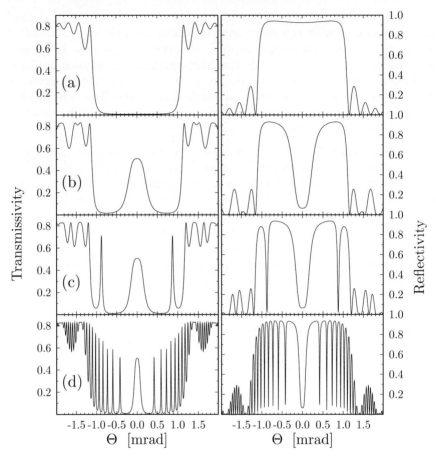

Fig. 4.10. Angular dependence of transmission (left panels) and reflection (right panels) for an x-ray Fabry-Pérot resonator made of two single crystal plates of α-Al$_2$O$_3$. The radiation is a plane monochromatic wave ($E = E_0 = 14.3148$ keV) at near-normal incidence to the reflecting (0 0 0 30) atomic planes. The reflecting (0 0 0 30) atomic planes in both crystal plates are in perfect registry relative to each other, i.e., $\boldsymbol{Hu} = 2\pi n$. The thicknesses of the crystal plates $d = \tilde{d} = 50$ µm, and their temperature $T = 300$ K. The vacuum gap between the plates is taken to be (a) $d_{\rm g} = 0$, (b) $d_{\rm g} = 27.2$ µm, (c) $d_{\rm g} = 81.6$ µm, (d) $d_{\rm g} = 462.4$ µm. Negative Θ signifies the change of the azimuthal angle ϕ to $\phi + \pi$.

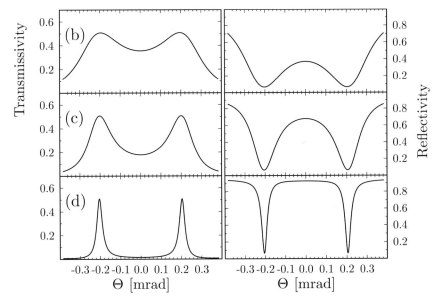

Fig. 4.11. Angular dependence of transmission (left panels) and reflection (right panels) for the x-ray Fabry-Pérot resonator made of two single crystal plates of α-Al$_2$O$_3$. The vacuum gap between the plates is taken to be (b) $d_g = 27.2$ μm, (c) $d_g = 81.6$ μm, (d) $d_g = 462.4$ μm. In contrast to Fig. 4.10, the x-ray energy is changed by $\delta E = 0.3$ meV ($E = E_0 + \delta E$). Negative Θ signifies the change of the azimuthal angle ϕ to $\phi + \pi$.

interest to figure out how broad remains the angular region of high transmissivity of the x-ray Fabry-Pérot resonator for offset angles $\Theta \simeq 200$ μrad. The results of numerical simulations shown in Fig. 4.11 elucidate these matters.

By way of example, the same sapphire resonator is used as for Fig. 4.10. The energy of x rays is now slightly changed by $\delta E = 0.3$ meV to $E = E_0 + \delta E$, so that the maximum transmissivity is observed at $\Theta = \pm 200$ μrad. With this exception the curves shown in Fig. 4.11 were calculated using the same set of parameters as for Fig. 4.10. For convenience, the angular dependence in Fig. 4.11 are shown for a smaller angular range. The results of the calculation are given only for non-zero gaps, which are labeled as before by (b), (c) and (d), respectively.

As in the previous case, the width of the angular range of high transmissivity decreases with increasing gap width. The absolute values of the angular widths are smaller. For example, for a gap of $d_g = 462.4$ μm - panel (d) - the angular width is 30 μrad which is already comparable with a typical angular divergence of x rays available at third generation synchrotron radiation facilities.

The above example shows that in principle it is also possible, provided Θ is not too large, to operate the x-ray Fabry-Pérot resonator off exact normal

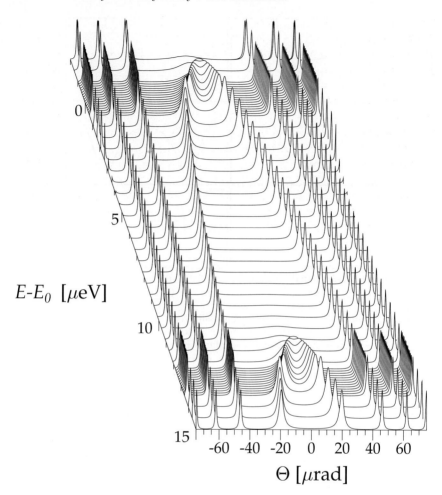

Fig. 4.12. Two-dimensional (E, Θ) plot of transmissivity of x rays through a system of two ideally parallel single crystal plates of α-Al$_2$O$_3$ spaced a distance of 5 cm apart. Crystal plates are 50 μm thick. Crystal surfaces are parallel to the (0001) atomic planes. X rays with an energy E in the vicinity of $E_0 = 14.3148$ keV incident upon the crystal plates almost normal to the atomic planes. The angle of deviation from normal incidence is Θ. The (0 0 0 30) Bragg reflection is used.

incidence without deteriorating its performance too much. In practice a decision, as to whether it is possible or not, depends mostly on the interplay of the angular divergence of the incident beam and the angular width of high transmissivity, which is eventually determined by the required energy width of the Fabry-Pérot resonances, see (4.66).

Two-Dimensional Picture. To gain more insight into the performance of the x-ray Fabry-Pérot resonator a two-dimensional (E, θ) plot of transmissivity is presented in Fig. 4.12. The same model, as for Figs. 4.10 and 4.11, of the sapphire x-ray resonator is used. It is a system of two ideally parallel 50 μm thick single crystal plates in perfect registry. But the gap between the plates is now chosen to be 5 cm.

Because of the rather large gap, the widths of the Fabry-Pérot resonances are significantly smaller than in the previous cases. The energy width of the resonances is about 1 μeV. Despite this fact these very sharp resonances can be selectively excited by a beam of 10 – 20 μrad angular divergence. For this however, the center of the radiation cone should be directed exactly perpendicular to the reflecting planes ($\Theta = 0$). If this condition is not fulfilled, i.e., the deviation of the cone axis from normal incidence is $\Theta \geq 20$ μrad, a broad distribution of the Fabry-Pérot resonances with different resonance energies will be excited. This example clearly demonstrates the advantages of the normal-incidence geometry for selective excitation of the resonances in x-ray Fabry-Pérot resonators.

4.3.8 Temperature Dependence

Another physical parameter, which influences the performance of x-ray Fabry-Pérot resonators is the temperature T. Its takes effect in two different ways. First, it changes the interplanar distance $d_H(T)$ of the reflecting atomic planes in the crystal mirrors and thus shifts the Bragg peak position. Second, it may change the relative position of the mirrors, and thus the gap width d_g. The latter happens because of thermal expansion (contraction) of the spacer, the part designed for holding the crystal mirrors at a fixed distance in space. The linear thermal expansion coefficient of the crystals $\beta_H = d_H^{-1}(\mathrm{d}d_H/\mathrm{d}T)$ in the direction of the diffraction vector $\boldsymbol{H}$, and the linear expansion coefficient of the supporting material $\beta_g = d_g^{-1}(\mathrm{d}d_g/\mathrm{d}T)$ are used in the subsequent considerations.

The change of the interplanar distance $d_H(T)$ with temperature shifts the region of total reflection, i.e. the window of observation. The shift on the energy scale is $\delta E = -E_H \beta_H \delta T / \sin\theta$, as derived from Bragg's law. Let us assume, that the energy E (or the glancing angle of incidence θ) of x rays follows the change of the temperature in such a way, that the "observation" point is always kept at the center of the region of total reflection. In such a "reference system" α_c and ϕ_R do not change with changing temperature. One can then use (4.37) or (4.38) for the subsequent analysis of the variation of the Airy phase with temperature. To simplify the analysis, we assume $\boldsymbol{H} \parallel \hat{\boldsymbol{z}}$, and $b = -1$. As a result we find

$$\frac{\mathrm{d}\varphi_A}{\mathrm{d}T} = 2\pi \frac{d_g}{d_H} \left[(\beta_g - \beta_H) - \beta_g \left(\chi_0' - \chi_g' \right) \right]. \tag{4.67}$$

The first term in the square brackets is due to the variation of the diffractive phase and the second is due to the refractive phase, respectively. Because of the smallness of χ'_0 and χ'_g the second term in (4.67) is, generally speaking, about 10^5 times smaller than the first one. In most cases it can be neglected, except for the case, where the first term is zero.

This happens if the thermal expansion coefficient of the material that holds the crystals and that of the crystals themselves are equal, i.e., $\beta_g = \beta_H$. Under this condition the diffractive phase does not change at all - the first term in the square brackets is zero. This situation is automatically realized if the two crystal mirrors are monolithically cut from a single crystal block such that the spatial lattice coherence between the two slabs is preserved. Independent of the temperature the reflecting atomic planes in both crystals are in perfect registry, i.e., $\boldsymbol{Hu} = 2\pi n$, and the diffractive phase is insensitive to the temperature. In this case only the refractive phase changes, however, only very slowly.

Thus if the thermal expansion coefficients of the material that holds the crystal mirrors and that of the crystals themselves are equal, the Fabry-Pérot resonances within the region of total reflection do not move. A small correction, given by the second term in (4.67) be can neglected. The resonances move together with the region of total reflection as a whole. This property may be very important for the design of tunable high energy resolution filters based on the x-ray Fabry-Pérot resonator.

4.3.9 Finite Number of Multiple Reflections

An infinite train of successively reflected waves was used to calculate the transmissivity and reflectivity of the resonator - (4.19)-(4.21). It is of interest to figure out how many reflections $m_{\min}$ are actually sufficient to obtain a perfect interference pattern.

To simulate a finite number of interfering rays the summation in (4.19) and in (4.20) is truncated at the term $m = m_{\min}$. Figure 4.13 shows the results of calculations of the energy spectra for different values of $m_{\min}$. As can be seen, the interference pattern in Fig. 4.13(b), calculated with 15 reflected rays, is practically indistinguishable from the interference pattern in Fig. 4.13(a), calculated with infinite number of reflected rays[5].

$m_{\min} = 15$ is not a universal number. It is specific for the given resonator. Numerical simulations show that the minimum number of required reflections is larger the higher is the reflectivity (or the thickness) of the mirrors, or equivalently, the higher the finesse of the device.

The effective number of reflected waves required for a practically perfect interference pattern of the x-ray Fabry-Pérot resonator can be ascertained analytically by using the time dependence derived in Sect. 4.3.6. So many reflections will be required, that the remainder do not contribute to the signal by more than $\approx 1\%$. From (4.61) this criterion is fulfilled after a time interval

---

[5] Intensities in the vicinity of the transmission peaks differ by at most 1%.

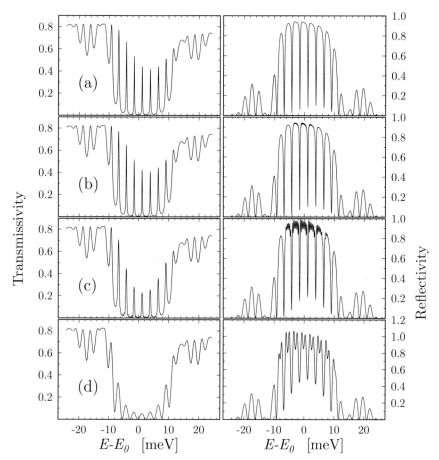

Fig. 4.13. Energy dependence of transmission (left panels) and reflection (right panels) for an x-ray Fabry-Pérot resonator made of two single crystal plates of α-Al$_2$O$_3$. The number of successively reflected rays contributing to the net signal is (a) infinite, (b) artificially limited to $m_{\min} = 15$, (c) 5, (d) 1. The gap between the crystal plates is $d_{\rm g} = 200$ μm. All other parameters are the same as used for calculation of the energy spectra in Fig. 4.7.

of $t^* \simeq 2\pi\tau$. The effective number of reflected waves is then $\mathcal{N}_r = t^*/t_{\rm f}$ or by using $\tau = t_{\rm f}\mathcal{F}/2\pi$ we find

$$\mathcal{N}_r \simeq \mathcal{F}. \tag{4.68}$$

The effective number of scattering events is obviously twice as much: $\mathcal{N}_s \simeq 2\mathcal{F}$. As follows from (4.68), the finesse $\mathcal{F}$ is also a measure of the effective number of reflections in the x-ray Fabry-Pérot resonator.

This analytical result is in agreement with the numerical simulations presented in Fig. 4.13. In this particular case the peak Bragg reflectivity for each mirror is $\simeq 0.8$. The finesse can be estimated as $\mathcal{F} \simeq 14$ and the effective number of rays contributing to the interference pattern is $\mathcal{N}_r \simeq 14$, in the sense of (4.68).

4.3.10 A Practical Expression for the Airy Phase

As was ascertained in the present chapter, the Airy phase changes with x-ray energy, incidence angle, temperature, gap width, etc. It is the variation of the Airy phase, which, after all, causes drastic changes in the transmissivity of an x-ray Fabry-Pérot resonator. As was also realized in the present chapter, x-ray Fabry-Pérot resonators operate favorably at near-normal incidence ($\Theta \ll 1$). Furthermore, the energy of x rays should be tuned to the region of total Bragg reflection, where Fabry-Pérot resonances have the smallest width. It would be practical to obtain an expression for the Airy phase under these conditions.

For this we shall use the general expression (4.41) and assume that at normal incidence to the reflecting atomic planes of the crystal mirrors the center of the region of total reflection is at an x-ray wavelength $\lambda_c = 2d_H/(1+w_H)$ or at an x-ray energy $E_c = hc/\lambda_c$, respectively. If we define relative deviations of x-ray energy from E_c as $\epsilon = (E - E_c)/E_c$, and take advantage of the fact that at small deviations from normal incidence $\sin\theta \simeq -\Theta^2/2$, then we find for the desired expression of the Airy phase:

$$\varphi_A = \boldsymbol{H}\boldsymbol{u} + 2\pi N_{g_0}\left(\epsilon - \frac{\Theta^2}{2} + w_H\right) + \phi_\mathcal{R}. \tag{4.69}$$

4.4 Imperfect X-Ray Fabry-Pérot Resonator

The performance of the perfect x-ray Fabry-Pérot resonator was discussed at length in the previous section. Here we consider the influence of possible imperfections, such as surface roughness of the crystal mirrors, the inhomogeneity of their thicknesses, errors in the parallelism of the atomic planes in the crystal mirrors, and last but not least, the difference in temperatures of the reflecting crystals.

To gain a better insight into the physical nature of the problem, it is advantageous to perform analytical calculations. The particular case of symmetric ($b = -1$) normal incidence ($\theta \simeq \pi/2$) Bragg diffraction is discussed.

4.4.1 Roughness of Internal Surfaces

The roughness of a surface may be characterized by the height h of the deviation of the surface from the mean value. Let us consider first the roughness of the internal surfaces of the x-ray Fabry-Pérot resonator. Due to the roughness of the internal surfaces, x rays hitting different points of the resonator sense different gap widths d_g. As was pointed out in the previous section the Airy phase, and thus the positions and widths of the Fabry-Pérot resonances, are very sensitive to d_g - (4.38). Thus the roughness of the internal surfaces, or more correctly, the inhomogeneous gap width, can blur the resonances. It is clear that the surface roughness does not affect the diffractive phase, as it does not change the coherence of the reflecting atomic planes. The roughness, however, influences the refractive phase.

If one considers two values of the gap width $d_g - h$ and $d_g + h$, respectively, then the phase difference associated with these two values can be estimated with (4.38) as

$$\delta\varphi_A = 2Kh\left(\chi'_0 - \chi'_g\right). \tag{4.70}$$

In order not to blur the resonances, the phase difference $\delta\varphi_A$ has to be less than $\Delta\varphi_A$, the width of the Fabry-Pérot resonances (4.35) on the φ_A scale. From (4.35) $\Delta\varphi_A = 2\pi/\mathcal{F}$. This yields for the admissible roughness (inhomogeneity of the gap width δd_g)

$$2h \equiv \delta d_g \ll \frac{2d_H}{|\chi'_0 - \chi'_g|}\frac{1}{\mathcal{F}}. \tag{4.71}$$

A very interesting consequence of the derived expression is that the performance of the x-ray resonator may be not influenced by surface roughness at all, if the gap is filled with a material of the same electric susceptibility as that of the crystals, i.e., if $\chi'_0 = \chi'_g$. This feature of the x-ray resonator was already pointed out by Steyerl and Steinhauser (1979), who have proposed filling the gap with liquids of sufficiently high electronic density but low x-ray absorption cross section. It should be noted that the low photo-absorption of the liquids is a very important issue. The photo-absorption in the gap, which is formally expressed via the imaginary part of the electric susceptibility χ''_g, decreases $\mathcal{R}$ and thus the finesse $\mathcal{F}$. This can be easily seen from (4.26) and (4.36).

The admissible surface roughness given by (4.71) could be relatively large, even for a vacuum gap $\chi'_g = 0$. For example, for the (0 0 0 30) Bragg reflection in Al_2O_3 $|\chi'_0| = 7.95 \times 10^{-6}$, $2d_H = 0.87$ Å, and by taking the finesse $\mathcal{F} = 30$, one obtains $2h \ll 0.36$ μm.

A roughness of $\simeq 0.1$ μm is routine for modern micro-technology. However, this concerns freely accessible surfaces. If the x-ray Fabry-Pérot resonator is monolithically cut out of a single crystal block, it could be a problem to manufacture internal surfaces with such a small roughness. It should be noted that the surface roughness required by (4.71) has to be preserved over the area of the crystal surfaces illuminated by the incident x rays.

The x-ray Fabry-Pérot resonator is less sensitive to the roughness of the external surfaces. What is actually important, is not the roughness itself, but the inhomogeneous mirror thickness due to the roughness of the internal and external surfaces.

4.4.2 Inhomogeneous Mirror Thickness

Varying mirror thickness results in a variation of the effective reflection coefficient $\mathcal{R}$ (4.26) of the mirrors, and in the variation of the phase jump $\phi_{\mathcal{R}}$ (4.27) upon Bragg reflection from the mirrors. The variation of the phase jump $\phi_{\mathcal{R}}$ changes the Airy phase (4.29) and shifts the Fabry-Pérot resonances; see right panels in Fig. 4.8. This effect clearly imposes some constraints on the admissible inhomogeneity of the mirrors' thicknesses. The limits are determined first of all by the sharpness of the Fabry-Pérot resonances. The requirement on the thickness homogeneity is not very stringent. For example, for the case of the $\Delta E = 72~\mu\mathrm{eV}$ broad resonance in Fig. 4.8(d) an error of $\delta d \simeq \delta \tilde{d} \simeq 10 \mu\mathrm{m}$ in the homogeneity of the mirror's thickness would be admissible.

4.4.3 Non-Parallel Mirrors

When the x-ray resonator is built of two separate single crystal plates, a major problem is to keep the reflecting planes in different mirrors parallel to each other. As will be shown the angle Φ which describes admissible deviations from parallel alignment, is very small despite the fact that Bragg back diffraction of a single crystal plate has a large angular width.

Following Kohn et al. (2000), the error in parallelism of the reflecting planes in the two crystals can be considered as a crystal lattice defect which leads to a displacement $\mathbf{u}(x, z)$ of the atoms from their ideal positions. Here we assume, as before, that the z-axis is normal while the x-axis and y-axis are parallel to the surface of one of the crystal plates. A rotation of the crystal plate around the y-axis at the position (x_0, z_0) by a small angle Φ results in the displacement field

$$\delta u_x = -(z - z_0)\,\Phi, \qquad \delta u_z = (x - x_0)\,\Phi. \tag{4.72}$$

Such a displacement leads to a broken coherence of the atomic planes in both crystals, which results in an additional shift of the diffractive component of the Airy phase:

$$\delta\varphi_{\mathrm{A}}(x) = \mathbf{H}\delta\mathbf{u}(x) = 2\pi s(x), \qquad s(x) = \frac{(x - x_0)}{d_H}\,\Phi. \tag{4.73}$$

The existence of a phase shift varying in space along the surface of the x-ray resonator leads to a variation of the transmissivity. The admissible angular error in parallelism which does not destroy the peaks of transmissivity is determined by the condition $\delta\varphi_{\mathrm{A}} < \Delta\varphi_{\mathrm{A}}$, where

$$\delta\varphi_{\mathrm{A}} = 2\pi \frac{l_x}{d_H}\,\Phi, \qquad \Delta\varphi_{\mathrm{A}} = \frac{2\pi}{\mathcal{F}}. \tag{4.74}$$

Here l_x is the linear dimension of the resonator surface illuminated with the incident x rays. Alternatively, l_x could be also interpreted as the width of the x-ray detector.

The admissible angular deviation from parallel adjustment becomes

$$\Phi < \frac{d_H}{l_x}\frac{1}{\mathcal{F}}. \qquad (4.75)$$

For $\mathcal{F} = 15$, $d_H = 0.43$ Å, and $l_x = 1$ mm one obtains $\Phi < 3 \cdot 10^{-9}$ rad. This is a very stringent condition. Equation (4.75) implies that it is favorable to work with beams of small cross section. For example, by reducing l_x to 10 μm the admissible angle increases to 0.3 μrad.

4.4.4 Temperature Gradients

It may happen that the crystal plates of the x-ray resonator have a slightly different temperature. Let the difference be δT. There are at least two deteriorating effects of such a temperature gradient on the performance of the resonator.

First, the interplanar distance d_H changes with crystal temperature. This shifts the location of the region of total reflection. From (2.108) the shift, in terms of the deviation parameter α, is $\delta\alpha = -4\beta_H(T)\delta T$. In the limiting case, if the temperature difference is too large, the regions in both crystal do not overlap and no multiple reflections and thus no interference effects take place. The width of the region of total reflection in weakly-absorbing and thick crystals is $\Delta\alpha = 4|P\chi_H|/\sqrt{|b|}$ (2.97). By using the condition $\Delta\alpha \gg |\delta\alpha|$ one obtains the following estimate for the tolerable temperature difference between the crystals:

$$\delta T \ll \frac{|\chi_H|}{\beta_H}. \qquad (4.76)$$

Here we used $b = -1$ for the asymmetry factor, and $|P| = 1$ for the polarization factor. For a sapphire x-ray resonator with the (0 0 0 30) Bragg reflection, the Fourier component of the electric susceptibility, and the linear thermal expansion coefficient are $|\chi_H| = 8.63 \times 10^{-7}$, and $\beta_H = 5.92 \times 10^{-6}$ K^{-1} at $T \approx 295$ K, respectively. Thus, we obtain in this particular case that the admissible temperature difference is $\delta T \ll 0.15$ K.

The second effect arises if the diffraction vector $\boldsymbol{H}$ is not parallel to the surface normal $\hat{\boldsymbol{z}}$, i.e., the asymmetry angle η is nonzero, as shown, e.g., in Fig. 2.2. As was discussed in Sect. 2.2 - (2.30) - the virtual reflecting planes are at an angle $\Psi = K|\alpha|\sin\eta/2H|\gamma_H|$ to the relevant atomic planes. If the temperature of the crystals is different, then the parameter α is different as well. As a result, the virtual reflecting planes in the crystals are not parallel. The angle between the planes is

$$\delta\Psi = \sin\eta\frac{K|\delta\alpha|}{2H|\gamma_H|} = \beta_H \tan\eta\, \delta T. \qquad (4.77)$$

4. Theory of X-Ray Fabry-Pérot Resonators

The last expression in (4.77) we have obtained by using the relation $\delta \alpha = -4\beta_H \delta T$ of (2.108), and by $\theta \approx \pi/2$, $K \simeq 2H$, $|\gamma_H| \simeq \cos \eta$.

Now one can use the result of the previous section, in which the admissible error in parallelism of the reflecting planes was estimated. Combining inequality (4.75) with (4.77) and taking $\Phi = \delta \Psi$ one arrives at the following admissible temperature difference in the crystals of the resonator:

$$\delta T < \frac{d_H}{l_x \mathcal{F}} \frac{1}{\beta_H \tan \eta}. \tag{4.78}$$

By using the numerical examples of the present and the previous section one obtains $\delta T < 0.3$ mK/$\tan \eta$. If $\tan \eta \approx 1$ ($\eta \approx 45°$), then one obtains a very demanding requirement: $\delta T < 0.3$ mK. The requirement becomes less stringent for small angles η. For example, if $\eta \leq 0.5°$ ($\tan \eta \leq 0.01$) then $\delta T < 30$ mK. Nonetheless, the admissible temperature difference is small and requires precise temperature control of the resonator crystals.

5. High-Resolution X-Ray Monochromators

5.1 Introduction

The fundamentals of the theory of x-ray diffraction in perfect crystals, which we have been extensively discussed in the previous chapters, will be applied in the remainder of this book to address the problem of designing highly accurate spectrometers for studying dynamics and the structure of condensed matter with atomic resolution.

A typical spectrometer consists of a monochromator selecting x rays with a very narrow spectral spread, and of an analyzer performing energy and momentum transfer analysis of the inelastically scattered x rays to determine the excitation spectra of a sample under study. Examples of spectrometers for inelastic x-ray scattering experiments are shown in Figs. 1.10, 1.13, and 1.15 on pages 13, 15, and 17, respectively. Nuclear resonant scattering experiments require only monochromators in most cases, as shown in Fig. 1.17 on page 20, since the resonant nuclei play the role of the built-in analyzers. Tunable high-resolution x-ray monochromators play the role of wavelength meters in the experiments on the precise measurement of the crystal lattice parameters, as depicted in Fig. 1.22 on page 25.

In this chapter, we shall concentrate on the design of x-ray monochromators. The design of x-ray analyzers will be considered in Chap. 6.

The performance of the monochromators is evaluated in terms of the relative spectral resolution $\Delta E/E$ (or the bandpass ΔE), the angular acceptance $\Delta\theta$, and the peak throughput T.

The physical problem to be studied determines the required energy resolution of the x-ray monochromator. For example, investigations of vibrational excitations in matter, such as phonon dispersion relations $\hbar\omega(\boldsymbol{Q})$, require the energy resolution of the whole instrument (monochromator and analyzer) to be better, often much better, than $\approx 1-5$ meV. Our main focus in this chapter will be on the design of the monochromators with very high energy resolution, i.e., with an energy bandpass $\lesssim 5$ meV.

The angular acceptance of the monochromators $\Delta\theta$ has to match the angular divergence $\varDelta\theta$ of x rays from the source. The modern storage ring based undulator x-ray sources are designed to provide x-ray beams with maximum divergence of a few tens of microradians. To be specific, the performance of all monochromators discussed in this chapter will be evaluated by assuming one and the same x-ray beam divergence of $\Delta\theta = 10$ μrad.

Most of the scientific work with x-ray photons benefits from high photon flux on the sample. Therefore, monochromator peak throughput T, the ratio of the number of photons emerging from the monochromator, to the number of photons incident upon the monochromator, with an energy of photons at the maximum of the monochromator spectral function, is an important characteristic of the instrument. For the peak throughput to be unambiguously defined, the angular spread of the incident x rays has to be always specified.

The quality of the monochromator can be quantified by the introduction of a cumulative characteristic, the figure of merit

$$M = \frac{T}{\Delta E}, \tag{5.1}$$

which is the ratio of the monochromator peak throughput T, calculated for an incident beam with a given angular divergence, to the spectral bandpass ΔE of the monochromator. It is clear, that the greater M is, the better is the monochromator quality.

We shall begin with the classification of the principles underlying the monochromatization of x rays. Then we shall discuss the design of different monochromator types.

5.2 Principles of X-Ray Monochromatization

Several physical principles can be used to select x rays with a small spectral spread. We shall discuss in the following four of them.

1. Bragg reflection off a perfect crystal is on its own a means of selecting x rays with a narrow spectral spread. As we know, the relative spectral width $\Delta E/E$ is to good accuracy a constant for a given Bragg reflection, and is in the range of $10^{-4} - 10^{-5}$ for low-indexed Bragg reflections, and can be as small as $10^{-9} - 10^{-10}$ for high-indexed Bragg reflections, as shown in Fig. 1.4 on page 7 (see also tables of Appendices A.3 and A.4). In agreement with (2.124) and (2.175), these numbers are determined eventually by the *effective* number of atomic planes N_e contributing to Bragg diffraction. The relative spectral width of a Bragg reflection is the reciprocal of N_e: $\Delta E/E \propto N_e^{-1}$. Monochromatization with a single Bragg reflection is addressed in detail in Sect. 5.3.

2. As was discussed in detail in Sect. 2.4.8, the intrinsic spectral and angular widths of Bragg reflections can be changed by applying an asymmetric scattering geometry, i.e., by using diffraction from crystal plates cut at a nonzero asymmetry angle to the reflecting atomic planes. Because of nonzero angular spread of incident x rays from real sources, asymmetric diffraction, however, does not offer on its own a means of selecting x rays with narrower

spectral spreads. A sequence of asymmetric Bragg reflections has to be used, to reduce in the first instance the angular spread, and then to obtain in the second step a smaller spectral spread. The simplest arrangement includes two crystals in the dispersive $(+,+)$ configuration, as in Fig. 3.1 on page 145.

Advances in monochromator design using this principle have been reviewed earlier by Matsushita and Hashizume (1983), and Toellner (2000). In Sect. 5.4 typical schemes of multiple-crystal monochromators will be presented, including those developed most recently.

3. The two aforementioned principles of monochromatization rely on the property of Bragg reflections to have a specific (very often a very small) relative energy width, and on the fact that it can be additionally reduced by using asymmetrically cut crystal optics. The effect of angular dispersion, discussed in Sect. 2.2.5, and Sect. 2.4.9, offers a means of monochromatization, which is independent of the intrinsic properties of Bragg reflections, such as intrinsic spectral width. The principle can be realized in the three-crystal $(+,+,-)$ and $(+,+,+)$ configurations, as discussed in Sects. 3.6 and 3.7, respectively. The collimator, dispersing element, and wavelength selector are the elements of the $(+,+,\pm)$ x-ray monochromator, as shown schematically in Fig. 5.1(b). This principle is illustrated in Fig. 5.1(a) with the help of the classical optical dispersing prism. The crystal acts as a dispersing element, which transforms

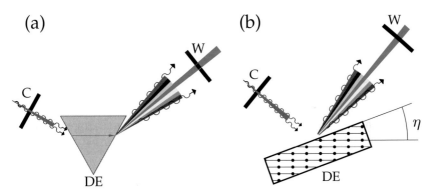

Fig. 5.1. Principle of the $(+,+,\pm)$ monochromator illustrated in (a) with the help of the classical optical prism used as dispersing element (DE). Other elements are a collimator (C) and a wavelength selector (W). Right panel (b): an asymmetrically cut crystal behaves like the optical prism dispersing the photons with different photon energies.

wavelength into angular dispersion. The same effect is brought about equally well by using both low-indexed and high-indexed Bragg reflections. The effect is applicable in any spectral range. In particular, for this reason, monochromators with a small spectral bandwidth $\Delta E/E \simeq 10^{-7} - 10^{-8}$ are feasible

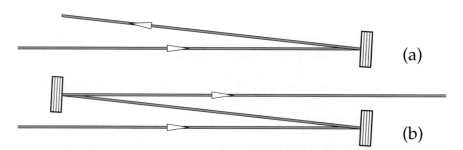

Fig. 5.2. (a) Schematic drawing of the single-bounce monochromator. (b) Two-bounce nondispersive monochromator with two equivalent crystals in the $(+,-)$ configuration. The two-bounce monochromator selects x rays with almost the same spectral spread as the single-bounce monochromator in panel (a), however, preserves the direction of the incident beam (in-line monochromator).

also in application to x rays with photon energies, as low as $2-5$ keV. Examples of the realization of this new concept are discussed in Sect. 5.5.

4. Another principle of monochromatization, which does not rely on the intrinsic spectral width of Bragg reflections, and which uses solely the high reflectivity of Bragg reflections, is multiple-wave interference. It can be implemented using Fabry-Pérot interference filters. X rays with extremely small spectral spread can be filtered in this way. Examples of realization of this concept are discussed in Sect. 5.6.

5.3 Single-Bounce and $(+,-)$-Type Monochromators

Spectral bandpass, angular acceptance, reflectivity, and some other characteristics of single-bounce monochromators, as well as the choice of the monochromator crystals will be discussed in this section.

A schematic drawing of the single-bounce monochromator is shown in Fig. 5.2(a). A particular case is displayed with the glancing angle of incidence to the reflecting atomic planes close to $\pi/2$.

5.3.1 Spectral Bandpass

Intrinsic Spectral Bandpass. For the monochromator design, it is of primary importance that for a given Bragg reflection the relative spectral width

$$\left(\frac{\Delta\lambda}{\lambda_c}\right)_{\mathrm{intrinsic}} = \epsilon_H^{(s)} \tag{5.2}$$

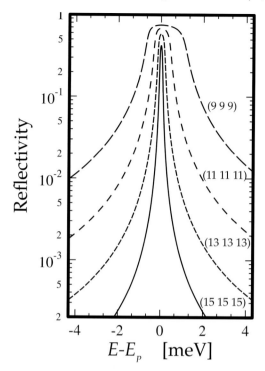

Fig. 5.3. Energy dependence of the reflectivity in Si in backscattering with $(h\,h\,h), h = 2n+1$ type Bragg reflections. Calculations are performed by the dynamical theory of x-ray diffraction for infinitely thick crystals in the two-beam approximation. The energy width decreases with increasing Miller indices $(h\,h\,h)$ of the Bragg reflections: $(9\,9\,9)$: $\Delta E = 1.9$ meV, $(11\,11\,11)$: $\Delta E = 0.82$ meV, $(13\,13\,13)$: $\Delta E = 0.35$ meV, and $(15\,15\,15)$: $\Delta E = 0.15$ meV. The peak reflectivity drops from 0.74 to 0.42, respectively. The photon energy E_P, at which the reflectivity reaches its peak value, is given for each reflection in the table of Appendix A.3.

is with good accuracy a constant, in the symmetric scattering geometry. The quantity $\epsilon_H^{(s)}$ given by (2.119) is independent of the glancing angle of incidence, and of the photon wavelength[1].

As follows from Fig. 1.4 on page 7, and the tables of Appendices A.3 and A.4 providing data on the spectral widths of Bragg reflections in Si and α-Al$_2$O$_3$ crystals, respectively, the relative spectral width $\Delta\lambda/\lambda_c$ (or equivalently $\Delta E/E_c$) is in the range from $\approx 10^{-4}$ to $\approx 10^{-5}$ for low-indexed Bragg reflections. For high-indexed Bragg reflections $\Delta\lambda/\lambda_c$ can be as small as $10^{-9} - 10^{-10}$ [2]. The choice of the Bragg reflection is determined by the required *relative* or *absolute* spectral resolution of the crystal monochromator.

---

[1] It is the momentum transfer, or diffraction vector $\boldsymbol{H}$, which is conserved for a given Bragg reflection. Since with good accuracy, $\epsilon_H^{(s)}$ (2.123) is almost entirely a function of $\boldsymbol{H}$, therefore, the relative spectral width is also conserved. For more details see the discussion in Sect. 2.4.7.

[2] Higher-indexed Bragg reflections are characterized by a larger value of $|\boldsymbol{H}|$. Because of the finite size of atoms, and the finite amplitudes of thermal vibrations, the atomic form factor $f_n^{(0)}(\boldsymbol{H})$ (2.6), and the Debye-Waller factor $g_n(\boldsymbol{H})$ (2.8) become smaller with increasing $|\boldsymbol{H}|$. As a result, the structure factor F_H (2.5) of the crystal unit cell, and the scattering amplitude from one atomic plane become also smaller. This requires more atomic planes N_e for the maximum reflectivity to be achieved. A larger number of reflecting planes always means higher spectral resolution $\Delta E/E \propto 1/N_e$, see (2.124).

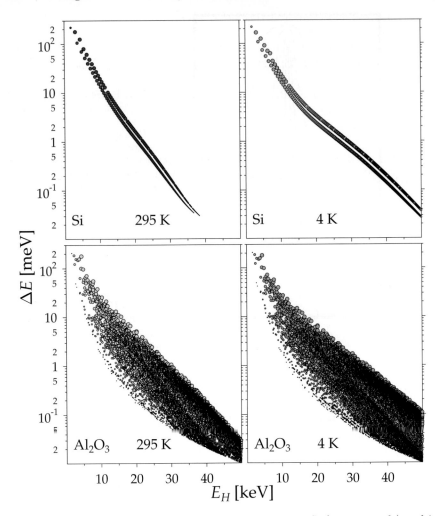

Fig. 5.4. Energy width ΔE of Bragg back-reflections in Si (upper panels) and in α-Al$_2$O$_3$ (lower panels) at $T = 295$ K (left) and at $T = 4$ K (right), respectively. Results for all allowed reflections are shown with Bragg energies E_H in the range from 0 to 50 keV and reflectivity larger than 10%. The diameter of each point is proportional to the peak reflectivity of the corresponding back reflection. The smallest point corresponds to a reflectivity of 10% and the largest to 95%.

The invariance of the relative spectral width for a given Bragg reflection implies, in particular, that the smallest energy width ΔE of the Bragg reflection is achieved at the smallest photon energy, i.e., is achieved in backscattering at the Bragg energy $E_H = H\hbar c/2$ (2.100).

Figure 5.3 shows the energy-dependent reflectivity of selected Si Bragg back-reflections, calculated according to the dynamical theory.

5.3 Single-Bounce and (+,−)-Type Monochromators

Figure 5.4 shows the energy width ΔE of all allowed Bragg back reflections in Si (upper panel) and in α-Al$_2$O$_3$ (lower panel) with Bragg energies E_H in the range from 0 to 50 keV and peak reflectivities exceeding 10%. The results of calculations at two different temperatures, $T = 295$ K (left panels), and at $T = 4$ K (right panels), are presented. Calculations are performed using the dynamical theory of x-ray diffraction for infinitely thick crystals. Atomic and crystal data (coordinates of atoms, thermal parameters, lattice parameters, anomalous scattering factors, etc.) were used, as given in Appendices A.1 and A.2, respectively.

A general property is easy to observe: the energy width decreases almost exponentially with E_H. That is, if a smaller bandpass is required, then a higher indexed Bragg reflection and photons with higher energy has to be used.

The two separate sets of points in the case of Si correspond to Bragg reflections with even $h+k+l = 4n$ (upper set), and odd $h+k+l = 2n+1$ (lower set) Miller indices. The separation is due to different values of the structure factors F_H (2.5) for odd and even reflections.

Figure 5.4 demonstrates an interesting property of α-Al$_2$O$_3$ crystals. For a given Bragg energy E_H, Bragg reflections are available with energy widths both larger and smaller, as compared to those of Si crystals. This property is due to strong structure factor variations resulting from the more complicated structure of α-Al$_2$O$_3$. As a result, it is easier to find a Bragg reflection with a desired spectral width in α-Al$_2$O$_3$.

Geometrical Spectral Broadening. The quantity $\epsilon_H^{(s)}$ is the intrinsic contribution of the Bragg reflection to the spectral resolution of the crystal monochromator. However, it is not the only one, which determines its spectral resolution.

Real sources produce x-ray beams with a finite angular divergence. Due to this, the glancing angle of incidence has a spread, which we shall denote as $\Delta\theta$. Variation of the incidence angle results in Bragg reflection in a shifted spectral range – as can be seen in Fig. 5.5 (see also Figs. 1.3 and 2.12) - and thus causes the broadening of the spectral dependence of the reflection, which eventually degrades the performance of the crystal monochromator, in terms of the spectral bandpass. This kind of broadening, which is often called geometrical broadening, is given by

$$\left(\frac{\Delta\lambda}{\lambda_c}\right)_{\mathrm{geometrical}} = \frac{\Delta\theta}{\tan\theta_c}, \qquad \text{for } 0 < \theta_c < \pi/2,$$

$$\left(\frac{\Delta\lambda}{\lambda_c}\right)_{\mathrm{geometrical}} = \Delta\theta\,(\pi/2 - \theta_c), \qquad \text{for } \theta_c \lesssim \pi/2, \qquad (5.3)$$

$$\left(\frac{\Delta\lambda}{\lambda_c}\right)_{\mathrm{geometrical}} = \frac{(\Delta\theta)^2}{2}, \qquad \text{for } \theta_c = \pi/2,$$

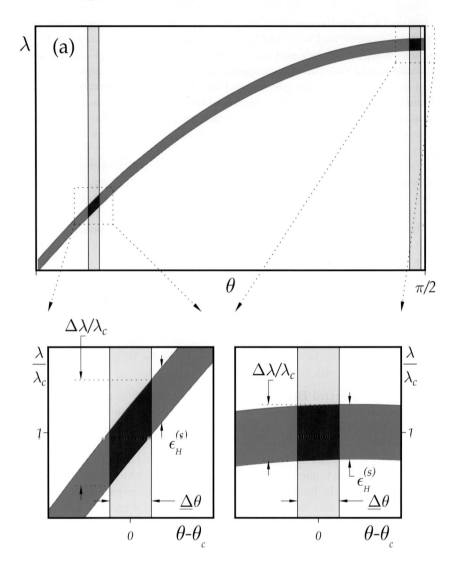

Fig. 5.5. DuMond diagram illustrating the influence of the angular divergence of x-ray beams on the spectral bandpass of the single-bounce monochromator, in different regions of glancing angles of incidence θ. Marked in green (dark grey) is the Bragg reflection region. $\underline{\Delta\theta}$ is the angular spread of x rays in the incident beam. The area in the boxes marked-off by dotted lines are shown below on enlarged scales. See text for other notations and details.

as can be readily derived from Bragg's law. This contribution may be very significant, especially for Bragg reflections with a small intrinsic width $\epsilon_H^{(s)}$.

Fortunately, it decreases with increasing glancing angle of incidence θ_c. As can be seen from (5.3), and Fig. 5.5, the situation changes dramatically in backscattering geometry, i.e., in the vicinity of $\pi/2$, where the geometrical broadening is the smallest. Remarkably, the Bragg back-reflection with a relative spectral width as small as $\epsilon_H^{(s)} = 10^{-9}$ is still not blurred off if the x-ray beam is at exact normal incidence, i.e., $\theta_c = \pi/2$, and it has an angular divergence of $\underline{\Delta\theta} \simeq 20$ μrad, a typical divergence of x-ray beams from modern undulator based sources.

Total Spectral Bandpass. The total spectral bandpass of the single-crystal monochromator can be estimated using Fig. 5.5 as

$$\frac{\Delta\lambda}{\lambda_c} = \left(\frac{\Delta\lambda}{\lambda_c}\right)_{\text{intrinsic}} + \left(\frac{\Delta\lambda}{\lambda_c}\right)_{\text{geometrical}}. \quad (5.4)$$

Finite beam divergence blurs the intrinsic spectral width, unless it is less than the intrinsic spectral width of the Bragg reflection.

A two-crystal monochromator, shown in Fig 5.2(b), with the crystals in the $(+,-)$ nondispersive configuration has the same spectral bandpass as the single crystal monochromator in Fig 5.2(a). Indeed, as we have ascertained earlier – see (3.27) and Fig. 3.6 – the second crystal in the nondispersive configuration does not change the spectrum of x rays. It regains only their direction. This fact is often used to design a monochromator, which preserves the initial beam direction, the so-called in-line monochromator.

To decide whether a single-crystal monochromator (or a double-crystal $(+,-)$ monochromator) is suitable for a high resolution spectrometer, one has to check: (i) is a perfect crystal available with an appropriate Bragg reflection? (ii) is the geometrical broadening sufficiently small? (iii) is the scattering geometry as in Fig 5.2(a) or equivalently Fig 5.2(b) suitable?

In the following, we shall assume that the backscattering geometry is suitable, and the geometrical broadening is sufficiently small, e.g., due to close proximity of glancing angles to $\pi/2$.

We have to address then only the availability of crystals with required Bragg reflections. In particular, the following issues are important to consider: degree of crystal perfection, crystal reflectivity, access to specific photon energies (tunability), and the influence of multiple-beam diffraction effects.

5.3.2 Choice of Crystals

The relative spectral width of a Bragg reflection is the reciprocal of the effective number of the periodically arranged atomic planes contributing to Bragg reflection, see (2.124). Crystal perfection in terms of periodicity of the atomic planes is therefore the first and most important criterion for a crystal to be used as an x-ray monochromator.

Among commercially available choices, silicon single crystals have the highest degree of perfection. Their use in high resolution crystal x-ray optics is well established. The instruments with the highest resolution are achieved by using silicon single crystals. Silicon remains the first choice in the realization of today's x-ray optical instruments.

The possibility of obtaining very high energy resolution (in the meV range) with silicon single crystals in backscattering was first demonstrated in the experiment of Graeff and Materlik (1982). Verbeni et al. (1996) have demonstrated a single-bounce backscattering monochromator with a relative energy resolution of $\Delta E/E = 1.8 \times 10^{-8}$ ($\Delta E = 0.45$ meV, $E = 25.7$ keV). Figure 5.6 shows some recent examples of measuring the energy width of Bragg reflections from silicon crystals in extreme backscattering geometry (Baron et al. 2001b). The experimental results demonstrate that one may obtain essentially theoretical widths for such Bragg reflections. In particular, a relative spectral width $\Delta E/E = 2.3 \times 10^{-8}$ was measured for the (13 13 13) Bragg reflection at $E = 25.7$ keV. It is about twice as much as the theoretical value for one crystal (see the data in Appendix A.3). This is, however, reasonable, because the measurements were performed using two identical crystals in the $(+,-)$ nondispersive configuration configuration, as in Fig. 5.2(b), and by changing the temperature of one of the two crystals. As we know from the discussion in Sect. 2.4.14, achieving the smallest theoretical value requires a perfect crystal with a thickness of $d \simeq 10 d_e(0)$. This corresponds in this particular case to a 6 mm of perfect crystal with $d/d_H \simeq 2.5 \times 10^8$ perfectly arranged atomic planes. The measurement of theoretical widths is a clear demonstration of a very high quality of the available silicon crystals[3].

Despite their very high quality, the application of silicon crystals is favorable not in all cases. For example, due to the highly symmetric crystal structure of Si, access to specific photon energies via Bragg back-reflections is very limited. Also, the reflectivity of silicon crystals at normal incidence is low because of, first, low Debye temperature and thus small Debye-Waller factors, and second, because of multiple-beam diffraction effects. Crystals with i) lower symmetry, ii) high Debye temperature and iii) low photo-absorption, such as hexagonal Be, BeO or rhombohedral sapphire (α-Al_2O_3) crystals, hexagonal and rhombohedral polytypes of silicon carbide (4H-SiC, 6H-SiC, 15R-SiC, etc) may be often a better choice. In the following, along with applications of silicon we shall discuss the applications of sapphire crystals, since very large – up to 50 cm in diameter (Schmid et al. 1994, Khattak and Schmid 2001, Dobrovinskaya et al. 2002) – sapphire crystals are commercially

---

[3] Still, crystal defects in silicon are traceable even in crystals of the best available quality (Tuomi et al. 2001, Tuomi 2002). Because of this, and because of residual carbon and oxygen impurities, the relative variation of the interplanar distances in silicon is about $\Delta d_H/d_H \simeq 10^{-8}$ within the same silicon sample of highest quality (Bergamin et al. 1999, Basile et al. 2000). At present, this number most likely determines a limit of homogeneity (within the same sample), and reproducibility (from sample to sample) of the silicon lattice parameter. As a consequence, this number determines also a limit for the achievable relative spectral bandpass $\Delta\lambda/\lambda = \Delta d_H/d_H$ of silicon based crystal monochromators.

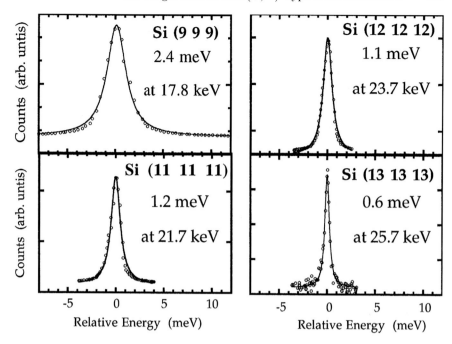

Fig. 5.6. Energy scans of one flat silicon crystal against another in a $(+,-)$ nondispersive configuration showing the energy resolution achieved with $(h\ h\ h)$ type Bragg reflections (Baron et al. 2001b). The results confirm that one may obtain essentially theoretical widths for high-indexed Bragg reflections from silicon crystals.

available, and their crystal quality is the best of the aforementioned materials (see Appendix A.7).

5.3.3 Access to Specific Photon Energies (Tunability)

Two different types of experiments (spectrometers) can be distinguished. In the first case, no preferred value for the photon energy is required. For example, to this class belong experiments on inelastic x-ray scattering, which use non-resonant Thomson scattering from atomic electrons to study vibrational dynamics in condensed matter. In the second case, photons with a specific energy are required. This is the case of experiments using nuclear resonant scattering or resonant inelastic x-ray scattering (RIXS). The energy of the nuclear resonance, or the energy of the $K(L, M, ..)$-absorption edge, define the required photon energy.

In the first case, any reflection with the desired spectral width can be used. In the second case, both the spectral width, and the Bragg energy are important parameters. This imposes constraints for using particular crystals as backscattering monochromators[4].

Si. It is difficult to find back-reflections in silicon crystals with Bragg energies matching some specific values, e.g., energies of nuclear resonances (see the data in Appendix A.5). Silicon crystallizes in a cubic structure. In crystals with a cubic unit cell the interplanar distance is $d_H = a/\sqrt{h^2 + k^2 + l^2}$ (A.1) where a is the lattice parameter, and h, k, l are Miller indices. It is easy to see that reflections with different Miller indices may often have the same value of d_H and accordingly the same Bragg energy $E_H = hc/(2d_H)$. The Bragg energies in cubic crystals are therefore highly degenerate and the number of different Bragg energies is rather low. Typically, one finds only one Bragg energy value in an interval of $\simeq 400$ eV width. This is illustrated in Fig. 5.7(Si), which shows the spectrum of Bragg energies in Si crystals. Matching an x-ray energy to the nearest Bragg energy is equivalent to tuning of the lattice parameter, which for practical reasons means a variation of the temperature.

For example, taking into account the thermal expansion coefficient β_H in Si (see Appendix A.1), the variation of the Bragg energy with temperature, given by (2.109), can be estimated as $dE_H/dT \simeq -0.04$ eV/K for $E_H \simeq 15$ keV. To shift E_H by 100 eV, the temperature would have to be changed by an unrealistic 2600 K. Therefore, Si single crystals allow backscattering only in limited regions of the x-ray spectrum.

α-Al$_2$O$_3$. This problem is less pronounced in the case of crystals with lower crystal symmetry, for which the interplanar distances d_H and hence Bragg energies $E_H = hc/(2d_H)$ are much less degenerate. Sapphire (α-Al$_2$O$_3$) has a rhombohedral crystal lattice, see Appendix A.2. Sapphire single crystals allow exact backscattering with a density of reflections of at least one per $\simeq 15$ eV in the $10-25$ keV range of the photon spectrum, and even more often for harder x rays, see Fig. 5.7(Al$_2$O$_3$). By heating or cooling α-Al$_2$O$_3$ by no more than 100 K from room temperature one can fulfill the backscattering condition for any x-ray energy above 10 keV. Sapphire single crystals allow exact Bragg backscattering for x rays in the 10-50 keV spectral range with high reflectivity, small energy bandwidth and large angular acceptance (Shvyd'ko et al. 1998, Shvyd'ko 2002).

This statement is illustrated in Appendix A.6 using the example of Mössbauer photon energies. Table A.4 demonstrates that one single α-Al$_2$O$_3$ crystal can be used to monochromatize x rays in the energy range from 6 to 70 keV

---

[4] According to Bragg's law in backscattering $E = E_R^{(s)}(1 + \Theta^2/2)$ (2.121)-(2.122), there is only a very weak dependence of the photon energy on the incidence angle Θ. In backscattering geometry $\Theta \ll 1$, and therefore, it is almost impossible to match the desired photon energy by angle variation. Thus, backscattering monochromators can be designed only for photon energies $E_R^{(s)} = E_H(1 + w_H^{(s)})$ close to the crystal's Bragg energies $E_H = hc/(2d_H)$. The latter can be varied only in a small range by changing d_H with crystal temperature.

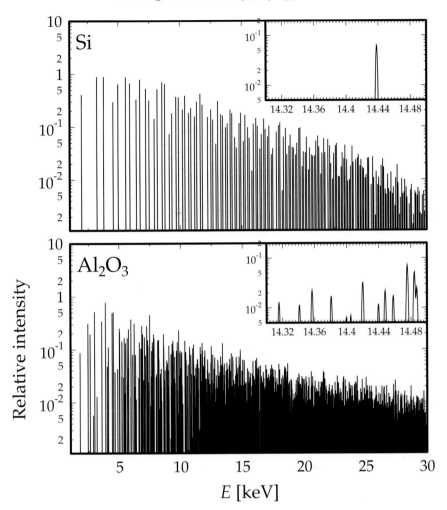

Fig. 5.7. Bragg backscattering spectra of Si and α-Al$_2$O$_3$ single crystals. Calculations are performed using the dynamical theory of diffraction for single crystal plates of 1 mm thickness. The height and width of the peaks result from the convolution of the energy profile of the theoretical reflectivity curve with the energy profile of the incident radiation, which is assumed to be 2 eV broad. Peaks corresponding to n-times degenerate Bragg energies are taken to have n-times greater height. The insets show expanded spectra in the range 14.3-14.5 keV.

with meV and sub-meV bandwidths. Coarse energy tuning can be performed by switching to another Bragg reflection. Fine energy tuning is accomplished by changing the crystal temperature. According to this table, sapphire crystals at T_H=371.6 K will reflect x rays matching the 14.4125 keV Mössbauer transition in ^{57}Fe nuclei using the (1 3 $\bar{4}$ 28) back-reflection with an angular

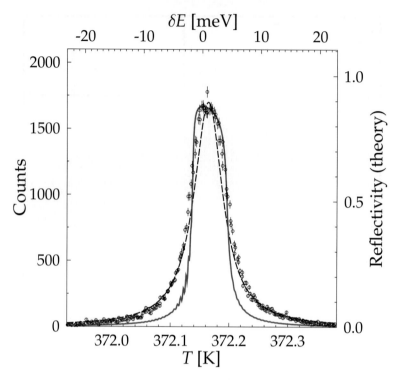

Fig. 5.8. Reflectivity of α-Al$_2$O$_3$ (circles) vs. crystal temperature T at almost exact backscattering ($\Theta = 0.2$ mrad) (Shvyd'ko and Gerdau 1999). The upper scale is the photon energy variation equivalent to the temperature change. The solid line is the calculation using the dynamical theory for an ideal crystal of 1 mm thick α-Al$_2$O$_3$, and an incident monochromatic plane wave. The dashed line is a Lorentzian with a width of 6.2 meV.

acceptance of 1.0 mrad, an energy bandwidth of 5.9 meV, and 87% reflectivity. Figure 5.8 shows the measured reflectivity of the 14.4125 keV Mössbauer photons from the (1 3 $\bar{4}$ 28) planes in α-Al$_2$O$_3$ vs. crystal temperature. Obviously, the expected theoretical width of the reflection was nearly reached, demonstrating that the sapphire crystals are of sufficient quality.

Single-bounce sapphire backscattering monochromators (as in Fig. 5.2a) have been used in the nuclear resonant scattering experiments, exploiting the 14.4 keV nuclear resonance of ^{57}Fe, the 25.65 keV resonance of ^{161}Dy, and some other nuclear resonances (Shvyd'ko et al. 1998, 2001, 2002).

Figure 5.9 shows by way of example the setup with a backscattering α-Al$_2$O$_3$ crystal used as a high-energy-resolution meV-monochromator in experiments on nuclear resonant scattering with 25.65 keV photons from ^{161}Dy nuclei. The Bragg energy of the (3 2 $\bar{5}$ 52) back-reflection in sapphire at $\simeq 374.6$ K fits to the nuclear transition energy of ^{161}Dy, see Table A.4. The

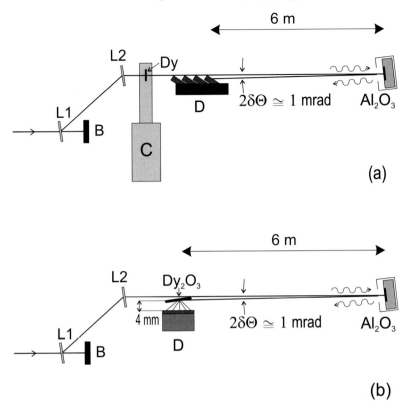

Fig. 5.9. Example experimental setup with a backscattering α-Al$_2$O$_3$ crystal used as a high-energy-resolution meV-monochromator in experiments on nuclear resonant scattering with 25.65 keV photons from ^{161}Dy nuclei (Shvyd'ko et al. 2001): a beam from the undulator (not shown) passes through the high-heat-load monochromator consisting of two diamond crystals C(220) in $(+,-)$ arrangement in Laue scattering geometry (L1 and L2). B - beamstop. Further monochromatization to meV-bandwidth is achieved by Bragg backscattering from a sapphire single crystal (α-Al$_2$O$_3$), the (3 2 $\bar{5}$ 52) reflection. (a) Setup for nuclear forward scattering (NFS), (C) a cryostat with a Dy-metal foil as sample, (D) detector: a stack of four avalanche photo diodes (APD). (b) Setup for nuclear resonant inelastic scattering with a Dy$_2$O$_3$ sample and a single APD as detector (D).

calculated energy bandwidth of the reflection is 0.66 meV. The backscattering energy and the energy of the nuclear transition can be perfectly matched by tuning the crystal temperature. From the temperature dependence of the lattice parameters of sapphire (see Appendix A.2) it follows that the temperature variation of the Bragg energy of the (3 2 $\bar{5}$ 52) back reflection is $dE/dT = -166$ meV/K. For proper performance of the monochromator, mK-temperature-control is required (Lucht 1998).

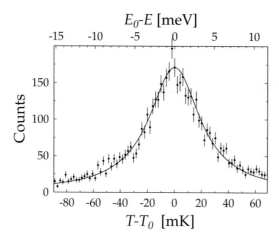

Fig. 5.10. Number of delayed 25.65 keV photons vs. the temperature of the sapphire crystal as measured with the NFS setup - Fig. 5.9(a). $T_0 = 374.624(40)$ K. The energy scale is calculated using the conversion factor 0.17 meV/mK valid for the $(3\ 2\ \bar{5}\ 52)$ reflection in α-Al_2O_3. (Shvyd'ko et al. 2001)

The setup for nuclear forward scattering (NFS) measurements is shown in Fig. 5.9(a). The beam passes through the sample, which is a foil of Dy metal placed in a cryostat.

The setup for nuclear resonant inelastic scattering measurements is shown in Fig. 5.9(b). As distinct from the case of the NFS setup, the sample has to be placed downstream of the sapphire monochromator. A Dy_2O_3 powder sample was used. A detector was placed close to the sample to detect the isotropic nuclear resonant fluorescence.

While measuring NFS time spectra the energy (temperature) of the sapphire monochromator is fixed at the energy of the nuclear resonance. While measuring nuclear resonant inelastic scattering spectra, the photon energy selected by the monochromator is scanned across the nuclear resonance by changing the temperature of the sapphire crystal.

The performance of the sapphire monochromator was tested with the help of the NFS setup. 25.65 keV Mössbauer photons emitted by ^{161}Dy nuclei (in the Dy metal foil) in forward direction with delay after excitation are reflected from the α-Al_2O_3 single crystal, and counted as a function of the crystal temperature (Fig. 5.10). The width of the curve is 46 mK, corresponding to an energy width of 7.8 meV. It is $\simeq$ 10 times larger than the expected theoretical width. The main reason for this is a relatively high dislocation density of $\simeq 3 \times 10^3$ cm^{-2} in the sapphire single crystal available at the time of the experiment.

The lack of sapphire crystals with a sufficient degree of perfection has not yet allowed to achieve experimentally sub-meV resolution with sapphire backscattering monochromators. So far the narrowest spectral bandwidth of a Bragg back-reflection in α-Al_2O_3 was measured to be 2.5 meV using the $(1\ 1\ \bar{2}\ 45)$ reflection at 21.630 keV (Alp et al. 2000a) (see also Appendix A.7, and Fig. A.11(b)). Attempts to improve the sapphire crystal quality are ongoing.

5.3 Single-Bounce and $(+,-)$-Type Monochromators   231

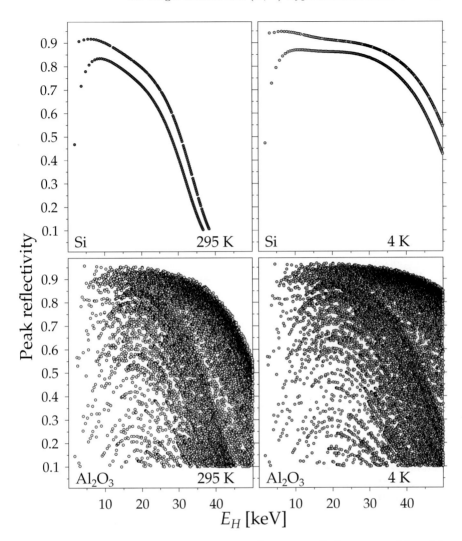

Fig. 5.11. Peak reflectivity of Bragg back-reflections in Si (upper panels) and in α-Al$_2$O$_3$ (lower panels) at $T = 295$ K (left) and at $T = 4$ K (right), respectively. All allowed reflections are shown with the Bragg energy E_H in the range from 0 to 50 keV and reflectivity exceeding 10%.

5.3.4 Reflectivity

A very important characteristic is the crystal reflectivity. Peak reflectivity values for all allowed Bragg back-reflections in Si (exceeding the 10% level), and with Bragg energies E_H in the range from 0 to 50 keV are shown in Fig. 5.11(Si). Calculations were performed using the dynamical theory with the crystal parameters as in Appendix A.1. The crystal was assumed to be infinitely thick. The upper left panel shows the results of calculations at a crystal temperature of $T = 295$ K. It is clearly seen that a reflectivity of more than 90% is achievable at room temperature for x rays only in the range from 3 to 8 keV. The crystal reflectivity drops very quickly with photon energy.

The low reflectivity at high energies, i.e. for reflections from atomic planes with small interplanar distance d_H is a consequence of the relatively large amplitudes of thermal vibrations in Si (large B-factor (2.8)), and therefore small Debye-Waller factors and scattering amplitudes, leading to the requirement of such a large number of reflecting atomic planes that photo-absorption becomes an important competing process drastically reducing the reflectivity. Silicon crystals are too "soft" to reflect hard x rays effectively (Debye temperature $\Theta_D \simeq 530$ K, see Appendix A.1). One could partly improve the situation by cooling the crystal. This is demonstrated by the results of calculations at crystal temperature $T = 4$ K in the upper right panel of Fig. 5.11(Si).

In contrast to silicon, sapphire is a "hard" material ($\Theta_D \simeq 900 - 1000$ K, see Appendix A.2). We may expect to observe higher values for the Bragg reflectivity. Calculated peak reflectivity (exceeding the 10% level) in α-Al$_2$O$_3$ for all allowed Bragg back-reflections with Bragg energies E_H in the range from 0 to 50 keV are shown in Fig. 5.11(Al$_2$O$_3$). The left panel shows the results of calculations at a crystal temperature of $T = 295$ K.

As distinct from the case of Si, reflections with a peak reflectivity of more than 90% are predicted in a broad spectral range from 5 to 35 keV. For Si this is not possible even at $T = 4$ K. The much higher peak reflectivity is first of all due to the smaller amplitudes of thermal vibrations in α-Al$_2$O$_3$. At 4 K the high reflectivity range extends even further, up to 45 keV, as can be seen in the right panel of Fig. 5.11(Al$_2$O$_3$).

An instructive example of Bragg reflectivity calculations in α-Al$_2$O$_3$ for high-energy photons is shown in Fig. 5.12. For practical reasons we have chosen the photon energy $E = 67.413$ keV, which coincides with the energy of the nuclear resonance in ^{67}Ni. Nuclear resonance spectroscopy with ^{67}Ni is attractive for studies of vibrational dynamics of the Ni-containing proteins, for studies of the magnetism of Ni-containing materials, etc. The (14 12 $\overline{26}$ 122) Bragg back-reflection in sapphire matches the nuclear resonance energy[5] (see Table A.4). The energy dependence of the reflectivity has been calculated for sapphire crystals of different thicknesses. As the calculations show, a perfect sapphire crystal of an "intermediate" thickness of 10 mm

---

[5] At this point it should be noted that presently the energy of the nuclear resonance in ^{67}Ni is determined with insufficient accuracy to identify exactly the suitable Bragg reflections in α-Al$_2$O$_3$. For example, the most recent data of the Nuclear Data Centrum at Brookhaven National Laboratory www.nndc.bnl.gov/nndc/nudat/levform.html provide the value $E = 67.413(3)$ keV. The experiment, in which recently the nuclear resonance in ^{67}Ni was for the first time excited with synchrotron radiation, has given the value $E = 67.419(8)$ keV (Wille et al. 2003).

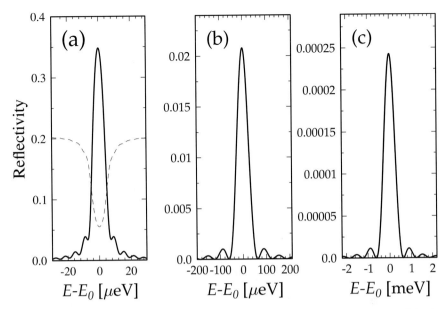

Fig. 5.12. Energy dependence of Bragg reflectivity from α-Al$_2$O$_3$ crystals of thickness $d = 100$ mm (a), $d = 10$ mm(b), and $d = 1$ mm (c). The dashed line in (a) shows the energy dependence of transmissivity. X rays are at normal incidence to the (14 12 $\bar{2}$6 122) atomic planes. Sapphire is at a temperature of $T = 50$ K. $E_0 \simeq 67.414$ keV. Calculations were performed in the two-beam approximation of the dynamical theory.

may serve as a monochromator selecting 67.413 keV photons with a spectral spread of 80 μeV – Fig. 5.12(b). The reflectivity of such a monochromator is about 2 %. Since at this photon energy the absorption length in sapphire is $d_{\rm ph} \simeq 63$ mm, one can use thicker crystals to achieve higher reflectivity. One can achieve up to 40 % reflectivity with 100 mm thick crystals. The theoretical spectral width is only a few micro-eV – Fig. 5.12(a).

5.3.5 Figure of Merit

In Sect. 5.1, we have introduced a cumulative characteristic, the figure of merit $M = T/\Delta E$ (5.1), to quantify the quality of monochromators.

Figure 5.13 shows the calculated values of the figure of merit for single-bounce back-reflection monochromators. The figure of merit for the monochromators using Bragg reflection from Si crystals are shown in the upper panels of Fig. 5.13, and those for the monochromators using Bragg reflection from α-Al$_2$O$_3$ crystals are shown in the lower panels. The left-hand panels show the results of calculations at a crystal temperature of $T = 295$ K, and the right-hand - at $T = 4$ K (right), respectively. The graphs demonstrate the

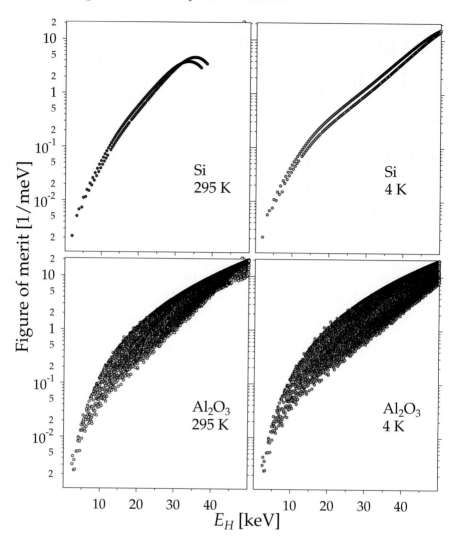

Fig. 5.13. Figure of merit $M = T/\Delta E$ of the single-bounce back-reflection monochromators, using Bragg reflection from Si crystals (upper panels) and from α-Al$_2$O$_3$ crystals (lower panels) at a crystal temperature of $T = 295$ K (left) and at $T = 4$ K (right), respectively. All allowed reflections are shown with the Bragg energy E_H in the range from 0 to 50 keV and reflectivity exceeding 10%.

very fast growing quality of the single-bounce backscattering monochromators, measured in terms of M, with an increasing of the Bragg energy E_H of the Bragg reflections.

5.3.6 Reflectivity in Exact Backscattering. Effects of Multiple-Beam Diffraction

Crystals with Diamond Structure. As was originally pointed out by Steyerl and Steinhauser (1979), and discussed in detail in Sect. 2.5, Sect. 2.6, and Sect. 2.7, Bragg back-reflections for crystals with diamond structure, like Si, Ge, and GaAs, are always accompanied by other simultaneous reflections. The (111) and (220) reflections are the only exceptions to the rule. The last column of the table in Appendix A.3 shows the number of accompanying reflections N_a, which are excited in silicon crystals at normal incidence to the (hkl) atomic planes.

For example, along with the (12 4 0) back-reflection another 22 reflections arise simultaneously in Si. It is a 24-beam diffraction case. Approaching normal incidence to the (12 4 0) planes, new reflection channels for x rays open up. One expects a reduced reflectivity in the backscattering channel and a complicated energy dependence. Detailed experimental and theoretical studies of the (12 4 0) backscattering case in Si, including detailed analysis of the accompanying reflections were presented by Sutter et al. (2001).

Table 5.1. Diffraction vectors $\boldsymbol{B}_n$ and $\boldsymbol{L}_n$ of the conjugate pairs of accompanying reflections arising simultaneously with the back-reflection with diffraction vector $\boldsymbol{N} = (12\ 4\ 0)$: $\boldsymbol{N} = \boldsymbol{B}_n + \boldsymbol{L}_n$. Assuming $\boldsymbol{N}$ to be normal to the crystal surface, $\boldsymbol{B}_n$ denote the diffraction vectors of the Bragg-case reflections, and $\boldsymbol{L}_n$ - of the Laue-case reflections. P_n labels the basal plane of the reflection pair $\boldsymbol{B}_n$, $\boldsymbol{L}_n$. The asterisk labels those reflections, which generate waves propagating parallel to the surface.

n	$\boldsymbol{B}_n$ (h,k,l)	$\boldsymbol{L}_n$ (h,k,l)	P_n
1	(8 $\bar{4}$ 0)*	(4 8 0)*	P_1
2	(8 8 0)	(4 $\bar{4}$ 0)	P_2
3	(12 0 0)	(0 4 0)	P_3
4	(6 8 $\bar{2}$)	(6 $\bar{4}$ 2)	P_4
5	(12 2 $\bar{2}$)	(0 2 2)	P_5
6	(6 4 6)	(6 0 $\bar{6}$)	P_6
7	(8 2 $\bar{6}$)	(4 2 6)	P_7
8	(8 2 6)	(4 2 $\bar{6}$)	P_8
9	(6 4 $\bar{6}$)	(6 0 6)	P_9
10	(12 2 2)	(0 2 $\bar{2}$)	P_{10}
11	(6 8 2)	(6 $\bar{4}$ $\bar{2}$)	P_{11}

The diffraction vectors of the accompanying reflections arising with the back-reflection with diffraction vector $\boldsymbol{N} = (12\ 4\ 0)$ are listed in Table 5.1. Figure 5.14 shows a fragment of the reciprocal space in the immediate vicinity

236   5. High-Resolution X-Ray Monochromators

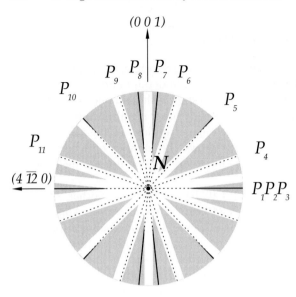

Fig. 5.14. The basal planes P_n (Table 5.1), seen as black solid and dotted lines, are projected onto the plane perpendicular to diffraction vector N of the back-reflection. The area where multiple beam diffraction takes place is shown as white stripes perpendicular to P_n, cf. Fig. 2.37.

of diffraction vector N projected onto the plane perpendicular to N. Projections of the basal planes P_n of the reflection pairs B_n, L_n (Table 5.1) are seen as black solid and dotted lines. All the basal planes intersect at different angles in a line parallel to N.

Based on the kinematic approximation, discussed in Sect. 2.5.6, and graphically illustrated by Fig. 2.37, the area where multiple beam diffraction takes place is mapped *qualitatively* by the white stripes radiating from the plot center and perpendicular to P_n. The stripe widths scale with the angular widths of the relevant Bragg reflections calculated in the two-beam approximation.

In the simplest case of only one pair of accompanying reflections arising with a back-reflection, the four-beam diffraction case, the angular width of the region of multiple-beam excitation and thus of the low reflectivity into backscattering channel was calculated in Sect. 2.6 to be a few μrad-wide, provided the direction of incidence was varied *in the basal plane* of the conjugate pair of the accompanying reflections, and about mrad-wide - in the direction *perpendicular to the basal plane*, see Figs. 2.43 and 2.44. Therefore, there is always one favorable direction (in the basal plane) in which one can escape the region of multiple-beam excitation in the shortest possible distance, and one unfavorable (perpendicular to the basal plane) direction of the angular variation for a single pair of accompanying reflections. In the 24-beam diffraction case presented graphically in Fig. 5.14, there are 11 conjugate pairs of accompanying reflections, and respectively 11 basal planes. Therefore, it is almost impossible to find one preferable direction in which the angular variation will be favorable. As a consequence, the angular width of the region of

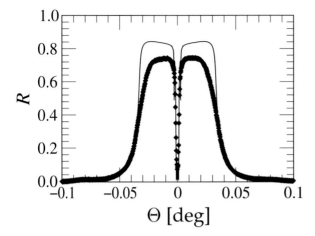

Fig. 5.15. Angular dependence of the reflectivity at near-normal incidence to the (12 4 0) atomic planes in Si of the 14.438 keV photons with 2 meV spectral bandwidth (Sutter et al. 2001). The black points show the experimental data, the solid curve shows simulations with the program NBMSIM authored by Colella (1974). Θ is the angular deviation from exact backscattering about the (4 $\bar{1}2$ 0) reciprocal lattice vector.

multiple-beam excitation and thus of the low reflectivity into backscattering channel is much larger that the angular width of the two-beam diffraction.

Figure. 5.3.6 shows the reflectivity of a Si single crystal into the backscattering channel for incoming x rays as a function of the angular deviation from normal incidence to the (12 4 0) atomic planes (Sutter et al. 2001). The dependence displays a mrad-wide region of high reflectivity which, however, collapses while approaching normal incidence at $\Theta = 0$. The angular range of the low reflectivity is as expected very broad, about 0.2 mrad broad. The reflectivity at normal incidence is nearly zero.

In addition, while approaching normal incidence the energy dependence of the reflectivity experiences dramatic changes, as demonstrated in Fig. 5.16. At $\Theta = 0.3$ mrad off normal incidence the energy dependence shows a few meV-wide narrow single peak, Fig. 5.16(D). At normal incidence the spectrum changes drastically and displays a triple peak structure, Fig. 5.16(A), which is due to multiple-beam diffraction[6].

These results definitively demonstrate that Si single crystals are very unfavorable for use as *exact backscattering* x-ray mirrors, because of their low reflectivity into the backscattering channel.

---

[6] The angular dependence shown in Fig. 5.3.6 was measured with x rays of 2 meV spectral bandwidth (Sutter et al. 2001). The spectra shown in Fig. 5.16 were measured with x rays having a larger energy spread of 7 meV. It is by this reason the reflectivity at normal incidence in Fig. 5.16(E) is not zero.

238 5. High-Resolution X-Ray Monochromators

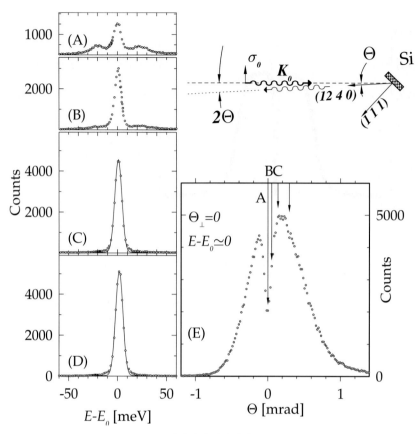

Fig. 5.16. Energy dependence (A),(B),(C),(D) and angular dependence (E) of the reflectivity of a Si single crystal for incoming x rays close to normal incidence to the (12 4 0) atomic planes (Shvyd'ko and Gerdau 1999). Scheme of the experimental setup for their observation is shown at the top right of the drawing. The angular divergence of the x rays is $\simeq 20 \times 20$ μrad^2. Θ and $\Theta_\perp$ are two complementary angular deviations from normal incidence to (12 4 0) in the (1 $\bar{3}$ 2) plane and in the plane perpendicular to it, respectively. Here is $\Theta_\perp = 0$. The x rays are monochromatized to an energy band of 7 meV width. A,B,C and D in (E) mark the Θ angular positions at which the energy spectra (A),(B),(C) and (D) are measured. $E_0 = 14.438$ keV.

Sapphire. We know from the discussion in Sect. 2.5 that, in contrast to crystals with a cubic lattice, in crystals with hexagonal or rhombohedral lattice structures many *multiple-beam free* Bragg back-reflections ($hkil$) can be found. Such cases in α-Al$_2$O$_3$ are listed in Appendix A.4.

In the following examples the energy dependence of the reflectivity under exact Bragg backscattering conditions was studied in α-Al$_2$O$_3$ crystals of different thickness. The (0 0 0 30) back reflection was used. Its Bragg energy is $E_H = 14.315$ keV at $T = 295$ K. The expected peak reflectivity for a thick crystal is 95%, the energy width $\Delta E = 13.1$ meV, and the extinction length $d_e(0) = 16$ μm. The dynamical theory calculations for this Bragg reflection in thin and thick crystals are shown in Figs. 2.27 and 2.26, respectively.

To measure the energy dependence, a monochromator was employed, which provided photons with meV spectral spread. Sapphire crystals with low dislocation density were used, see Appendix. A.7. Figures 5.17 and 5.18 show the energy dependence of the reflectivity under exact Bragg backscattering conditions. The thicknesses of the samples used to measure the reflectivity curves in Figs. 5.17(a), 5.18, 5.17(b), and Fig. 5.17(c) were 0.039(1), 0.063(1), 0.084(1) mm, and 2.5 mm, respectively. The first three values of the thickness were obtained by theoretical fits of the measured reflectivity curves. The measured peak reflectivities are 0.68, 0.87, 0.92, and 0.95, respectively. In agreement with the dynamical theory, the experimental results show that the thicker the crystal the higher the peak reflectivity R and the smaller is the observed energy width.

Another feature observed in Figs. 5.17(a), 5.17(b), and 5.18 are the intensity oscillations in the tails of the reflection curves. The thicker the sample, the shorter is the period of these oscillations. The oscillations in the 2.5 mm thick crystal are too fast to be resolved. All these observations are in a good agreement with the dynamical theory calculations for perfect crystals.

As was mentioned in Sect. 2.4.12, the intensity oscillations on the tails of the Bragg reflection curves are due to the interference of the radiation fields reflected from the front and rear surfaces of a parallel sided crystal, the thickness oscillations[7]. The thickness of the samples can be directly determined with (2.176) from the period of oscillations δE as $d = hc/2\delta E(1+w_H)$. Alternatively, an absolute energy calibration of the monochromator can be performed if the crystal thickness is known[8].

---

[7] Thickness oscillations have been observed earlier for the angular dependence of the Bragg reflection curves by Batterman and Hilderbrandt (1967) and Batterman and Hilderbrandt (1968). They have been observed as a function of the crystal thickness by Hashizume et al. (1970). These oscillations should not be confused with those originating from the Pendellösung effect, which takes place in the Laue scattering geometry, see, e.g., (Pinsker 1978). The oscillating angular dependence of Laue-case rocking curves has been used by Teworte and Bonse (1983) and Deutsch and Hart (1985) for very precise determination of the structure factors F_H in silicon.

[8] The thickness oscillations are observed both on the energy (wavelength) and the angular scale. In the sense of (2.168)-(2.170), they are described universally via the $\tilde{\alpha}$ parameter which contains both the energy (wavelength) and angular variables. With todays technology, angular measurements are extremely precise. The aforementioned universality could be therefore employed to carry out very precise wavelength calibrations of monochromators or wavelength-meters. To do so, the Bragg reflectivity curve should be measured twice. First - as a function of incidence angle Θ, and second - as a function of x-ray wavelength λ. These curves could be translated one into the other using (2.169)-(2.170) and thus the relative wavelength (energy) scale calibration could be performed. If the Bragg

240   5. High-Resolution X-Ray Monochromators

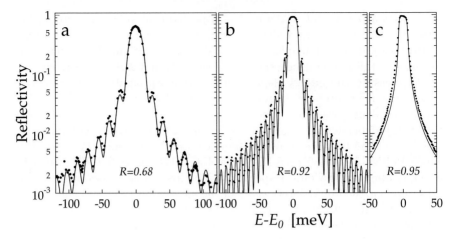

Fig. 5.17. Bragg reflectivity vs x-ray energy E of parallel-sided α-Al$_2$O$_3$ crystals of different thicknesses (Shvyd'ko 2002): (a) $d = 0.039$ mm , (b) $d = 0.084$ mm, and, (c) $d = 2.5$ mm. X rays are at normal incidence to the (0 0 0 30) atomic planes, $E_0 = 14.315$ keV. Solid lines are dynamical theory calculations for the incident plane wave packet with an energy spread of (a) 6 meV and (b-c) 2 meV.

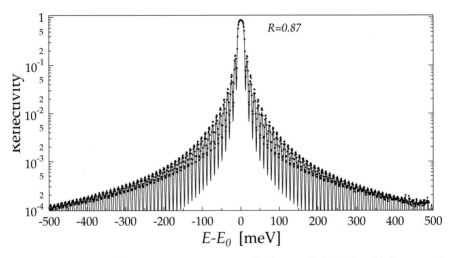

Fig. 5.18. Bragg reflectivity vs x-ray energy E of a parallel-sided α-Al$_2$O$_3$ crystal of thicknesses $d = 0.063$ mm (Shvyd'ko 2002). X rays are at normal incidence to the (0 0 0 30) atomic planes, $E_0 = 14.315$ keV. The solid line is a dynamical theory calculation for the incident plane wave packet with an energy bandwidth of 2 meV.

---

energy is known the relative calibration could be converted to an absolute one. The backscattering geometry is most favorable for such measurements, due to the Θ^2 stretching of the angular scale.

These results demonstrate that sapphire single crystals can be effectively used as *exact backscattering* x-ray mirrors of high reflectivity.

5.4 Multi-Bounce (+, +)-Type Monochromators

5.4.1 Introduction

The contribution to the total spectral width of the crystal monochromator resulting from the angular spread $\Delta\theta$ of the propagation directions of the incident waves, the geometrical broadening, can be reduced not only using extreme backscattering geometry, as was argued in the previous section, but also by reducing the angular spread $\Delta\theta$ itself, see (5.3) and (5.4). For this, additional collimating optics have to be used.

The most popular solution is to use an additional collimator crystal installed ahead of the monochromator crystal in the (+, +) dispersive configuration, as shown in Fig. 3.1, or Fig. 5.19(a). The principle of achieving a small relative spectral bandpass $\Delta\lambda/\lambda$ in successive reflections from two crystals in the (+, +) configuration is graphically explained by the DuMond diagram in Figs. 3.3 and 3.4. The first crystal limits the angular spread[9] of the propagation directions, and thus limits the spectral spread of the waves reflected from the second crystal, used as the monochromator. The theory presented in Sect. 3.3, based on DuMond diagram analysis, can be used to describe the performance of the (+, +)-type monochromators. A formal expression for the relative spectral bandpass $\Delta\lambda/\lambda$ of the (+, +) monochromator in the approximation of non-absorbing crystals is given by (3.13). Equation (3.19) is a particular case of (3.13) valid for two equivalent reflections[10], albeit with different asymmetry factors $b_1 = 1/b_2$. The angular acceptance of the two successive reflections is given by (3.17), or by (3.20), for the pair of equivalent reflections having asymmetry factors $b_1 = 1/b_2$.

According to (3.13), and (3.19), a small relative spectral spread $\Delta\lambda/\lambda_c$ is achieved in the first place by using Bragg reflections with small relative spectral widths given by $\epsilon_1^{(s)}$, and $\epsilon_2^{(s)}$.

A distinctive feature of the (+, +)-type monochromators consists in the possibility of varying the spectral bandpass and angular acceptance of the monochromator, although over a restricted range, by choosing properly the

---

[9] Hereafter the angular spread in the dispersion plane is meant. Small deviations of the propagation direction of the incident wave from the dispersion plane, i.e., small deviations of the azimuthal angle ϕ from 0 or π, do no impact significantly the monochromator performance, because of the $\cos\phi$ dependence.

[10] At this point the remark in footnote 3 on page 149 should be recalled: $\Delta\lambda/\lambda_c$ given by (3.13), and (3.19) presents an estimate of the upper bound for the relative spectral bandwidth of two reflections in the (+, +) configuration. The exact value has to be evaluated numerically using the exact expressions of the dynamical theory. The exact value is a factor 2 to 1.5 smaller than the one calculated by (3.13), or (3.19), in the framework of the DuMond diagram analysis.

asymmetry parameters b_1 and b_2 of the first and the second crystals, respectively, as follows from (3.13) and (3.19).

Unlike the single-bounce monochromators, the relative spectral bandpass $\Delta\lambda/\lambda_c$ is practically independent of the angular divergence of the incident beam. In fact, only those waves are reflected from the two crystals, whose propagation directions fit into the angular range given by the angular acceptance $\Delta\theta$ of the monochromators (3.17), or (3.20). The angular acceptance $\Delta\theta$ is a very important characteristic of monochromators. Only if $\Delta\theta$ is greater than the angular spread $\underline{\Delta\theta}$ of the incident radiation, could throughput of the monochromator be high.

Since the $(+,+)$-type monochromator design is no longer restricted to using extreme backscattering geometry (although proximity to backscattering geometry for the second crystal, the monochromator crystal, is always desirable for the same reason of reducing sensitivity to the angular spread and thus reducing the geometrical broadening) it is easier to tune monochromators of this type to a specific wavelength (photon energy) by changing angles of incidence and the relative angle between the crystals.

The angle ψ_{12} between the diffraction vectors $\boldsymbol{H}_1$ and $\boldsymbol{H}_2$, see Fig. 3.1, determines (along with the crystal parameters, such as interplanar distances d_1, and d_2, for the relevant reflecting atomic planes) the wavelength λ_c at the center of the monochromator spectral distribution function. It is given by (3.14)-(3.15). The wavelength (photon energy) variation is accomplished by changing the relative angle ψ_{12}, as given by (3.16).

The crystal peak reflectivity is never 100%, because of photo-absorption. Therefore, the cumulative peak reflectivity from all the monochromator crystals, which we term peak throughput, is always smaller than that of the single-crystal monochromator. Another evident disadvantage of the multiple-crystal monochromators is their increased design complexity.

A review of designs and applications of $(+,+)$-type monochromators covering the period until 1983 was published by Matsushita and Hashizume (1983). More recent advances were reviewed in the article of Toellner (2000). The field is developing so quickly, that even this recent survey is not completely up-to-date. The reader will find in this chapter an account of "old" as well as most recent designs of high-resolution monochromators using the $(+,+)$ crystal arrangement. In fact, not only two-crystal monochromators will be discussed. Also, four-crystal configurations such as $(+,+,-,-)$ and $(+,-,-,+)$ will be considered, which basically use the principle of the two-crystal $(+,+)$ monochromator. As we shall see, using the additional crystals offers the prospect of further improving the monochromator performance.

5.4.2 Symmetric $(+,+)$ and $(+,-,-,+)$ Monochromators

In the simplest case, see Fig. 5.19(a), the Bragg reflections from both crystals are symmetric, i.e., the asymmetry factors $b_1 = b_2 = -1$, and the relative spectral widths of the reflections $\epsilon_1^{(s)} = \epsilon_2^{(s)} = \epsilon_H^{(s)}$ are identical. In this case,

5.4 Multi-Bounce (+, +)-Type Monochromators

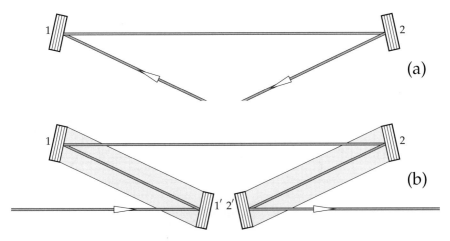

Fig. 5.19. (a) Symmetric (+, +)-type monochromator. (b) In-line symmetric (+, −, −, +)-type monochromator using the same principle as (a), which, however, redirects the beam to the forward direction.

the bandpass of the (+, +) symmetric monochromator equals $\Delta\lambda/\lambda_c = \epsilon_H^{(s)}$ for any photon energy and glancing angle of incidence, as follows from (3.19). By choosing Bragg reflections with an appropriate value of $\epsilon_H^{(s)}$ one can achieve a monochromator with a desired small bandpass.

Faigel et al. (1987) have demonstrated a monochromator of this kind with an energy bandpass $\Delta E = 5$ meV for $E = 14.4$ keV x rays ($\Delta E/E \simeq 3 \times 10^{-7}$), using the (10 6 4) Bragg reflection from symmetrically cut silicon crystals. In fact, the authors have used a four-crystal monochromator as in Fig. 5.19(b). The purpose of the additional two crystals 1' and 2' was to redirect the beam to the forward direction. The crystals 1 and 1', as well as 2 and 2' are in the (+, −) nondispersive configuration, respectively. The monochromator is termed a (+, −, −, +) symmetric in-line monochromator. The DuMond diagram analysis predicts the same spectral bandwidth and angular acceptance as for the (+, +)-type monochromator, with only the crystals 1 and 2. Numerical simulations using the exact expressions of the dynamical theory predict lower throughput, because of the additional Bragg reflections.

Hastings et al. (1991), and van Bürck et al. (1992) have used a monochromator of this type for the first observation of nuclear resonant forward scattering (NFS) of 14.4 keV synchrotron radiation from ^{57}Fe nuclei. The meV-monochromator was crucial for reducing the spectral spread of the photons and thus reducing the load on the timing detector and electronics. See Sect. 1.2 on some more details about NFS experiments.

The monochromator has a narrow energy bandpass, however it has also a small angular acceptance. From (3.20) we can estimate for the monochro-

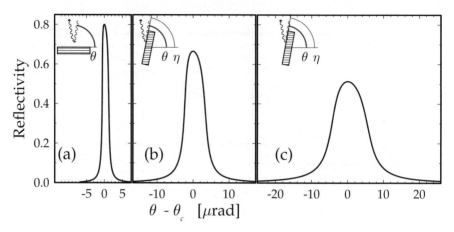

Fig. 5.20. Reflectivity of silicon crystals, with the (5 7 9) Bragg reflection, as a function of the glancing angle of incidence θ for different values of the asymmetry angle η, and the asymmetry factor b, respectively: (a) $b = -1$, $\eta = 0$; (b) $b = -1/10$, $\eta = 78.3°$; (c) $b = -1/30$, $\eta = 79.7°$. The center of the reflection curve is at photon energy $E_P = 14.4125$ keV, and the glancing angle of incidence is $\theta_c = 80.40°$. Asymmetry factor $b = \sin(\eta - \theta_c)/\sin(\eta + \theta_c)$ (2.113) with azimuthal angle $\phi_c = \pi$.

mator with the (10 6 4) Bragg reflections $\Delta\theta = \epsilon_H^{(s)} \tan\theta_c \simeq 1.8$ μrad, which is unfortunately smaller than the angular divergence of the synchrotron radiation at practically all modern existing sources.

5.4.3 Nested $(+, +, -, -)$ Monochromators

To increase the angular acceptance, while preserving the small spectral bandpass, one may use an asymmetrically cut first crystal, the collimator crystal. The reflectivity curves in Fig. 5.20 illustrate this statement. This figure shows examples of calculations of the reflectivity of silicon crystals with the (5 7 9) Bragg reflection as a function of the glancing angle of incidence for different values of the asymmetry angle η, i.e., the asymmetry factor b. A scattering geometry with the azimuthal angle $\phi = \pi$ resulting in $|b| < 1$ has to be used, to increase the angular acceptance, as discussed in Sect. 2.4.8. The angular width increases with the asymmetry angle from 1.8 μrad in Fig. 5.20(a) to 10 μrad in 5.20(c).

Figure 5.21(a) shows one possible design of the $(+,+)$-type monochromator with an asymmetrically cut first crystal. The crystals 1 and 2 compose the actual $(+,+)$ monochromator. Because of the dispersive configuration of the crystals, the beam direction is changed after two reflections. Because of the asymmetric reflection from the first crystal, the beam cross-section can be significantly increased. Therefore, two additional crystals are used of the same type as 2 and 1, i.e., the crystals 3 and 4 in Figure 5.21(a), which are

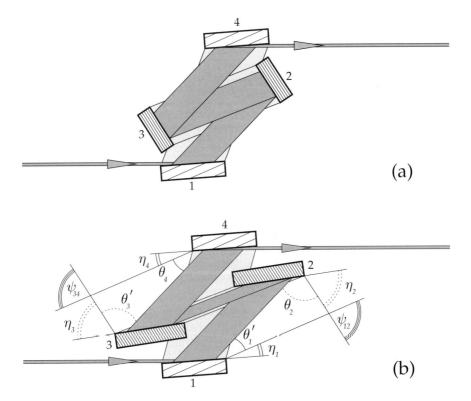

Fig. 5.21. Schemes of the $(+,+,-,-)$-monochromators, also known as nested in-line monochromators. The crystals 1 and 2 are in the $(+,+)$ configuration composing the actual monochromator. The function of the crystals 3 and 4 is to regain the incident beam direction, cross-section, and divergence. The reflecting atomic planes and thus Bragg reflections used in crystal 3 are exactly alike as in crystal 2. The same is valid for crystals 1 and 4. The atomic planes are parallel to those in the counterpart crystal.
(a) Crystals 1 and 4 are cut asymmetrically: $\eta_1 = \eta_4 \neq 0$. Notations are shown in Fig. 5.21(b).
(b) Also crystals 2 and 3 are cut asymmetrically: $\eta_2 = \eta_3 \neq 0$.

inverted with respect to the first two crystals. Their purpose is to regain the beam direction, cross-section, and divergence[11]. The same sets of reflecting atomic planes and thus the same Bragg reflections are used in crystals 1 and 4. The same set of reflecting atomic planes are used in crystals 2 and 3.

---

[11] The use of additional two crystals, the third and the fourth, preserves the beam direction and size at a cost of reduced monochromator throughput and increased design complexity. The reduced throughput results from reflections from additional crystals with less than unit reflectivity.

Figure 5.21(b) shows a more general design with crystals 2 and 3 also cut asymmetrically. The crystals 2 and 3 are set in the nondispersive configuration with respect to each other, with the atomic planes parallel to those in the counterpart crystal. The same is valid for crystals 1 and 2. The whole four-crystal configuration can be denoted as $(+,+,-,-)$. Although the Bragg reflections used in the nondispersive reflection pairs are exactly alike, the relevant asymmetry factors have to be different. For the entrance and exit waves to be symmetric, the following relations have to be fulfilled:

$$b_4 = 1/b_1, \qquad b_3 = 1/b_2.$$

Under these conditions, the results of the DuMond diagram analysis of the two-crystal $(+,+)$-case presented in Sect. 3.3 can be applied to this particular $(+,+,-,-)$-case.

The four crystal configuration increases the design complexity as the position and orientation of four instead of just two crystals have to be controlled. Since the reflecting atomic planes in the first and the fourth, as well as the third and the fourth crystals have to be set parallel to each other, it is favorable to manufacture nondispersive crystal pairs 1 and 4 as well as 2 and 3, each from a single crystal monolith, as shown in Fig. 5.22. Each monolith has a channel-cut. The resulting two inside parallel faces are used as the nondispersive $(+,-)$ reflection pairs of the monochromator. The two channel-cut crystals are nested within each other. The end result is a four-Bragg-reflection device, which is often referred to as nested in-line monochromator, that only requires as much angular control as a two-crystal device.

Spectral Bandpass and Angular Acceptance. The Bragg reflection and the asymmetry factor b_1 have to be chosen for the first crystal in such a way, that the following two conditions are fulfilled. First, its angular acceptance $\simeq \epsilon_1^{(s)} \tan \theta_1 / \sqrt{|b_1|}$ (2.125)-(2.126) has to be larger than the incident beam divergence. Second, the angular spread $\simeq \epsilon_1^{(s)} \tan \theta_1' \sqrt{|b_1|}$ (2.154) of the propagation directions of the waves leaving the first crystal, and, correspondingly, impinging upon the second crystal has to be small enough not to cause too large geometrical broadening in reflection from the second crystal. The Bragg reflection for the first crystal (in fact only the $\epsilon_1^{(s)}$ value of the Bragg reflection is important) and the asymmetry factor $|b_1| < 1$ are chosen along these requirements. The Bragg reflection of the second crystal has to be chosen according to the required spectral resolution of the monochromator, i.e., the Bragg reflection with an appropriate $\epsilon_2^{(s)}$ must be used.

Spectral bandpass, and angular acceptance of the monochromator can be estimated using (3.13), and (3.17), respectively, with the data for $\epsilon_1^{(s)}$ and $\epsilon_2^{(s)}$ taken, e.g., from the tables in Appendices A.3 and A.4.

The first working model of the high-resolution, large-angular-acceptance monochromator for hard x rays was demonstrated by Toellner et al. (1992), and Mooney et al. (1994). The design was proposed earlier by Ishikawa et al. (1992).

5.4 Multi-Bounce (+,+)-Type Monochromators 247

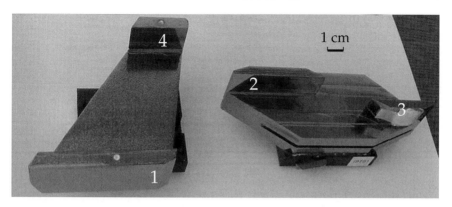

Fig. 5.22. Photograph of two silicon single crystal monoliths used in the four-Bragg-reflection (+, +, −, −) nested monochromator (monochromator #3 in Table 5.2) with two pairs of Bragg reflections (4 4 0) and (5 7 9), respectively (courtesy of T. Toellner). Each monolith has a channel cut. In the monochromator, the two crystal monoliths are nested within each other. The resulting two inside parallel faces of each monolith labeled as 1 and 4 on the left, and 2 and 3 on the right are used as the nondispersive (+, −) asymmetrically cut crystal pairs of the monochromator - Fig. 5.21 . The largest dimension of the monoliths is about 13 cm.

The crystals employed by Toellner et al. (1992) and Mooney et al. (1994) had parameters as given in Table 5.2[12], monochromator #1. The bandpass of the monochromator was measured to be $\Delta E = 6.7$ meV (FWHM) for $E = 14.4$ keV x rays ($\Delta E/E = 4.6 \times 10^{-7}$). The angular acceptance was 21 μrad, which matches the angular divergence of the beams delivered at most of the existing synchrotron radiation facilities.

The energy bandpass of the monochromator can be narrowed further by selecting Bragg reflections with a smaller relative spectral width $\epsilon_2^{(s)}$. The choice is certainly limited, since only those reflections can be used, whose Bragg energy is smaller than the energy of x-ray photons: $E_H < E$. Monochromator #2 presented in Table 5.2 differs from monochromator #1 by the use of the (5 7 9) Bragg reflection, instead of (12 2 2), with smaller relative energy width, c.f. the data in Appendix A.3. This results in a smaller energy bandpass for monochromator #2 as compared to #1.

Another way to reduce the monochromator bandpass is to also cut the second and the third crystals asymmetrically. The DuMond diagrams in Fig. 3.3, as well as (3.13) suggest that the crystals have to be cut and oriented in such a way that the asymmetry factors $|b_2| = 1/|b_3| > 1$. Figure 5.23 shows exam-

---

[12] Table 5.2, as well the subsequent Tables 5.3, 5.4, 5.5, and 5.6, provide in the sub-rows denoted as (e), (t), and (d) three different sets of the monochromator parameters: (e) as measured in the cited experiment, (t) calculated using the exact equations of the dynamical theory, and (d) calculated using the DuMond diagram technique with (3.13) and (3.17). The calculations were performed by the author of the book.

Table 5.2. Characteristics of the four-bounce $(+,+,-,-)$ monochromators designed for different photon energies E_P, and parameters of the silicon crystals composing the monochromators. The isotope symbol in brackets indicates the nucleus whose resonance energy coincides with the photon energy E_P. The monochromator schemes are shown in Fig. 5.21.
Crystal parameters: crystal's number n, Bragg reflection (hkl), glancing angle of incidence θ at which the photons with energy E_P are reflected, asymmetry angle η, absolute value of the asymmetry factor $|b|$.
Monochromator parameters: energy bandpass ΔE (FWHM), angular acceptance $\Delta\theta$ (FWHM), peak throughput T are given, as (e) measured in the experiment, (t) calculated by using the exact equations of the dynamical theory, and (d) calculated using the DuMond diagram technique with (3.13) and (3.17). The energy bandpass ΔE, and the peak throughput T are calculated for an x-ray beam with 10 μrad angular divergence. See footnote 3 on page 149 for the reasons for the discrepancies between the (t) and (d) values.

monochromator		monochromator crystals					monochromator			
#	E_P [keV]	n	(hkl)	θ_n [deg]	η_n [deg]	$\|b_n\|$	ΔE [meV]	$\Delta\theta$ [μrad]	T	
1	14.412 (^{57}Fe)						6.7			(e)
	(Toellner et al. 1992)	1,4	(4 2 2)	22.82	20	1/14,14	5.6	21	0.65	(t)
	(Mooney et al. 1994)	2,3	(10 6 4)	77.53	0	1,1	11.4	23	1	(d)
2	14.412 (^{57}Fe)									(e)
		1,4	(4 2 2)	22.82	20	1/14,14	3.8	21	0.55	(t)
		2,3	(5 7 9)	80.40	0	1,1	8	23	1	(d)
3	14.412 (^{57}Fe)						2.2			(e)
		1,4	(4 4 0)	26.62	24.7	1/23,23	2.1	17	0.45	(t)
	(Toellner 1996)	2,3	(5 7 9)	80.40	71	3,1/3	4.6	23	1	(d)
4	21.541 (^{151}Eu)						0.94			(e)
		1,4	(4 4 0)	17.44	16	1/22,22	0.72	11	0.38	(t)
	(Toellner 2000)	2,3	(15 11 3)	86.75	0	1,1	1.6	14	1	(d)
5	22.494 (^{149}Sm)						0.9			(e)
		1,4	(8 0 0)	23.95	20	1/10,10	1.0	3.9	0.17	(t)
	(Barla et al. 2004)	2,3	(16 8 8)	83.94	0	1,1	1.9	4.2	1	(d)
6	23.880 (^{119}Sn)						1.1			(e)
	(Toellner 2000)	1,4	(4 4 4)	19.34	18	1/26,26	0.84	7.2	0.28	(t)
	(Barla et al. 2000)	2,3	(12 12 12)	83.45	0	1,1	1.6	8.9	1	(d)
7	25.651 (^{161}Dy)						0.52			(e)
		1,4	(6 2 0)	16.345	15	1/22,22	0.46	8.1	0.15	(t)
	(Baron et al. 2001a)	2,3	(22 4 2)	87.23	0	1,1	0.9	9	1	(d)

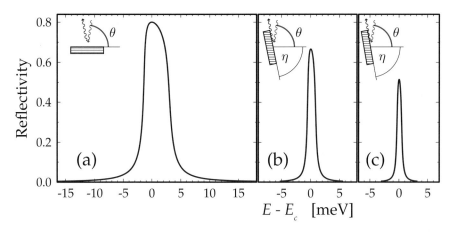

Fig. 5.23. Reflectivity of silicon crystals with the (5 7 9) Bragg reflection as a function of photon energy for different values of the asymmetry angle η, and the asymmetry factor b, respectively: (a) $b = -1$, $\eta = 0$; (b) $b = -10$, $\eta = 78.3°$; (c) $b = -30$, $\eta = 79.7°$. The center of the reflection region is at photon energy $E_P = 14.4125$ keV, and glancing angle of incidence $\theta_c = 80.40°$. Asymmetry factor $b = \sin(\eta + \theta_c)/\sin(\eta - \theta_c)$ (2.113) with azimuthal angle $\phi_c = 0$

ples of calculations of the reflectivity curves demonstrating how the spectral width decreases with increasing asymmetry angle η, and the asymmetry factor. Shown are the calculations of the reflectivity of silicon crystals with the (5 7 9) Bragg reflection. The monochromator with asymmetrically cut crystals 2 and 3 is shown schematically in Fig. 5.21(b). The properties of a functioning monochromator with all the crystals cut asymmetrically are given in Table 5.2, monochromator #3 (Toellner 1996). Monochromator #3 differs from monochromator #2 in the first place by the use of asymmetrically cut crystals 2 and 3, resulting immediately in a narrowed spectral bandpass.

The angular acceptance of the monochromators #1 – #3 in Table 5.2 does not differ significantly. Because of a relatively high angular acceptance, monochromators of the type #1 – #3 in Table 5.2 can be efficiently used not only at the undulator based x-ray sources providing beams with less than 20 μrad angular divergence, but also at the bending magnet beamlines delivering beams with somewhat larger angular divergence.

Tunability. The multi-bounce monochromators operate at glancing angles of incidence not necessarily very close to 90°. Due to this, it is easier than in the case of the single-bounce backscattering monochromator discussed in Sect. 5.3 to find suitable Bragg reflections for a monochromator with desired energy (wavelength). Table 5.2 presents examples of the $(+,+,-,-)$-type monochromators designed for different photon energies.

The center of the spectral distribution function of the $(+,+,-,-)$ nested monochromator is given by (3.14). Fine tuning of the center of the monochromator function in steps of meV or sub-meV is performed by changing the angle between the reflecting atomic planes of the crystal pair 1, 4 with respect to 2, 3. Formally, this is the angle ψ_{12} between the diffraction vectors $\boldsymbol{H}_1$ and $\boldsymbol{H}_2$, as shown in Fig. 3.1. The tuning is described by (3.16). In practice it is achieved by mounting the two single crystal monoliths, as, e.g., shown in Fig. 5.22, on high-precision rotation stages, and changing the relative angular position of the relevant atomic planes in angular steps of typically from 1 μrad to 10 nrad.

Throughput. One of the most important monochromator characteristics is peak throughput. The peak throughput of a monochromator is determined by the product of the peak reflectivities of the constituent monochromator crystals (for an incident beam with a given angular divergence). The crystals and Bragg reflections have to be chosen such that the peak crystal reflectivity is high. Striving for improved spectral resolution, and increased angular acceptance by increasing the asymmetry of the Bragg reflections, as discussed above, leads, however, to reduced peak reflectivity of the single crystals, as shown in Figs. 5.23 and 5.20, and to the reduced peak throughput of the monochromator as a whole. The optimum monochromator design is always a compromise between the conflicting requirements of small spectral bandpass and high peak throughput.

For real x-ray beams, which always possess a finite angular divergence $\underline{\Delta\theta}$, the monochromator peak throughput is also decided by the smallness of the ratio of $\underline{\Delta\theta}$ to the angular acceptance $\Delta\theta$ of the monochromator.

Angular Acceptance at High Photon Energies. The angular acceptance of the $(+,+,-,-)$ monochromators under discussion is determined in the first place by the angular acceptance of the Bragg reflection from the first crystal, which one can readily understand after inspection of (3.17).

The relatively high angular acceptance is achievable; however, it is not always so. The problem springs up with increasing photon energy. The monochromators designed for higher photon energies often have insufficient angular acceptance, as compared to the angular divergence of incident x rays. This is, e.g., the case of monochromator #5 in Table 5.2. In such cases, the application of additional collimating optics is necessary. For this purpose, Baron et al. (1999) have proposed using compound refractive lenses as introduced by Snigirev et al. (1996). The performance of high-energy resolution monochromators in combination with compound refractive lenses was tested by Chumakov et al. (2000).

Compact Design. At photon energies $E > 25$ keV the angular acceptance of Bragg reflections becomes so small that obtaining high energy resolution and at the same time preserving large angular acceptance and high peak throughput of the monochromators becomes problematic. A low-indexed Bragg reflection from the first crystal, and a high-indexed Bragg reflection with a glancing angle of incidence very close to 90° from the second crystal

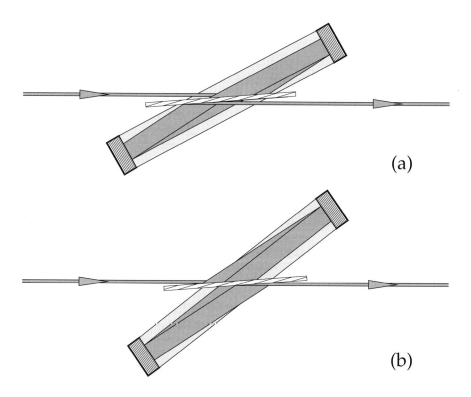

Fig. 5.24. Compact design of the $(+,+,-,-)$ monochromator (a), and of the $(+,-,+,-)$ monochromator (b).

have to be used to cope with the problem. The latter, however, is inconsistent with the monochromator design shown in Fig. 5.21

To circumvent this difficulty, Baron et al. (2001a) have proposed an alternative compact design, optimized for Bragg back-reflections very close to backscattering. The essential idea is to use, instead of the channel-cut crystal for Bragg reflections 1 and 4, a very thin crystal plate, as shown in Fig. 5.24. It couples the x-ray beam into the channel-cut backscattering crystal. A crystal plate, a few micrometers thick, is already sufficient to achieve high reflectivity using low-indexed Bragg reflections, provided the crystal is cut asymmetrically with an asymmetry parameter $|b_1| \ll 1$. The beam back-reflected from the channel-cut crystal is transmitted through the thin crystal plate with low losses, as it enters the crystal at a large angle to the crystal surface[13]. Fig. 5.24(a) shows the $(+,+,-,-)$-type monochromator equivalent to those discussed in the present section. Fig. 5.24(b) shows a $(+,-,+,-)$ four-

---

[13] The crystal plate implemented by Baron et al. (2001a) was 0.8 mm thick, absorbing 50% of the radiation. Theoretically, the peak throughput of the monochro-

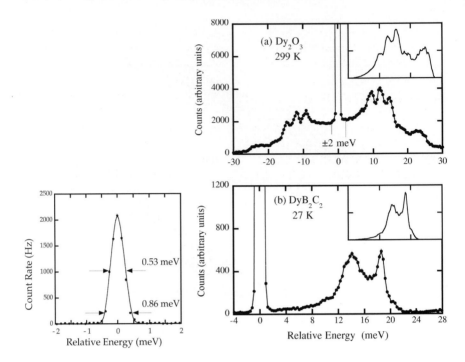

Fig. 5.25. Left: Energy-resolution function for the four-bounce Si $(+,-,+,-)$ monochromator for 25.65 keV x rays, as in Fig. 5.24(b) (monochromator #7, Table 5.2). The solid line is the calculation according to the dynamical theory of x-ray diffraction.
Right: Spectra of nuclear resonant inelastic scattering in (a) Dy_2O_3 powder near room temperature and (b) a DyB_2C_2 crystal at 27 K. Elastic peak heights are 1×10^5 and 4×10^4, respectively. Lines are to guide the eye. Insets show the derived partial phonon densities of states (30 meV full scale in each case). (Baron et al. 2001a)

bounce monochromator, which was implemented by Baron et al. (2001a). It has almost the same characteristics as the $(+,+,-,-)$-type monochromator Fig. 5.24(a), provided the Bragg reflections from the channel-cut crystal are close to backscattering.

The design was demonstrated in an almost theoretically perfect high-resolution monochromator providing photons in a 0.52 meV bandpass at 25.65 keV. The monochromator was used to measure inelastic nuclear resonant scattering from phonons in ^{161}Dy containing samples, shown in Fig. 5.25.

Crystal Surface Flatness. The monochromator crystals must be of the highest quality, i.e., they must have the lowest density of crystal defects.

---

mator (monochromator #7, Table 5.2) can be increased by almost a factor of two by using a much thinner plate of $\lesssim 0.2$ mm.

Along with this evident requirement for the quality of the crystal material, there is a stringent requirement for the crystal surfaces to be extremely flat. We distinguish the long length-scale variation of flatness ($\gg 100\lambda$), the slope error, and the short length-scale variation ($\lesssim 100\ \lambda$), the surface roughness. Only the influence of the slope error will be addressed here.

The slope error leads to local variations of the asymmetry angle η_n ($n = 1, 2$) and thus of the asymmetry factor b_n. This in turn causes the local variation of the spectral bandpass $\Delta\lambda/\lambda_c$ (3.13), and of the peak position λ_c (3.14)-(3.15) of the spectral function. Eventually, these factors result in the integral broadening of the spectral bandpass.

From (3.14)-(3.15) we find for the relative variation of the peak position λ_c of the monochromator spectral function with asymmetry angles η_n of the monochromator crystals:

$$\left|\frac{\delta\lambda_c}{\lambda_c}\right| \simeq \sum_{n=1,2} \frac{A_n\, w_n^{(s)}\, |\delta\eta_n|}{2\sin 2\theta_{c_n}}, \qquad A_n = \left(\frac{\lambda_c}{2\sin\psi_{12}}\right)^2 \left(\frac{1}{d_n^2} + \frac{\cos\psi_{12}}{d_1 d_2}\right). \tag{5.5}$$

In the derivation of (5.5) we used $b_1 = \sin(\eta_1 - \theta_{c_1})/\sin(\eta_1 + \theta_{c_1})$, $b_2 = \sin(\eta_2 + \theta_{c_2})/\sin(\eta_2 - \theta_{c_2})$, and assumed that both crystals are cut strongly asymmetrically: $\eta_n \simeq \theta_{c_n}$ ($\eta_n < \theta_{c_n}$).

Similarly, from (3.13) we find for the variation of the relative spectral bandpass $\Delta\lambda/\lambda_c$ of the monochromator with asymmetry angles η_n of the monochromator crystals:

$$\left|\delta\left(\frac{\Delta\lambda}{\lambda_c}\right)\right| \simeq \sum_{n=1,2} \frac{B_n\, \epsilon_n^{(s)}\, |\delta\eta_n|}{2\sin 2\theta_{c_n}} \qquad B_1 = \frac{\tau_1}{\sqrt{|b_1|}}, \quad B_2 = \tau_2 \sqrt{|b_2|}. \tag{5.6}$$

The net broadening of the monochromator spectral function is the sum $|\delta\lambda_c/\lambda_c| + |\delta(\Delta\lambda/\lambda_c)|$ of both contributions. In this respect, it should be noted that the constants A_n and B_n are of the same order of magnitude, however, $w_n^{(s)} \gg \epsilon_n^{(s)}$. Therefore, typically $|\delta\lambda_c/\lambda_c| \gg |\delta(\Delta\lambda/\lambda_c)|$, i.e., the variation of the peak position broadens the monochromator spectral function much more than the variation of the bandpass.

In each particular case, the admissible variation $|\delta\eta_n|$ of the asymmetry angles can be estimated with the help of (5.5) and (5.6). Assuming $w^{(s)} \approx 10^{-5}$, we can estimate that an asymmetry angle variation of $|\delta\eta_n| \lesssim 10^{-4}$ will cause a relative broadening of the order of $10^{-8} - 10^{-9}$. This corresponds to an absolute broadening of $150 - 15\ \mu$eV for 15 keV photons.

Artificial channel-cut crystal. The manufacturing of the internal surfaces of the channel-cut crystals is a delicate process. Therefore, it is often beneficial to manufacture independently flat crystal pairs with very good surfaces and then assemble them into an artificial channel-cut crystal.

254   5. High-Resolution X-Ray Monochromators

Fig. 5.26. Picture of the weak-link mechanism that allows positioning of two crystals with better than 50 nrad angular resolution (Toellner and Shu 2001). See also Fig. 5.27 and text for the description of the parts.

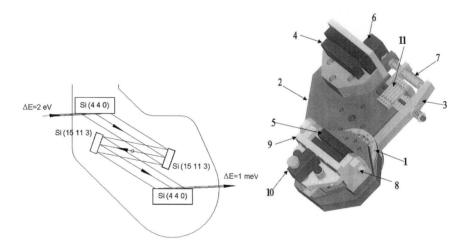

Fig. 5.27. Left: Schematic of a four-crystal in-line $(+,+,-,-)$ monochromator in a nested configuration. Right: The design of an artificial channel-cut crystal assembly employing a weak-link mechanism (Fig. 5.26) for positioning of the two crystals (4) and (5), which are used to reflect x-rays with the (4 4 0) Bragg reflections. See text for details. (Shu et al. 2001, 2003).

Some designs require channel-cut crystals to be too big. In this case producing an artificial channel-cut crystal is also favorable.

The difficulty is, however, that the reflecting atomic planes in the two flat crystals should be adjusted to be parallel with a very high accuracy, which itself is determined by the angular acceptance of the Bragg reflection, i.e., very often with sub-microradian accuracy.

Shu et al. (2001, 2003) have designed and constructed a miniature overconstrained weak-link mechanism that allows positioning of two crystals with better than 50 nrad angular resolution and nanometer linear driving sensitivity, see Fig. 5.26. The precision and stability of this structure allow the user to align or adjust an assembly of crystals to achieve the same performance as does a single channel-cut crystal. The authors termed this assembly "artificial channel-cut crystal".

Figure 5.27 shows the scheme of a four-crystal in-line $(+,+,-,-)$ monochromator with nested configuration (left), and the design of an artificial channel-cut crystal assembly (right) employing a weak-link mechanism (Fig. 5.26) for positioning of the two crystals (4) and (5), which are used to reflect x-rays with the (4 4 0) Bragg reflections. The weak-link mechanism that allows positioning of two crystals contains two sets of stacked thin-metal weak-link modules used in the driving mechanism: one is a planar-shaped, high-stiffness, high-stability weak-link mechanism acting as a planar rotary shaft (1), and the other one is a weak-link mechanism acting as a linear stage (11) to support a piezo translator. Both weak-link mechanisms have two modules mounted on each side of the base plate (2). A sine-bar (3) is installed on the center of the planar rotary shaft for the pitch alignment between the two (4 4 0) single crystals (4) and (5). Two linear drivers are mounted serially on the base plate to drive the sine-bar. The rough adjustment is performed by a piezo-augmented screw driver (6) with a 20-30 nm step size. A closed-loop controlled piezo translator (7) with strain sensor provides 1 nm resolution for the pitch fine alignment. A pair of commercial flexure bearings (8) is mounted on one of the crystal holders (9), and another piezo-augmented screw driver (10) provides the roll alignment for the crystal.

5.4.4 Asymmetric $(+,+)$ Two-Crystal Monochromators

As (3.13) demonstrates, the smallest relative spectral bandpass of the $(+,+)$-type monochromator is achieved in the first place by using Bragg reflections with the smallest intrinsic relative spectral widths $\epsilon_1^{(s)}$ and $\epsilon_2^{(s)}$. Secondly, highly asymmetric Bragg diffraction has to be used from the crystals with asymmetry factors $|b_1| \ll 1$, and $1/|b_2| \ll 1$. The first requirement is met automatically, if one and the same Bragg reflection with the smallest possible intrinsic relative spectral width $\epsilon_1^{(s)} = \epsilon_2^{(s)} = \epsilon^{(s)}$ is chosen for both monochromator crystals[14]. Since $|b_1|$ and $1/|b_2|$ have also to be minimized, they can be chosen equally small: $|b_1| = 1/|b_2|$. In this case the simplified expressions

---

[14] For a given photon energy E, this is very often the Bragg reflection, for which the energy of exact backscattering is the nearest to E, however, smaller than E. For example, in silicon for $E = 14.4125$ keV this is the (5 7 9) Bragg reflection, cf. the table in Appendix A.3.

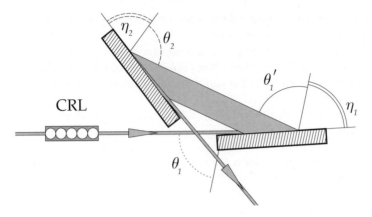

Fig. 5.28. Two-crystal $(+,+)$ monochromator with asymmetrically cut crystals. The asymmetry factors are usually chosen to be $|b_1| \simeq 1/|b_2| \ll 1$ to achieve the smallest spectral bandpass and the largest angular acceptance.
Here, $b_1 = \sin(\eta_1 - \theta_1)/\sin(\eta_1 + \theta_1)$ (obtained from (2.113) using azimuthal angle $\phi_1 = \pi$), and $b_2 = \sin(\eta_2 + \theta_2)/\sin(\eta_2 - \theta_2)$ (obtained from (2.113) using azimuthal angle $\phi_2 = 0$).
To match the angular divergence of the incident x-ray beam and the angular acceptance of the monochromator, and thus to increase the peak throughput of the monochromator collimating optics, such as compound refractive lens (CRL), can be installed upstream of the monochromator.

can be used to estimate the relative spectral bandpass of the monochromator $\Delta E/E = \epsilon^{(s)}\sqrt{|b_1|}$ (3.19), and of its angular acceptance $\Delta\theta \simeq \epsilon^{(s)}/\sqrt{|b_1|}$ (3.20).

The aforementioned requirements lead to a simple monochromator design. The scheme of the monochromator is shown in Fig. 5.28. The energy of the transmitted photons is changed in a way, which is usual for all types of $(+,+)$ monochromators, i.e., by changing the angle $\psi_{12} = \theta'_1 + \theta_2$ between the reflecting atomic planes in the monochromator crystals, see Figs. 5.28, 3.1, and Eq. (3.16). This scheme was implemented by Chumakov et al. (1996b), Toellner et al. (1997), Chumakov et al. (2000), and Toellner et al. (2001) in a series of monochromators, whose characteristics are given in Table 5.3. Using this, in principle, very simple scheme, it was possible to achieve a much higher energy resolution than that obtained with the four-crystal in-line $(+,+,-,-)$ monochromator in a nested configuration at the same photon energy. This can be seen from a comparison of the monochromator characteristics given in Tables 5.2 and 5.3.

Figure 1.19 on page 22 shows the instrumental function of the asymmetric $(+,+)$ monochromator for 14.412 keV x rays with 0.65 meV bandpass (FWHM), and examples of the spectra of nuclear resonant inelastic scattering from ^{57}Fe nuclei in $(NH_4)_2Mg^{57}Fe(CN)_6$ measured with the help of

5.4 Multi-Bounce (+,+)-Type Monochromators

Table 5.3. Characteristics of the two-bounce (+,+) monochromators - Fig. 5.28 - designed for different photon energies E_P, and parameters of the silicon crystals composing the monochromators. The isotope symbol in brackets indicates the nucleus whose resonance energy coincides with the photon energy E_P.
Crystal parameters: crystal's number n, Bragg reflection (hkl), glancing angle of incidence θ at which the photons with energy E_P are reflected, asymmetry angle η, absolute value of the asymmetry factor $|b|$.
Monochromator parameters: energy bandpass ΔE (FWHM), angular acceptance $\Delta\theta$ (FWHM), peak throughput T, as (e) measured in the experiment, or (t) calculated by using the exact equations of the dynamical theory, or (d) calculated using the DuMond diagram technique with (3.19) and (3.20). The energy bandpass ΔE, and the peak throughput T are calculated for an x-ray beam with 10 μrad angular divergence. See footnote 3 on page 149 for the reasons for the discrepancies between the (t) and (d) values.

monochromator		monochromator crystals				monochromator			
#	E_P [keV]	n	(hkl)	θ_n [deg]	η_n [deg] $\quad |b_n|$	ΔE [meV]	$\Delta\theta$ [μrad]	T	
1	14.412 (^{57}Fe)					1.65			(e)
		1	(5 7 9)	80.40	75.4 1/4.7	1.48	3.7	0.19	(t)
	(Chumakov et al. 1996b)	2	(5 7 9)	80.40	75.4 4.7	2.1	4.1	1	(d)
2	14.412 (^{57}Fe)					0.92			(e)
		1	(5 7 9)	80.41	79.5 1/22	0.8	7.8	0.28	(t)
	(Toellner et al. 1997)	2	(5 7 9)	80.40	78.4 10.4	1.20	8.9	1	(d)
3	14.412 (^{57}Fe)					0.47			(e)
		1	(5 7 9)	80.40	79.87 1/36	0.47	9.4	0.18	(t)
	(Chumakov et al. 2000)	2	(5 7 9)	80.42	79.87 36	0.76	11	1	(d)
4	23.880 (^{119}Sn)					0.14			(e)
		1	(12 12 12)	83.45	83.0 1/30	0.10	1.3	0.02	(t)
	(Toellner et al. 2001)	2	(12 12 12)	83.45	83.0 30	0.14	1.5	1	(d)

this monochromator. The monochromator parameters are close to those of monochromator #3 in Table 5.3.

Figure 5.29 shows another example: the instrumental function of the asymmetric (+,+) monochromator (the dashed line) with 0.85 meV bandpass (FWHM) for 14.412 keV, and the spectra of nuclear resonant inelastic scattering from ^{57}Fe nuclei in myoglobin, an oxygen-storing protein. The monochromator parameters are close to those of monochromator #2 in Table 5.3.

Using the same scheme, Toellner et al. (2001) have demonstrated a crystal monochromator with $\Delta E/E \simeq 6 \times 10^{-9}$, monochromator #4 in Table 5.3. The monochromator has an energy bandpass of $\Delta E = 140$ μeV and operates

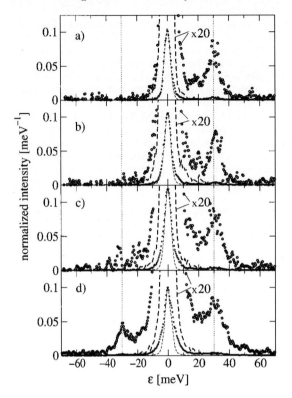

Fig. 5.29. Spectra of nuclear resonant inelastic scattering from ^{57}Fe in met-myoglobin measured at a) $T = 50$ K, b) 170 K, c) 235 K, and d) 300 K, respectively, as a function of photon energy E deviation $\varepsilon = E - E_M$ from the Mössbauer resonance in ^{57}Fe at $E_M = 14.412$ keV. Data are normalized and enlarged by a factor of 100.(Achterhold et al. 2002)
The dashed line presents the 0.85 meV broad (FWHM) spectral function of the $(+, +)$ two-crystal Si monochromator, with monochromator parameters close to those of monochromator #2 in Table 5.3.

with 23.880 keV x rays, see Fig. 5.30. Such a high energy resolution is already almost at the limit determined by the available quality of silicon single crystals, as discussed in the footnote on page 224.

The price for the high energy resolution is a relatively low peak throughput of monochromators of this type, which can be seen from the comparison of the corresponding numbers given in Tables 5.3 and 5.2.

The transformation of the two-crystal $(+, +)$ monochromator into the four-crystal one, as shown in Fig. 5.31(a), with the purpose of regaining the incident beam direction, would be therfore impractical, because of the very low peak throughput. A more practical solution is to use an additional crystal with Bragg reflection having a high reflectivity, and a Bragg angle close to $\pi - \theta_1 - \theta_2$ (Chumakov et al. 2000), see Fig. 5.31(b).

There are two reasons for the low throughput of the $(+, +)$ monochromators under discussion in this section. First, the requirement of using Bragg reflections with small intrinsic relative spectral width $\epsilon^{(s)}$ leads automatically to small angular acceptances, although the angular acceptance is increased by a factor $1/\sqrt{|b_1|}$. Installing collimating optics, such as a compound refractive lens (CRL), as shown in Fig. 5.28, helps to increase the throughput (Chumakov et al. 2000).

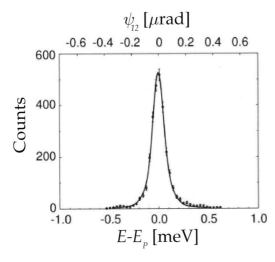

Fig. 5.30. Energy-resolution function for the two-crystal Si $(+,+)$ monochromator #4, Table 5.3. $E_P \simeq 23.880$ keV. The solid line is a simulation using the theory of dynamical diffraction. The full-width at half-maximum is 140 μeV. Energy tuning is performed by changing the angle $\psi_{12} = \theta'_1 + \theta_2$, see Figs. 5.28 and 3.1. (Toellner et al. 2001)

Second, the requirement of minimizing the parameters $|b_1|$, and $1/|b_2|$ results in another problem, which is difficult to avoid. The minimization is achieved by increasing the asymmetry angles $\eta_1 \to \theta_1$ and $\eta_2 \to \theta_2$, see Fig. 5.28, and thus minimizing $\eta_1 - \theta_1$ and $\eta_2 - \theta_2$. According to the results of the dynamical theory calculations shown in Figs. 5.20 and Figs. 5.23 the peak reflectivity of each crystal drops with increasing asymmetry angle. The peak throughput of the monochromator decreases as well, as the results of calculations in Fig. 5.32(a) show. The calculations are performed, assuming a well collimated incident beam with an angular divergence of 1 μrad, which is smaller than the angular acceptance of the monochromators.

The monochromator spectral throughput in Fig. 5.32(b) is calculated for a more realistic incident x-ray beam with an angular divergence of 10 μrad. A comparison of the spectral functions in Figs. 5.32(a) and (b) shows that in the case of the monochromator with the smallest asymmetry angle (largest $\sqrt{|b_1|}$)

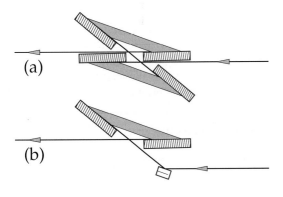

Fig. 5.31. (a) The two-crystal monochromator in Fig. 5.28 is transformed here into a four-crystal monochromator, in order to regain the primary beam direction. (b) With an additional crystal (using a low-indexed Bragg reflection with high reflectivity) the beam emerging from the monochromator is directed into the primary beam direction.

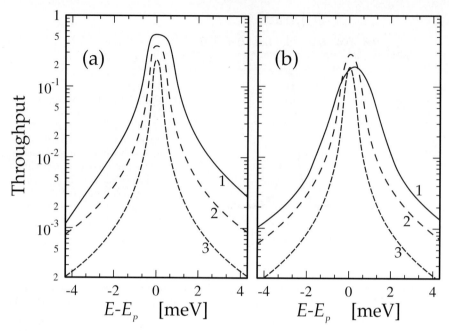

Fig. 5.32. Throughput of the two-crystal $(+,+)$-type monochromator as a function of the photon energy E. The curves labeled by 1, 2, and 3 were calculated for different values of the asymmetry angles η_1 and η_2 of the monochromator crystals (see Fig. 5.28). The asymmetry angles increase from case 1 to 3. The asymmetry angles and other monochromator crystal parameters are those of monochromators #1, #2, and #3 in Table 5.3.
(a) Calculations for an x-ray beam with 1 μrad angular divergence.
(b) Calculations for an x-ray beam with 10 μrad angular divergence.

the main factor, which reduces the peak throughput is insufficient angular acceptance of the monochromator. On the contrary, for the monochromator with the crystals cut with the largest asymmetry angle, the reduction of the peak throughput results from the greatly reduced peak reflectivities of the monochromator crystals.

Therefore, the parameter $\sqrt{|b_1|}$, and $1/\sqrt{|b_2|}$, respectively, cannot be chosen extremely small, not only because of technical problems in manufacturing the crystals, but also because of the aforementioned problem. There is an elegant way to avoid this problem. For this, however, a four-crystal $(+,-,-,+)$ scheme has to be used as described in the next section.

5.4.5 Asymmetric $(+,-,-,+)$ Monochromators

Principle. The principle and performance of the four-crystal asymmetric $(+,-,-,+)$ monochromators can be understood via the DuMond diagram analysis developed in Section 3.5.

5.4 Multi-Bounce (+, +)-Type Monochromators   261

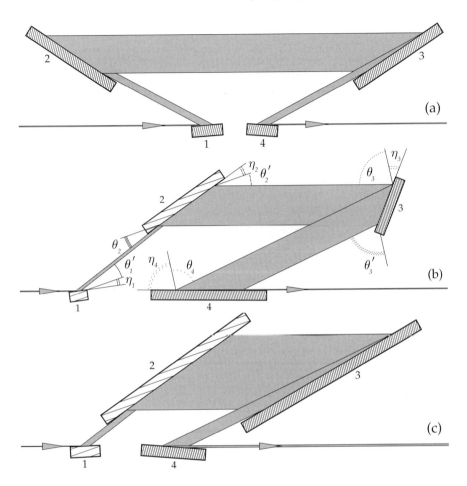

Fig. 5.33. Schematic drawings of four-crystal (+, −, −, +) monochromators with asymmetrically cut crystals: (a) corresponds to monochromators #1 and #4 of Table 5.4, (b) corresponds to monochromator #2 of Table 5.4, and (c) corresponds to monochromator #3 of Table 5.4.

A schematic view of four successive asymmetric Bragg reflections from crystals in the (+, −, −, +) configuration is shown in Fig. 3.7 on page 155. Schematic drawings of the four-crystal (+, −, −, +) monochromators, of different designs, which will be discussed in the present section, are shown in Fig. 5.33.

The basic principle of achieving high spectral resolution with the four-crystal (+, −, −, +) monochromators is in fact the same as for the two-crystal (+, +) monochromators: Bragg diffraction from *reflectors* in the *dispersive* configuration. The major distinction between these two schemes is in the kind of the reflectors employed. While for the two-crystal (+, +) monochro-

Table 5.4. Characteristics of the four-bounce $(+,-,-,+)$ monochromators designed for different photon energies E_P, and parameters of the silicon crystals composing the monochromators. The isotope symbol in brackets indicates the nucleus whose resonance energy coincides with the photon energy E_P. The monochromator schemes are shown in Fig. 5.33.
Crystal parameters: crystal's number n, Bragg reflection (hkl), glancing angle of incidence θ at which the photons with energy E_P are reflected, asymmetry angle η, absolute value of the asymmetry factor $|b|$.
Monochromator parameters: energy bandpass ΔE (FWHM), angular acceptance $\Delta\theta$ (FWHM), peak throughput T are given, as (e) measured in the experiment, (t) calculated by using the exact equations of the dynamical theory, and (d) calculated using the DuMond diagram technique with (3.28) and (3.29). The energy bandpass ΔE, and the peak throughput T are calculated for an x-ray beam with 10 μrad angular divergence. See footnote 3 on page 149 for the reasons for the discrepancies between the (t) and (d) values.

	monochromator	monochromator crystals				monochromator		
#	E_P [keV]	n	(hkl)	θ_n [deg]	η_n [deg] $\quad \|b_n\|$	ΔE [meV]	$\Delta\theta$ [μrad]	T
1	14.412 (^{57}Fe)					0.11		(e)
		1,2	(5 7 9)	80.40	78.4 $\quad$ 1/10.4	0.093	5.4	0.096 (t)
	(Yabashi et al. 2001)	3,4	(5 7 9)	80.40	78.4 $\quad$ 10.4	0.14	6.1	1 (d)
2	14.412 (^{57}Fe)					1.0		(e)
		1	(0 0 4)	18.47	11.0 $\quad$ 1/3.8	0.85	10.1	0.37 (t)
		2	(0 0 4)	18.47	16.0 $\quad$ 1/13.1	1.7	18	1 (d)
		3	(10 6 4)	77.53	34.25 $\quad$ 1.35			
	(Toellner et al. 2002)	4	(10 6 4)	77.53	76.85 $\quad$ 36.4			
3	14.412 (^{57}Fe)							
		1	(0 0 4)	18.47	16.0 $\quad$ 1/13.1	0.43	14	0.40 (t)
		2	(0 0 4)	18.47	14.0 $\quad$ 1/6.9	1.4	33	1 (d)
		3	(10 6 4)	77.53	74.0 $\quad$ 7.7			
		4	(10 6 4)	77.53	74.0 $\quad$ 7.7			
4	9.405 (^{83}Kr)					1.0		(e)
		1	(0 0 8)	76.15	72.15 $\quad$ 1/7.7	0.91	24	0.46 (t)
		2	(0 0 8)	76.15	70.75 $\quad$ 1/5.8	1.8	33	1 (d)
	(Toellner and Shu 2001)	3	(0 0 8)	76.15	70.75 $\quad$ 5.8			
	(Toellner et al. 2002)	4	(0 0 8)	76.15	72.15 $\quad$ 7.7			

mators the reflector is an isolated crystal, i.e., crystal 1 or 2 in Fig. 5.28, in the case of the four-crystal $(+,-,-,+)$ monochromator, this is a cascade of reflecting crystals. The most commonly employed cascade consists of two asymmetrically cut crystals using the same Bragg reflection. For example, crystals 1 and 2 in Figs. 5.33(a), (b), and (c) compose the first cascade, while crystals 3 and 4 – the second one. Of principal importance for achieving high spectral resolution with the $(+,-,-,+)$ monochromators is the fact that the crystals in each cascade are arranged in the *nondispersive* $(+,-)$ or $(-,+)$ configuration, respectively.

The main effect of the cascade of the nondispersively arranged asymmetrically cut crystals is to produce a change in the width of the Bragg reflection region (as seen in the phase space of the waves emerging after the cascade of two Bragg reflections) proportional to factor $|b_2|\sqrt{|b_1|}$, as graphically shown in Fig. 3.5 (Sect. 3.4), and expressed by Eq. (3.24). As opposed to this, single asymmetric Bragg reflection changes the width by a factor proportional only to $\sqrt{|b_1|}$, which is the *square root* of the asymmetry factor $|b_1|$, see Sect. 3.3 and Eq. (3.10). The additional multiplier in the case of the two cascaded reflections, the asymmetry factor $|b_2|$ of the *first power*, may enhance tremendously the effect of the cascade of the asymmetric Bragg reflections on the width, as compared to the effect of the asymmetric Bragg reflection from an isolated crystal.

As a consequence, the relative spectral width of the reflection from the two dispersively arranged nondispersive crystal cascades is $\Delta E/E = \epsilon^{(s)}|b_2|\sqrt{|b_1|}$, as given by (3.32), instead of $\Delta E/E = \epsilon^{(s)}\sqrt{|b_1|}$, which is valid for the relative spectral width of the reflection from two dispersively arranged isolated crystals, see (3.19). Here we assume that the same Bragg reflections are used in the cascades, and the asymmetry factors are $b_4 = 1/b_1$, and $b_3 = 1/b_2$, with $|b_1|, |b_2| \ll 1$. If this is not the case, the more general equation (3.28) has to be used to estimate the relative spectral bandpass of the monochromator. The basic principles are exactly alike.

The $(+,-,-,+)$ scheme was proposed and implemented for use in high-resolution monochromators by Yabashi et al. (2001). The design with four crystals using equivalent Bragg reflections with the asymmetry factors $b_4 = b_3 = 1/b_2 = 1/b_1$ was realized. The crystal parameters of the monochromator are given in Table 5.4, monochromator #1. The schematic drawing of the monochromator is shown in Fig. 5.33(a). The monochromator was designed for x rays with a photon energy of 14.412 keV. The application of the dispersively arranged nondispersive crystal cascades has allowed significant reduction of the spectral bandpass of x-ray monochromators. The spectral function with a width as small as 114 μeV was measured, as shown in Fig. 5.34[15]. The monochromator was applied to measure the duration of

---

[15] Narrowing of the energy bandpass while reducing the beam cross-section, as demonstrated in Fig. 5.34, is a clear indication of the influence of defects, e.g., periodicity of the reflecting atomic planes, in silicon crystals on the monochromator performance. Despite the highest quality of the silicon crystals used, this effect is noticeable. This experiment demonstrates that obtaining a relative energy resolution of $\simeq 10^{-8}$ is already almost at the limit determined by the quality of available silicon single crystals. For some more discussion, see also the footnote on page 224.

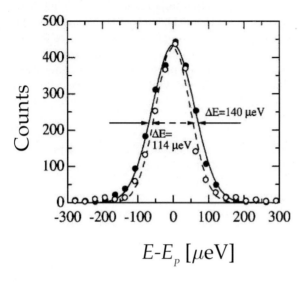

Fig. 5.34. Energy-resolution function for the four-crystal Si $(+,-,-,+)$ monochromator as given in Table 5.4, monochromator #1 (Yabashi et al. 2001). $E_p \simeq 14.412$ keV. The full-width at half-maximum is 140 μeV (incident beam cross-section 500×90 μm, closed circles) and is 114 μeV (beam cross-section 100×20 μm, open circles). The monochromator scheme is shown in Fig. 5.33(a).

synchrotron radiation pulses, pulse lengths about 30 ps, using an intensity interferometry technique (Yabashi et al. 2002).

Angular Acceptance and Throughput. While the spectral bandpass can be drastically reduced, the application of the cascades does not enhance in the same way the angular acceptance of the monochromator. Similar to the angular acceptance of the two-crystal $(+,+)$ monochromators, the angular acceptance $\Delta\theta$ of the four-crystal $(+,-,-,+)$ monochromators is determined in the first place by the angular acceptance of the Bragg reflection from the first crystal $\Delta\theta \simeq (\epsilon^{(s)}/\sqrt{|b_1|})\tan\theta_1$, see (3.33).

Because of the small angular acceptance the peak throughput may be low. In particular, the angular acceptance of the monochromator designed by Yabashi et al. (2001) is only $\simeq 5$ μrad. As a result for the incident x-ray beam with an angular divergence of 10 μrad the peak throughput is only 10%, see Table 5.4. A peak throughput of 20%, is expected for the incident non-divergent beam, according to the dynamical theory calculations. Application of additional collimating optics, like compound refractive lenses should lead to higher peak throughput.

In this respect, the $(+,-,-,+)$ scheme seems to be more beneficial for monochromatization of x rays beams with lower energy photons. For low-energy photons only low-indexed Bragg reflections with relatively large intrinsic relative spectral width $\epsilon^{(s)}$ can be used. This results in a relatively high angular acceptance of the monochromators. Additionally, the asymmetry factor enhancement due to the application of the cascade, may significantly reduce the spectral bandpass. This salient feature was demonstrated in the monochromator designed for 9.405 keV x rays (Toellner and Shu 2001). The monochromator parameters are given in Table 5.4, monochromator #4. The scheme of the monochromator is very similar to that shown in Fig. 5.21(a). The monochromator demonstrated an energy bandpass of 1.0 meV (Toellner et al. 2002), which is the best value achieved so far in this low-energy spectral range, and simultaneously has almost 50% peak throughput, see Table 5.4.

Unequal Cascades. So far we have discussed designs with the same Bragg reflection used in both cascades. One may think of designs, which use different Bragg reflections in different cascades.

For example, the application of a low-indexed reflection in the first cascade could increase the angular acceptance. The asymmetry factors $|b_1|$ and $|b_2|$ have to be chosen in such a way, that the beam after the first cascade has sufficiently small angular divergence. Using low-indexed reflections is advantageous, since as a rule they have higher peak reflectivity.

In the second cascade, high-indexed reflections have to be used to achieve the required spectral properties of the radiation. In this way one may design a device with excellent spectral properties and high peak throughput.

An example of monochromator design with unequal cascades is shown in Figs. 5.33(b). The monochromator crystal parameters and monochromator characteristics are given in Table 5.4, monochromator #2. The monochromator was designed and experimentally demonstrated by Toellner et al. (2002). The design was intentionally constrained to obtain a spectral bandpass of $\simeq 1$ meV, and unit magnification, as this suits most nuclear resonant scattering experiments, given the need for subsequent focusing of the exit x rays on samples.

The monochromator characteristics can be easily varied by changing the asymmetry angles, and Bragg reflections, giving great flexibility to the design and application of the $(+,-,-,+)$-type monochromators. For example, choosing somewhat different asymmetry angles, the spectral bandpass of monochromator #2 can be narrowed by a factor of two without changing the peak throughput, as the characteristics of monochromator #3 in Table 5.4 demonstrate (see also Fig. 5.33(c) for the monochromator scheme). The specified characteristics are based on dynamical theory simulations.

Transmission through unequal cascades can change the beam cross-section, angular divergence, etc., as compared to those of the incident beam. To preserve the beam cross-section and angular divergence one has to take care that $1/(b_1 b_2) \simeq b_3 b_4$. As Fig. 5.33(b) demonstrates, the cross-section of the emerging beam is smaller than that of the incident beam. On the contrary, the monochromator shown in Fig. 5.33(c) increases the beam cross-section. The angular divergence behaves in the opposite way. If the beam cross-section matters one can use additional collimating optics after the monochromator.

A comparison with the characteristics of the two-crystal $(+,+)$ monochromators in Table 5.3 shows that the $(+,-,-,+)$ monochromators with the same spectral bandpass have much higher peak throughput. For example, the $(+,-,-,+)$ monochromator #3 in Table 5.4 has a more than doubled

266 5. High-Resolution X-Ray Monochromators

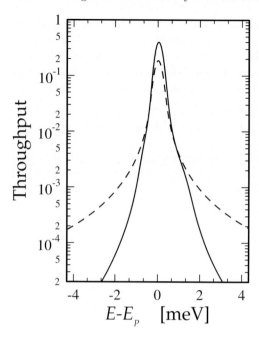

Fig. 5.35. Spectral function of the four-crystal $(+,-,-,+)$ monochromator (#3 in Table 5.4) – solid line – is compared with the spectral function of the two-crystal $(+,+)$ monochromator with almost the same bandpass (#3 in Table 5.3) – dashed line. The spectra are calculated according to the dynamical theory, for an x-ray beam with 10 μrad angular divergence in the dispersion plane.

throughput and slightly smaller energy bandpass, than the $(+,+)$ monochromator #3 in Table 5.3. The cause for the lower peak throughput of the two-crystal $(+,+)$ monochromators is highly asymmetric Bragg reflections. This is not required in the $(+,-,-,+)$ scheme, since the asymmetry factor enhancement works due to the application of the nondispersive reflection cascades.

It is not only the narrowness of the energy bandwidth, which is of critical importance to spectroscopic studies. The presence of steep wings in the monochromator spectral functions are very valuable for discriminating weak signals. Remarkably, the spectral function of the four-crystal $(+,-,-,+)$ monochromator has much steeper wings, because of the four reflections involved, than the spectral function of the same width (FWHM) of the two-crystal $(+,+)$ monochromator, as demonstrated in Fig. 5.35.

Technical Details. Operating four-crystal monochromators is complex, since this requires controlling the position and angular orientation of all four crystals. In the case of the four-crystal $(+,+,-,-)$ monochromator with nested configuration, the operation can be simplified, because the position and angular orientation of two channel-cut crystals, each containing two Bragg reflectors, has to be controlled, instead of four independent crystals, as discussed in Sect. 5.4.3.

The situation with the four-crystal $(+,-,-,+)$ monochromator is not as simple. Only if all four crystals are cut symmetrically, can each cascade of two nondispersively arranged crystals be manufactured as a single channel-cut

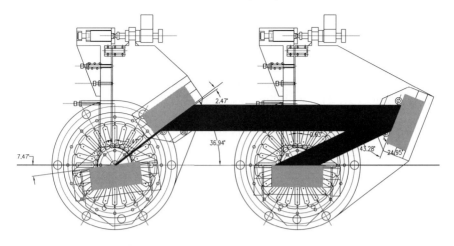

Fig. 5.36. Schematic drawing of the $(+,-,-,+)$ monochromator (monochromator #2, Table 5.4) with two pairs of nondispersively arranged asymmetrically cut crystals, assembled with the help of the weak-link mechanism (see Fig. 5.27) into the two dispersively arranged reflection cascades (Toellner and Shu 2001).

crystal, as shown in Fig. 5.19(b). In the case of interest, which is discussed in the current section, the crystals are cut asymmetrically. As a consequence, despite using the same Bragg reflection, the angle of reflection θ'_1 is not equal to the angle of incidence θ_2, considering the first cascade as an example, and, therefore, the reflecting atomic planes in the two asymmetrically cut crystals must not be parallel, see Fig. 5.33(b). The cascade of two nondispersively arranged asymmetrically cut crystals cannot be manufactured as a monolith channel-cut crystal.

As follows from (3.22)-(3.23), the atomic planes of the crystals in each cascade, which we shall number as cascade $k=1$, and $k=2$, respectively, must be canted by a small angle

$$\theta'_{2k-1} - \theta_{2k} = w^{(s)}_{2k} \tan\theta_{2k-1} \left(1/b_{2k} - b_{2k-1}\right)/2 \qquad (5.7)$$

from the parallel state, about the axis perpendicular to their dispersion planes, see Fig. 5.33. Bragg's law correction $w^{(s)}_{2k}$ is typically much less than 10^{-4}. Therefore, the canting angle is usually small. Bragg's law correction $w^{(s)}_{2k}$ is not only small, but also (almost) energy independent. For these reasons, there is only a weak dependence of the canting angle (5.7) on the photon energy, which is in the first place due to the variation with energy of the glancing angles of incidence θ_{2k-1} and θ_{2k} (not included explicitly in the expression for the canting angle), and (also indirectly) via the asymmetry factors b_{2k-1} and b_{2k}, respectively. In the first approximation, and for small photon energy variations, the canting angle may be regarded as a constant quantity.

Yabashi et al. (2001) controlled four crystals independently, each installed on a separate rotation stage. Toellner and Shu (2001) employed another procedure to operate the monochromator. The relative angular orientation of the crystal pairs in each cascade is adjusted using a weak-link mechanism analogous to the weak-link mechanism of the artificial channel-cut crystals, shown in Fig. 5.27. The weak-link mechanisms are used to adjust the relative orientation of the crystals in the cascades. The assembled cascades, being installed on two independent rotation stages with parallel rotation axes, are oriented relative to the incident beam and relative to each other. Figure 5.36 displays a schematic drawing of the monochromator (monochromator #2, Table 5.4) with two pairs of nondispersively arranged asymmetrically cut crystals each assembled and oriented using the weak-link mechanism for the two dispersively arranged reflection cascades.

Energy Tuning. Each monochromator is designed for a nominal photon energy E_P. It determines the Bragg reflections to be used and the relative angle $\psi_{23} = \theta'_2 + \theta_3$ between the reflecting atomic planes of the second and the third crystals, as discussed in Sect. 3.5. The photon energy can be tuned over a relatively small energy range by changing the angle ψ_{23}. The tuning is described by (3.31): $\delta E = E_P \, \delta\psi_{23}/\tau_{23}$. In practice this is achieved by rotation of the second cascade relative to the first one. The crystals in the cascades are assumed to be adjusted for maximum throughput.

5.4.6 Figure of Merit for $(+,+)$-Type Monochromators

Four different concepts of $(+,+)$-type monochromators were considered in the previous sections. Which of the presented concepts is the most favorable? Which one should be applied in a particular case? There are different basic and practical criteria for choosing the monochromator concept, for judging whether the concept merits attention: spectral bandwidth, peak throughput, angular acceptance, ease of fabrication, ease of operation and control, etc. Let us compare the monochromators designed for the same photon energy, $E = 14.412$ keV, using the cumulative characteristic, the figure of merit M (5.1).

The figure of merit for four-crystal $(+,+,-,-)$ nested monochromators, whose parameters and characteristics are given in Table 5.2, varies from $M \simeq 0.11$ meV^{-1} to $M \simeq 0.22$ meV^{-1} [16]. The figure of merit for the two-crystal $(+,+)$ monochromators given in Table 5.3 varies approximately in the range from $M \simeq 0.13$ meV^{-1} to $M \simeq 0.38$ meV^{-1}, i.e., attains values a factor of two higher.

The figure of merit for the four-crystal $(+,-,-,+)$ monochromators given in Table 5.4 is much higher. It varies in the range from $M \simeq 0.43$ meV^{-1} to $M \simeq 1$ meV^{-1}. The $(+,-,-,+)$ monochromators are not easy to build

---

[16] Here, the values of the peak throughput are calculated for an x-ray beam with a 10 µrad angular divergence, typical for modern synchrotron radiation sources.

and control; however, as the figure of merit informs us, they have the best performance among the discussed $(+,+)$-type monochromators: high peak throughput and narrow spectral bandpass.

Interestingly, the three-crystal $(+,+,\pm)$ monochromators, which will be discussed in Sect. 5.5 show even higher performance, as measured by the figure of merit M.

5.5 Multi-Bounce $(+,+,\pm)$-Type Monochromators

5.5.1 Principle

The common feature of both the single-bounce, and the $(+,+)$-type monochromators, which we have considered in the previous sections, is that their relative spectral bandpass is determined in the first place by the intrinsic relative spectral width $\epsilon_H^{(s)}$ of the Bragg reflections used. The spectral bandpass of the $(+,+)$-type monochromators can be additionally tailored by using the asymmetrically cut crystal optics ($\eta_H \neq 0$, $|b_H| \neq 1$). Still, $\epsilon_H^{(s)}$ remains the main formal parameter, which determines the spectral properties of the monochromators.

The intrinsic relative spectral width of Bragg reflections decreases with increasing magnitude of the diffraction vector $\boldsymbol{H}$ (with increasing Bragg energy), which is clearly seen in Fig. 1.4 on page 7, and from the data in the tables of Appendices A.3 and A.4. This means, that to achieve higher energy resolution, as a rule, higher indexed Bragg reflections have to be used. This in turn requires using high energy x-ray photons (bearing in mind the preference of Bragg reflections in backscattering for the monochromator crystals). To design a monochromator for x-ray photons in the low-energy region $\approx 2-10$ keV with a small bandpass, say about or smaller than 1 meV, becomes problematic, if we must rely on the traditional monochromator concepts.

In the present section we shall discuss a fundamentally different monochromator concept, which unlike the previously considered ones does not rely on the intrinsic spectral properties of Bragg reflections. As we shall see, this concept offers a means of designing monochromators with a sub-meV bandpass for x-ray photons also in the low-energy range. It relies on the *angular dispersion* of x rays in highly asymmetric reflection close to backscattering.

As was shown in Sect. 2.2.5, and Sect. 2.4.9, any asymmetrically cut crystal acts in Bragg diffraction as a dispersing element transforming wavelength into angular dispersion, with respect to the angle of reflection θ'. In backscattering, in the dispersion plane (ϕ_c and $\phi'_c = 0$ or π), the angular dispersion is described by $\delta\theta' \simeq -2(\delta\lambda/2d_H)\tan\eta$ (2.145), with η being the asymmetry angle between the diffracting atomic planes and the entrance crystal surface. A collimator, dispersing element, and wavelength selector compose the elements of the x-ray monochromator, as shown schematically in Fig. 5.1(b) on page 217. The principle is also illustrated in Fig. 5.1(a) with the help of the

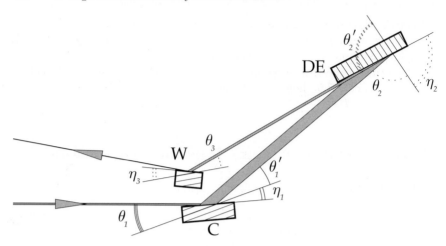

Fig. 5.37. Schematic view of the $(+,+,-)$ monochromator. C - collimator crystal, DE - dispersing crystal element, W - wavelength selector crystal.

classical optical dispersing prism. Slits can be used for the collimator and wavelength selector, as shown in Fig. 5.1(b). A more efficient way to realize the collimator and wavelength selector is, however, by applying asymmetrically cut crystals instead. The resultant design comprises three crystals as shown in Fig. 5.37 (see also Fig. 3.11 on page 161) with the second crystal as the dispersing element, the first crystal as the collimator, and the third crystal as the wavelength selector.

The center of the spectral region of Bragg reflection varies according to Bragg's law with the angle of incidence as $\delta\lambda_c/\lambda_c = \delta\theta_c/\tan\theta_c$. This is the reason for what we termed geometrical broadening in Sect. 5.3, see Eq. (5.3). Since the real beams always have a finite divergence, the effect of geometrical broadening interferes and may blur the effect of angular dispersion, unless the angular divergence of the beam incident upon the dispersing element is small enough. To minimize the influence of the geometrical broadening, and thus to mitigate the requirements imposed on the incident beam divergence, it is beneficial that the dispersing crystal element diffracts in backscattering. The least geometrical broadening is in exact backscattering $\theta \simeq \pi/2$. Therefore, as in the previously considered monochromator concepts, the backscattering geometry is the most favorable.

In Sect. 3.6 it was argued that the $(+,+,-)$ three-crystal configuration is the most favorable in terms of achieving the smallest spectral bandpass. Approaching backscattering (for the dispersing crystal element) the $(+,+,-)$ configuration, however, switches over to the $(+,+,+)$ configuration, as shown in Sect. 3.7. Since both configurations may occur, we shall term the three-crystal monochromators discussed here, as $(+,+,\pm)$ monochromators. The principle and performance of the three-crystal $(+,+,\pm)$ monochromators can

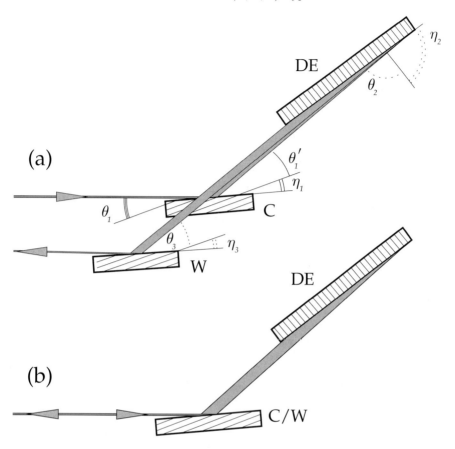

Fig. 5.38. Schematic view of the $(+,+,\pm)$ monochromator. C - collimator crystal, DE - dispersing crystal element, W - wavelength selector crystal. The dispersing crystal element is diffracting (a) in almost exact backscattering, (b) in exact backscattering. In the latter case the collimator and the wavelength selector is one and the same crystal.

be understood via the DuMond diagram analysis of x-ray diffraction from crystals in the $(+,+,-)$ and $(+,+,+)$ configurations developed in Sects. 3.6 and 3.7, respectively.

We shall consider in the following a particular case of the $(+,+,\pm)$ monochromators with the second crystal, the dispersing element, diffracting in almost exact backscattering, as shown in Fig. 5.38(a), or in exact backscattering, as shown in Fig. 5.38(b). Identical Bragg reflections are also used in this case for the collimator and the wavelength selector. An evident merit of using the dispersing element diffracting in (almost) exact backscattering is the (almost) antiparallel propagation directions of the incident and transmitted beams, allowing in-line arrangement of other optical elements of the whole spectrometer, whose main constituent is the monochromator.

5.5.2 Spectral Bandpass

If the second crystal of a $(+,+,\pm)$ three-crystal monochromator, the dispersing element, diffracts in backscattering ($\theta_2 \to \pi/2$), the expression for the monochromator spectral bandpass is especially simple, and is given by (3.41) or (3.47). It is most striking that the relative bandpass given by these equations is independent of the Bragg reflection from the dispersing crystal. It is determined solely by geometrical parameters, such as angular spread of the waves incident upon the dispersing crystal element, its asymmetry angle, and angular acceptance of the wavelength selector crystal. More specifically, the relative bandpass of the $(+,+,\pm)$ monochromator with the dispersing element diffracting in almost exact backscattering is equal to the average $(\Delta\theta'_1 + \Delta\theta_3)/2$ of the angular spread $\Delta\theta'_1$ of the radiation waves incident on the dispersing element and the angular acceptance $\Delta\theta_3$ of the wavelength selector crystal, divided by the factor $\tan\eta_2$. If the collimator and wavelength selector are equivalent, or in the extreme case are even one and the same crystal, as in Fig. 5.38(b), then $\Delta\theta'_1 = \Delta\theta_3 = \Delta\theta_1^{(s)}\sqrt{|b_1|} = \Delta\theta_3^{(s)}/\sqrt{|b_3|}$, and the expression for the relative bandpass of the $(+,+,\pm)$ monochromator simplifies to

$$\frac{\Delta E}{E} = \frac{\Delta\theta_1^{(s)}\sqrt{|b_1|}}{\tan\eta_2}, \tag{5.8}$$

which is equivalent to (3.48). We used here the fact that the angular width of the symmetric Bragg reflection from the first crystal collimator $\Delta\theta_1^{(s)} = \epsilon_1^{(s)} \tan\theta_1$ equals that from the third crystal, the wavelength selector: $\Delta\theta_3^{(s)} = \Delta\theta_1^{(s)}$. Additionally, we assumed that $b_1 = 1/b_3$, with

$$b_1 = \frac{\sin(\eta_1 - \theta_1)}{\sin(\eta_1 + \theta_1)}, \quad \text{and} \quad b_3 = \frac{\sin(\eta_3 + \theta_3)}{\sin(\eta_3 - \theta_3)},$$

obtained from (2.113) with azimuthal angle $\phi_1 = \pi$, and $\phi_3 = 0$, respectively.

As follows from (5.8), there are two ways of varying the relative energy bandpass of the $(+,+,\pm)$ monochromators.

First, it can be varied by changing the asymmetry angle η_2. The smallest $\Delta E/E$ is achieved for the largest possible value of $\tan\eta_2$. Since we do not want to excite additional waves in the crystal, and thus diminish the monochromator throughput, the upper limit for $\tan\eta_2$ is set by the condition $|\gamma_H|^2 \gg |\chi_0|$, as explained in Sect. 2.2.9. This condition ensures that the angle of the reflected beam relative to the surface is much larger than the critical glancing angle of incidence. In our specific case of $\theta_2 \simeq \pi/2$ this condition can be written as $\eta_2 \lesssim \pi/2 - \sqrt{2w_H^{(s)}}$. Here we have used the fact that in backscattering $|\chi_0| \simeq 2w_H^{(s)}$ (2.112). This leads to $\tan\eta_2 \lesssim 1/\sqrt{2w_H^{(s)}}$. The Bragg's law correction $w_H^{(s)}$ is typically $10^{-4} - 10^{-6}$ (see Appendices A.3 and A.4). Therefore, $\tan\eta_2 \approx 100 - 500$ are in principle possible.

The second way of varying the relative energy bandpass of the $(+,+,\pm)$ monochromators is, according to (5.8), to change the angular spread $\Delta\theta_1' = \Delta\theta_1^{(s)} \sqrt{|b_1|}$ of the waves incident upon the dispersing element, and simultaneously to change the acceptance $\Delta\theta_3 = \Delta\theta_3^{(s)}/\sqrt{|b_3|}$ of the wavelength selector crystal, for which we have assumed $\Delta\theta_3 = \Delta\theta_1'$. By selecting even a low-indexed Bragg reflection, and cutting properly the collimator and wavelength selector crystals, it is easy to make $\Delta\theta_1^{(s)} \sqrt{|b_1|} \approx 10^{-6}$.

According to the above estimated values of $\tan\eta_2 \approx 100$, and $\Delta\theta_1^{(s)} \sqrt{|b_1|} \approx 10^{-6}$, and using (5.8), it follows that a relative spectral bandpass as small as $\Delta E/E \approx 10^{-8}$ (and even smaller) for the $(+,+,\pm)$ monochromators is feasible.

The relative energy bandpass $\Delta E/E$ is Bragg reflection and photon energy independent. It is valid both for the monochromators designed for x rays in the high-energy and in the low-energy regions. Photons with an energy as low as ≈ 2 keV can be considered[17].

Since the relative spectral bandwidth $\Delta E/E$ given by (5.8) is independent of Bragg reflection and photon energy (a very beneficial feature of the $(+,+,\pm)$ monochromators!) the energy width ΔE is smaller for lower photon energy E. This means, the spectral properties of the $(+,+,\pm)$ monochromators tend to become better with decreasing photon energy, as opposed to the case of the single-bounce and $(+,+)$-type monochromators, whose spectral properties tend to improve with increasing photon energy.

Table 5.5 gives examples of $(+,+,\pm)$-type monochromators, with their parameters and characteristics, designed for x-ray photons with an energy of $E_P \simeq 9.13$ keV[18].

---

[17] The 2 keV limit is determined here, as it is the typical smallest energy of the photons, which can be diffracted by crystals, such as Si or α-Al$_2$O$_3$, as can be verified by inspection of the tables in Appendices A.3 and A.4.

[18] The (0 0 8) Bragg reflection with Bragg energy $E_H = 9.13$ keV is chosen for the dispersing crystal element in the presented monochromator design. This choice is dictated by two reasons. The first reason is a small number (four) of accompanying reflections, which can be excited simultaneously with the (0 0 8) Bragg reflection in exact backscattering. The second reason is the fact that all four accompanying reflections can be effectively suppressed. All accompanying reflected waves propagate in this particular case perpendicular to the incident beam. Therefore, one conjugate pair of accompanying reflections can be suppressed by a judicious choice of the polarization of the incident wave, as discussed in Sect. 2.5.5. The second pair is suppressed due to asymmetric Bragg diffraction in backscattering with the incident wave propagating at a small angle to the entrance crystal surface, see Sect. 2.7. The choice of the (0 0 8) Bragg reflection is not unique.

Table 5.5. Characteristics of the three-bounce $(+,+,\pm)$ monochromators, shown schematically in Fig. 5.38, and parameters of the silicon crystals composing the monochromators. E_P is the design value of the photon energy.
Crystal parameters: crystal number n, Bragg reflection (hkl), glancing angle of incidence θ at which the photons with energy E_P are reflected, asymmetry angle η, absolute value of the asymmetry factor $|b|$.
Monochromator parameters: energy bandpass ΔE (FWHM), angular acceptance $\Delta\theta$ (FWHM), peak throughput T, figure of merit M, as (t) calculated by using exact equations of the dynamical theory, or (d) calculated by using the DuMond diagram technique with (3.40) and (3.42). The energy bandpass ΔE, and the peak throughput T are calculated for an x-ray beam with 10 μrad angular divergence. See footnote 3 on page 149 for the reasons for the discrepancies between the (t) and (d) values.

monochromator		monochromator crystals				monochromator				
#	E_P [keV]	n	(hkl)	θ_n [deg]	η_n [deg]	$\|b_n\|$	ΔE [meV]	$\Delta\theta$ [μrad]	T	M [meV^{-1}]	
1	9.13155										
		1	(0 2 2)	20.71	14.78	1/5.6	3.9	20	0.78	0.20	(t)
		2	(0 0 8)	89.80	84.80	1.1	7.3	49	1		(d)
		3	(0 2 2)	20.71	14.78	5.6					
2	9.13155										
		1	(0 2 2)	20.71	14.78	1/5.6	0.83	18	0.78	0.94	(t)
		2	(0 0 8)	89.80	88.80	1.4	1.7	49	1		(d)
		3	(0 2 2)	20.71	14.78	5.6					
3	9.13155										
		1	(0 2 2)	20.71	14.78	1/5.6	0.25	15	0.74	3.0	(t)
		2	(0 0 8)	89.80	89.60	3.1	0.56	49	1		(d)
		3	(0 2 2)	20.71	14.78	5.6					

Figure 5.39 shows results of evaluations of the throughput as function of the photon energy for the $(+,+,\pm)$ monochromators, whose parameters are given in Table 5.5. The throughput is calculated for three different values of the asymmetry angle η_2 of the second crystal, the dispersing element. The energy bandpass changes with the asymmetry angle from $\Delta E = 3.9$ meV for monochromator #1, to $\Delta E = 0.83$ meV for monochromator #2, and to $\Delta E = 0.25$ meV for monochromator #3. While the energy bandpass changes drastically, the peak throughput in the three cases shown in Fig. 5.39 is almost constant, varying slightly from $T = 0.78$ to 0.74, respectively. This salient feature of the $(+,+,\pm)$ monochromators will be discussed in more detail below.

5.5 Multi-Bounce $(+, +, \pm)$-Type Monochromators 275

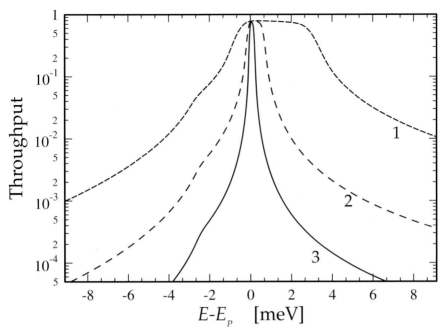

Fig. 5.39. Throughput of the three-crystal $(+, +, \pm)$ monochromator as a function of the photon energy E, calculated for three different values of the asymmetry angle η_2 of the second crystal, the dispersing crystal element. $1 : \eta_2 = 84.8°$, $2 : \eta_2 = 88.8°$, and $3 : \eta_2 = 89.6°$. Other crystal parameters are given in Table 5.5. The energy bandpass changes with the asymmetry angle from $\Delta E = 3.9$ meV (curve 1), to $\Delta E = 0.83$ meV (curve 2), and to $\Delta E = 0.25$ meV (curve 3). The throughput is calculated for an x-ray beam with 10 μrad angular divergence.

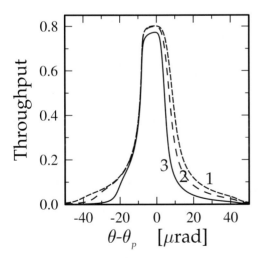

Fig. 5.40. Throughput of the three-crystal $(+, +, \pm)$ monochromators similar to Fig. 5.39, however now calculated as a function of the glancing angle of incidence θ (here θ is actually the glancing angle of incidence θ_1 on the first crystal, see Fig. 5.38). Incident x rays have a spectral distribution bandwidth equal to the bandpass ΔE of the monochromator, as given in the caption of Fig. 5.39. The angular acceptance of the monochromators changes from $\Delta \theta = 20$ μrad (1), to $\Delta \theta = 18$ μrad (2), and to $\Delta \theta = 15$ μrad (3), respectively.

Table 5.6. Characteristics of the three-bounce $(+, +, \pm)$ monochromators, shown schematically in Fig. 5.38, and parameters of the silicon crystals composing the monochromators. E_P is the design value of the photon energy.
Crystal parameters: crystal's number n, Bragg reflection (hkl), glancing angle of incidence θ at which the photons with energy E_P are reflected, asymmetry angle η, absolute value of the asymmetry factor $|b|$.
Monochromator parameters: energy bandpass ΔE (FWHM), angular acceptance $\Delta\theta$ (FWHM), peak throughput T, figure of merit M, as (t) calculated by using exact equations of the dynamical theory, or (d) calculated by using the DuMond diagram technique with (3.40) and (3.42). The energy bandpass ΔE, and the peak throughput T are calculated for an x-ray beam with 10 μrad angular divergence. See footnote 3 on page 149 for the reasons for the discrepancies between the (t) and (d) values.

monochromator		monochromator crystals				monochromator				
#	E_P [keV]	n	(hkl)	θ_n [deg]	η_n [deg]	$\|b_n\|$	ΔE [meV]	$\Delta\theta$ [μrad]	T	M [meV^{-1}]	
1'	9.13155										
		1	(0 2 2)	20.72	19.71	1/37	1.5	96	0.63	0.4	(t)
		2	(0 0 8)	89.80	84.80	1.1	2.8	126	1		(d)
		3	(0 2 2)	20.70	19.71	37					
2'	9.13155										
		1	(0 2 2)	20.72	19.71	1/37	0.32	86	0.63	2.0	(t)
		2	(0 0 8)	89.80	88.80	1.4	0.65	126	1		(d)
		3	(0 2 2)	20.70	19.71	37					
3'	9.13155										
		1	(0 2 2)	20.72	19.71	1/37	0.09	65	0.61	6.7	(t)
		2	(0 0 8)	89.80	89.60	3.1	0.22	126	1		(d)
		3	(0 2 2)	20.70	19.71	37					

Figure 5.41 shows the results of similar calculations for the same parameters of the dispersing crystal element. However, in this situation the asymmetry angle η_1 of the collimator crystal and the asymmetry angle η_3 of the wavelength selector crystal are being increased, as given in Table 5.6. These changes result in the decreased asymmetry factors $|b_1|$ and $1/|b_3|$, respectively. The final effect is consistent with expectations based on Eq. (5.8): the energy bandpass of the monochromators becomes smaller. The energy bandpass is $\Delta E = 1.5$ meV for monochromator #1', $\Delta E = 0.35$ meV for monochromator #2', and $\Delta E = 0.09$ meV for monochromator #3', see Fig. 5.41. While the energy bandpass changes drastically, the peak throughput in all the three cases again is almost constant, varying slightly from $T = 0.63$ to 0.61, respectively.

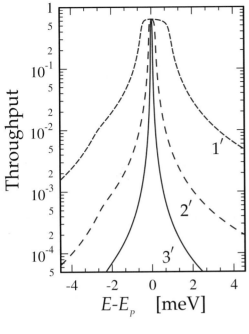

Fig. 5.41. Throughput of the three-crystal $(+,+,\pm)$ monochromator as in Fig. 5.39, however, now the asymmetry angle η_1 of the collimator and the asymmetry angle η_3 of the wavelength selector crystals are changed (compare the crystal parameters in Table 5.5 and Table 5.6) to increase the angular acceptance, and simultaneously, to reduce the spectral bandpass. The latter now becomes $\Delta E = 1.5$ meV ($1'$), $\Delta E = 0.32$ meV ($2'$), and $\Delta E = 0.09$ meV ($3'$).

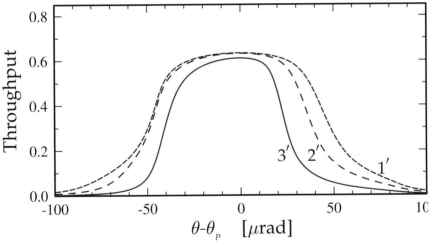

Fig. 5.42. Throughput of the three-crystal $(+,+,\pm)$ monochromators, similar to Fig. 5.41, however, now calculated as a function of the glancing angle of incidence θ. The x-ray beam is assumed to have a width of the spectral distribution equal to the bandpass of the corresponding monochromator, as given in the caption of Fig. 5.41. The angular acceptance of the monochromators changes from $1'$: $\Delta\theta = 96$ μrad, to $2'$: $\Delta\theta = 86$ μrad, and $3'$: $\Delta\theta = 65$ μrad, respectively.

5.5.3 Angular Acceptance

In the Dispersion Plane. According to the arguments presented at the end of Sect. 3.6, the angular acceptance $\Delta\theta$ of the three-crystal $(+,+,\pm)$

monochromators with respect to the incident waves, is determined in the first place by the angular acceptance of the Bragg reflection from the first crystal: $\Delta\theta \simeq \Delta\theta_1^{(s)}/\sqrt{|b_1|}$ (3.42). Comparison with (5.8), in particular, leads to the conclusion that the collimator crystal cut at an asymmetry angle η_1, and oriented to the incident beam such that the asymmetry factor $|b_1| \ll 1$, increases the angular acceptance of the $(+, +, \pm)$ monochromator, and simultaneously narrows its spectral bandpass!

The angular acceptance $\Delta\theta$ of the three-crystal $(+, +, \pm)$ monochromators is independent of the asymmetry angle η_2 of the dispersing crystal element. This means that by varying η_2 one can change the energy bandpass of the monochromator without changing its angular acceptance - a very attractive feature for x-ray monochromators. This prediction is based on the DuMond diagram analysis of Sect. 3.6.

The exact dynamical theory calculations in principle confirm this prediction, however, not completely. They reveal a weak dependence of the angular acceptance of the $(+, +, \pm)$ monochromators on the asymmetry angle of the dispersing crystal element. Examples of calculations shown in Figs 5.40 and 5.42 demonstrate that the angular width narrows with increasing asymmetry angle η_2. However, the narrowing is by far not as drastic as the changes, which occur simultaneously to the energy bandwidth. While the angular width becomes smaller by a factor of a maximum ≈ 1.5, the energy bandwidth reduces by a factor ≈ 15.

In the Plane Perpendicular to the Dispersion Plane. A wave propagating parallel to the dispersion plane is described by the wave vector $\boldsymbol{K}_0 = K(\pm\cos\theta, 0, -\sin\theta)$ (2.25) with zero y'-component (see Fig. 2.4 on page 48), and with $\phi = 0$, or $\phi = \pi$. The wave vector $\boldsymbol{K}_0$ of the incident radiation, which makes a small angle $|\delta\Phi| \ll 1$ with the dispersion plane, acquires a small y'-component $K\delta\Phi$. As a result, the wave vector $\boldsymbol{K}_0$ changes to $\boldsymbol{K}_0 = K[\cos(\theta + \delta\theta)\cos\delta\phi, \cos(\theta + \delta\theta)\sin\delta\phi, -\sin(\theta + \delta\theta)]$ (2.25). Being expressed in the same terms, the angular deviation $\delta\Phi$ takes the form $\delta\Phi = \cos(\theta + \delta\theta)\sin\delta\phi$.

Thus, if the glancing angle of incidence θ is *not* close to $\pi/2$, the small angular deviation $\delta\Phi$ off the dispersion plane results only in a small change of $\theta \rightarrow \theta + \delta\theta$, with $|\delta\theta| \ll 1$, and of the azimuthal angle of incidence $\phi \rightarrow \phi + \delta\phi$, with $|\delta\phi| \ll 1$, and therefore may result only in an insignificant shift of the spectral bandpass of a Bragg reflection in question.

However, if the glancing angle of incidence θ is close to $\pi/2$, $\delta\phi$ can no longer be considered as small, and the spectral changes as insignificant.

Applying these general considerations to our case, we conclude. The performance of the collimator and of the wavelength selector, having both $\theta \ll \pi/2$, is quite insensitive to the angular spread $\Delta\Phi$ of the propagation directions of the incident x-rays in the plane perpendicular to the dispersion plane. (We remind the reader, that the dispersion plane is one and the same for all three crystal elements of the $(+, +, \pm)$ monochromator.) On the contrary, even a small angular spread $\Delta\Phi$ may seriously impair the spectral characteristics of the dispersing element operating in extreme backscattering, and thus of the $(+,+,\pm)$ monochromator as a whole.

Yet, in one particular case the spectral properties of the dispersing element, and thus of the $(+,+,\pm)$ monochromator, is rather insensitive to the angular spread $\Delta\Phi$. As Fig. 2.17(a) on page 86 suggests (see also discussion of Fig. 2.17), in *exact* or in almost *exact* backscattering, when the projection of the wave vector $\boldsymbol{K}_0$ onto the (x', y') plane is in the vicinity of $(\Theta_R, 0)$, the angular spread $\Delta\Phi$ of the incident waves, resulting in the variations of the wave vector projections parallel to the y'-axis in Fig. 2.17(a), will produce no significant spectral broadening if

$$\Delta\Phi \lesssim \sqrt{\frac{\Delta E}{E}}. \tag{5.9}$$

Equation (5.9) gives also an estimate of the admissible angular acceptance of the $(+,+,\pm)$ monochromators (for angular deviations in the plane perpendicular to the dispersion plane) in almost *exact* backscattering geometry. For monochromator #3′ in Table 5.6 with an energy bandpass of $\Delta E = 0.09$ meV and $\Delta E/E_P \simeq 10^{-8}$, the admissible angular spread is thus $\Delta\Phi \simeq 0.1$ mrad, which is quite a large value.

5.5.4 Energy Tuning

If the second monochromator crystal, the dispersing element, diffracts in backscattering, the energy tuning of the $(+,+,\pm)$ monochromator can be performed by varying its (dispersing element) temperature, as is similarly done in the case of the single-bounce backscattering monochromators. The variation with crystal temperature of the energy of photons reflected in backscattering $\delta E/\delta T = E_H \beta_H$ (2.109) increases with Bragg energy E_H (β_H is the linear thermal expansion coefficient of the crystal in the direction of the diffraction vector $\boldsymbol{H}$). In this respect, a next attractive property of the $(+,+,\pm)$ monochromators has to be pointed out here.

As we know, a smaller bandpass is required, thus a Bragg reflection with higher Bragg energy E_H has to be employed, for the case of the single-bounce monochromators. This has important implications for the temperature control of the monochromator crystal. It becomes more and more demanding, since the energy variations become more and more sensitive to smaller and smaller changes in the temperature.

Unlike the single-bounce monochromators, the energy bandpass of the $(+,+,\pm)$ monochromators can be tailored by changing the asymmetry angle of the dispersing crystal element, i.e., without changing the photon energy. Since monochromators with very small energy bandpass are also feasible at small photon energies, thus using Bragg reflections with small E_H, temperature control is not as troublesome as for the single-bounce monochromators (with the same energy bandpass).

5.5.5 Throughput, Figure of Merit, Spectral Function

Throughput. As was already mentioned in Sect. 5.5.2, while the spectral bandpass of the $(+,+,\pm)$ monochromators can be efficiently varied by changing the asymmetry angle η_2 of the dispersing element, the angular acceptance and the peak throughput do not vary as significantly with η_2. If the dispersing crystal element diffracts in almost exact backscattering, the variations of the throughput are not significant at all, as the examples of calculations shown in Figs 5.39 and 5.41 demonstrate.

This salient property of the $(+,+,\pm)$ monochromators is explained in the following way. The angular dispersion, which underlies the operation of the $(+,+,\pm)$ monochromators depends directly on the asymmetry angle η_2 of the dispersing element, resulting in equation (5.8) for the energy bandpass. As opposed to this, the expressions for the crystal reflectivity in Bragg diffraction (see Sect. 2.4) contain the asymmetry factor as a formal parameter. This, however, varies marginally with η_2 in almost exact backscattering, being almost constant and very close to $b_2 \simeq -1$.

Figure of Merit. The energy bandpass of the $(+,+,\pm)$ monochromators can be made very narrow, with at the same time very high throughput. Therefore, the figure of merit, the cumulative characteristic of the monochromator quality introduced in Sect. 5.4.6, can take very large values in the case of $(+,+,\pm)$ monochromators. The figure of merit of the $(+,+,\pm)$ monochromators presented in Table 5.5 and Table 5.6 attains values as high as $M \simeq 6.7$ meV^{-1}, i.e., a factor $\simeq 6.7$ greater than the figure of merit of the best $(+,+)$-type monochromators.

Spectral Function. As was already mentioned, it is not only the narrowness of the energy bandwidth, which is of critical importance to spectroscopic studies. Monochromator spectral function wings with a steep slope are very valuable as well. Figure 5.43 shows the spectral function of a three-crystal $(+,+,\pm)$-type monochromator compared with the spectral functions of other monochromators, all having (almost) the same spectral bandwidth $\Delta E \simeq 0.8 - 0.9$ meV (FWHM). The solid line presents the spectral function of the $(+,+,\pm)$ three-crystal Si monochromator, with the parameters of monochromator #2 in Table 5.5. It is asymmetric because of the asymmetric form of the reflectivity curves of the crystals involved in Bragg scattering. The wings of the spectral function have a steeper slope, than the wings of the functions representing the single-bounce backscattering monochromator (dotted line), and of the two-bounce $(+,+)$ monochromator (dashed-dotted line), with the same bandwidth. However, they are less steep than the wings of the spectral function of the four-crystal $(+,-,-,+)$ monochromator. Such behaviour is consistent with intuitive expectations: the more crystals that are involved in the reflection process the steeper are the wings.

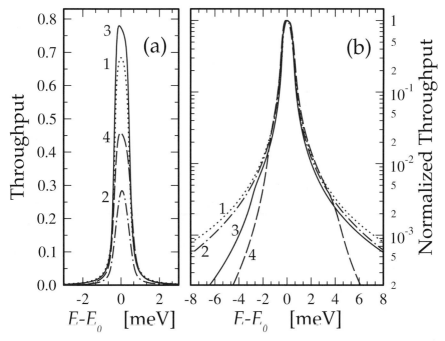

Fig. 5.43. (a) Spectral functions of different types of monochromators. (b) The same spectral functions are shown in logarithmic scale and normalized to the peak throughput.
Dynamical theory calculations for an incident x-ray beam with 10 μrad angular divergence. The spectral functions of the monochromators with (almost) the same bandwidth ΔE (FWHM) are compared:
(1) Single-crystal backscattering monochromator with the (11 11 11) Bragg reflection from silicon, $\Delta E = 0.82$ meV, $E_0 = 21.7$ keV.
(2) Two-crystal $(+,+)$ monochromator (#2 in Table 5.3), $\Delta E = 0.80$ meV, $E_0 = 14.4$ keV.
(3) Three-crystal $(+,+,\pm)$ monochromator (#2 in Table 5.5), $\Delta E = 0.83$ meV, $E_0 = 9.1$ keV.
(4) Four-crystal $(+,-,-,+)$ monochromator (#4 in Table 5.4), $\Delta E = 0.91$ meV, $E_0 = 9.4$ keV.

5.5.6 Technical Details

As usual, the experimental realization of the $(+,+,\pm)$ monochromators is connected with diverse technical problems.

Attenuation in the Collimator. One of the problems becomes evident by inspection of the monochromator design shown in Fig. 5.38(a). The beam reflected from the dispersing element again strikes the crystal collimator. Because of photo-absorption, the beam intensity at the wavelength selector can be greatly attenuated. The problem can be solved at least in one of two ways.

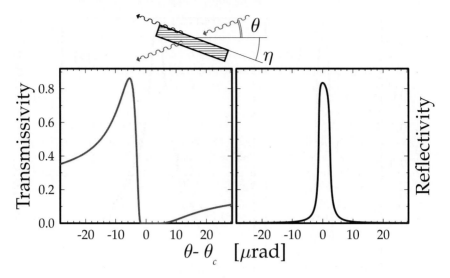

Fig. 5.44. Angular dependence of transmission (left), and reflection (right) of x rays with $E = 9.13$ keV photon energy in Bragg diffraction by a silicon crystal, the (220) Bragg reflection. Crystal plate thickness $d = 100$ μm, $\theta_c = 20.7°$, asymmetry angle $\eta = 19.7°$, asymmetry factor $b = -37$ (the same parameters as the collimator crystal of the monochromators presented in Table 5.6).
At $\theta - \theta_c \simeq -6$ μrad the crystal is highly transparent. Such transparency will have the collimator crystal (C) in the $(+,+,\pm)$ monochromator shown in Fig. 5.38(a) set in this angular position with respect to x rays propagating from the dispersing element (DE) to the wavelength selector (W).

First, the collimator crystal can be made very thin $\simeq 1 - 10$ μm. This thickness on the one hand is sufficient to achieve maximum reflectivity for the primary beam, e.g., with the (2 2 0) Bragg reflection from the collimator crystal. On the other hand, since the beam reflected from the dispersing element reenters the crystal collimator at a quite large angle of $\approx \theta_1 + \eta_1 \simeq 40°$, the photo absorption for this beam is minimal, less then 90 %. Such thin crystals, however, are difficult to handle.

To have a small attenuation level, and simultaneously thicker, easy to handle collimator crystals, the effect of anomalous transmission in Bragg scattering geometry, a kind of Borrmann effect (Pinsker 1978, Authier 2001), can be additionally used. This is the second among the two aforementioned possible solutions. Figure 5.44 explains briefly how it works according to the principles of the dynamical diffraction theory. The reflectivity and transmissivity curves are shown calculated for x rays in Bragg diffraction by the collimator crystal for different angles of incidence of the beam propagating from the dispersing element to the wavelength selector. As can be seen the transmission maximum is achieved at a glancing angle of incidence $\theta = \theta_c - \delta\theta$, which is $\delta\theta \simeq 6$ μrad smaller than that for the reflection maximum at $\theta = \theta_c$. For a 100 μm thick crystal the peak transmissivity is as high as $\simeq$ 90 %. Thus if Bragg diffraction in the dispersing element is set $\delta\theta/2$ off exact backscattering then only $\simeq$ 10 % is absorbed in the collimator, and $\simeq$ 90 % of the radiation will reach the wavelength selector (in the peak of monochromator throughput). Clearly, the thinner the collimator crystal the higher is the transmissivity.

Length of the Dispersing Element. Because of the asymmetric reflections from the first two crystals, the size of the beam footprint on the dispersing element is by a factor of $\tan\eta_2/b_1$ magnified with respect to the size of the primary beam incident on the collimator crystal, see Fig. 5.38. For example, in the extreme case of the $(+,+,\pm)$ monochromator with a very small bandpass of $\Delta E = 90$ μeV, the monochromator #3′ in Table 5.6, the magnification factor is $\simeq$ 5000. For the $(+,+,\pm)$ monochromator with an energy bandpass of $\Delta E = 1.5$ meV, the monochromator #1′ in Table 5.6, the magnification factor is $\simeq$ 400. This means, either the primary beam size has to be small, or the length of the dispersing element has to be large.

5.5.7 Summary

This section will be a brief summary of the distinctive properties and merits of the $(+,+,\pm)$ monochromators mentioned above.

1. The relative spectral bandpass $\Delta E/E$ is independent of Bragg reflection and the photon energy. The smaller the photon energy E the smaller is the bandpass ΔE. The spectral properties of the $(+,+,\pm)$ monochromators thus tend to become better with decreasing photon energy.

2. The monochromator spectral bandpass can be varied efficiently by changing the asymmetry angle of the dispersing element. Thus, the bandpass can be varied without changing the nominal photon energy of the monochromator (unlike the single-bounce backscattering monochromator).

3. Changing the monochromator spectral bandpass by varying the asymmetry angle of the dispersing element (as above) does not, however, change significantly the throughput and angular acceptance of the monochromator. In other words, the spectral characteristics of the monochromator can be significantly improved, without worsening appreciably its throughput and angular acceptance. This is a unique monochromator property.

4. The temperature control and thus the energy tuning of the $(+,+,\pm)$ monochromators is technically not very demanding, as long as the monochromators are designed and operated with x-ray photons in the low-energy region $(2-10 \text{ keV})$.

5. The figure of merit of the $(+,+,\pm)$ monochromators considered can attain values as high as $M \approx 7 \text{ meV}^{-1}$.

5.6 µeV-Resolution Fabry-Pérot Interference Filters

In the previous section a monochromator concept was presented, which does not rely on intrinsic spectral properties of Bragg reflections. The effect of angular dispersion underlies the concept.

Another principle of monochromatization, which does not rely on intrinsic spectral properties of Bragg reflections, and which uses solely the high reflectivity of crystals in Bragg diffraction, is *multiple-wave interference*. It is implemented as a Fabry-Pérot resonator or, in this case it is better to say, as a Fabry-Pérot interference filter. X rays with extremely small spectral spread can be filtered in this way. The theory of x-ray Fabry-Pérot resonators (interference filters) was discussed in detail in Chap. 4. One example of realization of this monochromator concept is discussed in this section.

The higher the reflectivity of the mirrors the larger is the finesse of the resonator and thus the sharper are the Fabry-Pérot resonances. This rule applies to Fabry-Pérot resonators in any spectral range.

High peak reflectivity for the electromagnetic radiation in the x-ray spectral range is achieved by Bragg diffraction from single crystals. The crystal reflectivity is defined by two competing effects: Bragg diffraction and photo-absorption. As was discussed in Sect. 2.4.3, the reflectivity of a non-absorbing crystal with thickness larger than the extinction length approaches 100%. However, since photo-absorption can never be avoided, 100% reflectivity exists only in the theory. The larger the ratio $d_e(0)/d_{\mathrm{ph}}$ of the extinction length to the photo-absorption length the closer is the peak reflectivity to the 100% limit. Theoretically the highest reflectivity can be achieved by using such crystals as diamond (C) or beryllium (Be).

Good quality beryllium crystals, with almost perfect crystal lattices, are not yet available. Good diamond single crystals are already available and their quality is steadily improving (Ishikawa et al. 2000, Sellschop et al. 2000).

Bragg back-reflection from diamond crystals can clearly be influenced by multiple-beam diffraction effects, resulting in considerably reduced reflectivity, in a similar way to silicon crystals belonging to the same space group. One has to employ different ways of suppression of the accompanying reflections. For example, this can be achieved by directing the incident beam slightly off normal incidence to the reflecting atomic planes (responsible for backscattering). If there is a small number of accompanying reflections, then an angular deviation of $\approx 5-10$ µrad can be sufficient, as demonstrated in Sect. 2.6.3 (see also Fig. 2.43 on page 138) by an example of 4-beam diffraction. Such a small angular deviation does not yet impair significantly the angular acceptance of x-ray Fabry-Pérot resonators. When excited with incident radiation slightly off exact backscattering the resonators have still very good performance, as was discussed in Sect. 4.3.7 and illustrated in Fig. 4.11 on page 205.

Figure 5.45 shows calculated transmission spectra of Fabry-Pérot interference filters built from diamond single crystals for x rays with photon energy

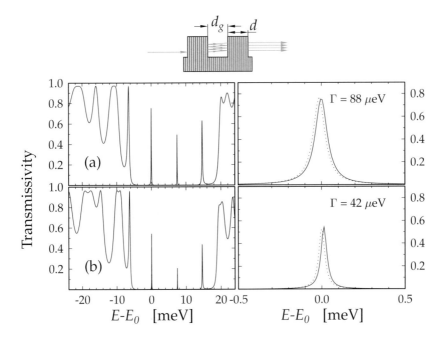

Fig. 5.45. Energy dependence of transmission through a Fabry-Pérot interference filter made of two single crystal plates of diamond - left panels. The right panels show the parts of the same spectra in a narrower energy range. The vacuum gap between the plates is taken to be $d_g = 64.0$ μm. The dotted line shows the sensitivity of the resonances to a change of the gap by 0.02 μm. The reflecting atomic planes in both crystal plates are in perfect registry relative to each other, i.e., $\boldsymbol{H}\boldsymbol{u} = 2\pi n$ (see Sect. 4.3.4 for details). The thicknesses of the crystal plates are (a) $d = \tilde{d} = 50$ μm, (b) $d = \tilde{d} = 60$ μm, and their temperatures are $T = 300$ K. The incident radiation is a plane monochromatic wave at normal incidence ($\delta\theta = 0$) to the (444) reflecting atomic planes. $E_0 = 12.040$ keV.

$E \simeq 12$ keV. Two different spectra shown in Figs. 5.45(a) and (b) are the results of calculations for resonator crystal mirrors of two different thicknesses: (a) $d = \tilde{d} = 50$ μm, and (b) $d = \tilde{d} = 60$ μm, respectively.

In the case of the resonator with crystal mirrors of thickness $d = \tilde{d} = 50$ μm - Fig. 5.45(a) - the Fabry-Pérot resonance at $E = E_0$ has a spectral width $\Gamma = 88$ μeV, and a peak transmissivity (throughput) as high as 75%. The deviation from 100% is only because of photo-absorption in diamond. The relative spectral width for this Fabry-Pérot resonance is $\epsilon_{\rm FP} = \Gamma/E = 7.3 \times 10^{-9}$. As follows from (4.66) the angular acceptance of the interference filter employing this resonance is $\Delta\theta \simeq 2\sqrt{\epsilon_{\rm FP}} = 170$ μrad. All the numbers, the resonance spectral width, angular acceptance, and peak transmission (or throughput), are exceptionally good for this instrument when used as a monochromator[19].

In the case of the resonator with crystal mirrors of thickness $d = \tilde{d} = 60$ μm - Fig. 5.45(b) - the Fabry-Pérot resonance at $E = E_0$ has a spectral width of $\Gamma = 42$ μeV, a relative spectral width of $\epsilon_{\rm FP} = \Gamma/E = 3.5 \times 10^{-9}$, an angular acceptance $\Delta\theta \simeq 120$ μrad, and a peak transmissivity of 53%.

The figure of merit of these two Fabry-Pérot resonators, being used as monochromators, are $M \simeq 8.5$ meV^{-1} (a), and $M \simeq 12.6$ meV^{-1} (b), respectively. These numbers are about two times higher than those of the $(+,+,\pm)$ monochromators, the monochromators with the highest values of the figure of merit discussed so far[20].

It is a technical challenge to realize such a resonator. For the resonator to be robust in operation, it is most beneficial to carve out the back-reflecting mirrors in a monolithic diamond crystal, as shown, e.g., in the inset of Fig. 5.45. In this case the very precise adjustment of the mirrors and keeping them in the parallel setting at a fixed distance would be not necessary. The difficult problem is to manufacture the gap between the mirrors, which has to be a few tens of micrometers wide. But probably the most difficult problem is to make the gap with 10 nm precision homogeneous over the area illuminated by the incident x rays. The homogeneity is necessary to prevent blurring off the very sharp resonances. The dotted lines in the right panels of Figure 5.45 show the results of calculations demonstrating how the transmission resonance alters its position as a result of a minute variation of the gap by 20 nm.

One of the interesting applications of such μeV filters could be x-ray holography requiring not only transverse but also longitudinal coherence. Future x-ray sources based on free-electron lasers driven by linacs will provide radiation with full transverse coherence, and about 1 mm^2 in cross-section (LCLS 1998, Materlik and Tschentscher 2001). The above described interference filters could be used to obtain photons with a coherence length of about 1 mm, and thus completely coherent in a volume of 1 mm^3.

---

[19] Another outstanding feature of diamond Fabry-Pérot resonators demonstrated by the spectra in Figs. 5.45(a) and (b) is a finesse of $\simeq 100$, which is exceptionally large, as compared to x-ray Fabry-Pérot resonators built from other crystals. The finesse achievable with sapphire backscattering mirrors is about $15-25$. For silicon crystals it is typically about a factor of two less.

[20] Obviously, such μeV interference filters have to be used in combination with meV monochromators, which have first to select one of the Fabry-Pérot resonances. This may make the figure of merit smaller.

6. High-Resolution X-Ray Analyzers

6.1 Introduction

Along with the monochromator, the x-ray analyzer is another major component for inelastic x-ray scattering instrumentation. Examples of spectrometers for inelastic x-ray scattering experiments are shown in Figs. 1.10, 1.13, and 1.15 on pages 13, 15, and 17, respectively. The performance of the analyzers, as for the monochromators, is evaluated in terms of spectral resolution (or bandpass), angular acceptance, and throughput.

Both instruments (monochromator and analyzer) being used in the same spectrometer must have in the optimum case approximately the same energy resolution. As was already mentioned, investigations of vibrational excitations in matter, such as phonon dispersion relations $\hbar\omega(\boldsymbol{Q})$, require the energy resolution of the whole instrument (monochromator and analyzer) to be better, often much better, than $\approx 1 - 5$ meV.

In contrast to the monochromator, which has to monochromatize x rays emerging from an undulator in a cone with an opening angle of $\lesssim 20$ μrad, the analyzer has to accept the radiation emerging from the sample in a cone with a much larger opening angle ≈ 5 mrad, while providing energy filtering with an equally good resolution. This is a challenging problem.

The angular acceptance of the analyzer is altogether not a simple issue. On the one hand, it has to be as large as possible to gather as many photons as possible scattered by the sample. On the other hand, the maximum angular acceptance of the analyzer is limited by the desired momentum transfer resolution in scattering from the samples. For acoustic phonons, one needs as much momentum resolution as possible, especially if the speed of sound is high. On the other hand, for low dispersion optical modes, one can afford lower resolution. The required momentum-resolution is sample- and experiment-dependent. It is usually required to be $\Delta Q \approx 0.5$ nm^{-1}, which is a compromise between physical requisites and photon flux limitation[1]. Assuming a photon energy $E \simeq 20$ keV, which corresponds to a wavelength of $\lambda \simeq 0.06$ nm, we may find that the angular acceptance of the analyzer has

---

[1] Typical count-rates in the experiments on inelastic x-ray scattering are $\simeq 1$ counts/s and even less at modern synchrotron radiation sources. For details on the estimation of the photon intensities see, e.g., (Sinn 2001).

to be $\varUpsilon^2 = 5 \times 5$ mrad^2, where $\varUpsilon$ is estimated as $\varUpsilon = \Delta Q/K \approx 5$ mrad, with $K = 2\pi/\lambda$. No single Bragg reflection at $E \simeq 20$ keV has such a large angular acceptance even in the backscattering geometry. Typical values are $\lesssim 0.5$ mrad, as can be found in the tables of Appendices A.3 and A.4.

We shall consider in this chapter two different types of x-ray analyzers.

The first type is an analyzer composed of a great number (≈ 10000) of small ($\approx 1 \times 1$ mm^2) crystals attached to a spherical substrate. Single Bragg reflection in backscattering from these crystals is used. The energy resolution of such analyzers improves as the higher indexed Bragg reflections, and thus photons with higher energy ($\gtrsim 20$ keV) are used.

The second type is an analyzer, whose spectral properties on the contrary, becomes better as the photon energy decreases ($\lesssim 10$ keV). It is composed of a collimating mirror and the $(+,+,\pm)$-type monochromator.

At the end of each section, devoted to a certain analyzer, we shall discuss layouts of spectrometers for inelastic x-ray scattering with very high energy resolution.

6.2 Spherical Crystal Analyzers

6.2.1 Historical Background

To perform the energy analysis with meV resolution of the photons emerging from the sample in a large solid angle of $\simeq 5 \times 5$ mrad^2, Bragg reflection from a spherically bent crystal wafer in backscattering geometry can be used. The curved analyzer gathers x rays from the sample, the source of the secondary waves, selects photons in a narrow energy band, and delivers them to the focal point where the detector is placed. However, bending introduces elastic strain in the crystal wafer, which immediately deteriorates the intrinsic energy resolution of the Bragg reflection. To minimize the energy broadening due to strain Moncton (1980) and Dorner and Peisl (1983), have proposed dicing the silicon wafers into small segments. The segments are kept oriented by leaving a very thin back-wall. An analyzer with $\simeq 1 \times 1$ mm^2 crystal segments providing 10 meV energy resolution has been achieved in this way for 13.8 keV x rays using the Si(7 7 7) Bragg reflection (Dorner et al. 1986). The first measurement with x rays of the phonon dispersion relations in Be crystals was accomplished by Burkel et al. (1987).

To eliminate the strain completely Masciovecchio et al. (1996b) have developed a procedure for mounting a large amount of ≈ 12000 small perfect crystal segments, obtained from the same silicon wafer, on a spherical substrate. An energy resolution of $\Delta E = 3.2$ meV for 17.8 keV x rays using the Si(9 9 9) Bragg reflection, $\Delta E = 1.4$ meV for 21.75 keV x rays using the Si(11 11 11), and $\Delta E = 1.0$ meV for 25.70 keV x rays using the Si(13 13 13) reflection was attained (Masciovecchio et al. 1996a, Verbeni et al. 2003), as illustrated in Fig. 6.1. The higher resolution was achieved not only due to

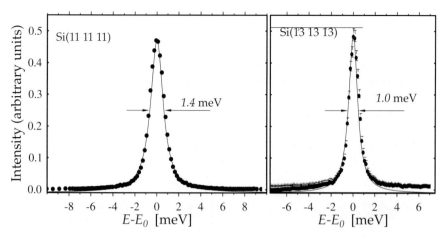

Fig. 6.1. Spectral resolution functions of inelastic x-ray scattering spectrometers, as measured by elastic scattering of x-rays from a plexiglass sample at momentum transfer $Q = 10$ nm^{-1}. The spectrometer consists of the monochromator and the spherical crystal analyzer using the Si($h\ h\ h$) Bragg reflection in backscattering, as shown schematically in Fig. 1.13 on page 15. The analyzer is operated with an opening of 3×9 mrad^2 (horizontal and vertical, respectively), and a bending radius of 6.5 m. The energy scans are performed by varying the temperature of the monochromator and keeping the analyzer temperature constant. (Masciovecchio et al. 1996b,a, Verbeni et al. 2003).

the new mounting procedure, but also due to further improvements in the fabrication procedure, optimization of the construction parameters, overall temperature control of the analyzer in the 1 mK range, and increased radius of the sphere, as compared to the first inelastic x-ray scattering instrument.

Further studies and developments aimed at achieving better performance in terms of energy resolution, throughput, and stability, were reported by Sinn (2001), Baron et al. (2001b), Sinn et al. (2001, 2002), and Alp et al. (2001a).

The crystal segments are glued either on a static spherical substrate (Masciovecchio et al. 1996b), or on a planar substrate, which is then dynamically bent to a spherical shape with a given radius (Sinn et al. 2002, 2004). Figure 6.2 shows a photograph of the array of ≈ 8000 perfect crystal segments glued on the planar substrate, which has to be dynamically bent to spherical shape (Sinn et al. 2002).

With such spherical focusing analyzers integrated into an inelastic x-ray scattering spectrometer measurements were performed of sound-wave propagation in water (Sette et al. 1995, 1998, Krisch et al. 2002), in liquid metals (Sinn et al. 1997, Scopigno et al. 2002), in glasses Scopigno et al. (2004), of phonon dispersion relations in single crystals (d'Astuto et al. 2002, Shukla et al. 2003), and in alloys (Wong et al. 2003), see Fig. 1.16 on page 18, to name

Fig. 6.2. Left: spherical crystal analyzer - an array of ≈ 8000 perfect crystal segments, each 1×1 mm^2 small, obtained by dicing the same silicon wafer, 4 mm thick and 10 cm in diameter (courtesy of H. Sinn). The crystal segments are glued either on a static spherical substrate (Masciovecchio et al. 1996a), or on a planar substrate, which is dynamically bent to a spherical shape with a given radius (Sinn et al. 2002, 2004). Right: silicon crystal segments on the glass substrate, shown with magnification (Sinn et al. 2002).

only few. A representative selection of high-pressure inelastic x-ray scattering experiments was given by Krisch (2003).

Figure 6.3 shows one of the recent examples of inelastic x-ray scattering spectra measured in liquid Al_2O_3 at a temperature of 2323 K for different momentum transfers Q of x-ray photons. The excitation spectra show a well-defined triplet structure of the anti-Stokes, elastic, and Stokes components. Analysis of the experimental data gave new insights into the microscopic dynamics of liquids (Sinn et al. 2003).

In the following sections we shall discuss the principle of spherical focusing analyzers, and the ultimate energy resolution which can be achieved with such optical devices.

6.2.2 Flat Crystal Analyzer

Before tackling the difficult subject of spherical focusing analyzers consisting of very many flat crystals, it is desirable first to simplify the problem. We start with a simple type of crystal analyzer in the form of a flat crystal plate with a surface area of $p_A \times p_A$ and a thickness of d_A placed at a distance $R_A \gg p_A$ from the sample, which is assumed to be a source of spherical waves, whose spectrum has to be analyzed. The angular dimension $\Delta \varUpsilon$ of the crystal as seen from the sample is

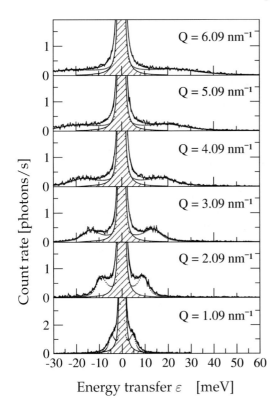

Fig. 6.3. Inelastic x-ray scattering spectra in liquid Al$_2$O$_3$ measured at a temperature of 2323 K for different momentum transfer Q of x-ray photons. The theoretical fits are shown as solid lines. The spectral resolution function, with an energy width of 1.8 meV (FWHM), is shown hatched.(Sinn et al. 2003)

$$\Delta \Upsilon \simeq \frac{p_A}{R_A}. \tag{6.1}$$

The possible scattering schemes are shown in Fig. 6.4. Symmetric Bragg reflection from the crystal with diffraction vector $\boldsymbol{H}$, and with a relative spectral bandwidth $\Delta E/E = \epsilon_H$ is assumed. We remind the reader, that ϵ_H can be considered as a universal parameter for the chosen Bragg reflection (assuming low photo-absorption), which with high accuracy is independent of incidence angle and x-ray energy.

We shall focus on the backscattering geometry. There are two reasons for this.

First, the smallest energy bandwidth $\Delta E = \epsilon_H E$ of the selected Bragg reflection is realized for x-ray photons reflected at normal (or close to normal) incidence to the reflecting atomic planes, where E takes a minimal value. Eventually, it is the ΔE value that determines the most suitable Bragg reflection. It can be selected from the tables of Bragg reflections in Appendices A.3 and A.4. As an example, the Si(2 2 20) Bragg reflection is characterized by an energy width $\Delta E = 0.87$ meV, an energy of photons reflected in backscattering $E = 23.06$ keV, an angular acceptance $\Delta \theta = 0.35$ mrad, and a peak reflectivity of $R = 0.75$.

292   6. High-Resolution X-Ray Analyzers

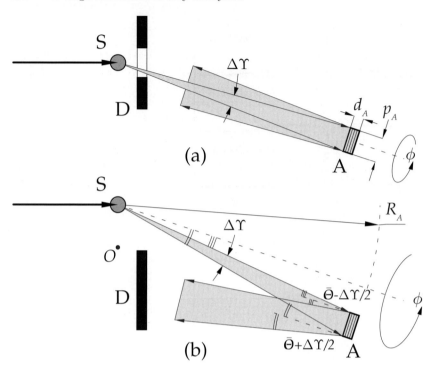

Fig. 6.4. Schematic drawing detailing the performance of the flat crystal analyzer - A. The incident monochromatic beam strikes the sample - S. The radiation scattered from the sample hits the analyzer. The analyzer is at a distance R_A from the sample in the exact backscattering geometry (a), and in off-exact backscattering geometry (b) with an average incidence angle $\bar{\Theta}$ to the analyzer. The radiation reflected from the analyzer is recorded by the x-ray detector - D. The crystal analyzer size measures $p_A \times p_A \times d_A$. $\Delta\Upsilon$ is the angular dimension of the flat crystal analyzer seen from the sample.

Second, and may be the most important reason, the geometrical broadening, the broadening of the spectral dependence of reflection due to the finite angular divergence of the incident beam, is minimal in backscattering, as was discussed in Sect. 5.3.1, and expressed by (5.3).

The admissible angular dimension $\Delta\Upsilon$, the admissible linear dimensions p_A, and d_A of the flat crystal analyzer in the exact and off-exact backscattering geometries are the important issues, which we shall consider next.

The crystal thickness d_A has to be chosen to be at least ten times the extinction length $d_A \gtrsim 10 d_e(0)$ to ensure maximal peak reflectivity, minimal energy width of the Bragg reflection, and an absence of thickness oscillations on the wings of the reflection curve, as is discussed in detail in Sect. 2.4.14 and illustrated in Fig.2.28 on page 109. For example, for the Si(2 2 20) Bragg reflection the extinction length is $d_e(0) = 0.24$ mm, which requires using flat crystals with a thickness of $d_A \gtrsim 2.4$ mm.

The transverse dimension p_A of the crystal, and the angular dimension $\Delta\Upsilon$ are limited by the condition of preservation of the intrinsic energy resolution of the Bragg reflection in the chosen scattering geometry. If $\overline{\Theta}$ is the average incidence angle to the reflecting atomic planes, then the variation of incidence angle Θ (assuming a point-like source of waves emerging from the sample) is in the range from $\overline{\Theta} + \Delta\Upsilon/2$ to $\overline{\Theta} - \Delta\Upsilon/2$, as shown in Fig. 6.4(b). Variation of the incidence angle shifts the spectral range of the Bragg reflection and thus causes geometrical broadening of the spectral dependence of reflection - as can be seen in Fig. 5.5 on page 222 - which eventually degrades the performance of the analyzer.

The geometrical broadening can be formally described in the following way. Let $R(\epsilon - \epsilon_c)$ be the crystal reflectivity as a function of the relative deviation $\epsilon = (E-E_H)/E_H$ (2.105) of the x-ray photon energy E from Bragg's energy E_H (2.100). The relative energy width of the reflection curve $R(\epsilon - \epsilon_c)$ is ϵ_H. The center of the reflection curve ϵ_c is given in backscattering by $\epsilon_c = \Theta^2/2 - \alpha_c/4$, in the sense of (2.104) and (2.96). Omitting the term $\alpha_c/4$, which gives a constant contribution, we may write $R(\epsilon - \epsilon_c) = R(\epsilon - \Theta^2/2)$. Finally, the energy dependence of the reflectivity of the analyzer crystal is obtained by averaging this expression over the directions of incidence of the photons, i.e., over incidence angle Θ and azimuthal angle ϕ:

$$\overline{R}(\epsilon, \overline{\Theta}) = \frac{1}{Z} \int_{\Theta_1}^{\Theta_2} \int_{\phi_1}^{\phi_2} R\left(\epsilon - \Theta^2/2\right) \Theta \, d\Theta \, d\phi. \tag{6.2}$$

Here $Z = \int_{\Theta_1}^{\Theta_2} \int_{\phi_1}^{\phi_2} \Theta \, d\Theta \, d\phi$ is the normalization factor.

We shall demand that $\Delta\Upsilon$ and p_A are chosen in such a way that the variation of the incidence angles over the surface of the analyzer crystal does not cause essential extra broadening as compared to the intrinsic width of the chosen Bragg reflection. This requirement can be quantitatively expressed as

$$\Delta\epsilon = f\epsilon_H, \qquad f < 1. \tag{6.3}$$

That is, the additional relative broadening $\Delta\epsilon$ may not exceed the intrinsic relative width ϵ_H of the Bragg reflection by a factor of more than $f < 1$.

Exact Backscattering. To calculate the spectral function of the reflectivity (6.2) in the exact backscattering geometry shown in Fig. 6.4(a) we have to use $\Theta_1 = 0$, $\Theta_2 = \Delta\Upsilon/2$, $\phi_1 = 0$, and $\phi_2 = 2\pi$. Assuming that $R(\epsilon)$ has a rectangular form with a width of ϵ_H (which is a fairly good approximation for low absorbing thick crystals), we obtain after the averaging procedure in (6.2) that the maximal geometrical contribution to the broadening is

$$\Delta\epsilon = \left(\frac{\Delta\Upsilon}{2\sqrt{2}}\right)^2, \qquad (\overline{\Theta} = 0). \tag{6.4}$$

294   6. High-Resolution X-Ray Analyzers

Using (6.4) together with (6.3), yields for the admissible angular dimension of the crystal analyzer in the exact backscattering geometry

$$\Delta \Upsilon \simeq 2\sqrt{2f\,\epsilon_H}, \qquad (\overline{\Theta}=0). \tag{6.5}$$

Equations (6.5) and (6.1) can be also used to calculate the admissible transverse linear dimension p_A of the flat crystal analyzer.

Table 6.1 shows how the admissible angular and linear dimensions ($\Delta \Upsilon$, p_A, d_A) of a silicon analyzer crystal change with Bragg reflection. The meaning of the quantities h_D, l_D will be explained later in Sect. 6.2.4, while discussing a ring detector.

Table 6.1. Bragg back-reflections (hkl) in silicon and the admissible angular/linear dimensions ($\Delta \Upsilon$, p_A, d_A) of the flat silicon analyzer crystals, and the parameters of the ring detector (h_D, l_D) in the *exact* backscattering geometry. Bragg back-reflections are listed in the table, which are accompanied by a maximum of $N_a = 6$ parasitic Bragg reflections. The angular and linear dimensions are calculated by using (6.5),(6.1), and (6.12) with $f = 0.25$, $R_A = 10$ m, and $\Upsilon = 5$ mrad. $d_A = 10 d_e(0)$. Crystal temperature $T = 300$ K in all cases except those labeled by (*), for which $T = 120$ K.

(hkl)	E_P	ΔE	$\Delta E/E$	R	N_a	$\Delta \Upsilon$	p_A	d_A	h_D	l_D
	keV	meV	$\times 10^{-8}$			mrad	mm	mm	mm	cm
(1 9 9)	14.57	4.18	28.7	0.79	4	0.76	7.6	0.5	4.8	95.
(2 6 12)	15.48	4.69	30.3	0.85	6	0.78	7.8	0.4	4.9	98.
(8 8 8)	15.82	4.33	27.4	0.85	6	0.74	7.4	0.5	4.7	93.
(0 6 14)	17.39	2.99	17.2	0.83	2	0.59	5.9	0.7	3.7	74.
(0 0 16)	18.26	2.46	13.5	0.82	4	0.52	5.2	0.9	3.3	65.
(6 10 12)	19.10	2.04	10.7	0.81	6	0.46	4.6	1.0	2.9	58.
(4 10 14)	20.16	1.62	8.0	0.80	6	0.40	4.0	1.3	2.5	51.
(0 2 18)	20.67	1.45	7.0	0.79	6	0.37	3.7	1.5	2.4	47.
(2 2 20)	23.06	0.87	3.8	0.75	6	0.27	2.7	2.4	1.7	35.
(0 6 22)	26.03	0.45	1.7	0.68	6	0.19	1.9	4.7	1.2	23.
*(10 12 18)	27.20	0.95	3.5	0.86	6	0.26	2.6	2.2	1.7	33.
*(16 16 16)	31.64	0.47	1.5	0.82	6	0.17	1.7	4.5	1.1	22.
*(0 0 32)	36.54	0.20	0.5	0.74	4	0.10	1.0	10.6	0.7	13.

Those Bragg back-reflections in silicon with Miller indices (hkl) are listed in Table 6.1, which have a narrow energy width $\Delta E \lesssim 5$ meV, and, which are accompanied by a maximum of $N_a = 6$ parasitic Bragg reflections at normal incidence to the (hkl) atomic planes. The better the energy resolution required, the higher the photon energy E_P must be used. This in turn requires thicker analyzer crystals with smaller transverse dimensions. Importantly, however, in all the cases the transversal dimension p_A is greater than 1 mm, i.e., greater than a practical lower limit on p_A, which is considered to be $1 - 0.5$ mm (Alp et al. 2001a, Sinn 2001, Baron et al. 2001b). The values in Table 6.1 apply to the allowed relative energy broadening $f = 0.25$, and radius of the sphere $R_A = 10$ m. The p_A value can be easily scaled up or down for spheres of different radius.

Let us now examine, how the requirements for the analyzer crystal dimensions change by a transition to the off-exact backscattering geometry.

Off-Exact Backscattering. To calculate the spectral function of the reflectivity (6.2) in the off-exact backscattering geometry shown in Fig. 6.4(b) we have to use $\Theta_1 = \overline{\Theta} - \Delta\Upsilon/2$, $\Theta_2 = \overline{\Theta} + \Delta\Upsilon/2$, and $\phi_2 = -\phi_1 \approx \Delta\Upsilon/2$. The minimum average incidence angle $\overline{\Theta}$ is not zero. It is determined by the requirement that the source and its image are separated in space. Since the image is twice as large as the analyzer crystal size, the average incidence angle $\overline{\Theta} \geq \Delta\Upsilon/2$. In practice it is not possible to get closer than $\overline{\Theta}_{\min} \approx 3\Delta\Upsilon/2$ (Sinn 2001). Under this condition, $\Theta \, d\Theta$ in (6.2) can be approximated by $\simeq \overline{\Theta} \, d\Theta$. Performing averaging in (6.2), we may obtain for the geometrical contribution to the relative energy broadening in this scattering geometry

$$\Delta\epsilon \simeq \overline{\Theta}\Delta\Upsilon, \qquad (\overline{\Theta} \geq \overline{\Theta}_{\min}). \tag{6.6}$$

The admissible angular dimension of the analyzer is then equal to:

$$\Delta\Upsilon \simeq \frac{f\,\epsilon_H}{\overline{\Theta}}, \qquad (\overline{\Theta} \geq \overline{\Theta}_{\min}). \tag{6.7}$$

For the average incidence angle $\overline{\Theta}_{\min} \approx 3\Delta\Upsilon/2$, the admissible angular dimension of the analyzer crystal is

$$\Delta\Upsilon \simeq 0.8\sqrt{f\,\epsilon_H}, \qquad (\overline{\Theta} = \overline{\Theta}_{\min}). \tag{6.8}$$

It is a factor $\simeq 3.5$ smaller than the admissible angular dimension of the crystal in the exact backscattering geometry, as given by (6.5), and decreases rapidly with deviation of $\overline{\Theta}$ from $\overline{\Theta}_{\min}$ - (6.7).

Comparison of (6.4) and (6.8) shows that the geometrical broadening $\Delta\epsilon$ takes its smallest, and the admissible angular/linear dimensions $\Delta\Upsilon$, p_A of the analyzer crystal take their largest values in the exact backscattering geometry.

The $\Delta\Upsilon$, and p_A values in Table 6.1 being reduced by a factor of ≈ 3.5 give the admissible angular and linear dimensions of the flat analyzer in the off-exact backscattering geometry at the minimum average incidence angle $\overline{\Theta}_{\min}$. For example, in the extreme case of the (0 0 32) Bragg reflection, which offers a very high energy resolution of 0.2 meV, the linear dimension p_A reduces from 1 mm (in the exact backscattering geometry) to 0.3 mm in the scattering geometry with an average incidence angle of $\overline{\Theta}_{\min} \simeq 0.05$ mrad.

296   6. High-Resolution X-Ray Analyzers

Table 6.2 shows some other examples of the admissible angular and linear dimensions of the analyzer crystal segments. They are calculated for the (hhh) family of Bragg reflections in silicon at a fixed average incidence angle $\overline{\Theta} = 0.15$ mrad, a practical value for today's instruments.

Table 6.2. Bragg reflections (hhh) in silicon and the admissible angular/linear dimensions $(\Delta \Upsilon, p_A, d_A)$ of the flat crystal silicon analyzer. N_a is the maximum number of accompanying Bragg reflections, which can be excited at near normal incidence to the (hhh) atomic planes. The angular and linear dimensions are calculated by using (6.7) with $f = 0.25$, $R_A = 10$ m, $\overline{\Theta} = 0.15$ mrad (source-image distance $2R_A \overline{\Theta} = 3$ mm). $d_A = 10 d_e(0)$. Crystal temperature $T = 300$ K.

(hhh)	E_P	ΔE	$\Delta E/E$	R	N_a	$\Delta \Upsilon$	p_A	d_A
	keV	meV	$\times 10^{-8}$			mrad	mm	mm
(9 9 9)	17.793	1.94	10.9	0.75	24	0.1817	1.817	1.1
(11 11 11)	21.747	0.82	3.8	0.69	24	0.0628	0.628	2.6
(12 12 12)	23.725	0.75	3.2	0.73	30	0.0527	0.527	2.8
(13 13 13)	25.702	0.35	1.4	0.59	24	0.0227	0.227	6.2
(15 15 15)	29.656	0.14	0.5	0.43	54	0.0079	0.079	15.4

Source Size Contribution. So far it was assumed, that the sample is a point-like source of spherical waves. This assumption is justified, if the transverse dimension s of the part of the sample illuminated by the incident beam is much smaller than the dimension of the analyzer crystal: $s \ll p_A$. Only in this case can one use relation (6.1). Otherwise,

$$\Delta \Upsilon \simeq \frac{1}{R_A} \sqrt{s^2 + p_A^2}, \tag{6.9}$$

has to be used to estimate the admissible transversal linear crystal dimension p_A from its admissible angular dimension $\Delta \Upsilon$, as given by (6.5) or (6.7).

Choice of the Crystal and Bragg Reflection. The choice of the crystal material and Bragg reflection for the analyzers are other important issues, which have to be addressed. A Bragg reflection free of accompanying reflections has to be used in the *exact* backscattering geometry. Sapphire (α-Al$_2$O$_3$) crystals offer a great variety of *multiple-beam free* Bragg back-reflections with energy widths from a few milli- to a few micro-eV, as listed in the Table of Appendix A.4. However, the quality of present day sapphire crystals is so far not sufficient for achieving sub-meV resolution in practice, see Appendix A.7.

Therefore, silicon still remains the first choice in realization of optical instruments with the highest resolution.

However, as already discussed many times in this book, Bragg reflections at normal incidence in silicon are always accompanied by a large, often a very large, number of parasitic reflections, which impairs the crystal reflectivity and its spectral characteristics. At this point, one may ask whether the use of silicon as the analyzer crystal is altogether justified in the exact backscattering geometry? The short answer is: if the angular dimension $\Delta \Upsilon \times \Delta \Upsilon$ of the analyzer crystal, is much larger than the angular region of the influence of the accompanying reflections, then the use of the exact-backscattering geometry can be justified. As the example in Fig. 2.44 on page 139 shows, one conjugate pair of accompanying reflections affects Bragg backscattering in a very narrow angular band as compared to the whole angular region of the back-reflection. Another one or two conjugate pairs of accompanying reflections will increase this unfavorable angular region, however not significantly. Therefore, Bragg back-reflections with a relatively small number $N_{\mathrm{A}} \lesssim 6$ of accompanying reflections could be used for crystal analyzers in the exact backscattering geometry. Such reflections are given in Table 6.1.

It is worth noting here that Bragg reflections in silicon of the family $(0\ 0\ 2^m)$, $m = 2, 3..$ like $(0\ 0\ 16)$ and $(0\ 0\ 32)$ of Table 6.1 are an interesting special case. By a judicious choice of the scattering geometry the number of accompanying reflections can be reduced from four to two. To explain how it works, it is essential to note that all four accompanying waves propagate perpendicular to the direction of the incident and back-reflected waves. Therefore, one conjugate pair of accompanying reflections can be suppressed by directing the polarization vector of the incident wave parallel to the basal plane of this reflection pair, as is discussed in detail in Sect. 2.5.5.

Also the $(2\ 2\ 20)$ Bragg reflection in silicon is an interesting special case. The waves excited by the conjugate pair of the $(\bar{6}\ \bar{6}\ 12)$ and $(8\ 8\ 8)$ accompanying Bragg reflections propagate almost normal to the incident and back-reflected waves, and therefore can be suppressed by directing the polarization vector of the incident wave parallel to the basal plane of this reflection pair.

6.2.3 Focusing Array of Flat Crystals

A single crystal flat analyzer cannot deliver photons in the required solid angle of $\simeq 5 \times 5$ mrad^2. The problem could be solved by using curved crystal analyzers. Due to the technical reasons discussed in Sect. 6.2.1, it is more favorable to use a two-dimensional array of many small flat crystal segments fixed on a sphere of radius R_{A}, as shown in Fig. 6.5. The analyzer segments receive x rays from the source, and each segment simultaneously delivers its component to the focus.

The results of the analysis performed in the previous section can be used to estimate the admissible linear and angular dimensions of the individual

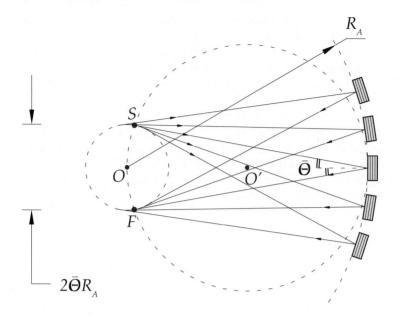

Fig. 6.5. Scheme of gathering and focusing x rays by the crystal analyzer consisting of the two-dimensional array of flat crystals on a sphere of radius R_A centered at O. The best focusing (apparent focus at point F) is achieved by placing the sample - a source of secondary spherical waves - at point S on a circle of radius $R_A/2$ centered at O' - Rowland's circle.

crystal segments. However, there are additional issues, which have to be addressed, namely, the position of, first - the sample, second - the focus, third - the detector.

Given that the flat crystals are placed on a sphere of radius R_A centered at O, as shown in Fig. 6.5, the rays with the angle of incidence $\overline{\Theta} \ll 1$ at the center of each flat crystal are tangent to a circle with radius $\overline{\Theta} R_A$ centered also at O. The reflected rays are tangent to the same circle. They are not crossing in one point. Therefore, *perfect* focusing can never be achieved even for the rays hitting the crystal center. Still, the best gathering and focusing will occur by placing the sample and the detector at the intersection points of the circle with radius $\overline{\Theta} R_A$, and the circle with radius $R_A/2$ centered at O', i.e., by placing at points S and F, respectively. The circle of radius $R_A/2$ is the locus of points S and F for different incidence angles $\overline{\Theta}$. It is known in optics as Rowland's circle, see, e.g., (Born and Wolf 1999).

Each crystal segment is flat. Thus, for this reason, the focusing is not perfect, as is illustrated in Fig. 6.6. The rays reflected at the edges of the crystal segments do not converge to the apparent focus F, which results in a nonzero source image size. Given $\Delta \Upsilon$ is the angular dimension of each crystal

6.2 Spherical Crystal Analyzers   299

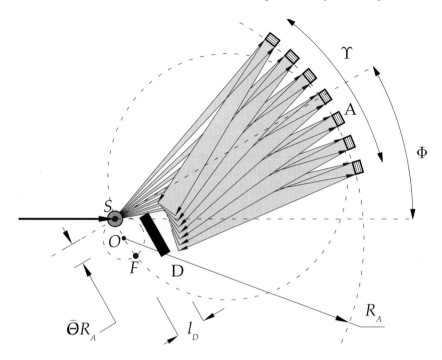

Fig. 6.6. Schematic drawing detailing the performance and design concept of the spherical crystal analyzer - A - a two-dimensional array of flat single crystals on a sphere of radius R_A in backscattering geometry. The incident monochromatic beam strikes the sample placed at point S. The radiation scattered from the sample with an average scattering angle Φ hits the analyzer. The signal from the analyzer is recorded by the x-ray detector - D.
The angular acceptance Υ of the analyzer, which is typically $\approx 5-10$ mrad, and the size of the flat crystal elements are shown greatly exaggerated for clarity.
The meaning of the points S, F, and O is the same as in Fig. 6.5.

segment at a distance $\simeq R_A$ from the point-like source, the source image dimension is $s' \simeq 2R_A \Delta \Upsilon \simeq 2p_A$. Thus, the source image dimension is twice as large as the transversal dimension p_A of the crystal segments. In fact, the only immediate consequence of this non-perfect focusing is that the detector active area should measure at least $2p_A \times 2p_A$. In the exact backscattering geometry, the non-perfect focusing may be even advantageous, as is explained in the next paragraph.

High energy resolution and large angular acceptance require using extreme backscattering geometry with average incidence angles $\overline{\Theta} < 1$ mrad. The distance between the source and focus points $SF \approx 2\overline{\Theta}R_A$ is very small. For example, for the cases presented in Table 6.2 the source-focus distance SF is 3 mm. In the experiments on inelastic x-ray scattering it is often very important to have more space around the sample to accommodate sample environments like an oven, a cryostat, pressure cells, etc. To gain more space, it is favorable to shift the detector from the apparent focus point F towards the analyzer, as shown in Fig. 6.6. The maximum sample-detector distance can be estimated by the formula

$$l_\mathrm{D} \approx \frac{\overline{\Theta} R_\mathrm{A}}{\varUpsilon}, \qquad \left(\overline{\Theta} \geq \overline{\Theta}_\mathrm{min}\right). \qquad (6.10)$$

It is derived from the requirement that the detector does not intercept the waves propagating towards and back from the analyzer. If $\overline{\Theta} = \overline{\Theta}_\mathrm{min} \approx 3\varDelta\varUpsilon/2$, which in practice is usually the most interesting case (due to larger p_A values), then by using (6.8) we obtain

$$l_\mathrm{D} \approx 1.2 R_\mathrm{A} \frac{\sqrt{f\,\epsilon_H}}{\varUpsilon}, \qquad \left(\overline{\Theta} = \overline{\Theta}_\mathrm{min}\right), \qquad (6.11)$$

as an estimate for the maximum sample-detector distance in the off-exact backscattering geometry with the incidence angle $\overline{\Theta}_\mathrm{min}$.

6.2.4 Exact Backscattering Analyzer and Ring Detector

The difficulty in realizing the focusing crystal analyzer is in the production stage, where a silicon wafer is diced into segments and glued onto a substrate and then bent to a spherical shape with a given radius (Masciovecchio et al. 1996b, Sinn et al. 2002). The practical lower limit on the transverse size p_A of the crystal segments is around $0.5 - 1$ mm (Alp et al. 2001a, Sinn 2001, Baron et al. 2001b). It is determined by modern technological possibilities, and by the minimum source size. The analysis performed in Sect. 6.2.2 shows that the exact backscattering geometry is most beneficial, since it offers considerably larger admissible transverse dimension p_A of the crystal segments. Comparison of Tables 6.1 and 6.2 shows that, because of the limitation on p_A, achieving sub-meV resolution in the off-exact backscattering geometry is much more difficult, than in the exact backscattering geometry.

What is not clear, how to realize the analyzer in the exact backscattering geometry in practice? The detector intercepts the photons propagating towards the analyzer. The most evident solution is using the semitransparent detector, as in the experiments on exact backscattering described in (Shvyd'ko et al. 1998). However, it is wasteful, since even in the optimal setting only 25% of the photons scattered by the sample can be recorded: 50% are transmitted through the detector to the analyzer, and 50% of the reflected photons from the analyzer are detected.

A more effective solution offers a ring detector as shown in Fig. 6.7. A small hole at the center of the detector lets the radiation through to the analyzer. Due to non-perfect focusing, the radiation reflected from the analyzer may illuminate an area larger than the detector hole, and thus most of the photons could be recorded. Two conditions have to be fulfilled for this

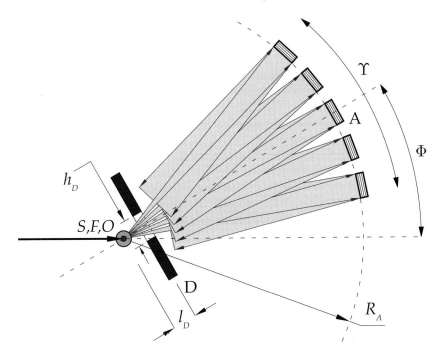

Fig. 6.7. Schematic drawing detailing the performance and design concept of the spherical crystal analyzer - A - a two-dimensional array of flat single crystals on a sphere of radius R_A, similar to Fig. 6.6; however, now the analyzer's flat crystal elements are set in the *exact* backscattering geometry. A ring detector - D - with a hole of diameter h_D at a distance l_D from the sample is used.
The angular acceptance of the analyzer Υ, which is typically $\approx 5-10$ mrad, and the size of the flat crystal elements are shown greatly exaggerated for clarity. Because of this the true relation $l_D \gg h_D$ is also distorted.
For other notations see the legend to Figs. 6.5 and 6.6.

procedure to be acceptable. First, the opening of the detector h_D and the source image size $2p_A$ should be chosen in such a way, that the area of the detector which is illuminated by the photons reflected from the analyzer is much larger than the area of the opening: $4p_A^2 \gg h_D^2$. To be specific, we shall demand that 90% of the radiation can be recorded, i.e., $4p_A^2 = 10h_D^2$, and thus $h_D \simeq 0.63 p_A$. Second, the detector should be placed at a distance l_D relatively close to the sample, so that the angular dimension of the opening is not smaller than the required angular acceptance Υ of the analyzer: $h_D/l_D \simeq \Upsilon$. The conditions are usefully summarized in the following expressions, which can be used to estimate the optimal detector opening h_D and its distance l_D from the sample:

$$h_D \simeq 0.63 p_A, \qquad (6.12)$$

$$l_{\mathrm{D}} \simeq \frac{0.63}{\Upsilon} p_A \simeq 1.8 R_A \frac{\sqrt{f \epsilon_H}}{\Upsilon}. \qquad (6.13)$$

To obtain the last expression in (6.13), (6.1) and (6.5) were used.

The h_D and l_D values calculated by using (6.12) and (6.13) for the ring detector integrated into the spherical analyzer are given in Table 6.2. The calculations were performed for different Bragg reflections. The values are reasonable even in the extreme case of the Si(0 0 32) Bragg reflection, offering a very high energy resolution of 0.2 meV.

At this point it is interesting to compare the maximum sample-detector distance permissible in the exact and off-exact backscattering geometries. They are given by (6.13), and (6.11), respectively. Comparison shows that for the same Bragg reflection, and with other things being equal, the permissible maximum sample-detector distance l_D is greater in the exact backscattering geometry.

To summarize the discussion of this section, the admissible crystal segment size p_A, and the sample-detector distance l_D are greater in the exact backscattering geometry. This makes the use of the spherical crystal analyzers in the exact backscattering geometry being superior to their use in the off-exact backscattering geometry. The exact backscattering geometry is also attractive due to its axial symmetry.

Although many arguments are in favor of the exact backscattering geometry, so far the off-exact backscattering geometry was realized in the experiments on inelastic x-ray scattering. However, the necessity of achieving higher energy resolution, and higher throughput of the focusing crystal analyzers, will eventually enforce using spherical crystal analyzers also in the exact-backscattering geometry. Some technical problems like ring detectors have to be solved in advance of this. No doubt, technological advances in semiconductor x-ray detectors will pave the way also for ring detectors.

6.2.5 Inelastic X-Ray Scattering Spectrometers

Thus far, we have addressed separately the problems of designing highly accurate monochromators and analyzers with meV-resolution. There are several possibilities of how these optical elements can be joined into one spectrometer, and each of these possibilities has its own merits. So far, we have discussed only one type of analyzers, the single-bounce backscattering analyzer. Therefore, the difference between the spectrometers may basically arise due to application of different types of monochromators.

Backscattering. If the throughput of the spectrometer is of primary importance (for achieving higher count-rates) then a single bounce backscattering monochromator is the best choice. The scheme of such a spectrometer is shown in Fig. 1.13 on page 15. One may also find a drawback in this arrangement, as the sample is too close to the incident beam, and there is little space for sample environment. The problem can be circumvented by installing, e.g.,

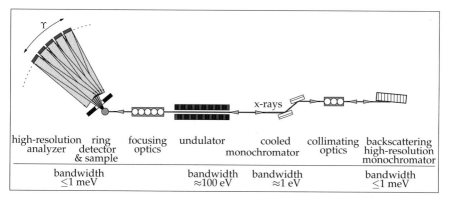

Fig. 6.8. Layout of the exact backscattering beamline with an inelastic x-ray scattering spectrometer. The cooled monochromator, and the single bounce backscattering monochromator are installed downstream of the undulator. The sample and the analyzer are installed upstream of the undulator. The *exact backscattering* analyzer is shown combined with the ring detector.

a pair of non-dispersively arranged crystals, to shift the beam reflected from the backscattering monochromator crystal further off the incident beam, see, e.g., (Baron et al. 2001b).

Exact Backscattering. Another solution of this problem is shown schematically in Fig. 6.8. The spectrometer is split up into two parts, one on the upstream and the other on the downstream sides of the undulator. Downstream of the undulator, a high-heat-load (cooled) monochromator, and an exact backscattering high-resolution monochromator are installed. The x rays produced by the undulator travels to the right. They are reflected from the backscattering monochromator crystal, travel back through the cooled monochromator, and the undulator, and arrive on its upstream side, where a sample and the high-resolution analyzer are installed. At least three questions have to be addressed, crucial for this special spectrometer arrangement.

First, access to the x-rays produced by the undulator is required both on its upstream and on the downstream sides. Recently, the feasibility of such an exact Bragg backscattering beamline, was demonstrated (Alp et al. 2004).

Second, x-rays may travel large distances of about 100 m, from the undulator source to the sample. Therefore, preserving the small cross-section of the beam using collimating optics (e.g., compound refractive lens), and focusing on the sample (e.g., by a compound refractive lens, or a focusing mirror) is a must.

Third, such a Bragg reflection has to be chosen for the exact backscattering monochromator, which is either not accompanied by excitation of simultaneous reflections, or the number of these accompanying reflections is relatively small, so that it is possible to suppress them by cutting the crystal asymmetrically, as discussed in Sect. 2.7, and shown schematically in Fig. 6.8.

304   6. High-Resolution X-Ray Analyzers

In-Line. The last but not least, a multi-bounce in-line monochromator can be used, as shown in Fig. 1.15 on page 17. In addition to having more free space for the sample environment, there are some other benefits of using multi-bounce in-line monochromators. First, energy tuning with the multi-bounce monochromators can be performed much faster (changing angle of incidence is quicker than changing the crystal temperature). Second, the spectral function of a multi-bounce monochromator has much steeper wings. A drawback is a smaller throughput as compared to the single-backscattering monochromators.

Some other possible layouts of inelastic x-ray scattering spectrometers will be presented in the next section, after discussing multi-bounce $(+,+,\pm)$-type flat crystal analyzers.

6.3 Multi-Bounce $(+,+,\pm)$-Type Flat Crystal Analyzers

6.3.1 High-Energy or Low-Energy Photons?

Single Bragg reflection from a crystal in backscattering has the advantage of being a simple and robust means for monochromatization, and spectral analysis of x rays. But, this way of monochromatization also has an apparent drawback. The higher energy resolution is required, the higher indexed Bragg reflections and thus photons with higher energy have to be used (Fig. 5.4, page 220). The use of x rays with high photon energies, however, is unfavorable in experiments on inelastic x-ray scattering, for at least three reasons.

First, modern x-ray sources - undulator based sources at storage electron rings or at linacs - generate more photons in the low-energy range ($E \lesssim 10$ keV) than in the high-energy range ($E \gtrsim 20$ keV). This statement is illustrated in Fig. 6.9 by a theoretical undulator x-ray spectrum. An example undulator is considered designed for a so-called high-energy synchrotron radiation facility (here both high energy electrons generating x rays in undulators, and high energy x-ray photons are implied). According to this example, a factor of 7.5 more photons can be generated with energy $E = 9.1$ keV than with $E = 21.7$ keV[2]. This factor can vary depending on the undulator design and the energy of the electrons, but it is always greater than one. This factor, being representative of undulator sources at high-energy synchrotron radiation facilities, becomes much higher for undulator sources at

---

[2] We are comparing the photon flux at the energies 9.1 keV and 21.7 keV, which are the Bragg energies of the (0 0 8) and (11 11 11) reflections in silicon. As will be clear from the following, these are example back-reflections used in two conceptually different designs for x-ray analyzers, which are, however, comparable in terms of the energy bandpass ($\simeq 1$ meV). The (11 11 11) Bragg reflection is used in the single-bounce spherical focusing crystal analyzer design, discussed in Sect. 6.2. The (0 0 8) Bragg reflection is used in this section, in the example design of the multi-bounce $(+,+,\pm)$ type analyzer.

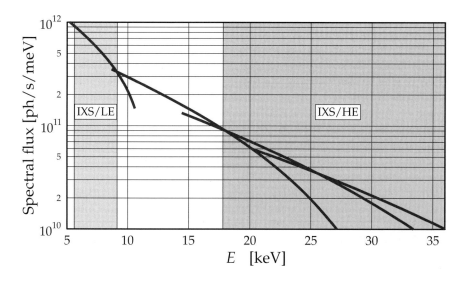

Fig. 6.9. Spectral flux of x-ray photons from an undulator as a function of photon energy E. Highlighted areas show spectral ranges for the low-energy (LE) and high-energy (HE) type spectrometers for inelastic x-ray scattering (IXS).
The parameters of the undulator are given in the legend to Fig. 1.9 on page 12. Different lines correspond to the 1st, 3rd, 5th, and 7th harmonic of the undulator spectrum, see Fig. 1.9. Calculations by Franz (2003) with the program SPECTRA (Tanaka and Kitamura 2001).

low-energy facilities. In this respect, it is also worth noting that the x-ray free electron lasers of the first generation (LCLS 1998, Materlik and Tschentscher 2001), the most powerful forthcoming x-ray sources, will be lasing in the low-energy range. Taking into account that the experiments on inelastic x-ray scattering still suffer from low photon intensities scattered even at modern synchrotron radiation facilities (typical count-rates are $\simeq 1$ counts/s and less) the aforementioned arguments are serious drawbacks to the use of high-energy photons, and they speak in favour of using low-energy photons.

Second, to achieve the same momentum resolution ΔQ in experiments on inelastic x-ray scattering with x-ray photons of a higher energy, the analyzer has to be operated with a reduced solid angle of acceptance Υ^2 ($\Upsilon = \Delta Q/K$, $K = \hbar c/E$). This leads immediately to reduced count-rates. We estimate that for this reason, we will get another factor of 5.6 higher count-rate in the case of using photons with an energy of $E = 9.1$ keV than with $E = 21.7$ keV.

Due to both factors, one may altogether expect in our particlar example a gain of $\simeq 42$ in the count-rate with the spectrometer, which uses the 9.1 keV photons, as opposed to that using the 21.7 keV photons[3]. Here the

---

[3] Clearly, there are also disadvantages to using low-energy photons. One is the higher photo-absorption, and therefore, lower penetration depth. For exam-

same energy and momentum resolution, and the same monochromator's and analyzer's throughput is assumed.

Third, in addition to the decline in the photon flux, the throughput of single-bounce backscattering analyzers also decreases rapidly with photon energy. This happens because, on the one hand, the peak reflectivity of Bragg reflections drops, and, on the other hand, the crystal segments have to be made of smaller size, which reduces the analyzer's effective working area (the number of necessary cuts increases, see Fig 6.2). For these reasons, the efficiency, in terms of count-rate, of the x-ray spectrometers using single-bounce backscattering analyzers decreases much faster than the increasing photon energy causes the narrowing of the analyzer's bandpass. It is very doubtful at all that a practically useful spectrometer with energy bandpass significantly smaller than 1 meV (which would require using photons with $E \gtrsim 30$ keV) can be realized with the single-bounce backscattering analyzers.

We may add one more argument in favour of using photons in the low-energy range: K-absorption edges of the important transition metals are precisely in this range (Thompson and Vaughan 2001). Proximity to atomic resonances would allow element selective measurements and enhanced scattering cross-sections.

Employing low-energy photons is, however, in conflict with the principles underlying single-bounce backscattering analyzers. New concepts and new solutions are required to exploit the potential of the low-energy photons.

One solution could be, for example, using α-Al$_2$O$_3$ or other non-cubic crystals instead of traditional Si crystals. As can be seen from Fig. 5.4, and the table of Appendix A.4 there exist Bragg back-reflections in sapphire in the low-energy range with small bandwidth $\Delta E \lesssim 10$ meV. One of them is the (0 $\bar{1}$ 1 22) Bragg back-reflection with a bandwidth of $\Delta E = 1.7$ meV ($E = 10.6$ keV). However, the Bragg reflections in the low-energy range with a narrow energy width have, as a rule, small peak reflectivity ($\lesssim 20 - 30\%$).

In the remainder of this section we shall discuss an alternative x-ray analyzer concept. As will become clear from the following discussion, analyzers of this type have three major beneficial features. The energy bandpass narrows with *decreasing* photon energy. The bandpass can be changed on demand from $\simeq 10$ meV to $\simeq 0.1$ meV *without changing* the photon energy. The peak throughput and angular acceptance can be kept almost *unchanged* while reducing the analyzer's bandpass.

6.3.2 Design Concept

The same optical instrument can be used as the monochromator and the analyzer. The sole distinguishing functional feature of the analyzer is the ability

---

ple, transmission through a 2 mm diamond pressure cell is 8 times higher at $E = 21.7$ keV than at $E = 9.1$ keV. The gain factor of 42 is reduced, but not annihilated. The low-energy spectrometer may still deliver a factor of 6 more photons than the high-energy one.

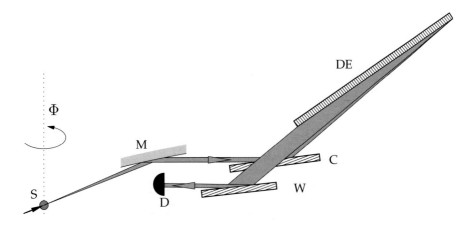

Fig. 6.10. Schematic drawing of the $(+,+,\pm)$ analyzer comprising of: two-dimensional paraboloidal mirror – M, collimator crystal – C, dispersing crystal element – DE, and wavelength selector – W. The incident monochromatic beam (its direction is shown by the arrow) impinges upon the sample – S. The dotted line indicates the common rotation axis of the analyzer and the detector – D – to access different momentum transfer in scattering from the sample.

to accept photons in a relatively large solid angle. Typically an opening angle of $\varUpsilon \approx 5 - 10$ mrad is required for the inelastic x-ray scattering experiments, as was discussed in Sect. 6.1. A monochromator can become an analyzer by installing ahead of it collimating optics, i.e., optical elements which gather x rays from the sample in the required solid angle and deliver them with a reduced angular spread to the monochromator. The basic design concept of the analyzer discussed in this section just uses a monochromator and collimating optics integrated into one instrument, an x-ray analyzer.

Our main focus in this section is on feasibility of the inelastic x-ray scattering experiments in the low-energy photon range. As we know, the refraction effects at the interface of two media gain in magnitude with decreasing photon energy (increasing wavelength), see (2.55). Therefore, for the purpose of collimating low-energy x-rays photons it seems appropriate to use refractive optics, such as: curved grazing incidence mirrors (Howells et al. 2000), curved graded multilayer mirrors (Schuster and Göbel 1995, Morawe et al. 1999, Protopopov et al. 2000), compound refractive lenses (Snigirev et al. 1996), or tapered glass capillaries (Kumakhov 1990, Bilderback et al. 1994).

Figure 6.10 demonstrates an example of how a collimating two-dimensional paraboloidal mirror and the $(+,+,\pm)$-type monochromator can be integrated into one instrument, an x-ray analyzer.

In the following, the two main optical components will be treated separately. We commence with the discussion of the monochromator component, which largely determines also the parameters of the collimating optics.

Table 6.3. Characteristics of the three-bounce $(+,+,\pm)$ monochromators shown schematically in Fig. 5.38(a), and parameters of the silicon crystals composing the monochromators. E_P is the design value of the photon energy. For the meaning of other notations see caption for Table 5.6 on page 276.
The presented versions are designed to have the same energy bandpass ΔE, but different angular acceptance $\Delta\theta$ suitable either for a stand-alone monochromator (#1), or for a monochromator integrated into the x-ray analyzer (#2).

monochromator		monochromator crystals				monochromator				
#	E_P [keV]	n	(hkl)	θ_n [deg]	η_n [deg]	$\|b_n\|$	ΔE [meV]	$\Delta\theta$ [μrad]	T	M [meV^{-1}]
1	9.13155									
		1	(0 2 2)	20.71	14.78	1/5.6	0.83	18	0.78	0.94 (t)
		2	(0 0 8)	89.80	88.80	1.4	1.7	49	1	(d)
		3	(0 2 2)	20.71	14.78	5.6				
2	9.13155									
		1	(0 2 2)	20.72	19.71	1/37	0.82	97	0.63	0.8 (t)
		2	(0 0 8)	89.90	87.10	1.1	1.5	127	1	(d)
		3	(0 2 2)	20.70	19.71	37				

6.3.3 Monochromator Component

In Chap. 5 we considered two different types of monochromators, designed for x rays with photon energies $E \lesssim 10$ keV, capable of achieving meV and sub-meV resolution. These are, first, the $(+,-,-,+)$-type monochromators, discussed in Sect. 5.4.5, and, second, the $(+,+,\pm)$-type monochromators discussed in Sect. 5.5. We shall be concerned in this section with $(+,+,\pm)$-type monochromators, for two main reasons. First, they usually have a higher peak throughput (for the same energy bandpass) as illustrated, e.g., in Fig. 5.43 on page 281. Second, they may have a larger angular acceptance, which is of primary importance.

Other distinguishing features of the three-crystal $(+,+,\pm)$ monochromators, as mentioned in Sect. 5.5, are, first, the ability to change the bandpass from $\simeq 10$ meV to $\simeq 0.1$ meV at a fixed photon energy, e.g., by changing the asymmetry angle of the dispersing element. Second, both the high peak throughput ($\simeq 0.8 - 0.6$) and the relatively high angular acceptance ($\simeq 0.1 - 0.07$ mrad), can be kept almost unchanged, while reducing the monochromator's bandpass, see Tables 5.5 and 5.6.

The aforementioned numbers correspond to the monochromators employing a silicon crystal with the (0 0 8) asymmetric Bragg reflection in backscattering, as the dispersing element. This back-reflection necessitates using 9.13 keV x-ray photons. The photon energy can be changed by choosing another Bragg reflection, and even choosing another crystal instead of silicon. However, to be specific, we shall focus in the following on this particular case. More specifically, we shall have in mind a monochromator whose parameters are given in Table 6.3 (monochromator #2) with an energy bandpass of $\Delta E = 0.82$ meV[4], and an angular acceptance in the dispersion plane of $\simeq 0.1$ mrad. The admissible angular acceptance of the monochromator in the plane perpendicular to the dispersion plane is $\simeq \sqrt{\Delta E / E_P}$ (5.9) in the exact backscattering scattering geometry, taking on a value of $\simeq 0.3$ mrad in the particular example under consideration.

In the optical design shown in Fig. 6.10, the version of the $(+,+,\pm)$ monochromator is chosen with the dispersing element diffracting x rays in *almost* exact backscattering, as in Fig. 5.38(a) on page 271. It has practically the same spectral characteristics and the same angular acceptance as the *exact* backscattering version, shown in Fig. 5.38(b), however, it requires using a more complicated three-crystal design instead of the two-crystal one. Nevertheless, the three-crystal design is advantageous in our particular case, as it allows separation in space of the incident and exit beams, and thus enables detection of the output signal without interfering with the incident beam[5].

Even though the angular acceptance of the $(+,+,\pm)$ monochromators can be as high as $\simeq 0.1 \times 0.3$ mrad^2, this number, being very good for the instrument used as a monochromator, is insufficient if it acts as the analyzer. Additional optics must be used to gather x rays from the sample in a solid angle of $\Upsilon^2 \approx 5 \times 5$ mrad^2, and deliver them with a reduced angular spread of $\lesssim 0.1 \times 0.1$ mrad^2 to the monochromator.

6.3.4 Collimating Optics

Curved mirrors are widely used in optics to focus and collimate photon beams (Hecht and Zajac 1974). Photons from a point source are collected and delivered to a point focus using elliptical mirrors. Paraboloid mirrors perfectly suit the divergent beam from a point source being condensed into a non-divergent parallel beam.

We shall discuss here only paraboloidal mirrors as an example of the possible realization of the collimating optics for the $(+,+,\pm)$ analyzer, although also elliptical mirrors are of substantial interest.

Figure 6.11 shows a reflection scheme off a paraboloidal mirror for x rays emerging from a point source placed in the focal point F of the mirror. The

---

[4] The single-bounce backscattering spherical focusing crystal analyzer, which uses the (11 11 11) symmetric Bragg back-reflection of the 21.7 keV photons from silicon, has the same energy bandpass, see Sect. 6.2. The parameters of monochromator #2 in Table 6.3 are deliberately chosen to allow for closer comparison between these two conceptually different analyzer designs.

[5] In the case of the three-crystal version of the $(+,+,\pm)$ monochromator schematically shown in Fig. 5.38(b), the peak throughput T given in Table 6.3 has to be reduced by a factor $0.8 - 0.9$ (depending of the collimator crystal thickness), to take into account the anomalous transmission in the collimator crystal, as discussed in Sect. 5.5.6.

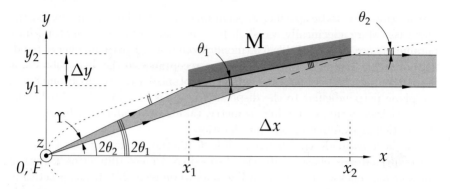

Fig. 6.11. Reflection of x rays off a paraboloidal mirror – M. The mirror surface is generated by the parabola $y^2 = 4fx$ rotated about the x-axis, with f being the focal length. X rays emerge from a point source placed in the focal point F of the mirror. For the assumed small glancing angles of incidence $\theta \ll 0.1$ rad, the vertex 0 of the parabola practically coincides with the focal point F.

mirror surface is generated by the parabola $y^2 = 4fx$ rotated about the x-axis, with f being the focal length. Assuming very small glancing angles of incidence $\theta \ll 0.1$ rad, as will be substantiated later, the vertex 0 of the parabola practically coincides with the focal point F. The figure also displays the parameters of paraboloidal mirrors and of the reflecting x rays.

The mirror parameters have to be optimized for obtaining larger angular acceptance $\varUpsilon^2$ ($\varUpsilon \gtrsim 5$ mrad), larger source-mirror distance x_1, and smaller vertical beam size Δy. The source-mirror distance is in fact the sample-mirror distance in the inelastic x-ray scattering experiments, which must be as large as possible to accommodate sample environments like an oven, a cryostat, pressure cells, etc. The vertical beam size is, on the contrary, must be kept small: $\Delta y \lesssim 1$ mm. Otherwise the dispersing crystal element of the $(+,+,\pm)$ monochromator may become unreasonably large[6].

One more important parameter is the admissible source size S. Nonzero source size results in an additional angular spread $\Delta\theta$ upon reflection from the mirrors, as it is graphically explained in Figure 6.12. With the help of this drawing it is easy to derive the following estimate for the admissible source size $S \simeq x_1 \Delta\theta$ for the given $\Delta\theta$ and the distance to the mirror point closest to the source. Since small glancing angles of incidence $\theta \ll 0.1$ rad are assumed, the distance is taken to be x_1. In the following examples $\Delta\theta = 100$ µrad was used, which equals the angular acceptance of the $(+,+,\pm)$ monochromator.

The angular acceptance of the monochromator also determines the admissible slope error, which should be $\ll \Delta\theta/2 = 50$ µrad (FWHM).

---

[6] To minimize the vertical beam size on the dispersing crystal element of the $(+,+,\pm)$ monochromator, it is more favourable to employ elliptical mirrors.

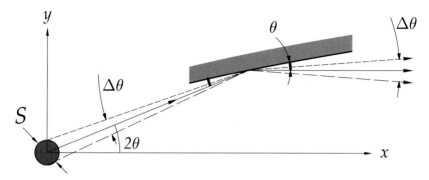

Fig. 6.12. Angular spread $\Delta\theta$ of x rays reflected off the paraboloidal mirror due to nonzero source size S. For small glancing angles of incidence θ one can use the approximate relation $S \simeq x_1 \Delta\theta$ to estimate the admissible values of the source size for the given $\Delta\theta$.

The mirror parameters (focal length, size, etc.) are also determined by the range of glancing angles of incidence with respect to the mirror surface (in which the photons can be reflected from the mirror). Figure 6.13 shows the angular dependence of reflectivity from flat mirrors coated with palladium (Pd) or platinum (Pt). As an example, x rays with a photon energy of $E = 9.13$ keV are again assumed. The optimal angular range of reflection is from 2 to 5 mrad. At larger glancing angles of incidence, the reflectivity becomes too small. Employing a glancing angle of incidence smaller than 2 mrad would require mirrors, which are too long.

Parameters of some mirrors are listed in Table 6.4.

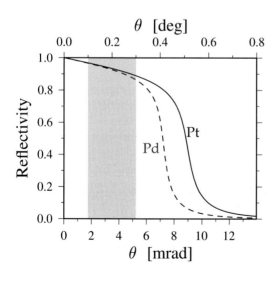

Fig. 6.13. Reflectivity of flat mirrors coated with palladium (Pd) or platinum (Pt) as a function of glancing angle of incidence θ for x-ray photons with an energy of $E = 9.13$ keV. Highlighted area indicates the optimal angular range for reflection with the paraboloidal mirror.

Table 6.4. Parameters of paraboloidal mirrors, and parameters of the reflecting x rays: f – focal length, θ_1 – largest glancing angle of incidence, $\Upsilon = 2(\theta_1 - \theta_2)$ – mirror's angular acceptance in the (x, y)-plane, Δx – mirror length, Δy – vertical size of the reflected beam, Δz - mirror width, in the z-axis direction, ensuring an angular acceptance of Υ in the (x, z)-plane, S – permissible x-ray source size (FWHM) at 0, for the angular spread of the reflected x rays to be smaller than $\Delta\theta = 100$ μrad, as explained in Fig. 6.12. For other notations see Fig. 6.11.

mirror #	f	θ_1	$\Upsilon = 2(\theta_1 - \theta_2)$	x_1	Δx	y_1	Δy	Δz	S
	μm	mrad (deg)	mrad (deg)	mm	mm	mm	mm	mm	μm
1	2.5	5.0 (0.29)	5.0 (0.29)	101	104	1.0	1.0	1	10
2	30	17.5 (1.0)	8.0 (0.46)	100	30	3.5	1.0	1	10
3	50	24 (1.35)	8.7 (0.5)	100	23	4.7	1.0	1	10

Mirror #1 in Table 6.4 assumes reflection with glancing angles of incidence in the range from 1.5 to 5 mrad (from the aforementioned Pd or Pt coatings). The angular acceptance is $\Upsilon = 5$ mrad. This number could be increased (while keeping Δy constant) by making the source-mirror distance x_1 smaller, but decreasing x_1 is not desirable, as was argued above. Besides, decreasing x_1 would also necessitate reducing source size, which already has to be quite small $\simeq 10$ μm.

Another way of enhancing Υ, is by shifting the range of glancing angles of incidence to higher values. This is, however, not possible with mirrors having usual single-layer coatings, as the range of high reflectivity is limited by $\theta \lesssim 5$ mrad, see Fig. 6.11.

An effective solution is to apply instead so-called multilayer mirrors, an artificial stratified structure, which reflects according to the principles of Bragg diffraction. For our purposes we need, however, a special multilayer mirror, with the period of the layers changing along the mirror's length (along the x-axis). Such mirrors with variable period, which are termed laterally graded multilayer mirrors, were introduced in the second half of 1990s, see, e.g., (Schuster and Göbel 1995, Morawe et al. 1999). The variation of the period is necessary, in order to match the changing glancing angle of incidence to the "Bragg angle" of the multilayer mirror, i.e., to the angle of peak reflectivity. The graded mirrors may have a reflectivity close to 90 %.

Mirrors #2 and #3 in Table 6.4 assume using such graded multilayer coatings. Application of such mirrors would allow for an increase of the angular acceptance $\varUpsilon$ almost by a factor of two, while keeping other parameters, such as source-mirror distance, and the vertical beam size, unchanged. Another advantage of larger glancing angles of incidence is the smaller mirror size.

Two-dimensionally curved paraboloidal mirrors are difficult to manufacture. The two-dimensional collimation can be alternatively achieved using instead a Kirkpatrick-Baez focusing system. Horizontal and vertical focusing are carried out by two separate orthogonal mirrors. Initially these are plane mirrors, which are dynamically bent to have a parabolic or elliptic shape. The basic principle for focusing x rays in orthogonal directions was originally suggested by Kirkpatrick and Baez (1948). Since then, the technology has been developed by many others, see, e.g., (Howells et al. 2000).

6.3.5 Inelastic X-Ray Scattering Spectrometers

Figures 6.14 and 6.15 show examples of how the high-resolution analyzers of the $(+, +, \pm)$ type can be integrated into an inelastic x-ray scattering spectrometer. The $(+, +, \pm)$-type backscattering monochromators are used as high-resolution monochromators in theses schemes.

Using the *exact* backscattering version of the $(+, +, \pm)$ monochromator, as in Fig. 6.15, has several benefits. Firstly, it consists of two rather than of three crystals (cf. Fig. 6.14) and therefore, it can be more easily controlled. For the same reason it has 10% to 20% higher throughput. Secondly, since the sample and the analyzer are installed in this case on the upstream side of the undulator, there is much more space for the sample environment. As was already mentioned in the discussion on the similar situation in Sect 6.2.5, the x rays may travel large distances from the undulator to the sample. Therefore, preserving their small cross-section requires collimating optics, e.g., compound refractive lenses.

Instead of using the backscattering $(+, +, \pm)$-type monochromators, as in Figs. 6.14 and 6.15, the in-line $(+, -, -, +)$-type monochromators, see Sect. 5.4.5, can be used alternatively. (Since the layout is obvious, we do not show it here graphically.) The benefits of using in-line monochromators are: firstly, sufficient space for the sample environment, secondly, faster energy tuning, and thirdly, spectral function of better shape (steeper wings), see Fig. 5.43. A disadvantage is a smaller throughput as compared to the $(+, +, \pm)$ monochromators (Fig. 5.43). And last but not least, switching to another bandpass would require changing all four monochromator crystals. In contrast to this, the $(+, +, \pm)$ monochromators would only require changing one crystal, the dispersing element.

There is one more major requirement. In any spectrometer using $(+, +, \pm)$ type analyzers, the application of focusing optics is mandatory, since the size S of the secondary source, has to be $\lesssim 10$ μm, see Table 6.4.

314   6. High-Resolution X-Ray Analyzers

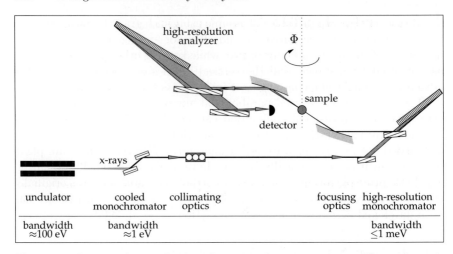

Fig. 6.14. Layout of an inelastic x-ray scattering spectrometer. The main components are the undulator x-ray source, the cooled monochromator, the high-resolution $(+,+,\pm)$ monochromator, the focusing optics, and the high-resolution $(+,+,\pm)$ analyzer.

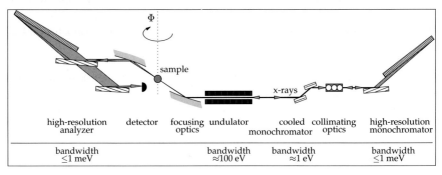

Fig. 6.15. Similar to Fig. 6.14, however, using the *exact* backscattering high-resolution monochromator, and with the sample and the high-resolution analyzer installed upstream of the undulator.

7. Towards Realizing X-Ray Resonators

X-ray Fabry-Pérot resonators (or interferometers) offer the potential for achieving much higher spectral resolution than is possible with the monochromators and analyzers discussed in the previous two sections. The physics underlying the x-ray Fabry-Pérot resonators was addressed in Chap. 4, where also the technical challenges of their implementation were investigated. In Chapt. 5.3 it was ascertained that sapphire is a proper material for Bragg back-reflecting mirrors. In the current chapter we present the concept and design of a prototype x-ray Fabry-Pérot resonator, and pilot experiments with such a device, following original publications (Shvyd'ko et al. 2003b,a, Shvyd'ko 2002). The primary aim of these studies was to test the theory, to test the principles, and to demonstrate the feasibility of achieving very high energy resolution with x-ray Fabry-Pérot resonators.

The concept of the prototype resonator is discussed in Sect. 7.1. In Sect. 7.2 technical details are described, such as: the procedure of manufacturing sapphire resonator mirrors, the design of a two-axis nanoradian goniometer for angular adjustment of the mirrors, the temperature control of the mirrors, and fast detectors. The experimental setup is described in Sects. 7.3 and 7.4. In Sect. 7.5 the results of the experiments with x-ray resonators are presented.

7.1 Concept of the Test Device

The first question to address is the design concept. What design is the best? There are many options. Should the mirrors be monolithically cut from a single crystal piece, or should the mirrors be independently adjustable? Both alternatives are of interest for future applications, but each of them has its own difficulties in realization.

Using one single crystal automatically ensures alignment of the reflecting atomic planes, and therefore reliable and robust operation of the device. It is, however, difficult to manufacture such a monolithic structure with inner surfaces of the mirrors having a good quality.

Realization of the second alternative, i.e., two independent mirrors, is not much easier, since the mirrors have to be aligned in parallel with nanoradian accuracy and be maintained aligned. However, it does have the advantage of allowing for better quality control of the sapphire mirrors. Secondly, it gives an additional opportunity to study the sensitivity to angular misalignment of the mirrors – an important issue for future applications. Since the first aim was a test of principles, a prototype device with independently adjustable mirrors was realized first, and it will be presented here.

The next problem with which one is confronted: is how to observe the Fabry-Pérot resonances? The first idea is to employ monochromatic x rays, with the bandwidth smaller than the Fabry-Pérot resonance width. The resonances are typically on the order 100 μeV in width, see Fig. 4.6 on page 193. This necessitates the use of x-ray monochromators with a bandpass $\lesssim$ 100 μeV, and this, although not impossible, see Sect. 5, is problematic. Thus, although the idea looks conceptually so simple, its implementation is not at all straightforward[1].

Instead of using monochromatic x-rays, and thus studying the energy dependence of transmission, one can use short radiation pulses to measure the time response of the resonator. As we have discussed in Sect. 4.3.6, the time response provides information on the width of the resonances and on the magnitude of the free spectral range. Furthermore, detection of the time response offers a possibility of direct observation of the multiple reflections from the mirrors. The longest time response (the largest number of multiple reflections) is expected for perfectly aligned mirrors. Therefore, observation of the time response and maximizing its duration (i.e., minimizing the decay rate) offers additionally a means of aligning the mirrors.

The duration of x-ray radiation pulses at modern synchrotron radiation facilities is typically 50 to 150 ps. The time resolution of the available x-ray detectors is in the same range (Baron 2000)[2]. To resolve the multiple reflections on the time scale, the back and forth time-of-flight between the mirrors should be larger than the pulse width and the detector time resolution. This dictates that the gap width d_g should be a few centimeters. For a resonator with a gap of $d_g = 5$ cm the time delay between two successively reflected beams is $t_f = 2d_g/c = 334$ ps. The free spectral range under these conditions is $E_f = 12.39$ μeV (4.49).

The Bragg reflection is chosen in line with the requirements of highest peak reflectivity, and of the Bragg energy being in a convenient spectral range. A good candidate is the (0 0 0 30) reflection in α-Al$_2$O$_3$. It is a *multiple-beam-free* Bragg back-reflection with a peak reflectivity of 95% (for infinitely thick crystal) and a Bragg energy of $E_H = 14.3148$ keV, see Appendix A.4. For the (0 0 0 30) Bragg back-reflection, the optimal thickness of the sapphire mirrors is $\simeq 40-70$ μm, as can be seen from the simulation of the transmission energy spectra shown in Fig. 4.8 on page 196. For a $d \simeq 55$ μm thick α-Al$_2$O$_3$ crystal

---

[1] One could use Mössbauer photons, which have neV bandwidth. This is certainly an interesting possibility, especially regarding the metrological applications of the Fabry-Pérot resonators, as illustrated in Fig. 1.24 on page 30.

[2] We do not consider here streak cameras which possess in fact better time resolution but very low efficiency.

7.1 Concept of the Test Device 317

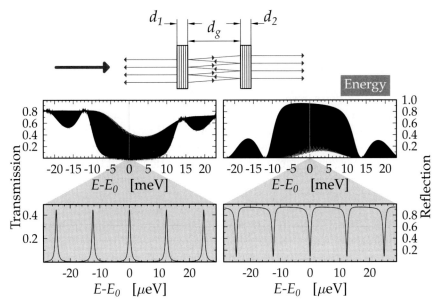

Fig. 7.1. Top: schematic of an x-ray Fabry-Pérot resonator. Center and bottom: the position of the Fabry-Pérot resonances on the x-ray energy scale in reflection (left), and transmission (right). The central parts on an expanded scale are shown in the bottom graphs. X rays are at normal incidence to the (0 0 0 30) atomic planes of α-Al$_2$O$_3$ single crystals used as the interferometer mirrors with $d_1 = d_2 = 55$ μm, separated by a gap of $d_g = 50$ mm. $E_0 = 14.315$ keV.

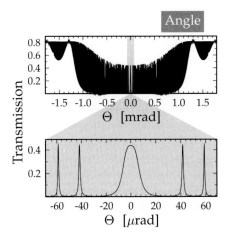

Fig. 7.2. The Fabry-Pérot resonances in transmission of Fig. 7.1(left) are shown now as a function of the incidence angle Θ, the angle of deviation from normal incidence to the (0 0 0 30) atomic planes. The central part on an expanded angular scale is shown in the bottom graph. The photon energy is $E_0 = 14.315$ keV.

plate, used as an example, the dynamical theory calculations give a peak reflectivity of $R \simeq 0.83$. The resonator with mirrors of such reflectivity has a finesse of $\mathcal{F} \simeq 16.8$. The expected resonance width $\Gamma = E_r/\mathcal{F}$ is $\simeq 0.73$ μeV and the response time $\tau = \hbar/\Gamma$ is $\simeq 0.89$ ns. The relative spectral resolution of the resonances $\epsilon_{FP} = \Gamma/E_H$ is $\simeq 5.5 \times 10^{-11}$. The angular acceptance of the Fabry-Pérot resonances at exact normal incidence $\Delta\theta = 2\sqrt{\epsilon_{FP}}$ (4.66) is $\simeq 15$ µrad.

The above estimates of the resonator parameters agree with those obtained from numerical dynamical theory calculations. Numerical calculations for x-ray Fabry-Pérot resonators with the (0 0 0 30) Bragg reflection have been presented in Sect. 4.3. In particular, a two-dimensional (E, Θ) plot of transmissivity of x rays through a resonator with two ideally parallel 50 µm thick sapphire crystal mirrors spaced a distance $d_g = 5$ cm apart, is shown in Fig. 4.12 on page 206. Figure 7.1 shows a schematic view of the test resonator and the position of the Fabry-Pérot resonances on the energy scale.

Figure 7.2 shows also the position of the Fabry-Pérot resonances, however, now as a function of the incidence angle Θ, the angle of deviation from normal incidence to the (0 0 0 30) atomic planes. It shows in particular that to selectively excite the extremely narrow ($\approx$ µeV-broad) Fabry-Pérot resonances in the test resonator, x-ray beams with an angular spread of $\lesssim 15-20$ µrad have to be used. Fortunately, today synchrotron radiation facilities routinely deliver such beams. The graph also shows how important it is to direct the incident beam as close as possible to $\Theta = 0$. If it deviates from normal incidence, for example, it is directed at an incidence angle of $\Theta \simeq 40$ µrad, this would immediately necessitate using x-ray beams with a much smaller angular spread of $\lesssim 1$ µrad, for selective resonance excitation (see also the two-dimensional (E, Θ) plot of transmissivity in Fig. 4.12). The larger the deviation from normal incidence, better collimated beams must then be used.

7.2 Technical Details

To implement the test Fabry-Pérot resonator one "merely" needs thin wafers of perfect sapphire with reflecting atomic planes aligned in parallel, and both wafers maintained at the same temperature. To observe the time response of the test device one needs x-ray detectors with a time resolution of $\simeq 100$ ps.

7.2.1 Manufacturing Sapphire Mirrors

Given the (0 0 0 30) Bragg reflection, the optimal thickness of the sapphire mirrors is $\simeq 40 - 70$ µm (Fig. 4.8, page 196). Such thin crystals are not easy to produce and to handle. Luckily, the mirrors have to be small, with an area of only about 10 mm^2, a little bit more than the cross-section of the incident x-ray beam.

Instead of manufacturing thin wafers, a technique of thinning down the center of a thick large sapphire disk can be employed. In this way, the thin mirror is framed by a thick ring and is safe to handle. The crystal surface is chemically polished, as described in Appendix A.7, in order to remove any possible defective layers or stresses introduced by the preceding mechanical treatment.

What should be also respected in the production, is the angle between the (0 0 0 30) reflecting atomic planes and the mirror surface, the asymmetry angle η. It has to be as small as possible to simplify the alignment of the atomic planes of both resonator crystal mirrors. As discussed in the next section, the angular range of the rotation stage used for this adjustment is 0.7 mrad (0.04°). Therefore, the asymmetry angle η has to be less than this value. This is a technological problem, which, however, can be circumvented by applying the following procedure. The resonator mirrors are produced as parts of the same sapphire monolith, which is then divided into parts. Due to this procedure the parts, i.e., the mirrors, have the same η. Taking advantage of this fact, the reflecting atomic planes of the mirrors can be oriented very close to the parallel state, by folding one of the mirrors back on itself, and rotating it in the plane of the surface at a definite relative azimuthal angle, as explained in more detail in (Shvyd'ko 2002).

7.2.2 Two-Axis Nanoradian Rotation Stage

The reflecting atomic planes of the crystal mirrors have to be parallel with sub-μrad or even nanoradian accuracy, as was discussed in Sect. 4.4, see (4.75). The higher the finesse, and the greater the incident beam cross-section, the more stringent is the requirement. Not only must the atomic planes be aligned for a short period; but they have also to be maintained in the aligned state with the aforementioned accuracy for a long time.

The range, over which the crystals have to be tilted, for the atomic planes to be aligned, is defined by the asymmetry angle η. Assuming $\eta \simeq 1$ mrad, to align the mirrors a two-dimensional rotation stage is needed with $\simeq 1$ nrad resolution and $\simeq 1$ mrad range (dynamic range 10^6).

The atomic planes of the mirrors M_1 and M_2 of the x-ray Fabry-Pérot interferometer are adjusted to be parallel to each other using a specially designed two-axis nanoradian rotation stage. A cross section of the stage is shown in Fig. 7.3; a perspective view with a cross section is shown in Fig. 7.4. As the name suggests, the smallest angular step of the stage is about 1 nrad. Flexural pivots instead of ball bearings are used to provide sub-nanometer repeatability. A flexured pivot is a frictionless linkage based on elastic deformation, also known as a weak-link mechanism. A general discussion of the monolithic rotation and translation mechanisms can be found, e.g., in the publications of Tanaka (1983) and Nakayama and Tanaka (1986).

The nanoradian stage is machined out of a single piece of invar. Invar, with a thermal expansion coefficient of less than 10^{-6} K^{-1}, is used to minimize instabilities due to temperature gradients and temperature fluctuations. Two-dimensional rotation of the mirror M_2 relative to M_1 is performed with the help of a two-dimensional circular notched flexure hinge – H. The rotation is activated by bending or releasing the leaf springs S_{F_μ} and S_{C_μ} cut out of the

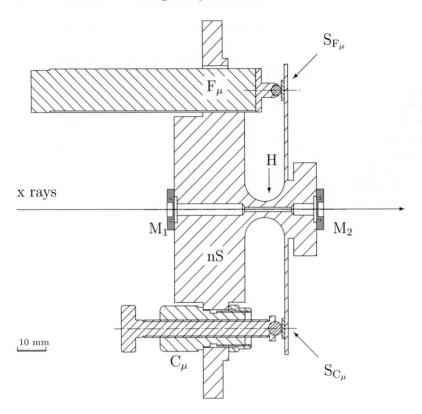

Fig. 7.3. Cross section of the monolithic nanoradian stage (nS) for parallel adjustment of the sapphire mirrors M_1 and M_2. Details see in text.

monolith (index μ takes values A or B for one of two mutually perpendicular rotation axis of the flexure hinge). This is brought about by the linear drives C_μ (a piezo-augmented screw driver) and F_μ (a piezo translator). Two pairs of leaf springs are used, one pair for each axis. One leaf spring in each pair is used for fine (F) and another for coarse (C) motion to reach the required angular resolution and angular range, respectively. In the initial position the leaf springs are pre-loaded, so that the driving points are shifted by 0.5 mm from their free state. The total range of the linear motion is $\simeq 1$ mm determined by the elastic deformation limit in invar.

By choosing the dimension (cross-section, radius) of the flexure hinge H and the dimension (length, thickness, width) of leaf springs S_{F_μ} and S_{C_μ} the specified parameters of angular rotation can be reached. The analytical simulation of the circular flexure hinge can be performed by using the formulae of Smith et al. (1997). The dimensions have been chosen to obtain a transmission factor of the linear to angular motion of 1 mrad/mm.

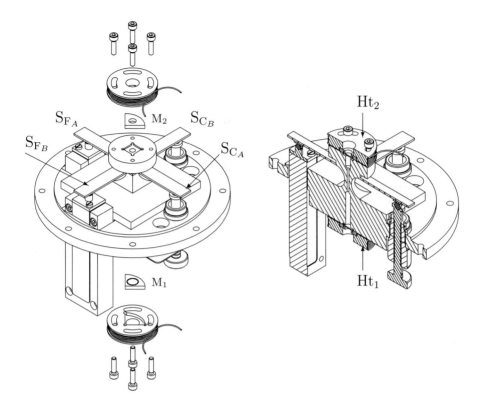

Fig. 7.4. Monolithic nanoradian stage for parallel adjustment of the sapphire mirrors M_1 and M_2. Left: perspective view. Right: perspective view with cross section. Details see in text.

The linear dependence of the tilt angle vs translation is observed (using an optical collimator) for both axes. Both axes show an approximately equal transmission factor of 700 ± 30 μrad/mm, a value very close to the designed value. The angular resolution of the nanoradian stage is 0.7 nrad. The maximal angular range is $\simeq 700$ μrad.

Both flat surfaces of the nanoradian stage to which the mirrors are attached are polished. In the unloaded state (the leaf springs are not loaded), the surfaces are parallel to each other with an accuracy of $\simeq 50$ μrad.

7.2.3 Temperature Control

The peak reflectivity shifts with crystal temperature on the x-ray energy scale at a rate of -0.083 meV/mK for the (0 0 0 30) reflection. The energy width of the reflection is $\gtrsim 13$ meV. Therefore, the temperature of both crystal mirrors has to be equal and stable within $\simeq 10$ mK, to ensure that their reflectivity peaks are not shifted apart by significantly more than 1 meV.

For this, a combination of passive and active temperature controls is applied. The nanoradian stage with the sapphire mirrors (the resonator) is mounted inside two concentric stainless steel sealed tanks. The inner tank, which houses the resonator, is filled with air for better temperature balance of the different parts of the resonator. The space between the outer and inner tanks is evacuated to thermally isolate the resonator from the environment. The walls of the internal tank are heated with surface heaters to a temperature slightly higher than the ambient temperature. The temperature is measured at one position with a thermo-resistor. At this point the temperature is maintained constant with 1 mK accuracy by using computer control (Lucht 1998).

The 2 mm thick rims of the sapphire mirrors are pressed against the flat surfaces of the nanoradian stage with the help of copper holders, which also incorporate wire heaters Ht_1 and Ht_2, see Fig. 7.4. A thermo-resistor is mounted in each copper holder to monitor the temperature. Its sensitive part is about 1 mm away from the correspondent sapphire mirror. The temperature of each copper holder in the immediate vicinity of the sensor is stabilized to an accuracy of 1 mK by using the heaters and the computer control.

7.3 Experimental Setup

Optical Elements. Figure 7.5 shows the schematic view of the setup used in the first experiments on x-ray resonators, which are presented in the following sections. X rays from an undulator are monochromatized at first to a ≈ 2 eV bandwidth with a two-bounce nondispersive $(+,-)$ high-heat-load monochromator. Both the undulator (at ≈ -40 m) and the monochromator (at ≈ -20 m) are not shown in Fig. 7.5. By using a four-bounce $(+,+,-,-)$ monochromator (see Sect. 5.4.3) the bandwidth of the x-ray spectrum is further narrowed down to $\simeq 2$ meV, a bandwidth, which is much narrower than the energy width of the (0 0 0 30) Bragg reflection. By tuning the meV-monochromator, the resonator can be excited in a desired region of its spectral function.

The vertical divergence of the beam (in the plane of Fig. 7.5) is $\sigma_v \simeq 20$ μrad. The horizontal divergence of the beam emerging from the undulator is $\sigma_h \simeq 40$ μrad. It can be narrowed down to $\sigma_h \simeq 20$ μrad by installing additionally a two-bounce $(+,-)$ Si(111) channel-cut crystal (horizontal scattering plane) upstream of the meV-monochromator, or by using a system of orthogonal slits (S). The same slit system is also used to change the beam cross-section.

Monochromatized and collimated x rays impinge upon the x-ray resonator (XR). The x-ray resonator comprises the sapphire (0 0 0 30) mirrors

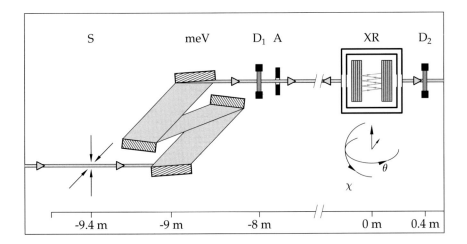

Fig. 7.5. Schematic view of experiments on x-ray resonators. X rays monochromatized to a bandwidth of 2 meV with a $(+,+,-,-)$ monochromator (meV) impinge upon an x-ray resonator (XR). Detectors D_1 and D_2 monitor the reflected and transmitted x rays. An aperture (A) is used as autocollimator to adjust exact backscattering from the resonator mirrors. The resonator is installed on a two-circle goniometer. S – slit system.

attached to the nanoradian stage. The stage together with the linear actuators, temperature sensors, heaters, etc., is placed inside the two concentric sealed tanks. An additional feature is an absorber, which can be inserted between the resonator mirrors by external control. This is helpful to distinguish between the back-reflection signals originating from the two mirrors of the resonator (while adjusting the resonator). The x-ray resonator is installed on a two-circle goniometer (χ, θ), enabling a variation of the incidence angle of the x rays.

Detectors. A semitransparent detector (D_1) with a time resolution of 0.7 ns is used to monitor the incident and reflected beams[3]. The time-of-flight $2L/c \simeq 53$ ns to the crystal and back to the detector, which are separated by a distance $L \simeq 8$ m, allows one to distinguish between the incident and reflected radiation pulses.

Another detector (D_2) with a much better time resolution of $\simeq 100$ ps is employed to monitor the radiation transmitted through the resonator[4].

---

[3] EG&G C30703F avalanche photo diode (APD) with a 100 μm thick sensitive Si wafer (Baron 1995) serves as the semitransparent detector ($\simeq 35\%$ absorption of the incident 14.3 keV photons).

[4] The S5343 APD from Hamamatsu Photonics is used with a 10 μm active layer (⌀1 mm), and a time resolution close to 100 ps (Kishimoto 1991, 1992).

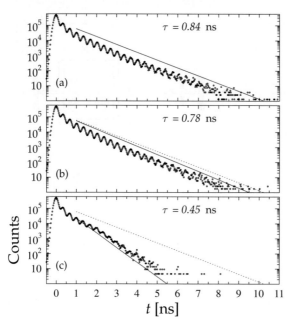

Fig. 7.6. Time response of an x-ray cavity with 55 μm thick sapphire mirrors for different relative angular settings of the resonator mirrors: (a) providing the longest response time, (b) $\Phi = 1.4$ μrad off, and (c) $\Phi = 15.4$ μrad off. The x-ray energy is tuned to the transmission minimum. The dotted lines indicate the decay rate with the largest response time $\tau = 0.84$ ns. The solid lines and the response time τ indicate the actual values.

With $\simeq 100$ ps synchrotron radiation pulses an overall duration of the time response of $\simeq 150$ ps (FWHM) is measured (with the resonator off the beam).

Preliminary Alignment of the Resonator Mirrors. An aperture (A) installed downstream of the semitransparent detector (D_1) is employed as an autocollimator for preliminary alignment (with μrad accuracy) of the resonator mirrors. The distance from the autocollimator to the resonator is about 8 m. By observing back-reflection through the aperture with an opening of 0.25 mm it is possible to adjust the reflecting atomic planes of both sapphire mirrors perpendicular to the incident beam with an accuracy of $\simeq 5 - 10$ μrad. The following procedure is used.

At first, with the help of the (χ, θ) two-circle goniometer the (0 0 0 30) atomic planes of the mirror M_2, see Fig. 7.3, are set perpendicular to the incident beam by observing backscattering through the aperture. The same procedure is repeated with the mirror M_1 [5]. The difference in the goniometer

---

[5] To distinguish between the signals originating from different mirrors, their temperatures and thus the peak reflectivities are detuned. Also, the absorber could be automatically inserted between the resonator mirrors.

positions gives the misalignment of the mirrors. The misalignment can be reduced to $\simeq 5-10$ μrad by tilting the mirror M_2 with the help of the nanoradian stage, as shown in Fig. 7.3, and checking the mirrors' orientation using the autocollimator technique.

A more accurate alignment is performed by observing the multiple reflections from the mirrors in the time dependence of transmission, as described below.

7.4 Adjustment of Mirrors

Fine Alignment. Light entering the resonator bounces back and forth between the mirrors producing multiple sub-waves emerging from the resonator cavity. A priori, it is obvious that the higher the reflectivity of the mirrors and the better the reflecting planes of both resonator mirrors are aligned,

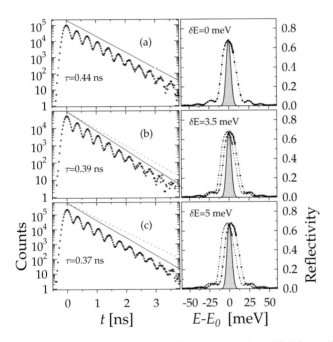

Fig. 7.7. Left: time response of an x-ray cavity with 39 μm thick sapphire mirrors. The temperature of the mirrors is (a) equal (with 10 mK accuracy), or differs by (b) 40 mK, and (c) 60 mK. The dotted lines indicate the time response with the largest response time $\tau = 0.44$ ns. The solid lines and the response times τ indicate the actual values.
Right: energy dependence of the reflectivity of both cavity mirrors (dots) and the spectrum of incident x-ray photons (solid line). δE is the mismatch of the peak reflectivities resulting from the temperature mismatch.

the more bounces take place. Their number can be measured in the most direct way by observing the time response in the forward direction of the resonator to the excitation with a short x-ray pulse. If the back-and-forth time-of-flight in the gap of the x-ray cavity is larger than the duration of the incident pulse and the time resolution of the detector, the number of bounces inside the cavity is easy to determine. Alternatively, the response time can be measured. The alignment of the mirrors can be improved by maximizing the number of observed bounces or maximizing the response time.

The time response of an x-ray cavity with 55 μm thick sapphire mirrors for different relative angular settings of the resonator mirrors is shown in Fig. 7.6. In particular, Fig. 7.6(a) shows the time dependence with the longest response time, corresponding to the resonator with best aligned mirrors. How is such a state reached? For this, a time window $t_1 \rightarrow t_1 + \Delta t$ ($t_1 > 0$) is selected in which the count rate is much lower than in the vicinity of $t = 0$. Only x-ray photons appearing in this time interval are counted. The angular position of the mirror M_2, see Figs. 7.3 and 7.4, is changed by the nanoradian stage to maximize the count rate in the selected time window. Then the window is chosen in the range $t_2 \rightarrow t_2 + \Delta t$, with $t_2 > t_1 + \Delta t$, and the angular position of the mirror M_2 is refined. The procedure is repeated several times until the count rate is still sufficient for further refinement. In this way, it is possible to adjust the mirrors with an accuracy of $\simeq 0.3$ μrad. This accuracy is limited by the brilliance of the third generation synchrotron radiation sources.

Equalizing Temperature. Figure 7.7 shows examples of what happens to the time response of the x-ray resonator if the temperatures of the resonator mirrors are not equal.

The temperature equality of the resonator mirrors is checked directly by measuring the energy dependence of Bragg reflectivity independently for each mirror, as explained later with help of Fig. 7.8 on page 327. The parameters of the computer control are chosen such that the peak reflectivities (or transmission minima) take place for practically the same photon energy.

7.5 Experiments on X-Ray Resonators

7.5.1 Reflectivity of Mirrors

α-Al$_2$O$_3$ single crystals were used as x-ray mirrors to reflect 14.315 keV x-rays backwards with the help of the (0 0 0 30) Bragg reflection. Crystals were grown by the heat-exchange method (Schmid et al. 1994). Dislocation-free samples were selected by white beam x-ray topography (Chen et al. 2001), see also Appendix A.7. The energy dependence of reflectivity and transmissivity in exact backscattering for two α-Al$_2$O$_3$ crystal plates picked out for the interferometer mirrors is shown in Fig. 7.8. The measurements were performed with x rays monochromatized to a bandwidth of 2 meV, and directed normal to the mirrors with an accuracy of 5 μrad.

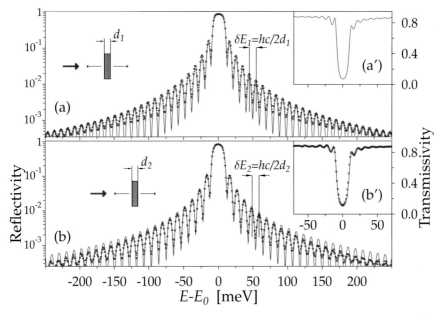

Fig. 7.8. Energy dependence of the reflectivity (a)-(b), and transmissivity (a')-(b'), of x rays at normal incidence to the (0 0 0 30) atomic planes of the two α-Al$_2$O$_3$ single crystals used as the interferometer mirrors: (a)-(a'): mirror 1, d_1 = 63.0(1) μm; (b)-(b'): mirror 2, d_2 = 58.6(2) μm thick. E_0 = 14.315 keV. Red lines are theoretical fits by using the dynamical theory of x-ray diffraction in exact backscattering. In the fit, the 2 meV energy width of the incident radiation is taken into account.

The energy dependence of reflectivity for mirror 1 in Fig. 7.8(a) and for mirror 2 in Fig. 7.8(b) look very similar. The peak reflectivity is observed practically at the same photon energy. This is achieved by equalizing the temperature of the mirrors. The peak reflectivities have almost the theoretical value, and are almost identical: $R = 0.84(2)$. The oscillations on the wings are, as we already well know, due to the interference of the waves reflected from the front and the rear crystal surfaces. A slightly different period of oscillations is observed in each mirror. The period is $\delta E_n = hc/2d_n(1 + w_H)$ (2.177). Neglecting the small Bragg's law correction $w_H = 4 \times 10^{-6}$ (see Appendix A.4), one can directly ascertain that the crystal thicknesses are $d_1 = 63.0(1)$ μm and $d_2 = 58.6(2)$ μm for mirrors 1 and 2, respectively.

At this point, it is important to memorize the fact that the mirrors have a slightly different thickness with a difference of $\delta d = d_1 - d_2 = 4.4(2)$ μm. Although it may sound strange, this small difference is of particular usefulness for proving the functioning of the interferometer, as it will become clear later (Fig. 7.12 on page 332).

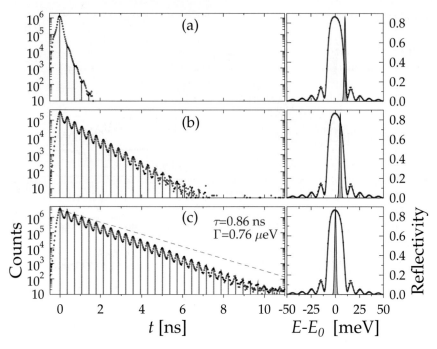

Fig. 7.9. Left: time response of the interferometer. The sharp green peaks are theoretical fits calculated by the theory of x-ray Fabry-Pérot interferometers (Sect. 4.3), without convolution with the time instrumental function.
Right: energy dependence of the reflectivity of a single α-Al$_2$O$_3$ mirror (red lines and dots) and the spectrum of incident x-ray photons (black line) normalized to the peak value of the reflection curve. The center of the spectral distribution E of the incident radiation pulses is detuned (a) $E - E_0 = 10$ meV off the peak reflectivity, (b) $E - E_0 = 5$ meV, and (c) tuned to the peak reflectivity of the mirrors: $E - E_0 = 0$, respectively. The spectral width of the incident radiation pulses is 2 meV, the pulse duration is 100 ps.
The 0.86 ns-decay-constant in (c) corresponds to 0.76 μeV-broad transmission Fabry-Pérot resonances.

A comparison of the energy dependencies both measured and calculated by the dynamical theory of x-ray diffraction (solid lines in Fig. 7.8) shows very good agreement, and thus demonstrates good performance of the selected back-reflecting x-ray mirrors. The mismatch in the amplitudes of the oscillations is attributed to a small thickness inhomogeneity of the mirrors. Built of such mirrors, a Fabry-Pérot interferometer should have a finesse of $\mathcal{F} = 18(3)$.

7.5.2 Time Response

X-Ray Resonator with $\simeq 60$ μm Thick Mirrors. The resonator composed of the mirrors presented above is excited by $\simeq 100$ ps radiation pulses. The left panels in Fig. 7.9 show the measured time response of the resonator in transmission (circles). The right panels in Fig. 7.9 show the energy dependence of the reflectivity (circles) of mirror 1, and the spectral distribution of the incident x-ray photons (solid line normalized to the maximum of the reflectivity curve). The Bragg reflectivity of the mirrors changes with the energy of the incident radiation. Far from the reflectivity maximum the incident radiation pulses traversed the resonator practically without interaction and any delay, as in Fig. 7.9(a). Apart from a few very weak multiple-reflection signals, this plot actually shows the instrument's time resolution function.

The closer the energy of x rays is to the reflectivity maximum the more significant the changes that occur. Multiple-reflection signals with descending strength are observed, appearing after every time-of-flight $t_f = 358(1)$ ps. These evenly spaced signals are a result of x-rays bouncing back and forth between the mirrors. The measured t_f gives the correct value for the gap between the mirrors $d_g = c/2t_f = 53.7(2)$ mm.

When the x-ray energy coincides with the reflectivity maximum, a train of more than 30 peaks is observed[6] due to $\simeq 60$ reflections, see Fig. 7.9(c). Such a long train of pulses is observed not only due to a high reflectivity of the mirrors, but also, because the reflecting atomic planes of the mirrors are set parallel with an accuracy of better than 0.35 μrad (by maximizing the intensity of the most delayed pulses in the train). The time response under these conditions is exponential. The longest decay time measured is $\tau \simeq 0.86(1)$ ns. Such a decay time corresponds to an energy width of $\Gamma \simeq 0.76(1)$ μeV for the transmission resonances. By using $t_f = 358(1)$ ps for the quantum beat period, and Eqs. (4.53)-(4.54), we derive for the finesse of the interferometer $\mathcal{F} = 2\pi\tau/t_f \simeq 15$, which matches the expected value 18(3), see discussion on page 328.

X-Ray Resonator with 39 μm Thick Mirrors. Figure 7.10 shows the measured time response of the x-ray resonator with another set of mirrors having smaller thicknesses of 39$\pm$1 μm. The measured peak reflectivity of the mirrors is $R = 0.69$, as shown in Fig. 7.10(right). The finesse of the resonator with mirrors of such reflectivity is expected to be $\mathcal{F} = 8.4$. The gap between the mirrors is set to $d_g = 52.5$ mm. One expects a quantum beat period of $t_f = 0.350$ ns, and a response time $\tau = 0.47$ ns, respectively.

When the center of the spectral distribution of the incident x-ray photons is tuned to the peak reflectivity of the mirrors, the time response is purely exponential– Fig. 7.10(left). Importantly, the use of thinner mirrors results in a shorter decay time of $\tau \simeq 0.44\pm0.02$ ns. This is consistent with the fact that thinner mirrors have a lower Bragg reflectivity, and thus a smaller finesse. The

---

[6] The number 30 is limited only by the intensity of the x-ray source used.

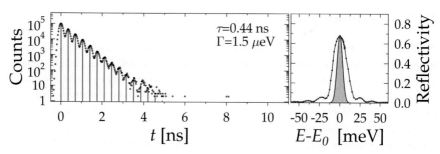

Fig. 7.10. Left: time response of an x-ray resonator with 39 μm thick sapphire mirrors (circles). The sharp green peaks are theoretical fits calculated by the theory of x-ray Fabry-Pérot interferometers (Sect. 4.3), without convolution with the time instrumental function. The peak at 4 ns is due to a spurious electron bunch in the storage ring.
Right: energy dependence of the reflectivity of the resonator mirrors (dots) and the spectrum of incident x-ray photons (solid line). The center E of the spectral distribution of the incident radiation pulses is tuned to the peak reflectivity of the mirrors: $E - E_0 = 0$. The spectral width of the incident radiation pulses is 7 meV, the pulse duration is 150 ps.
The 0.44 ns-decay-constant corresponds to 1.5 μeV-broad transmission Fabry-Pérot resonances.

observed decay time agrees very well with the estimated one. A decay time of $\tau \simeq 0.44 \pm 0.02$ ns corresponds to an energy width of $\Gamma \simeq 1.5$ μeV for the transmission resonances.

Comparison with Theory. The sharp vertical lines in the left panels of Fig. 7.9 and Fig. 7.10, which are in fact a set of very short ($\simeq 2$ ps) pulses, show the numerically calculated time spectra using the theory of x-ray Fabry-Pérot interferometers without any free parameter. The calculation procedure was described in Sect. 4.3.6. The shortness of the pulses is due to the interference of a very large number of Fabry-Pérot resonances simultaneously excited by a radiation spectrum much broader than the free spectral range. For example, in the case of the resonator corresponding to Fig. 7.9 a total number of $\Delta E/E_\mathrm{f} \simeq 160$ of Fabry-Pérot resonances are simultaneously excited by the $\Delta E = 2$ meV broad incident radiation spectrum. No convolution with the instrumental function (pulse width and detector resolution) was made. A very good agreement with respect to the amplitudes of the peaks between the experimental and calculated spectra is observed in almost all cases studied so far[7].

---

[7] To obtain agreement with the experimental spectrum in Fig. 7.9(c), a small correction to the energy of the incident photons was necessary. The best fit was achieved by assuming an excitation 3 meV off the reflectivity maximum. The discrepancy is attributed to a drift of the x-ray's energy spectrum during the two-hour measurement run. The dashed line in the left panel of Fig. 7.9(c) shows the envelope of the time response with $\tau = 1.1$ ns expected for excitation in the

The theoretical fit of the time spectrum in Fig. 7.9(c) demonstrates that the attained accuracy of 0.35 μrad in the angular alignment of the mirrors is sufficient to observe at least 30 successive x-ray beams (60 reflections) with intensities in agreement with the theory. In Sect. 4.3.9, it was shown that the interference of $\mathcal{N}_r \simeq \mathcal{F}$ successively reflected beams is sufficient for the formation of perfect Fabry-Pérot resonances. We observe $\simeq$ 30 beams, i.e., as $\mathcal{F} = 15$ – twice this number. One must, however, verify that these beams really interfere. The next experiment demonstrates the interference of the beams reflected from the resonator mirrors.

7.5.3 Interference Effects

Figures 7.11 and 7.12 show the energy dependences of the reflectivity and transmissivity of the resonator. The crystal thicknesses are $d_1 = 63.0(1)$ μm and $d_2 = 58.6(2)$ μm for mirrors 1 and 2, respectively as in Fig. 7.8. The atomic planes of both crystal mirrors of the interferometer are adjusted to be parallel to better than 0.35 μrad. In both figures, the same set of experimental data are shown by circles connected with dotted lines. However, the theoretical spectra shown by solid and dashed lines are calculated using two different models.

The solid red lines in Fig. 7.11 are the theoretical reflectivity and transmissivity (inset) of the first mirror, as in Fig. 7.8(a) and (a'), respectively. In other words, no interference between the mirrors is assumed. The dashed line in the inset is the product of the transmissivities through mirrors 1 and 2, i.e., describes similarly the transmissivity through a system of two noninteracting mirrors in Bragg backscattering. The period of the thickness oscillations observed in the experiment fits perfectly to that of the theoretical reflectivity of the first mirror (the theoretical curve is the same as in Fig. 7.8(a)). While the period fits perfectly, the theoretical curve in Fig. 7.11 does not agree with the experimental data in other respects. An additional periodical modulation - a beat-pattern - of the thickness oscillations with four nodal points is observed. Furthermore, the measured reflectivity in the antinodes of the beat-pattern is a factor of two in excess of the single crystal reflectivity.

The measured spectra in Fig. 7.12 agree almost perfectly with the spectra evaluated by the theory of x-ray Fabry-Pérot interferometers (see Sect. 4.3). The red solid lines in Fig. 7.12 are the spectra calculated for monochromatic x rays and then averaged over the 2 meV bandwidth of the incident radiation. No free parameters were used. Almost all the details, i.e., the beat-pattern in the reflection spectrum, the transmissivity of the system, are described by the theory. The transmissivity is higher than that of two independent mirrors – shown by the red dashed line in Fig. 7.11 – as it should be for a Fabry-Pérot interferometer.

---

maximum at $E - E_0 = 0$. In short preliminary measurements it was measured as $\tau = 1.05$ ns.

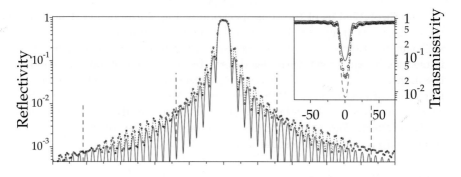

Fig. 7.11. Energy dependence of the reflectivity and transmissivity (inset) of the two-crystal interferometer. X rays are at normal incidence to the first crystal. The atomic planes of both crystals are adjusted to be parallel to better than 0.35 μrad. The solid red lines are the theoretical reflectivity and transmissivity (inset) of the first mirror, as in Fig. 7.8(a) and (a′), respectively. The dashed line in the inset is the product of the transmissivities through the mirror 1 and 2, i.e., describes the transmissivity through a system of two noninteracting mirrors in Bragg backscattering.

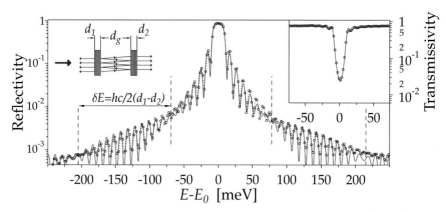

Fig. 7.12. Energy dependence of the reflectivity and transmissivity (inset) of the two-crystal interferometer. The reflectivity of the interferometer crystal mirrors is shown in Fig. 7.8. X rays are at normal incidence to the first crystal. The atomic planes of both crystals are adjusted to be parallel to better than 0.35 μrad. The solid red lines are theoretical spectra calculated by the theory of x-ray Fabry-Pérot interferometers (Sect. 4.3) and averaged over the 2 meV bandwidth of the incident radiation. Vertical dashed lines indicate the nodes of beatings in the energy dependence of the reflectivity.

The energy separation between the nodal points $\delta E = 140(3)$ meV corresponds to a quantity of length $hc/2\,\delta E = 4.4(2)$ μm, which agrees with the

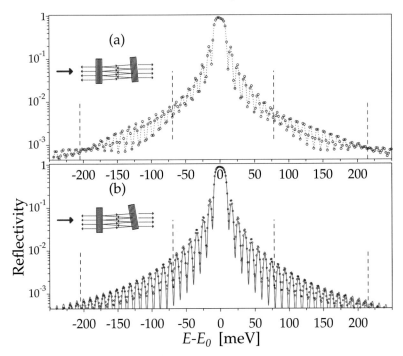

Fig. 7.13. Energy dependence of the reflectivity of the two-crystal interferometer. X rays are at normal incidence to the first crystal. The mirror 2 is tilted (a) by 3 μrad, (b) by 30 μrad from the parallel state. The solid red line in (b) is the theoretical reflectivity of the first mirror, the same as in Fig. 7.8(a) and in Fig. 7.11.

difference in the mirror thicknesses $\delta d = d_1 - d_2$ obtained from the fits of the spectra in Figs. 7.8(a) and 7.8(b). Therefore, the beat pattern appears due to the slightly different thickness of the crystal mirrors. The observed beating demonstrates unequivocally the interference of the waves reflected from the two mirrors, and therefore clearly proves the operation of a two-crystal interferometer.

The beat-pattern in Fig 7.12 can be easily destroyed by tilting the second crystal. Tilting by an angle of $\simeq 3$ μrad is enough to blur the nodes closest to the reflectivity maximum, as shown in Fig 7.13(a). With increasing tilt, the beat-pattern disappears completely – Fig 7.13(b) – and the reflectivity of the two-crystal system transforms to that of the first mirror, as in Fig 7.8(a).

The reported experimental result bears direct evidence of the successful performance of the two-crystal x-ray interferometer. This particular example demonstrates that x-ray Fabry-Pérot interferometers can be implemented in physical experiments requiring very high spectral resolution in the μeV range.

A. Appendices

A.1 Si Crystal Data

Space Group, Coordinates of Atoms. Silicon crystallizes in a cubic structure belonging to space group $Fd3m$ (O_h^7), No. 227 (Hahn 2002), the same as that for diamond, germanium, etc..

Si atoms in the unit cell are at the $8b$ positions (point symmetry $\bar{4}3m(T_d)$) with radius vector

$$\boldsymbol{r}_n = \boldsymbol{p}_k + \boldsymbol{t}_m,$$

where $n = k + 2(m-1)$, $k = 1, 2$, $m = 1, 2, 3, 4$,

$$\boldsymbol{p}_1 = \left(\frac{1}{2}, \frac{1}{2}, \frac{1}{2}\right), \qquad \boldsymbol{p}_2 = \left(\frac{3}{4}, \frac{3}{4}, \frac{3}{4}\right),$$

and the translation vectors

$$\boldsymbol{t}_1 = (0,0,0), \quad \boldsymbol{t}_2 = \left(0, \frac{1}{2}, \frac{1}{2}\right), \quad \boldsymbol{t}_3 = \left(\frac{1}{2}, 0, \frac{1}{2}\right), \quad \boldsymbol{t}_4 = \left(\frac{1}{2}, \frac{1}{2}, 0\right).$$

Lattice Parameter, Thermal Expansion. The distance between the atomic planes with Miller indices (hkl) in a cubic lattice is given by:

$$d_H = \frac{a}{\sqrt{h^2 + k^2 + l^2}}. \tag{A.1}$$

Standard Value at 22.5 °C
The following value of the lattice parameter of silicon $a = a_{\text{Si}}$, where

$$a_{\text{Si}} = 5.43102088(16) \times 10^{-10} \text{ m} \qquad \text{at } 22.500 \text{ °C}, \tag{A.2}$$

is accepted nowadays as a length standard for interatomic measurements (Mohr and Taylor 2000).

The linear thermal expansion coefficient of silicon measured in the temperature range from 21.0 °C to 23.5 °C by using a combined x-ray-optical interferometer (Bergamin et al. 1999) is equal to:

$$\beta(T) = \beta_0 + \beta_1 \Delta T, \qquad (A.3)$$

$$\beta_0 = 2.581(2) \times 10^{-6} \text{ K}^{-1} \qquad \beta_1 = 0.016(4) \times 10^{-6} \text{ K}^{-2},$$

where ΔT is the temperature deviation from 22.5 °C.

Temperature Dependence

Okada and Tokumaru (1984) have proposed the following empirical formula

$$\beta(T) = C_1 \{1 + \exp[-C_2(T - T_1)]\} + C_3 T, \qquad (A.4)$$

$$C_1 = 3.725 \times 10^{-6} \text{ K}^{-1}, \quad C_2 = 5.88 \times 10^{-3} \text{ K}^{-1},$$
$$C_3 = 5.548 \times 10^{-10} \text{ K}^{-2}, \quad T_1 = 124.0 \text{ K},$$

which fits experimental data for the linear thermal expansion coefficient in silicon in the temperature range 124 – 1500 K.

Anomalous Scattering Factors. The energy dependent parts of the structure factor (2.5)-(2.6) can be calculated for Si with f'_n and f''_n obtained from the library of anomalous scattering factors (Kissel and Pratt 1990, Kissel et al. 1995).

Debye-Waller Factor. The B-factor in the expression for the Debye-Waller factor (2.8) is $B = 0.4632$ Å^2 at ambient conditions (Deutsch and Hart 1985). This value corresponds to a Debye temperature of $\Theta_D = 530$ K. In the examples presented in this book, it is this quantity, which is used to calculate the Debye-Waller factors, in the framework of the Debye model (Ziman 1969, Ashcroft and Mermin 1976), at temperatures other than room temperature.

A.2 α-Al₂O₃ Crystal Data

Space Group. Sapphire (α-Al$_2$O$_3$) has a rhombohedral crystal lattice and belongs to the space group $R\bar{3}c$ (D_{3d}^6), No. 167 (Hahn 2002). It is usually specified with hexagonal axes of reference $\boldsymbol{a}, \boldsymbol{b}, \boldsymbol{c}$, where the c-axis is directed along the 3-fold symmetry axis, see, e.g., (Burns and Glazer 1978). There are two independent lattice parameters a and c ($a = b$). For convenience, one uses four Miller indices $(hkil)$, to specify atomic planes in the hexagonal basis, where the additional index i is redundant, and is given by $i = -(h+k)$.

Coordinates of Atoms. Coordinates of Al atoms in $12c$ positions of the hexagonal unit cell

$$\boldsymbol{r}_n = \boldsymbol{p}_m + \boldsymbol{t}_k,$$

where $n = m + 4(k-1)$, $k = 1, 2, 3$, $m = 1, 2, 3, 4$,

$$\boldsymbol{p}_1 = (0,0,z), \quad \boldsymbol{p}_2 = \left(0, 0, \frac{1}{2} - z\right), \quad \boldsymbol{p}_2 = (0, 0, -z), \quad \boldsymbol{p}_2 = \left(0, 0, \frac{1}{2} + z\right),$$

with $z = 0.35220(1)$ (Kirfel and Eichhorn 1990).

Coordinates of of O atoms in $18e$ positions of the hexagonal unit cell

$$\boldsymbol{r}_n = \boldsymbol{p}_m + \boldsymbol{t}_k,$$

where $n = m + 6(k-1)$, $k = 1, 2, 3$, $m = 1, 2, 3, 4, 5, 6$,

$$\boldsymbol{p}_1 = \left(x, 0, \frac{1}{4}\right), \quad \boldsymbol{p}_2 = \left(0, x, \frac{1}{4}\right), \quad \boldsymbol{p}_3 = \left(-x, -x, \frac{1}{4}\right),$$

$$\boldsymbol{p}_4 = \left(-x, 0, \frac{3}{4}\right), \quad \boldsymbol{p}_5 = \left(0, -x, \frac{3}{4}\right), \quad \boldsymbol{p}_6 = \left(x, x, \frac{3}{4}\right),$$

with $x = 0.30627(2)$ (Kirfel and Eichhorn 1990).

The translation vectors are

$$\boldsymbol{t}_1 = (0, 0, 0), \quad \boldsymbol{t}_2 = \left(\frac{2}{3}, \frac{1}{3}, \frac{1}{3}\right), \quad \boldsymbol{t}_3 = \left(\frac{1}{3}, \frac{2}{3}, \frac{2}{3}\right).$$

Lattice Parameters and Thermal Expansion. The distance between the atomic planes with Miller indices $(hkil)$ in a hexagonal lattice is given by:

$$d_H = \frac{ac}{\sqrt{(4/3)\, c^2\, (h^2 + k^2 + hk) + a^2 l^2}}. \tag{A.5}$$

The lattice parameters $a, b = a$, and c, and thermal expansion coefficients $\beta_a, \beta_b = \beta_a$, and β_c in HEMEX sapphire (Schmid et al. 1994) measured with a relative accuracy of better than 6×10^{-6} in a temperature range from

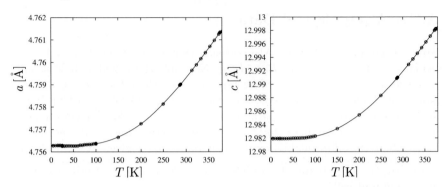

Fig. A.1. Lattice parameters a and c in α-Al$_2$O$_3$ (Lucht et al. 2003, Shvyd'ko et al. 2002). The solid lines are fits with functions (A.6) and the parameters from (A.9)

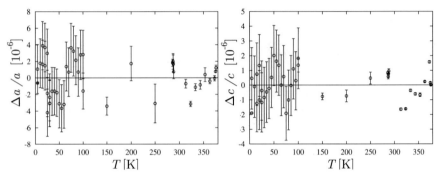

Fig. A.2. Deviation of the measured lattice parameters a and c in α-Al$_2$O$_3$ from the fit function (A.6) with parameters from (A.7).

$T_{\min} = 4.5$ K to $T_{\max} = 374$ K (Lucht et al. 2003, Shvyd'ko et al. 2002) can be approximated by the following empirical formulae (Lucht et al. 2003)

$$x(T) = \left(x_{44}T^4 + x_{04}\right) w(T) + (x_{11}T + x_{01})\left[1 - w(T)\right],$$

$$w(T) = \left(1 + \exp\frac{\sqrt{T} - \sqrt{\Theta_x}}{\sqrt{\Delta\Theta_x}}\right)^{-1}. \quad (A.6)$$

Here x takes the values a or c, respectively. The coefficients in (A.6) are equal to

$$\begin{array}{ll|ll}
a_{04} = 4.756274(4) \times 10^{-10} \text{ m} & c_{04} = 12.981943(4) \times 10^{-10} \text{ m} \\
a_{44} = 1.5(1.0) & \times 10^{-22} \text{ m K}^{-4} & c_{44} = 5.8(1.1) & \times 10^{-22} \text{ m K}^{-4} \\
a_{01} = 4.7501(4) & \times 10^{-10} \text{ m} & c_{01} = 12.9633(5) & \times 10^{-10} \text{ m} \\
a_{11} = 2.95(9) & \times 10^{-15} \text{ m K}^{-1} & c_{11} = 9.2(1) & \times 10^{-15} \text{ m K}^{-1} \\
\Theta_a = 193(30) & \text{K} & \Theta_c = 187(7) & \text{K} \\
\Delta\Theta_a = 1.0(3) & \text{K} & \Delta\Theta_c = 0.99(9) & \text{K}
\end{array}$$

$$(A.7)$$

The solid lines in Fig. A.1 are the functions (A.6) with the parameters from (A.7). shown together with the measured results for a and c. Figure A.2 shows the deviation of the measured results from the fit. The relative accuracy of the fit is about 4×10^{-6} which is close to the measurement uncertainty at low temperature, but worse than the uncertainty of the measurements at higher temperature. The polynomial functions of the 6th order can be used to approximate the experimental data with higher accuracy of about 1×10^{-6} in the $286 - 374$ K range, however, they are inapplicable for extrapolations (Shvyd'ko et al. 2002).

The linear thermal expansion coefficients in limiting cases of low and high temperatures:

$$\beta_x(T) = \left[\frac{4x_{44}}{x(0\text{ K})}\right] T^3 \quad \text{for } T \to T_{\min}$$

$$\beta_x(T) = \frac{x_{11}}{x(374\text{ K})} \quad \text{for } T_{\max} \geq T \gg \Theta_x. \tag{A.8}$$

From (A.8) and (A.7) we obtain the following expressions for the thermal expansion coefficients in the limiting cases:

$$\begin{aligned}
\beta_a &= 1.3(1.0) \times 10^{-12}\text{K}^{-1} \cdot (T/\text{K})^3 & \text{for } & T \to T_{\min} \\
\beta_c &= 1.79(35) \times 10^{-12}\text{K}^{-1} \cdot (T/\text{K})^3 & \text{for } & T \to T_{\min} \\
\beta_a &= 6.2(2) \times 10^{-6}\text{ K}^{-1} & \text{for } & T_{\max} \geq T \gg 200\text{ K} \\
\beta_c &= 7.07(8) \times 10^{-6}\text{ K}^{-1} & \text{for } & T_{\max} \geq T \gg 200\text{ K}.
\end{aligned} \tag{A.9}$$

White and Roberts (1983) have reported values for the thermal expansion coefficients in sapphire measured with a capacitance dilatometer, which are twice as small at low temperatures:

$$\begin{aligned}
\beta_a &= 0.5(1) \times 10^{-12}\text{K}^{-1} \cdot (T/\text{K})^3 & \text{for } & T \to 0 \\
\beta_c &= 0.75(10) \times 10^{-12}\text{K}^{-1} \cdot (T/\text{K})^3 & \text{for } & T \to 0.
\end{aligned} \tag{A.10}$$

Seel et al. (1997) have reported in agreement with White and Roberts (1983) a value of 5×10^{-12} K^{-1} for the thermal expansion of α-Al$_2$O$_3$ in the c-direction at 1.9 K. The measurements have been performed by modulating the temperature of a cryogenic optical resonator and observing the resulting change in resonator frequency. The reasons for the disagreement with Lucht et al. (2003) are not yet clear.

Anomalous Scattering Factors. The energy dependent parts of the structure factor (2.5)-(2.6) an be calculated for α-Al$_2$O$_3$ with f'_n and f''_n obtained from the library of anomalous scattering factors (Kissel and Pratt 1990, Kissel et al. 1995).

Debye-Waller Factors, Debye Temperature. For calculating the structure factor F_H (2.5) of the crystal unit cell, the knowledge of the Debye-Waller factors is required. This in turn requires the knowledge of the mean square displacement $\langle v_H^2 \rangle$ of the atoms from their equilibrium positions projected on the direction of the diffraction vector $\boldsymbol{H}$[1].

The mean square displacement values in α-Al_2O_3 at ambient conditions have been measured by Kirfel and Eichhorn (1990), Brown et al. (1992). The anisotropy of these parameters is small. To a fairly good approximation they can be assumed to be isotropic. By using the data of Kirfel and Eichhorn (1990), Brown et al. (1992) one can calculate the B-values (2.8) to be equal

$$B_{Al} = 0.195 \text{ Å}^2, \qquad B_O = 0.274 \text{ Å}^2.$$

These values are used to calculate the Debye-Waller factors at room temperature, in the examples presented in this book.

By using these B-values, the following effective Debye temperatures can be evaluated for the sub-lattice of aluminum, and for the sub-lattice of oxygen atoms:

$$\Theta_D(Al) = 890 \text{ K}, \qquad \Theta_D(O) = 995 \text{ K}.$$

These characteristic temperatures are used to calculate the Debye-Waller factors, in the framework of the Debye model (Ziman 1969, Ashcroft and Mermin 1976), at temperatures other than room temperature, in the examples presented in this book.

Thermal Conductivity. The thermal conductivity in α-Al_2O_3 attains its maximum value of $\simeq 200$ W cm^{-1}K^{-1} at $T \simeq 30$ K. This is the highest

---

[1] Lattice dynamics of sapphire, phonon dispersion relations, were studied by inelastic neutron scattering by Bialas and Stolz (1975) and Schober et al. (1993).

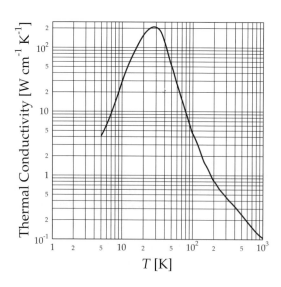

Fig. A.3. Thermal conductivity of sapphire (α-Al_2O_3). The recommended values for high purity synthetic sapphire single crystal with heat flow at 60° to the c-axis.(Touloukian et al. 1970)

thermal conductivity measured among mono-crystalline materials. The combination of high thermal conductivity and low thermal expansion at $T \simeq 30$ K is unique. It makes sapphire attractive also as a very robust crystal for high-heat-load monochromators.

A.3 Bragg Back-Reflections in Si

All allowed Bragg reflections (hkl) in silicon, and their parameters are listed in the following table. Equivalent reflections $(\tilde{h}\tilde{k}\tilde{l})$ are not included, for which $h^2 + k^2 + l^2 = \tilde{h}^2 + \tilde{k}^2 + \tilde{l}^2$. For the symmetric Bragg reflections the following parameters are given: the x-ray photon energy E_P at which the reflectivity reaches its peak value R at glancing angle of incidence $\theta = 90°$ to the reflecting atomic planes, the energy width ΔE, the relative energy width[2] $\Delta E/E_P$, the angular width[3] $\Delta\theta$ (calculated at x-ray energy $E = E_P$), the extinction length $d_e(0)$, and the Bragg's law correction $w_H^{(s)}$ (2.115) (calculated at x-ray energy $E = E_P$). These parameters were calculated in the framework of the *two-beam* dynamical theory of x-ray diffraction in perfect and thick silicon crystals at $T = 300$ K, with photo-absorption taken into account. Atomic and crystal data (coordinates of atoms, thermal parameters, lattice parameters, anomalous scattering factors, etc.) were used, as given in Appendix A.1. Only those reflections are listed whose peak reflectivity exceeds $R = 0.1$. The last column of the table shows the number of accompanying reflections N_a, which are excited in silicon crystals at normal incidence to the (hkl) atomic planes. The total number of the beams excited in the crystal is $N_a + 2$.

(hkl)	E_P	$w_H^{(s)}$	ΔE	$\Delta E/E_P$	$\Delta\theta$	R	$d_e(0)$	N_a
	keV	$\times 10^{-6}$	meV	$\times 10^{-8}$	μrad		μm	
(1 1 1)	1.9773	105	218	11025	15700	0.47	0.9	0
(0 2 2)	3.2287	47.2	178	5513	7370	0.91	1.1	0
(1 1 3)	3.7859	34.5	106	2799	7020	0.72	1.9	4
(0 0 4)	4.5659	23.8	125	2737	6400	0.91	1.8	4
(1 3 3)	4.9756	20.0	70.1	1408	4950	0.78	2.9	4
(2 2 4)	5.5920	15.8	79.7	1425	3070	0.92	2.5	2
(3 3 3)	5.9312	14.0	51.7	871	3930	0.81	4.0	6
(0 4 4)	6.4571	11.8	61.2	947	2690	0.92	3.3	10
(1 3 5)	6.7530	10.8	40.0	592	3290	0.82	5.2	10
(0 2 6)	7.2192	9.44	48.4	670	2410	0.92	4.2	2
(3 3 5)	7.4850	8.78	31.9	426	2830	0.83	6.5	4
(4 4 4)	7.9082	7.85	39.3	496	2190	0.91	5.2	6

[2] The quantity $\Delta E/E_P$ would be equal to $\epsilon_H^{(s)}$, the invariant quantity of Bragg reflections, introduced in (2.119)-(2.120), provided silicon was a non-absorbing crystal, see Sect. 2.4.3. Since silicon does not belong to a family of strongly absorbing crystals, the quantity $\Delta E/E_P$ in the table may be considered, with a certain degree of accuracy, equal to $\epsilon_H^{(s)}$, and used for $\epsilon_H^{(s)}$ to perform estimates with those equations throughout this book containing $\epsilon_H^{(s)}$.

[3] The angular width $\Delta\theta$ of a Bragg reflection at any glancing angle of incidence θ other than $90°$ can be calculated to a good accuracy as $\Delta\theta = \epsilon_H^{(s)} \tan\theta \simeq (\Delta E/E_P)\tan\theta$ in agreement with (2.126).

A.3 Bragg Back-Reflections in Si

(hkl)	E_P	$w_H^{(s)}$	ΔE	$\Delta E/E_P$	$\Delta\theta$	R	$d_e(0)$	N_a
	keV	$\times 10^{-6}$	meV	$\times 10^{-8}$	μrad		μm	
(1 1 7)	8.1516	7.39	26.0	318	2490	0.83	8.0	10
(2 4 6)	8.5419	6.72	32.4	379	2000	0.91	6.4	6
(1 3 7)	8.7677	6.38	21.6	246	2220	0.83	9.7	16
(0 0 8)	9.1316	5.87	27.1	296	1850	0.91	7.7	4
(3 3 7)	9.3432	5.61	18.1	193	2000	0.83	11.5	4
(0 6 6)	9.6856	5.21	22.9	236	1710	0.90	9.1	4
(5 5 5)	9.8853	5.00	15.4	155	1820	0.83	13.6	12
(0 4 8)	10.2095	4.69	19.6	191	1600	0.90	10.6	22
(1 1 9)	10.3991	4.52	13.2	126	1670	0.83	15.8	16
(4 6 6)	10.7078	4.26	17.0	158	1500	0.90	12.3	2
(1 3 9)	10.8888	4.12	11.4	104	1540	0.83	18.3	10
(4 4 8)	11.1839	3.90	14.8	132	1410	0.89	14.1	22
(1 7 7)	11.3573	3.78	9.99	87.9	1420	0.82	21.0	16
(0 2 10)	11.6406	3.60	13.0	111	1330	0.89	16.1	10
(1 5 9)	11.8073	3.50	8.79	74.4	1320	0.82	23.9	16
(3 5 9)	12.2407	3.25	7.78	63.5	1240	0.82	27.0	10
(2 4 10)	12.5040	3.11	10.2	81.5	1190	0.88	20.5	6
(1 1 11)	12.6593	3.04	6.93	54.7	1160	0.81	30.3	10
(0 8 8)	12.9140	2.92	9.10	70.4	1130	0.88	23.0	10
(1 3 11)	13.0645	2.85	6.21	47.5	1090	0.81	33.9	28
(0 6 10)	13.3115	2.75	8.18	61.4	1070	0.87	25.6	6
(3 3 11)	13.4575	2.69	5.59	41.5	1030	0.81	37.7	16
(0 0 12)	13.6974	2.59	7.38	53.8	1020	0.87	28.4	28
(7 7 7)	13.8393	2.54	5.05	36.4	976	0.80	41.7	12
(2 2 12)	14.0727	2.45	6.69	47.5	979	0.86	31.3	10
(3 5 11)	14.2109	2.41	4.58	32.2	926	0.80	45.9	22
(0 4 12)	14.4383	2.33	6.09	42.1	936	0.86	34.4	22
(1 9 9)	14.5731	2.29	4.18	28.6	880	0.79	50.4	4
(2 8 10)	14.7949	2.22	5.57	37.6	897	0.86	37.7	6
(1 1 13)	14.9264	2.18	3.82	25.5	838	0.79	55.1	28
(4 4 12)	15.1430	2.12	5.10	33.6	861	0.85	41.1	22
(1 3 13)	15.2716	2.08	3.51	22.9	800	0.79	60.1	28
(2 6 12)	15.4834	2.03	4.69	30.2	827	0.85	44.8	6
(3 3 13)	15.6091	1.99	3.23	20.6	764	0.78	65.3	10
(8 8 8)	15.8164	1.94	4.33	27.3	796	0.85	48.6	6
(1 5 13)	15.9395	1.91	2.98	18.7	732	0.78	70.8	22
(0 2 14)	16.1425	1.86	4.00	24.7	767	0.84	52.5	12
(1 9 11)	16.2631	1.83	2.76	16.9	702	0.77	76.5	22
(0 8 12)	16.4622	1.79	3.71	22.5	739	0.84	56.7	22
(3 9 11)	16.5805	1.76	2.56	15.4	673	0.77	82.5	16

(hkl)	E_P	$w_H^{(s)}$	ΔE	$\Delta E/E_P$	$\Delta\theta$	R	$d_e(0)$	N_a
	keV	$\times 10^{-6}$	meV	$\times 10^{-8}$	μrad		μm	
(2 4 14)	16.7758	1.72	3.45	20.5	713	0.84	61.0	14
(1 7 13)	16.8919	1.70	2.38	14.0	647	0.77	88.7	22
(4 8 12)	17.0836	1.66	3.21	18.7	689	0.83	65.6	46
(1 1 15)	17.1977	1.64	2.22	12.9	623	0.76	95.2	28
(0 6 14)	17.3860	1.60	2.99	17.2	666	0.83	70.3	2
(1 3 15)	17.4981	1.58	2.07	11.8	600	0.76	102	10
(9 9 9)	17.7934	1.53	1.94	10.9	578	0.75	109	24
(2 10 12)	17.9756	1.50	2.62	14.5	623	0.82	80.4	14
(1 5 15)	18.0840	1.48	1.81	10.0	558	0.75	117	40
(0 0 16)	18.2632	1.45	2.46	13.4	603	0.82	85.8	4
(3 5 15)	18.3699	1.44	1.70	9.25	539	0.74	124	22
(2 2 16)	18.5463	1.41	2.31	12.4	585	0.82	91.4	14
(5 11 11)	18.6514	1.39	1.60	8.58	520	0.74	132	10
(0 4 16)	18.8253	1.37	2.17	11.5	567	0.81	97.3	46
(1 7 15)	18.9288	1.35	1.50	7.92	503	0.74	141	28
(6 10 12)	19.1001	1.33	2.04	10.6	550	0.81	103	6
(3 7 15)	19.2021	1.31	1.42	7.40	487	0.73	150	16
(0 12 12)	19.3710	1.29	1.92	9.91	534	0.81	110	34
(1 1 17)	19.4717	1.28	1.34	6.88	471	0.73	159	22
(0 10 14)	19.6382	1.26	1.82	9.27	518	0.80	116	18
(1 3 17)	19.7375	1.24	1.26	6.38	457	0.72	168	46
(4 12 12)	19.9018	1.22	1.71	8.59	503	0.80	123	22
(1 9 15)	19.9998	1.21	1.19	5.95	443	0.72	178	16
(4 10 14)	20.1620	1.19	1.62	8.03	489	0.80	130	6
(1 5 17)	20.2587	1.18	1.13	5.58	429	0.72	188	34
(0 8 16)	20.4189	1.16	1.53	7.49	475	0.79	138	22
(3 5 17)	20.5143	1.15	1.07	5.22	416	0.71	199	22
(0 2 18)	20.6725	1.13	1.45	7.01	462	0.79	146	6
(5 9 15)	20.7668	1.12	1.01	4.86	404	0.71	210	16
(4 8 16)	20.9231	1.11	1.38	6.60	450	0.78	154	46
(1 7 17)	21.0163	1.10	0.96	4.57	392	0.70	222	34
(2 4 18)	21.1707	1.08	1.30	6.14	437	0.78	162	18
(1 11 15)	21.2628	1.07	0.91	4.28	381	0.70	234	28
(8 12 12)	21.4155	1.06	1.24	5.79	426	0.78	171	22
(3 11 15)	21.5065	1.05	0.86	4.00	370	0.69	246	22
(0 6 18)	21.6575	1.03	1.18	5.45	414	0.77	180	18
(11 11 11)	21.7475	1.02	0.82	3.77	360	0.69	260	24
(1 3 19)	21.9858	1.00	0.78	3.55	350	0.69	273	46
(4 6 18)	22.1335	0.99	1.06	4.79	393	0.77	199	14
(3 3 19)	22.2216	0.98	0.74	3.33	340	0.68	288	16

A.3 Bragg Back-Reflections in Si   345

(hkl)	E_P	$w_H^{(s)}$	ΔE	$\Delta E/E_P$	$\Delta\theta$	R	$d_e(0)$	N_a
	keV	$\times 10^{-6}$	meV	$\times 10^{-8}$	µrad		µm	
(8 8 16)	22.3677	0.97	1.01	4.52	382	0.76	210	22
(1 5 19)	22.4549	0.96	0.70	3.12	331	0.68	302	28
(0 14 14)	22.5995	0.95	0.96	4.25	373	0.76	221	16
(1 13 15)	22.6858	0.94	0.67	2.95	322	0.67	318	46
(0 0 20)	22.8290	0.93	0.91	3.99	363	0.75	232	28
(3 13 15)	22.9144	0.92	0.64	2.79	313	0.67	334	10
(2 2 20)	23.0561	0.91	0.87	3.77	354	0.75	243	6
(1 7 19)	23.1407	0.90	0.61	2.64	305	0.66	351	34
(0 4 20)	23.2811	0.89	0.83	3.57	345	0.74	256	70
(3 7 19)	23.3649	0.89	0.58	2.48	297	0.66	368	52
(0 10 18)	23.5039	0.88	0.79	3.36	336	0.74	268	10
(9 11 15)	23.5869	0.87	0.55	2.33	289	0.65	386	10
(12 12 12)	23.7246	0.86	0.75	3.16	328	0.73	282	30
(5 7 19)	23.8068	0.85	0.53	2.23	282	0.64	405	22
(2 6 20)	23.9432	0.84	0.72	3.01	319	0.73	295	22
(1 1 21)	24.0247	0.84	0.50	2.08	274	0.64	425	28
(1 3 21)	24.2407	0.82	0.48	1.98	267	0.63	446	34
(2 14 16)	24.3747	0.81	0.65	2.67	304	0.72	325	14
(1 13 17)	24.4547	0.81	0.46	1.88	261	0.63	467	46
(0 8 20)	24.5875	0.80	0.62	2.52	296	0.72	340	70
(1 5 21)	24.6669	0.79	0.44	1.78	254	0.62	489	40
(2 12 18)	24.7986	0.79	0.60	2.42	289	0.71	356	10
(3 5 21)	24.8773	0.78	0.42	1.69	248	0.61	513	28
(4 8 20)	25.0079	0.77	0.57	2.28	282	0.71	373	46
(1 11 19)	25.0859	0.77	0.40	1.59	241	0.61	537	22
(0 2 22)	25.2154	0.76	0.54	2.14	275	0.70	391	18
(1 7 21)	25.2928	0.76	0.38	1.50	235	0.60	562	52
(3 7 21)	25.4980	0.74	0.36	1.41	230	0.60	589	16
(2 4 22)	25.6254	0.74	0.50	1.95	262	0.69	428	22
(13 13 13)	25.7016	0.73	0.35	1.36	224	0.59	616	24
(0 16 16)	25.8280	0.72	0.47	1.82	256	0.69	448	10
(1 15 17)	25.9036	0.72	0.33	1.27	218	0.58	645	34
(0 6 22)	26.0290	0.71	0.45	1.73	250	0.68	469	6
(1 9 21)	26.1040	0.71	0.32	1.23	213	0.58	674	28
(4 16 16)	26.2285	0.70	0.43	1.64	244	0.67	490	46
(1 1 23)	26.3029	0.70	0.30	1.14	208	0.57	705	52
(4 6 22)	26.4264	0.69	0.41	1.55	238	0.67	513	26
(1 3 23)	26.5003	0.69	0.29	1.09	203	0.56	737	52
(0 12 20)	26.6229	0.68	0.40	1.50	232	0.66	536	46
(3 3 23)	26.6962	0.68	0.28	1.05	198	0.56	771	16

(hkl)	E_P	$w_H^{(s)}$	ΔE	$\Delta E/E_P$	$\Delta\theta$	R	$d_e(0)$	N_a
	keV	$\times 10^{-6}$	meV	$\times 10^{-8}$	μrad		μm	
(2 8 22)	26.8180	0.67	0.38	1.42	227	0.66	561	14
(1 5 23)	26.8907	0.67	0.27	1.00	193	0.55	806	22
(4 12 20)	27.0116	0.66	0.36	1.33	222	0.65	586	46
(1 11 21)	27.0838	0.66	0.26	0.96	189	0.54	843	52
(10 12 18)	27.2038	0.65	0.35	1.29	216	0.64	612	6
(3 11 21)	27.2756	0.65	0.24	0.88	184	0.54	881	28
(0 0 24)	27.3947	0.64	0.33	1.20	211	0.64	640	28
(1 7 23)	27.4660	0.64	0.23	0.84	180	0.53	920	46
(0 10 22)	27.5843	0.63	0.32	1.16	206	0.63	669	30
(1 15 19)	27.6551	0.63	0.23	0.83	176	0.52	961	40
(0 4 24)	27.7726	0.63	0.31	1.12	202	0.62	699	22
(3 15 19)	27.8429	0.62	0.22	0.79	172	0.51	1000	22
(2 14 20)	27.9596	0.62	0.29	1.04	197	0.62	730	18
(5 7 23)	28.0295	0.61	0.21	0.75	168	0.51	1050	28
(4 4 24)	28.1454	0.61	0.28	0.99	192	0.61	762	70
(1 9 23)	28.2148	0.61	0.20	0.71	164	0.50	1100	58
(2 6 24)	28.3300	0.60	0.27	0.95	188	0.60	796	14
(3 9 23)	28.3989	0.60	0.19	0.67	160	0.49	1140	28
(1 1 25)	28.5818	0.59	0.18	0.63	157	0.48	1190	22
(2 12 22)	28.6956	0.59	0.25	0.87	179	0.59	868	14
(1 3 25)	28.7636	0.58	0.17	0.59	153	0.47	1250	58
(0 8 24)	28.8766	0.58	0.24	0.83	175	0.58	906	22
(3 3 25)	28.9442	0.58	0.17	0.59	150	0.47	1300	16
(0 18 18)	29.0565	0.57	0.23	0.79	171	0.57	946	16
(1 5 25)	29.1237	0.57	0.16	0.55	146	0.46	1360	46
(0 16 20)	29.2353	0.57	0.22	0.75	168	0.57	987	94
(3 5 25)	29.3021	0.56	0.15	0.51	143	0.45	1420	64
(4 18 18)	29.4131	0.56	0.21	0.71	164	0.56	1030	18
(1 15 21)	29.4794	0.56	0.15	0.51	140	0.44	1480	22
(4 16 20)	29.5897	0.55	0.20	0.68	160	0.55	1080	46
(15 15 15)	29.6557	0.55	0.14	0.47	137	0.43	1540	54
(0 2 26)	29.7653	0.55	0.19	0.64	157	0.55	1120	22
(3 7 25)	29.8309	0.54	0.14	0.47	134	0.43	1610	28
(12 12 20)	29.9399	0.54	0.18	0.60	153	0.54	1170	22
(5 15 21)	30.0051	0.54	0.13	0.43	131	0.42	1680	28
(2 4 26)	30.1135	0.53	0.18	0.60	150	0.53	1220	22
(1 13 23)	30.1783	0.53	0.13	0.43	128	0.41	1750	58
(8 8 24)	30.2860	0.53	0.17	0.56	146	0.52	1270	22
(1 9 25)	30.3505	0.52	0.12	0.40	125	0.40	1830	34
(0 6 26)	30.4576	0.52	0.16	0.53	143	0.51	1330	14

A.3 Bragg Back-Reflections in Si

(hkl)	E_P	$w_H^{(s)}$	ΔE	$\Delta E/E_P$	$\Delta\theta$	R	$d_e(0)$	N_a
	keV	$\times 10^{-6}$	meV	$\times 10^{-8}$	μrad		μm	
(3 9 25)	30.5217	0.52	0.12	0.39	123	0.39	1910	22
(0 12 24)	30.6283	0.51	0.16	0.52	140	0.51	1390	70
(1 19 19)	30.6920	0.51	0.11	0.36	120	0.38	1990	22
(2 18 20)	30.7979	0.51	0.15	0.49	137	0.50	1450	22
(1 1 27)	30.8613	0.51	0.11	0.36	118	0.38	2080	70
(4 12 24)	30.9667	0.50	0.14	0.45	134	0.49	1510	46
(1 3 27)	31.0297	0.50	0.10	0.32	115	0.37	2170	28
(2 8 26)	31.1345	0.50	0.14	0.45	131	0.48	1570	22
(1 11 25)	31.1973	0.50	0.10	0.32	113	0.36	2260	52
(1 5 27)	31.3639	0.49	0.10	0.32	110	0.35	2350	70
(6 18 20)	31.4675	0.49	0.13	0.41	126	0.46	1710	6
(3 5 27)	31.5296	0.49	0.09	0.29	108	0.34	2450	22
(16 16 16)	31.6327	0.48	0.12	0.38	123	0.45	1780	6
(5 11 25)	31.6944	0.48	0.09	0.28	106	0.33	2560	34
(0 10 26)	31.7971	0.48	0.12	0.38	120	0.45	1860	38
(1 7 27)	31.8585	0.48	0.09	0.28	104	0.32	2670	58
(0 0 28)	31.9605	0.47	0.11	0.34	118	0.44	1940	52
(3 7 27)	32.0216	0.47	0.08	0.25	102	0.32	2780	28
(2 2 28)	32.1232	0.47	0.11	0.34	115	0.43	2020	18
(1 13 25)	32.1840	0.47	0.08	0.25	100	0.31	2900	22
(0 4 28)	32.2850	0.46	0.11	0.34	113	0.42	2100	82
(1 19 21)	32.3455	0.46	0.08	0.25	98	0.30	3020	58
(0 18 22)	32.4460	0.46	0.10	0.31	111	0.41	2190	10
(1 9 27)	32.5062	0.46	0.08	0.25	96	0.29	3150	40
(4 4 28)	32.6063	0.45	0.10	0.31	108	0.40	2280	46
(1 17 23)	32.6661	0.45	0.07	0.21	94	0.28	3280	58
(2 6 28)	32.7657	0.45	0.09	0.27	106	0.39	2380	38
(3 17 23)	32.8253	0.45	0.07	0.21	92	0.28	3420	40
(0 16 24)	32.9244	0.45	0.09	0.27	104	0.38	2480	22
(5 9 27)	32.9837	0.44	0.07	0.21	91	0.27	3560	34
(8 10 26)	33.0823	0.44	0.09	0.27	102	0.38	2580	14
(1 1 29)	33.1413	0.44	0.07	0.21	89	0.26	3710	34
(0 8 28)	33.2395	0.44	0.08	0.24	100	0.37	2690	70
(1 3 29)	33.2982	0.44	0.06	0.18	88	0.25	3870	58
(6 6 28)	33.3959	0.43	0.08	0.24	98	0.36	2810	10
(3 3 29)	33.4543	0.43	0.06	0.18	86	0.24	4030	40
(4 8 28)	33.5516	0.43	0.08	0.24	96	0.35	2920	94
(17 17 17)	33.6098	0.43	0.06	0.18	84	0.23	4200	36
(0 14 26)	33.7065	0.42	0.08	0.24	94	0.34	3050	18
(3 5 29)	33.7645	0.42	0.06	0.18	83	0.23	4370	70

(hkl)	E_P	$w_H^{(s)}$	ΔE	$\Delta E/E_P$	$\Delta\theta$	R	$d_e(0)$	N_a
	keV	$\times 10^{-6}$	meV	$\times 10^{-8}$	μrad		μm	
(1 21 21)	33.9185	0.42	0.06	0.18	82	0.22	4560	16
(2 10 28)	34.0144	0.42	0.07	0.21	90	0.32	3300	22
(1 7 29)	34.0718	0.42	0.06	0.18	80	0.21	4750	52
(8 16 24)	34.1672	0.41	0.07	0.20	89	0.31	3440	46
(1 13 27)	34.2244	0.41	0.05	0.15	79	0.20	4940	82
(0 2 30)	34.3194	0.41	0.07	0.20	87	0.30	3580	14
(3 13 27)	34.3763	0.41	0.05	0.15	78	0.20	5150	16
(8 8 28)	34.4710	0.41	0.06	0.17	86	0.30	3730	46
(1 17 25)	34.5276	0.40	0.05	0.14	76	0.19	5360	46
(2 4 30)	34.6218	0.40	0.06	0.17	84	0.29	3890	38
(1 9 29)	34.6782	0.40	0.05	0.14	75	0.18	5580	58
(0 12 28)	34.7720	0.40	0.06	0.17	82	0.28	4050	22
(3 9 29)	34.8282	0.40	0.05	0.14	74	0.18	5810	40
(0 6 30)	34.9216	0.40	0.06	0.17	81	0.27	4220	34
(5 17 25)	34.9775	0.39	0.05	0.14	73	0.17	6050	46
(4 12 28)	35.0705	0.39	0.06	0.17	80	0.26	4390	70
(5 9 29)	35.1262	0.39	0.05	0.14	72	0.16	6300	28
(4 6 30)	35.2188	0.39	0.05	0.14	78	0.25	4570	14
(1 15 27)	35.2742	0.39	0.04	0.11	70	0.16	6560	22
(1 1 31)	35.4217	0.38	0.04	0.11	69	0.15	6830	52
(0 22 22)	35.5135	0.38	0.05	0.14	76	0.24	4960	20
(1 3 31)	35.5685	0.38	0.04	0.11	68	0.15	7120	88
(0 20 24)	35.6600	0.38	0.05	0.14	74	0.23	5160	70
(3 3 31)	35.7147	0.38	0.04	0.11	67	0.14	7410	46
(2 14 28)	35.8058	0.38	0.05	0.14	73	0.22	5370	22
(1 5 31)	35.8604	0.38	0.04	0.11	66	0.13	7710	46
(4 20 24)	35.9511	0.37	0.05	0.14	72	0.21	5590	94
(3 5 31)	36.0054	0.37	0.04	0.11	66	0.13	8030	46
(0 10 30)	36.0957	0.37	0.05	0.14	71	0.21	5820	22
(7 15 27)	36.1498	0.37	0.04	0.11	65	0.12	8360	22
(1 7 31)	36.2937	0.37	0.04	0.11	64	0.12	8700	70
(4 10 30)	36.3834	0.36	0.04	0.11	68	0.19	6310	30
(1 17 27)	36.4370	0.36	0.04	0.11	63	0.11	9060	76
(0 0 32)	36.5263	0.36	0.04	0.11	67	0.18	6570	4
(3 17 27)	36.5798	0.36	0.04	0.11	62	0.11	9430	22
(2 2 32)	36.6687	0.36	0.04	0.11	66	0.18	6830	14
(5 7 31)	36.7220	0.36	0.03	0.08	61	0.10	9810	58
(0 4 32)	36.8106	0.36	0.04	0.11	65	0.17	7110	94
(2 12 30)	36.9519	0.35	0.04	0.11	64	0.16	7400	10
(4 4 32)	37.0927	0.35	0.04	0.11	63	0.16	7700	94

A.3 Bragg Back-Reflections in Si

(hkl)	E_P	$w_H^{(s)}$	ΔE	$\Delta E/E_P$	$\Delta\theta$	R	$d_e(0)$	N_a
	keV	$\times 10^{-6}$	meV	$\times 10^{-8}$	µrad		µm	
(2 6 32)	37.2329	0.35	0.04	0.11	62	0.15	8020	38
(12 12 28)	37.3726	0.35	0.04	0.11	62	0.14	8340	22
(2 20 26)	37.5118	0.34	0.03	0.08	61	0.14	8680	30
(0 8 32)	37.6505	0.34	0.03	0.08	60	0.13	9030	46
(0 14 30)	37.7886	0.34	0.03	0.08	59	0.13	9400	22
(4 8 32)	37.9263	0.34	0.03	0.08	58	0.12	9780	94
(2 18 28)	38.0635	0.33	0.03	0.08	58	0.12	10200	26
(12 20 24)	38.2001	0.33	0.03	0.08	57	0.11	10600	46
(2 10 32)	38.3363	0.33	0.03	0.08	56	0.11	11000	14

A.4 Two-Beam Bragg Backscattering Cases in α-Al$_2$O$_3$

All allowed *multiple-beam-free* Bragg back-reflections $(hkil)$ in α-Al$_2$O$_3$, and their parameters are listed in the following table. Equivalent reflections, e.g., $(ihkl)$, $(kihl)$, $(\bar{h}ki\bar{l})$, $(\bar{i}hk\bar{l})$, $(\bar{k}ih\bar{l})$ are not included. For the symmetric Bragg reflections the following parameters are given: the x-ray photon energy E_P at which the reflectivity reaches its peak value R at glancing angle of incidence $\theta = 90°$ to the reflecting atomic planes, the energy width ΔE, the relative energy width[4] $\Delta E/E_P$, the angular width[5] $\Delta\theta$ (calculated at x-ray energy $E = E_P$), the extinction length $d_e(0)$, and the Bragg's law correction $w_H^{(s)}$ (2.115) (calculated at x-ray energy $E = E_P$). The parameters were calculated in the framework of the *two-beam* dynamical theory of x-ray diffraction in perfect and thick α-Al$_2$O$_3$ crystals at $T = 300$ K, with photo-absorption taken into account. Atomic and crystal data (coordinates of atoms, thermal parameters, lattice parameters, anomalous scattering factors, etc.) were used, as given in Appendix A.2. Only those reflections are listed whose peak reflectivity exceeds $R = 0.1$ and $E_P \leq 40$ keV.

Two-beam Bragg diffraction in backscattering takes place at normal incidence to the reflecting planes with Miller indices $(hkil)$ obeying one of the reflection conditions:

$(0\ 0\ 0\ l)$: $l = 6n,\ n = 1, 2, 3, ...$
$(h\ h\ \overline{2h}\ 0)$: $h = 1, 3, 5, 9, 11, 15, ...$
$(0\ k\ \bar{k}\ 0)$: $k = 3(1 + 2n);\ k = 0, 1, 2, 3, ...$
$(0\ k\ \bar{k}\ l)$: $k = 1 - 6n;\ l = 2 + 6m;\ n = 0, 1, 2, 3, ..;\ m = 0, 1, 2, 3, ...$
  $k = -1 + 6n;\ l = 4 + 6m;\ n = 0, 1, 2, 3, ..;\ m = 0, 1, 2, 3, ...$

as derived from the table:

$(hkil)$	E_P	$w_H^{(s)}$	ΔE	$\Delta E/E_P$	$\Delta\theta$	R	$d_e(0)$
	keV	$\times 10^{-6}$	meV	$\times 10^{-8}$	μrad		μm
$(0\ 1\ \bar{1}\ 2)$	1.7817	248	206	11561	20500	0.14	1.5
$(0\ \bar{1}\ 1\ 4)$	2.4304	140	185	7611	12700	0.57	1.1
$(1\ 1\ \bar{2}\ 0)$	2.6054	122	123	4720	9920	0.53	1.6

---

[4] The quantity $\Delta E/E_P$ would be equal to $\epsilon_H^{(s)}$, the invariant quantity of Bragg reflections, introduced in (2.119)-(2.120), provided α-Al$_2$O$_3$ is a non-absorbing crystal, see Sect. 2.4.3. Since α-Al$_2$O$_3$ does not belong to the family of strongly absorbing crystals, the quantity $\Delta E/E_P$ in the table may be considered, with a certain degree of accuracy, equal to $\epsilon_H^{(s)}$, and used for $\epsilon_H^{(s)}$ to perform estimates with those equations throughout this book containing $\epsilon_H^{(s)}$.

[5] The angular width $\Delta\theta$ of a Bragg reflection at any glancing angle of incidence θ other than $90°$ can be calculated to a good accuracy as $\Delta\theta = \epsilon_H^{(s)} \tan\theta \simeq (\Delta E/E_P) \tan\theta$ in agreement with (2.126).

A.4 Two-Beam Bragg Backscattering Cases in α-Al$_2$O$_3$

$(hkil)$	E_P	$w_H^{(s)}$	ΔE	$\Delta E/E_P$	$\Delta\theta$	R	$d_e(0)$
	keV	$\times 10^{-6}$	meV	$\times 10^{-8}$	μrad		μm
(0 0 0 6)	2.8632	101	49.4	1725	8760	0.12	7.4
(0 1 $\bar{1}$ 8)	4.1031	49.4	63.5	1547	6750	0.59	3.3
(0 3 $\bar{3}$ 0)	4.5123	40.8	178	3944	4360	0.96	1.1
(0 $\bar{1}$ 1 10)	5.0032	33.1	96.9	1936	4580	0.88	2.1
(0 0 0 12)	5.7260	25.2	57.5	1004	5500	0.79	3.7
(0 1 $\bar{1}$ 14)	6.8476	17.6	43.0	627	3070	0.87	4.8
(0 $\bar{5}$ 5 2)	7.5807	14.3	4.53	59.7	1480	0.23	56.5
(0 5 $\bar{5}$ 4)	7.7588	13.7	39.9	514	2730	0.90	5.2
(0 $\bar{1}$ 1 16)	7.7814	13.6	23.0	295	2590	0.81	9.1
(3 3 $\bar{6}$ 0)	7.8154	13.5	55.1	705	2250	0.95	3.8
(0 $\bar{5}$ 5 8)	8.4337	11.5	6.94	82.3	1630	0.54	30.8
(0 0 0 18)	8.5889	11.1	35.1	408	2850	0.90	6.0
(0 5 $\bar{5}$ 10)	8.9064	10.3	21.6	242	1880	0.88	9.6
(0 1 $\bar{1}$ 20)	9.6610	8.77	22.9	237	1770	0.92	9.1
(0 $\bar{5}$ 5 14)	10.0589	8.09	22.5	223	1760	0.92	9.3
(0 $\bar{1}$ 1 22)	10.6048	7.27	1.73	16.3	804	0.23	147
(0 5 $\bar{5}$ 16)	10.7164	7.12	15.0	139	1720	0.88	14.0
(0 0 0 24)	11.4519	6.23	2.85	24.9	757	0.61	71.3
(0 $\bar{5}$ 5 20)	12.1502	5.53	8.38	68.9	1090	0.87	24.9
(0 1 $\bar{1}$ 26)	12.4970	5.23	5.14	41.1	1060	0.78	41.0
(0 5 $\bar{5}$ 22)	12.9133	4.89	2.53	19.6	875	0.61	84.5
(5 5 $\overline{10}$ 0)	13.0255	4.81	6.93	53.2	947	0.88	30.1
(0 $\bar{1}$ 1 28)	13.4449	4.51	6.50	48.4	1080	0.86	32.4
(0 9 $\bar{9}$ 0)	13.5365	4.45	11.3	83.5	1070	0.94	18.6
(0 0 0 30)	14.3148	3.98	13.1	91.5	1130	0.95	16.0
(0 $\bar{5}$ 5 26)	14.5075	3.87	5.93	40.9	1030	0.87	35.5
(0 5 $\bar{5}$ 28)	15.3316	3.47	2.25	14.7	634	0.74	93.4
(0 $\bar{1}$ 1 34)	16.2930	3.07	4.65	28.5	727	0.90	45.1
(0 $\overline{11}$ 11 2)	16.5721	2.97	0.90	5.45	440	0.53	238
(0 11 $\overline{11}$ 4)	16.6543	2.94	4.62	27.7	768	0.90	45.6
(0 $\overline{11}$ 11 8)	16.9792	2.82	1.44	8.48	524	0.70	147
(0 $\bar{5}$ 5 32)	17.0206	2.81	1.88	11.1	635	0.75	113
(0 0 0 36)	17.1777	2.76	1.97	11.5	722	0.77	108
(0 11 $\overline{11}$ 10)	17.2189	2.75	3.95	22.9	665	0.90	53.2
(0 $\overline{11}$ 11 14)	17.8423	2.56	4.08	22.9	689	0.91	51.5
(0 5 $\bar{5}$ 34)	17.8816	2.55	5.07	28.4	722	0.93	41.5
(0 1 $\bar{1}$ 38)	18.1943	2.46	3.19	17.5	657	0.89	66.1
(0 11 $\overline{11}$ 16)	18.2211	2.45	2.92	16.0	649	0.87	72.4
(0 $\overline{11}$ 11 20)	19.0996	2.23	2.89	15.1	571	0.90	72.7
(0 $\bar{1}$ 1 40)	19.1455	2.22	3.12	16.3	613	0.90	67.6

$(hkil)$	E_P	$w_H^{(s)}$	ΔE	$\Delta E/E_P$	$\Delta\theta$	R	$d_e(0)$
	keV	$\times 10^{-6}$	meV	$\times 10^{-8}$	µrad		µm
$(0\ 11\ \overline{11}\ 22)$	19.5939	2.12	0.31	1.56	250	0.30	775
$(0\ \overline{5}\ 5\ 38)$	19.6297	2.11	1.51	7.69	457	0.81	140
$(0\ 0\ 0\ 42)$	20.0407	2.03	1.57	7.83	575	0.81	135
$(0\ 5\ \overline{5}\ 40)$	20.5145	1.93	1.65	8.04	454	0.85	128
$(0\ \overline{11}\ 11\ 26)$	20.6793	1.90	1.73	8.37	501	0.85	122
$(0\ 1\ \overline{1}\ 44)$	21.0488	1.84	2.35	11.2	501	0.90	89.7
$(0\ 11\ \overline{11}\ 28)$	21.2656	1.80	1.38	6.49	436	0.83	153
$(0\ \overline{1}\ 1\ 46)$	22.0008	1.68	0.52	2.34	290	0.63	414
$(0\ \overline{5}\ 5\ 44)$	22.3012	1.64	2.64	11.8	517	0.93	79.8
$(0\ \overline{11}\ 11\ 32)$	22.5137	1.60	0.45	2.00	276	0.61	475
$(0\ 15\ \overline{15}\ 0)$	22.5608	1.60	1.75	7.76	434	0.89	120
$(0\ 0\ 0\ 48)$	22.9036	1.55	3.01	13.1	567	0.94	69.9
$(0\ 11\ \overline{11}\ 34)$	23.1716	1.51	2.04	8.80	461	0.91	103
$(0\ 5\ \overline{5}\ 46)$	23.2018	1.51	0.99	4.29	386	0.82	213
$(9\ 9\ \overline{18}\ 0)$	23.4458	1.48	1.47	6.27	397	0.89	144
$(0\ 1\ \overline{1}\ 50)$	23.9053	1.42	1.45	6.07	414	0.89	145
$(0\ \overline{11}\ 11\ 38)$	24.5457	1.35	1.06	4.32	358	0.86	199
$(0\ \overline{1}\ 1\ 52)$	24.8578	1.32	0.51	2.04	267	0.73	419
$(0\ \overline{5}\ 5\ 50)$	25.0151	1.30	0.86	3.43	321	0.84	246
$(0\ 11\ \overline{11}\ 40)$	25.2589	1.27	1.13	4.47	356	0.88	188
$(0\ \overline{17}\ 17\ 2)$	25.5867	1.24	0.81	3.18	335	0.83	261
$(0\ 17\ \overline{17}\ 4)$	25.6400	1.24	0.74	2.88	293	0.82	286
$(0\ 0\ 0\ 54)$	25.7666	1.22	0.55	2.13	248	0.78	384
$(0\ \overline{17}\ 17\ 8)$	25.8522	1.22	0.95	3.67	352	0.86	223
$(0\ 5\ \overline{5}\ 52)$	25.9268	1.21	0.77	2.96	319	0.83	276
$(0\ 17\ \overline{17}\ 10)$	26.0103	1.20	1.68	6.46	409	0.93	126
$(0\ \overline{17}\ 17\ 14)$	26.4271	1.16	0.76	2.89	291	0.84	277
$(0\ 17\ \overline{17}\ 16)$	26.6843	1.14	0.45	1.69	236	0.75	471
$(0\ \overline{11}\ 11\ 44)$	26.7303	1.14	1.26	4.71	358	0.91	168
$(0\ 1\ \overline{1}\ 56)$	26.7632	1.13	0.14	0.52	142	0.39	1610
$(0\ \overline{17}\ 17\ 20)$	27.2917	1.09	1.35	4.95	369	0.92	156
$(0\ 11\ \overline{11}\ 46)$	27.4862	1.08	0.39	1.42	228	0.74	544
$(0\ 17\ \overline{17}\ 22)$	27.6399	1.06	0.29	1.05	203	0.67	732
$(0\ \overline{1}\ 1\ 58)$	27.7160	1.06	0.98	3.54	322	0.89	215
$(0\ \overline{5}\ 5\ 56)$	27.7589	1.05	0.34	1.24	216	0.71	618
$(0\ \overline{17}\ 17\ 26)$	28.4197	1.01	0.28	0.98	186	0.68	762
$(0\ 0\ 0\ 60)$	28.6295	0.99	1.06	3.70	340	0.91	200
$(11\ 11\ \overline{22}\ 0)$	28.6560	0.99	0.69	2.42	268	0.87	305
$(0\ 5\ \overline{5}\ 58)$	28.6787	0.99	0.67	2.34	268	0.86	316
$(0\ 17\ \overline{17}\ 28)$	28.8490	0.98	0.79	2.73	297	0.88	269

A.4 Two-Beam Bragg Backscattering Cases in α-Al$_2$O$_3$

$(hkil)$	E_P	$w_H^{(s)}$	ΔE	$\Delta E/E_P$	$\Delta\theta$	R	$d_e(0)$
	keV	$\times 10^{-6}$	meV	$\times 10^{-8}$	μrad		μm
$(0\ \overline{11}\ 11\ 50)$	29.0331	0.96	0.61	2.11	260	0.85	345
$(0\ 1\ \overline{1}\ 62)$	29.6220	0.93	0.37	1.26	210	0.78	570
$(0\ \overline{17}\ 17\ 32)$	29.7810	0.92	0.09	0.30	110	0.34	2610
$(0\ 11\ \overline{11}\ 52)$	29.8223	0.91	0.34	1.16	205	0.77	615
$(0\ 17\ \overline{17}\ 34)$	30.2814	0.89	0.55	1.81	237	0.86	386
$(0\ \overline{5}\ 5\ 62)$	30.5247	0.87	0.46	1.49	228	0.83	465
$(0\ \overline{1}\ 1\ 64)$	30.5751	0.87	0.64	2.08	254	0.88	333
$(0\ \overline{17}\ 17\ 38)$	31.3454	0.83	0.59	1.88	249	0.88	360
$(0\ \overline{11}\ 11\ 56)$	31.4281	0.82	0.13	0.43	128	0.56	1610
$(0\ 5\ \overline{5}\ 64)$	31.4505	0.82	0.67	2.12	257	0.89	318
$(0\ 0\ 0\ 66)$	31.4924	0.82	0.23	0.72	169	0.71	934
$(0\ 17\ \overline{17}\ 40)$	31.9070	0.80	0.60	1.87	246	0.89	354
$(0\ 11\ \overline{11}\ 58)$	32.2434	0.78	0.46	1.43	216	0.86	460
$(0\ 1\ \overline{1}\ 68)$	32.4816	0.77	0.52	1.59	226	0.88	410
$(0\ \overline{17}\ 17\ 44)$	33.0840	0.74	0.39	1.17	196	0.85	547
$(0\ \overline{5}\ 5\ 68)$	33.3068	0.73	0.39	1.16	196	0.85	550
$(0\ \overline{1}\ 1\ 70)$	33.4349	0.73	0.27	0.82	171	0.79	775
$(0\ 17\ \overline{17}\ 46)$	33.6977	0.72	0.06	0.17	82	0.33	4000
$(0\ \overline{11}\ 11\ 62)$	33.8958	0.71	0.24	0.70	159	0.77	899
$(0\ 5\ \overline{5}\ 70)$	34.2372	0.69	0.18	0.54	140	0.73	1150
$(0\ 0\ 0\ 72)$	34.3554	0.69	0.12	0.34	112	0.61	1840
$(0\ \overline{23}\ 23\ 2)$	34.6063	0.68	0.23	0.66	158	0.78	925
$(0\ 23\ \overline{23}\ 4)$	34.6458	0.68	0.20	0.57	141	0.75	1080
$(0\ 11\ \overline{11}\ 64)$	34.7318	0.67	0.37	1.08	190	0.86	567
$(0\ \overline{23}\ 23\ 8)$	34.8031	0.67	0.27	0.77	168	0.81	790
$(0\ 23\ \overline{23}\ 10)$	34.9207	0.67	0.47	1.35	210	0.89	450
$(0\ \overline{17}\ 17\ 50)$	34.9709	0.66	0.34	0.96	182	0.85	632
$(0\ \overline{23}\ 23\ 14)$	35.2322	0.65	0.21	0.58	143	0.77	1030
$(0\ 1\ \overline{1}\ 74)$	35.3417	0.65	0.30	0.85	171	0.84	704
$(0\ 23\ \overline{23}\ 16)$	35.4256	0.65	0.12	0.34	111	0.64	1790
$(0\ 17\ \overline{17}\ 52)$	35.6288	0.64	0.08	0.22	92	0.51	2790
$(0\ \overline{23}\ 23\ 20)$	35.8853	0.63	0.39	1.09	191	0.88	542
$(0\ \overline{5}\ 5\ 74)$	36.1017	0.62	0.31	0.85	171	0.85	692
$(0\ 23\ \overline{23}\ 22)$	36.1508	0.62	0.09	0.25	100	0.58	2360
$(0\ \overline{1}\ 1\ 76)$	36.2952	0.62	0.04	0.12	70	0.31	5260
$(0\ \overline{11}\ 11\ 68)$	36.4213	0.61	0.26	0.72	158	0.83	811
$(0\ \overline{23}\ 23\ 26)$	36.7505	0.60	0.07	0.20	88	0.53	2880
$(0\ 23\ \overline{23}\ 28)$	37.0835	0.59	0.24	0.64	152	0.82	897
$(0\ 0\ 0\ 78)$	37.2183	0.59	0.32	0.85	171	0.87	667
$(0\ 11\ \overline{11}\ 70)$	37.2740	0.58	0.13	0.36	116	0.71	1590

$(hkil)$	E_P	$w_H^{(s)}$	ΔE	$\Delta E/E_P$	$\Delta\theta$	R	$d_e(0)$
	keV	$\times 10^{-6}$	meV	$\times 10^{-8}$	μrad		μm
$(0\ 25\ \overline{25}\ 2)$	37.6133	0.57	0.07	0.18	83	0.51	3280
$(0\ \overline{25}\ 25\ 4)$	37.6496	0.57	0.20	0.53	138	0.80	1060
$(0\ 17\ \overline{17}\ 58)$	37.6787	0.57	0.24	0.64	151	0.84	874
$(0\ 25\ \overline{25}\ 8)$	37.7945	0.57	0.09	0.25	97	0.63	2290
$(0\ \overline{23}\ 23\ 32)$	37.8131	0.57	0.04	0.09	61	0.27	6980
$(0\ \overline{25}\ 25\ 10)$	37.9028	0.57	0.22	0.58	143	0.82	958
$(0\ 25\ \overline{25}\ 14)$	38.1900	0.56	0.20	0.52	136	0.81	1070
$(0\ 1\ \overline{1}\ 80)$	38.2023	0.56	0.10	0.27	101	0.67	2080
$(0\ 23\ \overline{23}\ 34)$	38.2084	0.56	0.16	0.41	122	0.77	1350
$(0\ \overline{25}\ 25\ 16)$	38.3685	0.55	0.14	0.36	116	0.75	1510
$(0\ 25\ \overline{25}\ 20)$	38.7934	0.54	0.19	0.48	131	0.81	1140
$(0\ \overline{5}\ 5\ 80)$	38.9064	0.54	0.07	0.19	84	0.58	3000
$(0\ \overline{11}\ 11\ 74)$	38.9935	0.53	0.18	0.46	129	0.81	1180
$(0\ \overline{23}\ 23\ 38)$	39.0571	0.53	0.18	0.47	131	0.81	1150
$(15\ 15\ \overline{30}\ 0)$	39.0764	0.53	0.17	0.43	124	0.79	1270
$(0\ \overline{17}\ 17\ 62)$	39.1020	0.53	0.07	0.19	85	0.59	2920
$(0\ \overline{1}\ 1\ 82)$	39.1559	0.53	0.13	0.33	111	0.74	1630
$(0\ 23\ \overline{23}\ 40)$	39.5091	0.52	0.19	0.48	131	0.82	1130
$(0\ 25\ \overline{25}\ 26)$	39.5950	0.52	0.10	0.24	96	0.68	2210
$(0\ 17\ \overline{17}\ 64)$	39.8289	0.51	0.14	0.35	113	0.77	1520
$(0\ 5\ \overline{5}\ 82)$	39.8431	0.51	0.13	0.33	111	0.76	1600
$(0\ 11\ \overline{11}\ 76)$	39.8598	0.51	0.02	0.05	46	0.10	16200
$(0\ \overline{25}\ 25\ 28)$	39.9043	0.51	0.10	0.26	99	0.70	2050
$(0\ 0\ 0\ 84)$	40.0813	0.51	0.10	0.25	96	0.70	2120

A.5 Low-Lying Levels of Stable Isotopes

Selected parameters of low-lying excited states of stable isotopes with excitation energies below 100 keV are listed in the following table. The isotopes with "ground" state lifetime longer than 10^9 years are included in the table as well. They are labeled with (∗). The values of the energy of a nuclear excited state E, its spin J_e and parity P_e, the spin J_g and parity P_g of the ground state, the half-lifetime $T_{1/2}$ of the excited state, as well as the publication year of the data are included in the table.

The nuclear data were retrieved with the help of the level retrieval program of the Nuclear Data Centrum at Brookhaven National Laboratory www.nndc.bnl.gov/nndc/nudat/levform.html.

Isotope	E keV	J_e, P_e	J_g, P_g	$T_{1/2}$	publication year
^{40}K∗	29.8299(6)	3−	4−	4.24(9) ns	90
^{45}Sc	12.40(5)	3/2+	7/2−	0.318(7) s	92
^{57}Fe	14.4129(6)	3/2−	1/2−	98.3(3) ns	98
^{61}Ni	67.413(3)	5/2−	3/2−	5.34(16) ns	99
^{67}Zn	93.312(5)	1/2−	5/2−	9.16(3) µs	91
^{73}Ge	13.275(17)	5/2+	9/2+	2.95(2) µs	93
^{73}Ge	66.716(19)	1/2−	9/2+	0.499(11) s	93
^{73}Ge	68.752(7)	7/2+	9/2+	1.74(13) ns	93
^{83}Kr	9.4053(8)	7/2+	9/2+	154.4(1.1) ns	01
^{99}Ru	89.68(5)	3/2+	5/2+	20.5(1) ns	94
^{103}Rh	93.041(9)	9/2+	1/2−	1.11(3) ns	01
^{107}Ag	93.125(19)	7/2+	1/2−	44.3(2) s	00
^{109}Ag	88.0341(11)	7/2+	1/2−	39.6(2) s	99
^{119}Sn	23.871(8)	3/2+	1/2+	18.03(7) ns	00
^{121}Sb	37.133(8)	7/2+	5/2+	3.46(3) ns	00
^{125}Te	35.4922(5)	3/2+	1/2+	1.48(1) ns	99
^{127}I	57.608(11)	7/2+	5/2+	1.95(1) ns	96
^{129}I∗	27.80(2)	7/2+	5/2+	16.8(2) ns	96
^{129}Xe	39.578(2)	3/2+	1/2+	0.97(2) ns	96
^{131}Xe	80.1854(19)	1/2+	3/2+	0.48(3) ns	94
^{133}Cs	80.9974(13)	5/2+	7/2+	6.28(2) ns	95
^{145}Nd	67.22(2)	3/2−	7/2−	29.4(1.0) ns	93
^{145}Nd	72.50(1)	5/2−	7/2−	0.72(5) ns	93
^{149}Sm∗	22.507(6)	5/2−	7/2−	7.12(11) ns	94
^{151}Eu	21.541(3)	7/2+	5/2+	9.6(3) ns	97
^{153}Eu	83.36720(17)	7/2+	5/2+	0.793(17) ns	90
^{153}Eu	97.43103(17)	5/2−	5/2+	198(16) ps	90
^{154}Sm	81.976(18)	2+	0+	3.02(4) ns	98

Isotope	E keV	J_e, P_e	J_g, P_g	$T_{1/2}$	publication year
^{155}Gd	60.0087(10)	5/2−	3/2−	193(10) ps	94
^{155}Gd	86.5460(6)	5/2+	3/2−	6.50(4) ns	94
^{156}Gd	88.9666(14)	2+	0+	2.21(2) ns	92
^{157}Gd	54.533(6)	5/2−	3/2−	130(8) ps	96
^{157}Gd	63.917(5)	5/2+	3/2−	0.46(4) μs	96
^{158}Gd	79.510(2)	2+	0+	2.52(3) ns	96
^{158}Dy	98.9180(10)	2+	0+	1.66(3) ns	96
^{159}Tb	58.00(1)	5/2+	3/2+	53.6(1.4) ps	94
^{160}Gd*	75.26(1)	2+	0+	2.69(3) ns	96
^{160}Dy	86.7882(4)	2+	0+	2.026(12) ns	96
^{161}Dy	25.65136(3)	5/2−	5/2+	29.1(3) ns	00
^{161}Dy	43.8201(7)	7/2+	5/2+	0.83(6) ns	00
^{161}Dy	74.56668(5)	3/2−	5/2+	3.14(4) ns	00
^{162}Dy	80.6598(7)	2+	0+	2.20(3) ns	99
^{163}Dy	73.4448(4)	7/2−	5/2−	1.51(5) ns	00
^{164}Dy	73.392(5)	2+	0+	2.39(4) ns	01
^{164}Er	91.38(2)	2+	0+	1.47(3) ns	01
^{165}Ho	94.700(3)	9/2−	7/2−	22.0(4) ps	92
^{166}Er	80.577(7)	2+	0+	1.82(3) ns	92
^{167}Er	79.3221(13)	(9/2)+	7/2+	119(9) ps	00
^{168}Er	79.804(1)	2+	0+	1.88(2) ns	94
^{168}Yb	87.73(1)	2+	0+	1.47(3) ns	94
^{169}Tm	8.4103(3)	3/2+	1/2+	4.08(8) ns	91
^{170}Er	78.591(22)	2+	0+	1.891(23) ns	96
^{170}Yb	84.25468(8)	2+	0+	1.605(13) ns	96
^{171}Yb	66.721(7)	3/2−	1/2−	0.81(17) ns	92
^{171}Yb	75.878(5)	5/2−	1/2−	1.64(16) ns	92
^{171}Yb	95.272(5)	7/2+	1/2−	5.25(24) ms	92
^{172}Yb	78.7427(6)	2+	0+	1.65(5) ns	95
^{173}Yb	78.647(12)	7/2−	5/2−	46(5) ps	95
^{174}Yb	76.471(1)	2+	0+	1.79(4) ns	99
^{174}Hf*	90.985(19)	2+	0+	1.66(7) ns	99
^{176}Hf	88.351(24)	2+	0+	1.43(4) ns	98
^{178}Hf	93.180(1)	2+	0+	1.48(2) ns	94
^{180}Hf	93.326(2)	2+	0+	1.50(5) ns	94
^{181}Ta	6.238(20)	9/2−	7/2+	6.05(12) μs	91
^{183}W*	46.4839(4)	3/2−	1/2−	188(5) ps	92
^{183}W*	99.0793(4)	5/2−	1/2−	0.77(4) ns	92
^{187}Os	9.746(24)	3/2−	1/2−	2.38(18) ns	91
^{187}Os	74.33(3)	3/2−	1/2−	20(6) ps	91
^{187}Os	75.04(3)	5/2−	1/2−	2.16(16) ns	91

A.5 Low-Lying Levels of Stable Isotopes

Isotope	E	J_e, P_e	J_g, P_g	$T_{1/2}$	publication year
	keV				
^{189}Os	36.202(16)	1/2−	3/2−	530(30) ns	90
^{189}Os	69.537(15)	5/2−	3/2−	1.62(4) ns	90
^{189}Os	95.254(18)	3/2−	3/2−	230(30) ps	90
^{191}Ir	82.420(12)	1/2+	3/2+	4.08(7) ns	95
^{193}Ir	73.044(5)	1/2+	3/2+	6.09(15) ns	98
^{195}Pt	98.882(4)	3/2−	1/2−	170(19) ps	99
^{197}Au	77.351(2)	1/2+	3/2+	1.91(1) ns	95
^{201}Hg	26.269(20)	5/2−	3/2−	630(50) ps	94
^{201}Hg	32.138(16)	3/2−	3/2−	$\simeq 100$ ps	94
^{232}Th*	49.369(9)	2+	0+	345(15) ps	91
^{235}U*	13.040(2)	3/2+	7/2−	500(3) ps	93
^{235}U*	46.204(3)	9/2−	7/2−	≤ 60 ps	93
^{235}U*	51.701(1)	5/2+	7/2−	191(5) ps	93
^{236}U*	45.242(3)	2+	0+	234(6) ps	91
^{238}U*	44.91(3)	2+	0+	203(7) ps	91

A.6 α-Al₂O₃ as a Universal meV-Monochromator

In Sect. 5.3 it was argued that it is easy to find a matching Bragg back-reflection in α-Al$_2$O$_3$ for any desired x-ray energy above 10 keV. This statement is illustrated here using the example of Mössbauer photon energies.

Table A.4 displays nuclear transition energies E_γ for selected isotopes. Only those transition energies are given whose values are known precisely enough, and whose values do not exceed 70 keV[6]. For each selected energy we provide Miller indices $(hkil)$ of the atomic planes in α-Al$_2$O$_3$ which reflect photons of such an energy at normal incidence to $(hkil)$ planes. Also, the appendant crystal temperatures T_H are given at which the back-reflection takes place. The quantity T_H is referred to as the Bragg temperature, which applies to a selected set of atomic planes $(hkil)$ and to a selected x-ray energy. For each reflection the table provides the calculated angular width $\Delta\theta$, the energy bandwidth ΔE, the peak reflectivity R, the extinction length $d_e(0)$ (2.90) and the variation of the reflection peak position on the energy scale with crystal temperature dE/dT. The calculations were performed using the dynamical theory for an infinitely thick perfect crystal with the parameters given in Appendix A.2.

It is remarkable that even for x rays with relatively high energy, e.g., corresponding to the 46.4837 keV nuclear transition in ^{183}W, sapphire offers Bragg reflections with high reflectivity, with an angular acceptance much larger than the divergence of radiation from present day undulator based x-ray sources, and with an energy bandpass as low as 100 μeV. However, for this to be realized, a perfect sapphire crystal more than 5 mm thick and kept in an environment with mK-stable temperature has to be used. The crystal quality of the commercially available α-Al$_2$O$_3$ is addressed in Appendix A.7.

The higher the photon energy the easier it is to find a back-reflection with matching Bragg energy and temperature. Generally speaking, more than one set of reflecting atomic planes matches each selected photon energy. Data are presented for only one set with the highest reflectivity and with the Bragg temperature falling into the temperature range from 4 K to 400 K. In some cases the data for one more reflection are presented as well.

At high photon energies, however, a problem occurs - lower peak reflectivity. The solution is to cool down the monochromator crystal. Figure 5.11 on page 231 demonstrates how the crystal reflectivity increases with decreasing crystal temperature. The effect is most pronounced at higher photon energies. The other calculations shown in Fig. 5.12 on page 233 demonstrate that the peak reflectivity of sapphire at $T \simeq 50$ K may attain $\simeq 35$ % using some strong reflections for photon energies as high as ≈ 70 keV. Even a little bit higher reflectivity values are achieved at $T \simeq 4$ K. The extinction length is,

---

[6] The Bragg reflectivity of α-Al$_2$O$_3$ for x-ray photons with energy higher than 70 keV, at normal incidence to the reflecting atomic planes, becomes low, less than 30%, even at low temperatures. Therefore, we consider here only those transitions having $E_\gamma < 70$ keV.

Isotope	E_γ	$(hkil)$ hexagonal	T_H	$\Delta\theta$	ΔE	R	$d_e(0)$	$\dfrac{\mathrm{d}E}{\mathrm{d}T}$
	keV		K	µrad	meV		mm	$\dfrac{\mathrm{meV}}{\mathrm{mK}}$
^{181}Ta	6.2155	(2 1 $\bar{3}$ 10)	228	3355	59	0.89	.004	−0.027
^{169}Tm	8.4103	(1 5 $\bar{6}$ 2)	600	1322	4.0	0.31	.058	−0.067
^{73}Ge	13.275	(8 1 $\bar{9}$ 7)	337	1248	5.6	0.82	.038	−0.078
^{57}Fe	14.4125	(1 3 $\bar{4}$ 28)	371	1001	5.9	0.87	.036	−0.094
		(1 6 $\bar{7}$ 22)	161	719	1.9	0.63	.111	−0.037
^{151}Eu	21.5414	(11 2 $\bar{13}$ 24)	221	526	2.8	0.93	.075	−0.085
		(3 2 $\bar{5}$ 43)	287	324	0.56	0.64	.371	−0.125
^{149}Sm	22.496	(3 13 $\bar{16}$ 8)	134	597	2.7	0.93	.077	−0.036
^{119}Sn	23.8795	(1 9 $\bar{10}$ 40)	286	440	1.8	0.91	.120	−0.131
		(6 5 $\bar{11}$ 40)	286	353	1.1	0.85	.197	−0.131
^{161}Dy	25.6513	(14 5 $\bar{19}$ 0)	342	341	1.1	0.88	.197	−0.149
		(3 2 $\bar{5}$ 52)	374	301	0.66	0.80	.319	−0.166
^{129}I	27.77	(11 10 $\bar{21}$ 10)	344	296	0.82	0.87	.259	−0.162
^{40}K	29.8299	(5 15 $\bar{20}$ 26)	214	273	0.68	0.88	.310	−0.110
^{201}Hg	32.138	(2 17 $\bar{19}$ 36)	100	201	0.38	0.83	.559	−0.030
^{125}Te	35.4919	(11 6 $\bar{17}$ 68)	4	168	0.36	0.90	.575	−0.0001
^{121}Sb	37.1297	(15 13 $\bar{28}$ 14)	146	168	0.41	0.90	.552	−0.059
^{161}Dy	43.8211	(15 17 $\bar{32}$ 28)	155	93	0.10	0.77	2.05	−0.094
^{238}U	44.915	(8 25 $\bar{33}$ 4)	73	111	0.16	0.86	1.32	−0.019
^{183}W	46.4839	(7 12 $\bar{19}$ 82)	76	88	0.10	0.80	2.13	−0.028
^{67}Ni	67.413	(14 12 $\bar{26}$ 122)	90	21	0.008	0.28	37.4	−0.047

Table A.4. Nuclear transition energies E_γ for selected isotopes, and Bragg reflections $(hkil)$ together with appendant temperatures T_H in α-Al$_2$O$_3$, with back-reflection energy matching E_γ. For each reflection the table provides the calculated angular width $\Delta\theta$, the energy bandwidth ΔE, the peak reflectivity R, the extinction length $d_e(0)$ (2.90), and the variation of the reflection peak position on the energy scale with crystal temperature $\mathrm{d}E/\mathrm{d}T$.

however, very large - more than 30 mm, necessitating the use of $\gtrsim 100$ mm long, very high quality sapphire crystals.

A.7 Quality Assessment of α-Al$_2$O$_3$ Crystals

Introduction. The most commonly used crystals in x-ray optics are undoubtedly silicon, since their quality in terms of spatial periodicity of atoms is the best among the commercially available crystals. However, as was discussed in the present book, application of the non-cubic crystals, such as α-Al$_2$O$_3$, sapphire, in high-resolution x-ray optics (monochromators, resonators, etc.) is often advantageous, e.g., because of potentially higher reflectivity, and access to specific photon energies. Additionally, the combination of high thermal conductivity and low thermal expansion at low temperatures is unique, and makes sapphire attractive as a robust crystal for high-heat-load monochromators, see Appendix A.2. Obtaining sapphire crystals of high quality and their quality assessment is a high priority problem for high-resolution x-ray optics. It is the topic for discussion in this appendix.

Sapphire crystals grown by different techniques are commercially available. An overview of different sapphire crystal growth techniques can be found, e.g., in the book of Dobrovinskaya et al. (2002). The early experience of applications of sapphire in high-resolution x-ray optics (Shvyd'ko et al. 1998) have indicated that the crystals grown by the heat-exchange method (HEM) have on average a better quality[7]. For this reason they were studied later by Chen et al. (2001, 2003) by using x-ray topography. For comparison, samples grown by the modified Czochralski method were inspected as well. Most recent studies showed that the crystals grown by the Kyropoulos method may have a very good, dislocation free, quality (McNally et al. 2004). The results of the aforementioned studies are summarized in the present section.

Two x-ray diffraction techniques are used for the quality assessment of sapphire crystals: (i) reflectivity and energy width measurements of Bragg back-reflections, and (ii) white beam x-ray topography. The first technique is the most appropriate. However, it requires monochromatic x rays, and it is time consuming. X-ray topography, especially white beam backscattering x-ray topography, is much faster, requires minimum instrumentation, and as will be shown it gives adequate information.

Preparation of Samples. Before going into the details of the crystal quality studies, a few words about the preparation of the sapphire samples for x-ray measurements.

---

[7] The heat-exchange method has been developed to grow high-quality sapphire crystals of large size for industrial applications (Schmid et al. 1994, Khattak and Schmid 2001). Crystals of different quality, HEMEX, HEMLITE, etc., are available from the producer Crystals Systems Inc. (Salem, MA 01970, USA). The premium grade HEMEX sapphire is grown from ultra high purity material and selected by the producer using visible light optical techniques. However, visible light optical tests are not sufficient to characterize crystals destined for applications in x-ray crystal optics. For this one has to use x-ray diffraction techniques as well.

A.7 Quality Assessment of α-Al$_2$O$_3$ Crystals

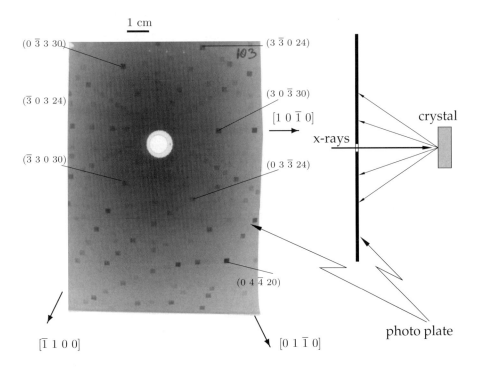

Fig. A.4. Schematic of the backscattering white beam x-ray topography technique. The topography spots corresponding to different Bragg reflections ($hkil$) are recorded on a high-resolution photo plate (left) in backscattering geometry (right). The distance from sample to photo plate is $\simeq 40 - 80$ mm. The spot size is determined by the incident beam cross-section.

The crystal surface is treated by chemical polishing in order to remove any possible defective layers or stresses which have been introduced due to preceding mechanical treatment (cutting, mechanical polishing, etc.). Typically 100 µm of the surface layer are removed. The polishing is performed in a 1:1 H$_2$SO$_4$: H$_3$PO$_4$ solution at 560 K, as described by Reisman et al. (1971). The temperature of the solution is controlled with an accuracy of 1 K. The removal rate in the (0 0 0 1) direction is about 8 µm/hour (for each side of the sample). The polishing in this solution is anisotropic. The removal rate, e.g., in the (0 4 $\bar{4}$ 14) or in the direction (0 3 $\bar{3}$ 16) is much higher, about $30 - 34$ µm/hour. More experimental details including the setup for chemical polishing was described by Lerche (2000).

White Beam X-Ray Topography. X-ray topography is a fast nondestructive technique, which can provide a map of the strain or defect distribution in crystals, such as dislocations, planar defects, stacking faults, domain walls in ferroelectric and magnetic materials, growth defects or large precipitates, see, e.g., (Bowen and Tanner 1998). The white beam backscat-

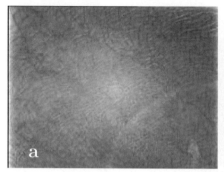 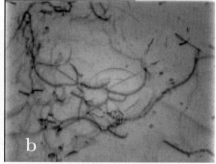

Fig. A.5. Back-reflection topographs of a sapphire wafer grown by the modified Czochralski method taken at regions with different dislocation densities: (a) Center part of the wafer, $(2\ 0\ \bar{2}\ 14)$ reflection ; (b) Edge part of the wafer, $(3\ 0\ \bar{3}\ 18)$ reflection (Chen et al. 2001). Sample area illuminated by x rays is 2.1×1.7 mm^2, the thickness of sample is about 0.3 mm.

tering x-ray topography technique was introduced by Tuomi et al. (1974). This technique is sensitive to strain fields extending over more than several micrometers. One can obtain depth information for defect distributions inside the sample by examining different reflections, as the penetration depths for different reflections differ from each other. It can provide depth information from several to several hundred micrometers from the sample surface.

White beam backscattering x-ray topography is recorded as per the usual Laue diagram method in backscattering geometry - Fig. A.4. Each spot on the photo plate[8] corresponds to a certain Bragg reflection $(hkil)$. The spot dimensions are given by the incident beam cross-section. Each single spot is a topograph. The spots are not equivalent, since Bragg reflections show different sensitivities to defects of different origin. Examples of such topographs are shown in Figs. A.6-A.9.

The geometrical resolution of white beam x-ray topography depends on a number of factors, including angular source size, the source to sample distance and the sample to photo plate distance. Synchrotron radiation facilities are ideally suited for high-resolution white beam x-ray topography experiments, since the aforementioned factors can be optimally chosen there. The resolution usually lies in the micrometer range. In addition, synchrotrons provide high a flux of x-ray photons over a broad spectral range. The typical time for taking one topograph at synchrotron radiation facilities is a few seconds.

Dislocation Distribution and Density. X-ray topographs show that dislocations exist in most of the sapphire crystal samples studied. However, in all samples discrete dislocation lines can be discerned. The dislocation den-

---

[8] The topographs are recorded on a high-resolution photo plate having an emulsion grain size of about 0.05 μm, and providing spatial resolution $\simeq 3$ lines/μm.

A.7 Quality Assessment of α-Al₂O₃ Crystals   363

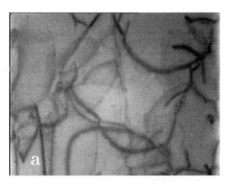
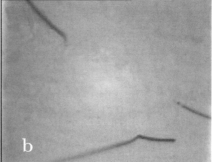

Fig. A.6. Back-reflection topographs of HEMEX sapphire wafers cut from different boules show different dislocation densities: (a) $\simeq 10^3$ cm^{-2}, (2 2 $\bar{4}$ 24) reflections; (b) much lower dislocation density, (3 0 $\bar{3}$ 18) reflection (Chen et al. 2001). Sample area illuminated by x rays is 2.1×1.7 mm^2.

Fig. A.7. Back-reflection topographs of HEMEX sapphire crystals with no dislocations, (2 $\bar{2}$ 0 16) reflections (Chen et al. 2001). The thickness of the sample is 0.04 mm. Sample area illuminated by x rays is 1.2×1.2 mm^2. These crystals have been used in the first experiments with the x-ray Fabry-Pérot interferometer, see Sect. 7.5.

Fig. A.8. (0 $\bar{1}$ 1 1) transmission section topography at a region in HEMEX sapphire with no dislocations (Chen et al. 2001). Sample thickness is 0.1 mm. The black spots in the image are due to defects in the photo plate.

sity[9] in the studied sapphire can vary significantly from sample to sample, especially between samples produced with different methods.

---

[9] Dislocation densities are calculated using the equation $\rho_\mathrm{d} = L/V$ where V is the volume of the sapphire wafer exposed to the x rays during the experiment and

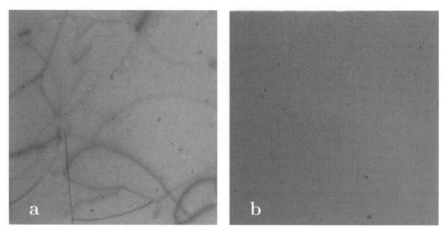

Fig. A.9. Back-reflection topographs of a Kyropoulos sapphire wafers (McNally et al. 2004). (a) Sapphire crystal grown with normal rate, (b) with reduced rate. The black spots are defects in the photo plate emulsion. Sample area illuminated by x rays is 2×2 mm^2.

Czochralski sapphire. The topograph of Fig.A.5a was taken from the center part of a wafer grown by the modified Czochralski method. The dislocation density is high: $\simeq 9 \times 10^4$ cm^{-2}. It can be observed from Fig.A.5a that two sets of dislocations exist in this wafer, which run roughly in two directions. An examination of the edge region of the same wafer indicates a higher crystal quality, as a lower dislocation density $\simeq 4 \times 10^3$ cm^{-2} was detected in its corresponding topographs - see Fig.A.5b.

HEMEX sapphire. The nominally highest quality HEMEX sapphire typically shows a dislocation density in the order of 10^3 cm^{-2} - see Fig.A.6a. In samples cut from a different boule a much lower dislocation density was detected - see Fig.A.6b. In some regions of these samples no dislocations were detected at all - see Fig.A.7. Pendellösung fringes were observed in the corresponding transmission section topographs[10] (Fig.A.8), which confirms the high quality of the sapphire wafer.

Kyropoulos sapphire. Figures A.9a and A.9b show topographs taken from two different sapphire crystals both grown by the Kyropoulos method[11]. The

---

L is the total dislocation line length in that volume. V is calculated using the extinction length and the beam size.

[10] White beam section topography is performed in transmission geometry. The beam is limited by a vertical slit of width approximately 20 μm and length 1.5 mm. The distance from sample to photo plate is usually set to 80 mm.

[11] The crystals were grown in the Institute for Single Crystals, Kharkov, Ukraine. The crystal wafer, whose topograph is shown in Fig. A.9a, was cut from a boule grown with a standard pulling up rate $\simeq 4$ mm/h. The crystal wafer, whose topograph is shown in Fig. A.9b, was cut from the boule grown with a reduced

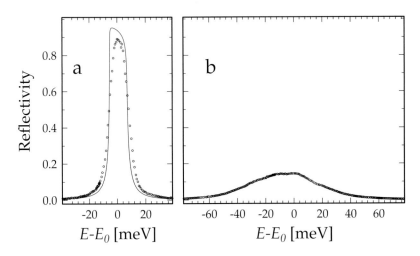

Fig. A.10. Energy dependence of the Bragg reflectivity of sapphire crystals for x rays at normal incidence to the (0 0 0 30) atomic planes, $E_0 \simeq 14.315$ keV (Chen et al. 2001). (a) HEMEX low dislocation density sapphire; its topograph is shown in Fig. A.6b. Circles are measured values, the solid line is the calculated profile from the dynamical theory of diffraction for perfect crystals (b) Czochralski grown sapphire with high dislocation density; its topograph is shown in Fig. A.5a.

crystal whose microstructure is mapped out in Fig. A.9a was grown with a standard pulling up rate $\simeq 4$ mm/h. The dislocation density is comparable with that one of the HEMEX crystals. The second crystal whose topograph is shown in Fig. A.9b was grown with a reduced pulling up rate $\simeq 0.1$ mm/h. No dislocation lines can be observed in the central area of 3×3 mm^2. This result demonstrates, that dislocation free sapphire crystals are feasible. Attempts to improve the quality of sapphire crystals are ongoing.

Bragg Reflectivity vs. Dislocation Density. Figure A.10 shows the energy dependence of the reflectivity of two sapphire crystals for x rays at normal incidence to the (0 0 0 30) atomic planes. The extinction length for this reflection is 16 μm. The x rays were monochromatized to an energy bandwidth of 7 meV width and scanned around $E_0 \simeq 14.315$ keV by using the $(+,+,-,-)$ nested monochromator described in detail in Sect. 5.4.3 and shown in Fig. 5.21. The reflectivity curve shown in Fig. A.10a was measured for the same region of the HEMEX α-Al$_2$O$_3$ crystal where few dislocations were detected - as shown in Fig. A.6b. The reflectivity curve shown in

---

pulling up rate $\simeq 0.1$ mm/h, and in an environment with a small temperature gradient of a few K/cm. After annealing at pre-melting temperature 1747 ± 5 K, the boule was cut into $\simeq 1$ mm thick wafers, and etched. The wafer with the lowest density of etch pits was selected.

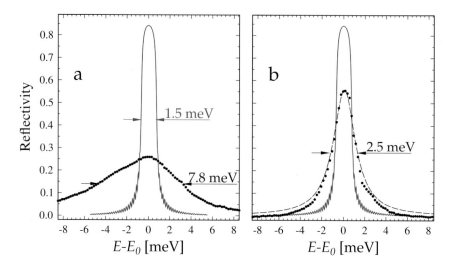

Fig. A.11. Energy dependence of the Bragg reflectivity of sapphire crystals for x rays at normal incidence to the (1 1 $\bar{2}$ 45) atomic planes, $E_0 \simeq 21.629$ keV (Alp et al. 2000a). (a) HEMEX sapphire with a $\simeq 10^3$ cm^{-2} dislocation density, its topograph is shown in Fig. A.6a; (b) HEMEX low dislocation density sapphire, its topograph is shown in Fig. A.6b. Circles are measured values, the solid line is the calculated profile from the dynamical theory of diffraction for perfect crystals. The dashed line in (b) is a fit with a Lorentzian.

Fig. A.10b was measured from the same region of the Czochralski grown α-Al$_2$O$_3$ crystal where the high dislocation density was detected - see Fig. A.5a. The HEMEX sapphire shows a high peak reflectivity of $\simeq$ 86% and a small width of 13 meV (FWHM) which fits well to the profile (solid line) calculated with the dynamical theory of x-ray diffraction in perfect crystals. The Czochralski grown sapphire shows a much lower peak reflectivity of $\simeq$ 14% and a large width of $\sim$ 50 meV (FWHM).

Another example is shown in Fig. A.11 where the reflectivity was measured for x rays of the higher energy $E_0 \simeq 21.630$ keV at normal incidence to the (1 1 $\bar{2}$ 45) atomic planes. The x rays were monochromatized to an energy bandwidth of 1 meV and scanned around 21.630 keV by using the high energy resolution monochromator shown in Fig. 5.27 (Shu et al. 2001).

The dynamical theory predicts a peak reflectivity of 84.2% and an energy width of 1.5 meV. The extinction length for this reflection is 144 μm, which is much larger than in the previous case. One should expect that a higher crystal quality is required to achieve theoretically predicted values of the reflectivity and energy width.

The reflectivity curve shown in Fig. A.11a was measured from the same region of the HEMEX α-Al$_2$O$_3$ crystal where a dislocation density of $\simeq$

10^3 cm^{-2} was detected - see Fig. A.6a. The reflectivity curve shown in Fig. A.11b was measured for the same region of the HEMEX α-Al$_2$O$_3$ crystal where few dislocations were detected (Fig. A.6b). The HEMEX sapphire with a dislocation density of $\simeq 10^3$ cm^{-2} shows a low peak reflectivity of $\simeq 25\%$ and a broad reflection curve with a width of $\simeq 7.8$ meV. The HEMEX sapphire with a lower dislocation density shows a peak reflectivity of $\simeq 55\%$ and a reflection curve - circles in Fig. A.11b - with a width of 2.5 meV. In both cases the theoretical values for a perfect crystal are not achieved. Yet, the HEMEX sapphire with lower dislocation density shows much better performance.

The above comparison of the reflectivity and topography results thus shows that white-beam x-ray topography provides information which correlates with the reflectivity measurements. This fact offers an argument for concluding that one can take advantage of the more rapid white-beam x-ray topography technique to assess the quality of the sapphire material for x-ray crystal optics.

A.8 Radiation Wavelength and Angle of Incidence for Exact Bragg Backscattering

The locus of points in (λ,θ) space that obey the exact backscattering condition (2.37) is shown by the solid line in Fig. 2.16 on page 84. The intersection of this line with the region of total Bragg reflection (a_0) presents the locus of points in (λ,θ) space for which total reflection in exact Bragg backscattering occurs.

Expressions (2.131)-(2.132) for the center of the region of *total exact Bragg back-reflection* (λ_R,θ_R) in the asymmetric scattering geometry are given in Sec. 2.4.8 without derivation. They are derived in this appendix.

To find the center of the region of *total exact Bragg back-reflection* in (λ,θ) space in the asymmetric scattering geometry, we shall solve for λ and θ the modified Bragg's law (2.114) together with the condition for exact backscattering (2.37).

We start with the expressions for the modified Bragg's law (2.114)-(2.115). By using (2.27) and (2.29) we obtain $(b-1)/2b = (\gamma_0 - \gamma_H)/2\gamma_0 = (H/2K)(\cos\eta/\gamma_0) = (\lambda/2d_H)(\cos\eta/\gamma_0)$. Further, we remind the reader that exact backscattering can be achieved only provided the wave vector of the incident wave, first, is lying in the dispersion plane, and, second, the azimuthal angle of incidence $\phi = 0$, as in Fig. 2.5(a_0). For this reason, and using (2.26) we have $\gamma_0 = \sin(\theta + \eta)$. In this way we find for the modified Bragg's law

$$\sin\theta = \frac{\lambda}{2d_H}\left(1 + w_H^{(s)}\frac{\lambda}{2d_H}\frac{\cos\eta}{\sin(\theta+\eta)}\right). \tag{A.11}$$

Combining this equation with the condition for exact backscattering (2.37)

$$\frac{\lambda}{2d_H} = -\frac{\cos(\theta+\eta)}{\sin\eta}, \tag{A.12}$$

we obtain the equation for the sought for glancing angle of incidence, denoted by θ_R:

$$\cos\theta_R \sin\eta \cos(\theta_R + \eta) = w_H^{(s)} \sin(\theta_R - \eta). \tag{A.13}$$

By using the following notations

$$\theta_R = \pi/2 - \Theta_R, \qquad x = \tan\Theta_R, \qquad t = \tan\eta, \tag{A.14}$$

$$A = t^2 - w_H^{(s)}, \qquad B = t(1 + 2w_H^{(s)}), \qquad \text{and} \qquad C = w_H^{(s)} t^2, \tag{A.15}$$

we can reduce (A.13) to a quadratic equation in $x = \tan\Theta_R$:

$$Ax^2 + Bx - C = 0. \tag{A.16}$$

In the symmetric scattering geometry, when $\eta = 0$ and thus $t = 0$, the solution of (A.16) is obvious: $\Theta_R = 0$. That is, exact backscattering occurs at normal incidence to the reflecting atomic planes.

Our main interest is in asymmetric scattering geometry. Let us assume a not very small asymmetry angle: $\eta \gg \sqrt{w_H^{(s)}}$. In this case we may write $A = t^2 > 0$. The quadratic equation (A.16) has two roots. Since the angle of incidence is positive, we select the positive root as a solution of our physical problem:

$$x = \frac{B}{2A}\left(\sqrt{1 + \frac{4AC}{B^2}} - 1\right) = \frac{2C/B}{1 + \sqrt{1 + 4AC/B^2}}. \tag{A.17}$$

Further, we know that $w_H^{(s)} \ll 1$, therefore, the quantity B can be presented with a high degree of accuracy as $B = t$. Finally, we obtain for the angle of incidence (at the center of the total reflection region) at which exact Bragg backscattering takes place:

$$\theta_R = \pi/2 - \Theta_R$$

$$\tan \Theta_R = \frac{2 w_H^{(s)}\, t}{1 + \sqrt{1 + 4 w_H^{(s)}\, t^2}}. \tag{A.18}$$

Combining (A.18) with (A.12) we obtain that the wave has to have the wavelength λ_R given by

$$\frac{\lambda_R}{2 d_H} = \cos \Theta_R \left(1 - \frac{1}{t} \tan \Theta_R\right), \tag{A.19}$$

to be reflected into itself.

If η is not too close to $\pi/2$, then the exact solution (A.18)-(A.19) for the incidence angle Θ_R and the wavelength λ_R of exact backscattering can be accurately approximated by

$$\Theta_R \simeq w_H^{(s)} \tan \eta \left(1 - w_H^{(s)} \tan^2 \eta\right), \tag{A.20}$$

$$\frac{\lambda_R}{2 d_H} \simeq 1 - w_H^{(s)} \left(1 - \frac{w_H^{(s)} \tan^2 \eta}{2}\right). \tag{A.21}$$

These expressions are obtained from the exact solution (A.18)-(A.19) by performing Taylor expansions, and keeping terms up to second order in $w_H^{(s)}$.

As Fig. 2.16 suggests the the maximum of the region of the total Bragg reflection is achieved at the incidence angle Θ_m, which strictly speaking does not coincide with Θ_R. To determine Θ_m we differentiate the modified Bragg's law (A.11) with respect to θ and impose the condition $d\lambda/d\theta = 0$ to determine the angle of incidence for the reflection with the greatest wavelength. As a result we obtain

$$\frac{\lambda}{2 d_H} = \frac{\sin(2\theta + \eta)}{\cos(\theta + \eta)}. \tag{A.22}$$

Combining this equation with the modified Bragg's law (A.11), and assuming that η is not too close to $\pi/2$, we obtain the expression for the incidence angle of Bragg reflection with greatest wavelength

$$\Theta_m \simeq w_H^{(s)} \, t \left(1 - 2w_H^{(s)} \, t^2\right). \tag{A.23}$$

As follows from (A.20) and (A.23) the difference between the angles Θ_R and Θ_m is small: $\Theta_R - \Theta_m = w_H^{(s)^2} \tan^3 \eta$. It most cases it can be neglected. Yet, the difference $\Theta_R - \Theta_m$ increases with η.

Bibliography

Achterhold, K., Keppler, C., Ostermann, A., van Bürck, U., Sturhahn, W., Alp, E. E. and Parak, F. G.: 2002, Vibrational dynamics of myoglobin determined by the phonon-assisted Mössbauer effect, *Phys. Rev. E* **65**, 051916.

Afanas'ev, A. M. and Kagan, Y.: 1967, The role of lattice vibrations in dynamical theory of x-rays, *Acta Cryst.* **A24**, 163.

Afanas'ev, A. M. and Kohn, V. G.: 1976, On the theory of simultaneous diffraction, *Acta Cryst.* **A32**, 308–310.

Afanas'ev, A. M. and Melkonyan, M. K.: 1983, X-ray diffraction under specular reflection conditions. Ideal crystals, *Acta Cryst.* **A39**, 207–210.

Alp, E. E., Moncton, D. E., Shu, D., Sutter, J., Sinn, H., Zhao, J., Sturhahn, W., Ramanathan, M., Haeffner, D., Shastri, S., Mills, D., Shenoy, G., Shvyd'ko, Y. V., Lerche, M., Wille, H. and Lucht, M.: 2004, An exact backscattering hard x-ray beamline at the APS, *to be published*.

Alp, E. E., Sinn, H., Alatas, A. and Shvyd'ko, Y. V.: 2000a, Bragg reflectivity studies of Al_2O_3 at 21.6 keV, *unpublished*.

Alp, E. E., Sinn, H., Alatas, A., Sturhahn, W., Toellner, T., Zhao, J., Sutter, J., Hu, M., Shu, D. and Shvyd'ko, Y. V.: 2001a, Source and optics considerations for new generation high-resolution inelastic x-ray spectrometers, *Nucl. Instrum. Methods Phys. Res. A* **467-468**, 617–622.

Alp, E. E., Sturhahn, W. and Toellner, T. S.: 2001b, Lattice dynamics and inelastic nuclear resonant x-ray scattering, *J. Phys.: Condens. Matter* **13**(34), 7645–7659.

Alp, E. E., Sturhahn, W., Sinn, H., Toellner, T., Hu, M., Sutter, J. and Alatas, A.: 2000b, Inelastic scattering of synchrotron radiation from electrons and nuclei for lattice dynamics studies, *AIP Conf. Proc.* **506**, 479.

Ashcroft, N. W. and Mermin, N. D.: 1976, *Solid State Physics*, Holt, Rinehart and Witson, New York.

Authier, A.: 2001, *Dynamical Theory of X-Ray Diffraction*, Vol. 11 of *IUCr Monographs on Crystallography*, Oxford University Press, Oxford, New York.

Ayvazyan, V., Baboi, N., Bohnet, I. and et al.: 2002a, Generation of GW radiation pulses from a VUV free-electron laser operating in the femtosecond regime, *Phys. Rev. Lett.* **88**, 104802.

Ayvazyan, V., Baboi, N., Bohnet, I. and et al.: 2002b, A new powerful source for coherent VUV radiation: Demonstration of exponential growth and saturation at the TTF free-electron laser, *Euro. Phys. J. D* **20**, 149–156.

Azaroff, L. V., Kaplow, R., Kato, N., Weiss, R. J., Wilson, A. J. C. and Yung, R. A.: 1974, *X-Ray Diffraction*, McGraw-Hill, Inc., New York.

Barbee, T. W. and Underwood, J. H.: 1983, Solid Fabry-Pérot étalons for x-rays, *Opt. Commun.* **48**, 161–166.

Barla, A., Rüffer, R., Chumakov, A. I., Metge, Plessel, J. and Abd-Elmeguid, M. M.: 2000, Direct determination of the phonon density of states in β-Sn, *Phys. Rev. B* **61**, R14881–R14884.

Barla, A., Sanchez, J. P., Haga, Y., Lapertot, G., Doyle, B. P., Leupold, O., Rüffer, R., Abd-Elmeguid, M. M., Lengsdorf, R. and Flouquet, J.: 2004, Pressure-induced magnetic order in golden SmS, *Phys. Rev. Lett.* **92**, 066401/1–4.

Baron, A. Q. R.: 1995, Report on the x-ray efficiency and time resolution of a 1 cm^2 reach through avalanche diode, *Nucl. Instrum. Methods Phys. Res. A* **352**, 665.

Baron, A. Q. R.: 2000, Detectors for nuclear resonant scattering experiments, *Hyp. Interact.* **125**, 29.

Baron, A. Q. R., Kohmura, Y., Ohishi, Y. and Ishikawa, T.: 1999, A refractive collimator for synchrotron radiation, *Appl. Phys. Lett.* **74**, 1492.

Baron, A. Q. R., Tanaka, Y., Ishikawa, D., Miwa, D., Yabashi, M. and Ishikawa, T.: 2001a, A compact optical design for Bragg reflections near backscattering, *J. Synchrotron Rad.* **8**, 1127–1130.

Baron, A. Q. R., Tanaka, Y., Miwa, D., Ishikawa, D., Mochizuki, T., Takeshita, K., Goto, S., Matsushita, T., Kimura, H., Yamamoto, F. and Ishikawa, T.: 2001b, Early commissioning of the SPring-8 beamline for high resolution inelastic x-ray scattering, *Nucl. Instrum. Methods Phys. Res. A* **467-468**, 627–630.

Bartlett, R. J., Trela, W. J., Kania, R. D., Hockaday, M. P., Barbee, T. W. and Lee, P.: 1985, Soft x-ray measurements of solid Fabry-Pérot étalons, *Opt. Commun.* **55**, 229–35.

Basile, G., Becker, P., Bergamin, A., Cavagnero, G., Franks, A., Jackson, K., Kuetgens, U., Mana, G., Palmers, E. W., Robbie, C. J., Stedman, M., Stümpel, J., Yacoot, A. and Zosi, G.: 2000, Combined optical and x-ray interferometry for high-precision dimensional metrology, *Proc. R. Soc. Lond. A* **456**, 701–729.

Basile, G., Bergamin, A., Cavagnero, G., Mana, G., Vittone, E. and Zosi, G.: 1994, Measurement of the Silicon (220) lattice spacing, *Phys. Rev. Lett.* **72**, 3133.

Batterman, B. W. and Cole, H.: 1964, Dynamical diffraction of x rays by perfect crystals, *Rev. Mod. Phys.* **36**, 681–717.

Batterman, B. W. and Hilderbrandt, G.: 1967, Observation of x-ray Pendellösung fringes in Darwin reflection, *Phys. Stat. Solidi* **23K**, 147–149.

Batterman, B. W. and Hilderbrandt, G.: 1968, X-ray Pendellösung fringes in Darwin reflection, *Acta Cryst.* **A24**, 150–157.

Bearden, J. A.: 1965, Selection of W K_{α_1} as the x-ray wavelength standard, *Phys. Rev.* **137**, B455–461.

Bearden, J. A.: 1967, X-ray wavelengths, *Rev. Mod. Phys.* **39**, 78–84.

Becker, P., Dorenwendt, K., Ebeling, G., Lauer, R., Lucas, W., Probst, R., Radermacher, H.-J., Reim, G., Seyfried, P. and Siegert, H.: 1981, Absolute measurements of the (220) lattice plane spacing in a silicon crystal, *Phys. Rev. Lett.* **46**, 1540.

Bergamin, A., Cavagnero, G., Mana, G. and Zosi, G.: 1999, Scanning x-ray interferometry and the silicon lattice parameter: towards 10^{-9} relative uncertainty?, *Euro. Phys. J. B* **9**, 225–232.

Bialas, H. and Stolz, H. J.: 1975, Lattice dynamics of sapphire (corundum). Part I: Phonon dispersion by inelastic scattering, *Z. Phys. B* **21**, 319–331.

Bilderback, D. H., Hoffman, S. A. and Thiel, D. J.: 1994, Nanometer spatial resolution achieved in hard x-ray imaging and Laue diffraction experiments, *Science* **263**(5144), 201–3.

Bonifacio, R., Casagrande, F., Cerchioni, G., de Salvo Souza, L., Pierini, P. and Piovella, N.: 1990, Physics of the high-gain FEL and superradiance, *La Rivista del Nuovo Cimento* **13**, 1–69.

Bonse, U. and Graeff, W.: 1977, X-ray and neutron interferometry, *in* H.-J. Queisser (ed.), *Topics in applied optics: X-ray optics*, Vol. 22, Springer, pp. 93–143. Berlin Heidelberg.

Bonse, U. and Hart, M.: 1965, An x-ray interferometer, *Appl. Phys. Lett.* **6**, 155.

Born, M. and Huang, K.: 1954, *Dynamical Theory of Crystal Lattices*, Clarendon, Oxford.

Born, M. and Wolf, E.: 1999, *Principles of Optics*, Cambridge University Press, Cambridge.

Bottom, V. E. and Carvalho, R. A.: 1970, Double crystal x-ray diffractometer using continuous radiation, *Rev. Sci. Instrum.* **42**, 196–199.

Bowen, D. K. and Tanner, B. K. (eds): 1998, *High Resolution X-ray Diffraction and Topography*, Taylor and Francis, London.

Brossel, J.: 1947a, Multiple-beam localized fringes: part I - Intensity distribution and localization, *Proc. Phys. Soc.* **59**, 224–234.

Brossel, J.: 1947b, Multiple-beam localized fringes: part II - Conditions of observation and formation of ghosts, *Proc. Phys. Soc.* **59**, 235–242.

Brown, A. S., Spackman, M. A. and Hill, R. J.: 1992, The electron distribution in corundum. Study of the utility of merging single crystal and powder diffraction data, *Acta Cryst.* **A49**, 513.

Brümmer, O., Höche, H. R. and Nieber, J.: 1979, X-ray diffraction in the Bragg case at Bragg angles of about $\pi/2$, *Phys. Stat. Solidi (a)* **53**, 565.

Burkel, E.: 1991, *Inelastic Scattering of X rays with Very High Energy Resolution*, Vol. 125 of *Springer Tracts in Modern Physics*, Springer, Berlin.

Burkel, E.: 2000, Phonon spectroscopy by inelastic x-ray scattering, *Rep. Prog. Phys.* **63**, 171–232.

Burkel, E., Dorner, B. and Peisl, J.: 1987, Observation of inelastic x-ray scattering from phonons, *Europhys. Lett.* **3**, 957–61.

Burns, C. A., Abbamonte, P., Isaacs, E. D. and Platzman, P. M.: 1999, Plasmons in lithium ammonia, *Phys. Rev. Lett.* **83**, 2390–2393.

Burns, G. and Glazer, A. M.: 1978, *Space Groups for Solid State Scientists*, Academic Press, New York.

Caliebe, W. A., Soininen, J. A., Shirley, E. L., Kao, C. and Hamalainen, K.: 2000, Dynamical structure factors of diamond and lif measured using inelastic x-ray scattering, *Phys. Rev. Lett.* **84**, 3907.

Caticha, A., Aliberty, K. and Caticha-Ellis, S.: 1996, A Fabry-Perot interferometer for sub-meV x-ray energy resolution, *Rev. Sci. Instrum.* **67**, 3380.

Caticha, A. and Caticha-Ellis, S.: 1982, Dynamical theory of x-ray diffraction at angles near $\pi/2$, *Phys. Rev. B* **25**, 971.

Caticha, A. and Caticha-Ellis, S.: 1990a, Dynamical theory of x-ray diffraction by thin crystals at angles near $\pi/2$, *Phys. Stat. Solidi (a)* **119**, 47.

Caticha, A. and Caticha-Ellis, S.: 1990b, A Fabry-Perot interferometer for hard x-rays, *Phys. Stat. Solidi (a)* **119**, 643.

Caticha, A. and Caticha-Ellis, S.: 1996, A thermal neutron interferometer of the Fabry-Perot type, *Phys. Stat. Solidi (a)* **153**, 29–46.

Chang, S.: 1998, Determination of x-ray reflection phases using N-beam diffraction, *Acta Cryst.* **A54**, 886–894.

Chen, W. M., McNally, P. J., Shvyd'ko, Y. V., Tuomi, T., Danilewsky, A. N., Kanatharana, J., Lowney, D., Lerche, M., Knuuttila, L. and Riikonen, J.: 2003, Dislocation analysis for heat-exchange method grown sapphire with white beam synchrotron x-ray topography, *J. Cryst. Growth* **252**, 113–19.

Chen, W. M., McNally, P. J., Shvyd'ko, Y. V., Tuomi, T., Lerche, M., Danilewsky, A. N., Kanatharana, J., Lowney, D., O'Hare, M., Knuuttila, L., Riikonen, J. and Rantamäki, R.: 2001, Quality assessment of sapphire wafers for x-ray crystal optics using white beam synchrotron x-ray topography, *Phys. Stat. Solidi (a)* **186**, 365–371.

Chukhovskii, F. N. and Förster, E.: 1995, Time-dependent x-ray Bragg diffraction, *Acta Cryst.* **A51**, 668–72.

Chumakov, A. I. and Rüffer, R.: 1998, Nuclear inelastic scattering, *Hyp. Interact.* **113**, 59.

Chumakov, A. I. and Sturhahn, W.: 1999, Experimental aspects of the inelastic nuclear resonance scattering, *Hyp. Interact.* **123/124**, 781.

Chumakov, A. I., Baron, A. Q. R., Rüffer, R., Grünsteudel, H., Grünsteudel, H. F. and Meyer, A.: 1996a, Nuclear resonance energy analysis of inelastic x-ray scattering, *Phys. Rev. Lett.* **76**, 4258.

Chumakov, A. I., Metge, J., Baron, A. Q. R., Grünsteudel, H., Rüffer, R. and Ishikawa, T.: 1996b, An x-ray monochromator with 1.65 meV energy resolution, *Nucl. Instrum. Methods Phys. Res. A* **383**, 642.

Chumakov, A. I., Rüffer, R., Leupold, O. and Sergueev, I.: 2003, Insight to dynamics of molecules with nuclear inelastic scattering, *Structural Chemistry* **14**, 109–119.

Chumakov, A. I., Rüffer, R., Leupold, O., Barla, A., Thiess, H., Asthalter, T., Doyle, B., Snigirev, A. and Baron, A. Q. R.: 2000, High-energy-resolution x-ray optics with refractive collimators, *Appl. Phys. Lett.* **77**, 31–33.

Colella, R.: 1974, Multiple diffraction of x-rays and the phase problem. Computational procedures and comparison with experiment, *Acta Cryst.* **A30**, 413–423.

Colella, R.: 1995, Multiple Bragg scattering and the phase problem in x-ray diffraction, *Comments Cond. Mat. Phys.* **17**, 175–215.

Colella, R. and Luccio, A.: 1984, Proposal for a free electron laser in the x-ray region, *Opt. Commun.* **50**, 41–44.

Compton, A. H. and Allison, S. K.: 1935, *X-Rays in Theory and Experiment*, van Nostrand, New York. Reprinted by van Nostrand, Princeton, 1963.

Cusatis, C., Udron, D., Mazzaro, I., Giles, C. and Tolentino, H.: 1996, X-ray back-diffraction profiles with a Si (111) plate, *Acta Cryst.* **A52**, 614–620.

Darwin, C. G.: 1914, The theory of x-ray reflexion, *Phil. Mag.* **27**, 315–333, 675–690.

d'Astuto, M., Giura, P., Krisch, M., Lorenzen, M., Mermet, A., Monaco, G., Requardt, H., Sette, F., Shukla, A. and Verbeni, R.: 2002, Phonon dispersion studies of crystalline materials using high-energy resolution inelastic x-ray scattering (IXS), *Physica B* **316-317**, 150–153.

Davis, T. J.: 1990, DuMond diagram mappings for multi asymmetric crystal monochromators, *Journal of X-Ray Science and Technology* **2**, 180–94.

Deslattes, R. D. and Henins, A.: 1973, X-ray to visible wavelength ratios, *Phys. Rev. Lett.* **31**, 972.

Deutsch, M. and Hart, M.: 1985, Electronic charge distribution in silicon, *Phys. Rev. B* **31**, 3846.

Deutsch, M. and Hart, M.: 1988, High-energy x-ray anomalous dispersion correction for silicon, *Phys. Rev. B* **37**, 2701.

Dierker, S. B., Pindak, R., Fleming, R. M., Robinson, I. K. and Berman, L.: 1995, X-ray photon correlation spectroscopy study of brownian motion of gold colloids in glycerol, *Phys. Rev. Lett.* **75**, 449–452.

Dobrovinskaya, L., Lytvynov, L. and Pishchuk, V.: 2002, *Sapphire and other corundum crystals*, Institute for Single Crystals (with the assistance of ISRA-BIT company), Ukraine-Charkiv.

Dorner, B.: 1982, *Coherent Inelastic Neutron Scatteing in Lattice Dynamics*, Vol. 93 of *Springer Tracts in Modern Physics*, Springer Verlag, Berlin.

Dorner, B. and Peisl, H.: 1983, An instrument with very high energy resolution in x-ray scattering, *Nucl. Instrum. Methods Phys. Res.* **208**, 587.

Dorner, B., Burkel, E. and Peisl, J.: 1986, An x-ray backscattering instrument with very high energy resolution, *Nucl. Instrum. Methods Phys. Res. A* **246**, 450–451.

DuMond, J. W. M.: 1937, Theory of the use of more than two successive x-ray crystal reflections to obtain increased resolving power, *Phys. Rev.* **52**, 872–883.

Ewald, P. P.: 1917, Zur Begründung der Kristalloptik. III. Die Kristalloptik der Röntgenstrahlen, *Ann. Physik* **54**, 519–597.

Fabry, C. and Pérot, A.: 1899, Theorie et applications d'une nouvelle methode de spectroscopie interferentielle, *Ann. Chim. Phys.* **7**, 115–144.

Faigel, G., Siddons, D. P., Hastings, J. B., Haustein, P. E., Grover, J. R., Remeika, J. P. and Cooper, A. S.: 1987, New approach to the study of the nuclear Bragg scattering of synchrotron radiation, *Phys. Rev. Lett.* **58**, 2699.

Firestone, R. B., Shirley, V. S., Chu, S. Y. F., Baglin, C. M. and Zipkin, J.: 1996, *Table of Isotopes*, John Wiley & Sons, Inc., New York.

Franz, H.: 2003, private communication.

Freund, A. and Schneider, J.: 1972, Two new experimental diffraction methods for a precise measurement of crystal perfection, *J. Cryst. Growth* **13-14**, 247–251.

Friedrich, W., Knipping, P. and von Laue, M.: 1913, Interferenzerscheinungen bei Röntgenstrahlen, *Ann. Physik* **41**, 971 – 988.

Gerdau, E. and de Waard, H. (eds): 1999/2000, *Nuclear Resonant Scattering of Synchrotron Radiation*, Baltzer. Special issues of the *Hyperfine Interact.*, Vol. 123-125.

Gerdau, E., Rüffer, R., Hollatz, R. and Hannon, J. P.: 1986, Quantum beats from nuclei excited by synchrotron radiation, *Phys. Rev. Lett.* **57**, 1141.

Gerdau, E., Rüffer, R., Winkler, H., Tolksdorf, W., Klages, C. P. and Hannon, J. P.: 1985, Nuclear Bragg diffraction of synchrotron radiation in yttrium iron garnet, *Phys. Rev. Lett.* **54**, 583.

Giles, C. and Cusatis, C.: 1991, Measurements of transmitted diffraction profiles on Bragg angles at $\pi/2$, *Appl. Phys. Lett.* **59**, 641.

Graeff, W.: 2002, Short x-ray pulses in a Laue-case crystal, *J. Synchrotron Rad.* **9**, 82–85.

Graeff, W. and Materlik, G.: 1982, Millielectron volt energy resolution in Bragg backscattering, *Nucl. Instrum. Methods Phys. Res.* **195**, 97.

Guichet, C. and Rasigni, G.: 1997, Considerations of the finesse and resolving power of Fabry-Pérot étalons for soft x rays, *Opt. Eng.* **36**, 2107.

Hahn, T. (ed.): 2002, *International Tables for Crystallography, Volume A: Space-group symmetry*, Kluwer Academic Publishers, (Dordrecht/Boston/London). 5th edition.

Hämäläinen, K. and Manninen, S.: 2001, Resonant and non-resonant inelastic x-ray scattering, *J. Phys. C: Solid St. Phys.* **13**, 7539–55.

Hart, M.: 1968, An angström ruler, *Brit. J. Appl. Phys.* **1**, 1405–1408.

Hashizume, H., Nakayama, K., Matsushita, T. and Kohra, K.: 1970, Variation of Bragg-case diffraction curves of x-rays from a thin silicon crystal with crystal thickness, *J. Phys. Soc. Japan* **29**, 806.

Hastings, J. B., Moncton, D. E. and Fujii, Y.: 1984, High energy-resolution inelastic scattering, *Proceedings of the 1984 Workshop on High Energy excitations in Condensed Matter.* Los Alamos.

Hastings, J. B., Siddons, D. P., van Bürck, U., Hollatz, R. and Bergmann, U.: 1991, Mössbauer spectroscopy using synchrotron radiation, *Phys. Rev. Lett.* **66**, 770.

Hecht, E. and Zajac, A.: 1974, *Optics*, Addison-Wesley Publishing Company, Inc., Massachusetts.

Hill, J. P., Kao, C.-C., Caliebe, W. A., Gibbs, D. and Hastings, J. B.: 1996, Inelastic x-ray scattering study of solid and liquid Li and Na, *Phys. Rev. Lett.* **77**, 3665.

Houston, W. V.: 1927, A compound interferometer for fine structure work, *Phys. Rev.* **29**, 478–84.

Howells, M. R., Cambie, D., Duarte, R. M., Irick, S., MacDowell, A. A., Padmore, H. A., Renner, T. R., Rah, S., and Sandler, R.: 2000, Theory and practice of elliptically bent x-ray mirrors, *Opt. Eng.* **39**, 2748–2762.

Hu, M. Y., Sinn, H., Alatas, A., Alp, E. E., Sturhahn, W., Wille, H.-C., Shvyd'ko, Y. V., Sutter, J. P., Bandaru, J., Haller, E. E., Ozhogin, V. I., Rodrigues, S., Colella, R., Kartheuser, E. and Villert, M. A.: 2003, The effect of isotopic composition on the lattice parameter of germanium, *Phys. Rev. B* **67**, 113306.

Ishikawa, T., Tamasaku, K., Yabashi, M., Gotob, S., Tanaka, Y., Yamazaki, H., Takeshita, K., Kimura, H., Ohashi, H., Matsushita, T. and Ohata, T.: 2000, 1-km beamline at SPring-8, *Proc. SPIE - Int. Soc. Opt. Eng.* **4145**, 1.

Ishikawa, T., Yoda, Y., Izumi, K., Suzuki, C. K., Zhang, X. W., Ando, M. and Kikuta, S.: 1992, Construction of a precision diffractometer for nuclear Bragg scattering at the photon factory, *Rev. Sci. Instrum.* **63**, 1015.

Izyumov, Y. A., Chernoplekov, N. A. and Iziumov, I. A.: 1994, *Neutron Spectroscopy*, Plenum Press.

Jackson, J. D.: 1967, *Classical Electrodynamics*, John Wiley & Sons, Inc., New York.

James, R. W.: 1950, *The Optical Principles of the Diffraction of X-Rays in Crystals*, G. Bell & Sons, Ltd., London.

Kagan, Y., Afanas'ev, A. M. and Kohn, V. G.: 1979, On excitation of isomeric nuclear states in a crystal by synchrotron radiation, *J. Phys. C: Solid St. Phys.* **12**, 615–631.

Kao, C., Caliebe, W. A., Hastings, J. B. and Gillet, J.: 1996, Resonant raman scattering in nio: Resonant enhancement of the charge transfer excitations, *Phys. Rev. B* **54**, 16361.

Kao, C., Hamalainen, K., Krisch, M., Siddons, D. P., Oversluizen, T. and Hastings, J. B.: 1995, Optical design and performance of the inelastic scattering beamline at the National Synchrotron Light Source, *Rev. Sci. Instrum.* **66**, 1699–1702.

Khattak, C. P. and Schmid, F.: 2001, Growth of the world's largest sapphire crystals, *J. Cryst. Growth* **225**, 572.

Kikuta, S., Imai, Y., Iizuka, T., Yoda, Y., Zhang, X.-W. and Hirano, K.: 1998, X-ray diffraction with a Bragg angle near to $\pi/2$ and its applications, *J. Synchrotron Rad.* **5**, 670.

Kim, Y. J., Hill, J. P., Burns, C. A., Wakimoto, S., Birgeneau, R. J., Ando, Y., Gog, T., Venkataraman, C. T. and Casa, D. M.: 2002, Resonant inelastic x-ray scattering study of $La_{2-x}Sr_xCuO_4$, *Phys. Rev. Lett.* **89**, 177003.

Kinosita, K.: 1953, Numerical evaluation of the intensity curve of a multiple-beam fizeau fringe, *J. Phys. Soc. Japan* **8**, 219.

Kirfel, A. and Eichhorn, K.: 1990, Accurate structure analysis with synchrotron radiation. The electron density in Al_2O_3 and Cu_2O, *Acta Cryst.* **A46**, 271.

Kirkpatrick, P. and Baez, A. V.: 1948, Formation of optical images by x-rays, *J. Opt. Soc. Am.* **38**, 766–774.

Kishimoto, S.: 1991, An avalanche photodiode detector for x-ray timing measurements, *Nucl. Instrum. Methods Phys. Res. A* **309**, 603.

Kishimoto, S.: 1992, High time resolution x-ray measurements with an avalanche photodiode detector, *Rev. Sci. Instrum.* **63**, 824.

Kissel, L. and Pratt, R.: 1990, Corrections to tabulated anomalous scattering factors, *Acta Cryst.* **A46**, 170–175. www-phys.llnl.gov/Research/scattering.

Kissel, L., Zhou, B., Roy, S., Gupta, S. K. S. and Pratt, R. H.: 1995, Validity of form-factor, modified-form-factor and anomalous-scattering-factor approximations in elastic scattering calculations, *Acta Cryst.* **A51**, 271–288.

Kohn, V. G., Kohn, I. V. and Manykin, E. A.: 1999, Diffraction of x-rays at a Bragg angle of $\pi/2$ (back reflection) with consideration of multi-wave effects, *JETP* **89**, 500.

Kohn, V. G., Shvyd'ko, Y. V. and Gerdau, E.: 2000, On the theory of an x-ray Fabry-Perot interferometer, *Phys. Stat. Solidi (b)* **221**, 597.

Kohra, K. and Kikuta, S.: 1967, A method of obtaining an extremely parallel x-ray beam by successive asymmetric diffractions and its applications, *Acta Cryst.* **A24**, 200.

Kohra, K. and Matsushita, T.: 1972, Some characteristics of dynamical diffraction at a Bragg angle of about $\pi/2$, *Z. Naturforsch. A* **27**, 484.

Kohra, K., Ando, M., Matsushita, T. and Hashizume, H.: 1978, Design of high resolution x-ray optical system using dynamical diffraction for synchrotron radiation, *Nucl. Instrum. Methods Phys. Res.* **152**, 161–166.

Krisch, M.: 2003, Status of phonon studies at high pressure by inelastic x-ray scattering, *J. Raman Spectroscopy* **34**, 628–632.

Krisch, M., Loubeyre, P., Ruocco, G., Sette, F., Cunsolo, A., D'Astuto, M., LeToullec, R., Lorenzen, M., Mermet, A., Monaco, G. and Verbeni, R.: 2002, Pressure evolution of the high-frequency sound velocity in liquid water, *Phys. Rev. Lett.* **89**, 125502.

Kumakhov, M. A.: 1990, Channeling of photons and new x-ray optics, *Nucl. Instrum. Methods Phys. Res. B* **48**, 283–286.

Kunz, C.: 2001, Synchrotron radiation: third generation sources, *J. Phys. C: Solid St. Phys.* **13**, 7499–7511.

Kushnir, V. I. and Suvorov, E. V.: 1986, Experimental observation of the half-degree angular interval of x-ray reflection in backscattering ($\theta \simeq \pi/2$) from a high-quality crystal, *JETP Lett.* **44**, 262.

Kushnir, V. I. and Suvorov, E. V.: 1990, X-ray backscattering on perfect crystals ($2\theta \simeq \pi$), *Phys. Stat. Solidi (a)* **122**, 391.

LCLS: 1998, Linac Coherent Light Source LCLS, Design Study Report, *Technical Report SLAC-R-521, UC-414*, Stanford Linear Accelerator Center, CA 94025 USA.

Lepetre, Y., Rivoira, R., Philip, R. and Rasigni, G.: 1984, Fabry-Pérot étalons for x-rays: construction and characterization, *Opt. Commun.* **51**, 127–130.

Lerche, M.: 2000, Investigations of sapphire single crystals: application for monochromatization and for exact Bragg backscattering of synchrotron radiation, Diploma thesis, University of Hamburg.

Liss, K.-D., Hock, R., Gomm, M., Waibel, B., Magerl, A., Krisch, M. and Tucoulou, R.: 2000, Storage of x-ray photons in a crystal resonator, *Nature* **404**, 371.

Lovesy, S. W.: 1986, *Theory of Neutron Scattering from Condensed Matter*, Vol. I and II, Oxford University Press.

Lucht, M.: 1998, High-precision temperature control: application for monochromatization and for exact Bragg backscattering of synchrotron radiation, Diploma thesis, University of Hamburg.

Lucht, M., Lerche, M., Wille, H.-C., Shvyd'ko, Y. V., Rüter, H. D., Gerdau, E. and Becker, P.: 2003, Precise measurement of the lattice parameters of Al_2O_3 in the temperature range 4.5 K–250 K using Mössbauer wavelength standard, *J. Appl. Cryst.* **36**, 1075–1081.

Masciovecchio, C., Bergmann, U., Krisch, M., Ruocco, G., Sette, F. and Verbeni, R.: 1996a, A perfect crystal x-ray analyzer with 1.5 meV energy resolution, *Nucl. Instrum. Methods Phys. Res. B* **117**, 339–340.

Masciovecchio, C., Bergmann, U., Krisch, M., Ruocco, G., Sette, F. and Verbeni, R.: 1996b, A perfect crystal x-ray analyzer with meV energy resolution, *Nucl. Instrum. Methods Phys. Res. B* **111**, 181–186.

Materlik, G. and Tschentscher, T. (eds): 2001, *TESLA. The superconducting Electron-Positron Linear Collider with an Integrated X-ray Laser Laboratory. Part V, The X-Ray Free Electron Laser*, DESY, Hamburg.

Matsushita, T. and Hashizume, H.: 1983, X-ray monochromators, *in* E. E. Koch (ed.), *Handbook on synchrotron radiation*, Vol. 1, North-Holland, Amsterdam, pp. 261–314.

McNally, P. J., O'Reilly, L. and Shvyd'ko, Y. V.: 2004, Kyropoulos grown sapphire crystals studied by white beam synchrotron x-ray topography, *(to be published)*.

Missalla, T., Uschmann, I., Förster, E., Jenke, G. and von der Linde, D.: 1999, Monochromatic focusing of subpicosecond x-ray pulses in the keV range, *Rev. Sci. Instrum.* **70**, 1288–99.

Mohr, P. J. and Taylor, B. N.: 2000, CODATA recommended values of the fundamental physical constants: 1998, *Rev. Mod. Phys.* **72**, 351–495.

Moncton, D. E.: 1980, High energy-transfer resolution, Vienna Synchrotron Summer School. Lecture Notes. Private communication *(unpublished)*.

Mooney, T. M., Toellner, T. S., Sturhahn, W., Alp, E. E. and Shastri, S. D.: 1994, High-resolution, large-angular-acceptance monochromator for hard x-rays, *Nucl. Instrum. Methods Phys. Res. A* **347**, 348.

Moos, H. W., Imbusch, G. F., Mollenauer, L. F. and Schawlow, A. L.: 1963, Tilted-plated interferometry with large plate separations, *Appl. Opt.* **2**, 817.

Morawe, C., Pecci, P., Peffen, J. C. and Ziegler, E.: 1999, Design and performance of graded multilayers as focusing elements for x-ray optics, *Rev. Sci. Instrum.* **70**, 3227–3232.

Mössbauer, R. L.: 1958, Kernresonanzfluoreszenz von Gammastrahlung in Ir^{191}, *Z. f. Physik* **151**, 124.

Mössbauer, R. L.: 1959, Kernresonanzabsorption von γ-Strahlung in Ir^{191}, *Z. Naturforsch. A* **14**, 211–216.

Mülhaupt, G. and Rüffer, R.: 1999, Properties of synchrotron radiation, *Hyp. Interact.* **123/124**, 13–30.

Murphy, J. B. and Pellegrini, C.: 1990, Introduction to the physics of FELs, *in* W. Colson, C. Pellegrini and A. Renieri (eds), *Laser Handbook*, Vol. 6, North Holland.

Nakayama, K. and Tanaka, M.: 1986, Monolithic rotation mechanism with nanoradian resolution, *in* S. Hosoya, Y. Ittaka and H. Hashizume (eds), *X-Ray Instrumentation for the Photon Factory: Dynamic Analyses of Micro-Structures in Matter*, KTK Scientific Publishers, pp. 277–282. Tokyo.

Nakayama, K., Hashizume, H., Kikuta, S. and Kohra, K.: 1973, Use of asymmetric dynamical diffraction of x-rays for multiple-crystal arrangements of the $(n_1, +n_2)$ setting, *Z. Naturforsch. A* **28**, 632–638.

Okada, Y. and Tokumaru, Y.: 1984, Precise determination of lattice parameter and thermal expansion coefficient of silicon between 300 and 1500 K, *J. Appl. Phys.* **56**, 314.

Okazaki, A. and Kawaminami, M.: 1973, Accurate measurements of lattice constants in a wide range of temperature: Use of white x-rays and double crystal diffractometry, *Japn. J. Appl. Phys.* **12**, 783.

Pinsker, Z. G.: 1978, *Dynamical Scattering of X rays in Crystals*, Springer, Berlin.

Polster, H. D.: 1969, Multiple beam interferometry, *Appl. Opt.* **8**, 522.

Protopopov, V. V., Shishkov, V. A. and Kalnov, V. A.: 2000, X-ray parabolic collimator with depth-graded multilayer mirror, *Rev. Sci. Instrum.* **71**, 4380–4386.

Reisman, A., Berkenblit, M., Cuomo, J. and Chan, S. A.: 1971, The chemical polishing of sapphire and MgAl spinel, *J. Electrochem. Soc.* **118**, 1653.

Renninger, M.: 1937, Umweganregung, eine bisher unbeachtete Wechselwirkungserscheinung bei Raumgitter-Interferenzen, *Z. f. Physik* **106**, 141–176.

Riese, D. O., Vos, W. L., Wegdam, G. H., Poelwijk, F. J., Abernathy, D. L. and Grübel, G.: 2000, Photon correlation spectroscopy: X rays versus visible light, *Phys. Rev. E* **61**, 1676.

Rogers, J. R.: 1982, Fringe shifts in multiple-beam Fizeau interferometry, *J. Opt. Soc. Am.* **72**, 638.

Röhlsberger, R., Gerdau, E., Rüffer, R., Sturhahn, W., Toellner, T. S., Chumakov, A. and Alp, E. E.: 1997, X-ray optics for μeV-resolved spectroscopy, *Nucl. Instrum. Methods Phys. Res. A* **394**, 251.

Rüffer, R. and Chumakov, A. I.: 2000, Nuclear inelastic scattering, *Hyp. Interact.* **128**, 255.

Rüffer, R., Gerdau, E., Hollatz, R. and Hannon, J. P.: 1987, Nuclear Bragg scattering of synchrotron radiation pulses in a single-reflection geometry, *Phys. Rev. Lett.* **58**, 2359.

Sachs, G. and Weerts, J.: 1930, Die Gitterkonstanten der Gold-Silberlegierungen, *Z. f. Physik* **60**, 481–490.

Saldin, E. L., Schneidmiller, E. A. and Yurkov, M. V.: 2000, *The Physics of Free Electron Lasers*, Advanced Texts in Physics, Springer, Berlin Heidelberg.

Saldin, E. L., Schneidmiller, E. A., Shvyd'ko, Y. V. and Yurkov, M. V.: 2001, X-ray FEL with a meV bandwidth, *Nucl. Instrum. Methods Phys. Res. A* **475**, 375–362.

Schmid, F., Khattak, C. P. and Felt, D. M.: 1994, Producing large sapphire for optical applications, *Am. Ceram. Soc. Bull.* **73**(2), 39.

Schober, H., Strauch, D. and Dorner, B.: 1993, Lattice dynamics of sapphire (Al_2O_3), *Z. Phys. B* **92**, 273–283.

Schülke, W.: 2001, Electronic excitations investigated by inelastic x-ray scattering spectroscopy, *J. Phys. C: Solid St. Phys.* **13**, 7557–91.

Schuster, M. and Göbel, H.: 1995, Parallel-beam coupling into channel-cut monochromators using curved graded multilayers, *J. Phys. D: Appl. Phys.* **28**, A270–A275.

Scopigno, T., Di Leonardo, R., Ruoccoand, G., Baron, A. Q. R., Tsutsui, S., Bossard, F. and Yannopoulos, S. N.: 2004, High frequency dynamics in a monatomic glass, *Phys. Rev. Lett.* **92**, 025503/1–4.

Scopigno, T., Filipponi, A., Krisch, M., Monaco, G., Ruocco, G. and F.Sette: 2002, High-frequency acoustic modes in liquid gallium at the melting point, *Phys. Rev. Lett.* **89**, 255506.

Seel, S., Storz, R., Ruoso, G., Mlynek, J. and Schiller, S.: 1997, Cryogenic optical resonators: A new tool for laser frequency stabilization at the 1 Hz level, *Phys. Rev. Lett.* **78**, 4741–4744.

Sellschop, J. P. F., Connell, S. H., Nilen, R. W. N., Detlefs, C., Freund, A. K., Hoszowska, J., Hustache, R., Burns, R. C., Rebak, M., Hansen, J. O., Welch, D. L., and Hall, C.: 2000, Synchrotron x-ray applications of synthetic diamonds, *New Diamond and Frontier Carbon Technology* **10**, 253.

Seto, M., Yoda, Y., Kikuta, S., Zhang, X. W. and Ando, M.: 1995, Observation of nuclear resonant scattering accompanied by phonon excitation using synchrotron radiation, *Phys. Rev. Lett.* **74**, 3828.

Sette, F., Krisch, M. H., Masciovecchio, C., Ruocco, G. and Monaco, G.: 1998, Dynamics of glasses and glass-forming liquids studied by inelastic x-ray scattering, *Science* **280**, 1550–1555.

Sette, F., Ruocco, G., Krisch, M., Bergmann, U., Masciovecchio, C., Mazzacurati, Signorelli, G. and Verbeni, R.: 1995, Collective dynamics in water by high energy resolution inelastic x-ray scattering, *Phys. Rev. Lett.* **75**, 850–853.

Shastri, S. D., Zambianchi, P. and Mills, D. M.: 2001a, Dynamical diffraction of ultrashort x-ray free-electron laser pulses, *J. Synchrotron Rad.* **8**, 1131–1135.

Shastri, S. D., Zambianchi, P. and Mills, D. M.: 2001b, Femtosecond x-ray dynamical diffraction by perfect crystals, *Proc. SPIE - Int. Soc. Opt. Eng.* **4143**, 69–77.

Shu, D., Toellner, T. S. and Alp, E. E.: 2001, Modular overconstrained weak-link mechanism for ultraprecision motion control, *Nucl. Instrum. Methods Phys. Res. A* **467-468**, 771–4.

Shu, D., Toellner, T. S. and Alp, E. E.: 2003, Redundantly constrained laminar structure as weak-link mechanisms, U.S. Patent granted No. 6,607,840.

Shukla, A., Calandra, M., d'Astuto, M., Lazzeri, M., Mauri, F., Bellin, C., Krisch, M., Karpinski, J., Kazakov, S. M., Jun, J., Daghero, D. and Parlinski, K.: 2003, Phonon dispersion and lifetimes in MgB_2, *Phys. Rev. Lett.* **90**, 095506/1–4.

Shvyd'ko, Y. V.: 1999, Nuclear resonant forward scattering of x-rays: Time and space picture, *Phys. Rev. B* **59**, 9132.

Shvyd'ko, Y. V.: 2002, *X-ray resonators and other applications of Bragg backscattering*, Habilitationsschrift. ISSN 1435-8085, DESY, Hamburg. www-library.desy.de/diss02.html (DESY-Thesis-2002-028).

Shvyd'ko, Y. V. and Gerdau, E.: 1999, Backscattering mirrors for x-rays and Mössbauer radiation, *Hyp. Interact.* **123/124**, 741–776.

Shvyd'ko, Y. V. and Smirnov, G. V.: 1990, Experimental study of time and frequency properties of collective nuclear excitations in a single crystal, *J. Phys.: Condens. Matter* **1**, 10563–10584.

Shvyd'ko, Y. V., Gerdau, E., Jäschke, J., Leupold, O., Lucht, M. and Rüter, H. D.: 1998, Exact Bragg backscattering of x-rays, *Phys. Rev. B* **57**, 4968.

Shvyd'ko, Y. V., Gerdau, E., Lerche, M., Lucht, M., Wille, H.-C., Alp, E. E. and Becker, P.: 2003a, Sapphire x-ray resonator. First experience and results, *Hyp. Interact. (C)* **5**, 25–28.

Shvyd'ko, Y. V., Gerken, M., Franz, H., Lucht, M. and Gerdau, E.: 2001, Nuclear resonant scattering of synchrotron radiation from ^{161}Dy at 25.61 keV, *Europhys. Lett.* **56**, 39.

Shvyd'ko, Y. V., Hertrich, T., van Bürck, U., Gerdau, E., Leupold, O., Metge, J., Rüter, H. D., Schwendy, S., Smirnov, G. V., Potzel, W. and Schindelmann, P.: 1996, Storage of nuclear excitation energy through magnetic switching, *Phys. Rev. Lett.* **77**, 3232.

Shvyd'ko, Y. V., Lerche, M., Jäschke, J., Gerdau, M. L. E., Gerken, M., Rüter, H. D., Wille, H.-C., Becker, P., Alp, E. E., Sturhahn, W., Sutter, J. and Toellner, T. S.: 2000, γ-ray wavelength standard for atomic scales, *Phys. Rev. Lett.* **85**, 495.

Shvyd'ko, Y. V., Lerche, M., Wille, H.-C., Gerdau, E., Alp, E. E., Lucht, M., Rüter, H. D. and Khachatryan, R.: 2003b, X-ray interferometry with micro-electronvolt resolution, *Phys. Rev. Lett.* **90**, 013904.

Shvyd'ko, Y. V., Lucht, M., Gerdau, E., Lerche, M., Alp, E. E., Sturhahn, W., Sutter, J. and Toellner, T. S.: 2002, Measuring wavelengths and lattice constants with the Mössbauer wavelength standard, *J. Synchrotron Rad.* **9**, 17–23.

Sinn, H.: 2001, Spectroscopy with meV energy resolution, *J. Phys.: Condens. Matter* **13**(34), 7525–7537.

Sinn, H., Alatas, A., Alp, E. E., Yavas, H., Nyman, A., Said, A., Shu, D., Sturhahn, W., Toellner, T. and Zhao, J.: 2004, A spherical bender for inelastic x-ray scattering, *(to be published)*.

Sinn, H., Alp, E., Alatas, A., Barraza, J., Bortel, G., Burkel, E., Shu, D., Sturhahn, W., Sutter, J., Toellner, T. and Zhao, J.: 2001, An inelastic x-ray spectrometer with 2.2 meV energy resolution, *Nucl. Instrum. Methods Phys. Res. A* **467-468**, 1545–1548.

Sinn, H., Glorieux, B., Hennet, L., Alatas, A., Hu, M., Alp, E. E., Bermejo, F. J., Price, D. L. and Saboungi, M.: 2003, Microscopic dynamics of liquid aluminum oxide, *Science* **299**, 2047–2049.

Sinn, H., Moldovan, N., Said, A. H. and Alp, E.: 2002, Development of a two-dimensional focusing faceted x-ray analyzer, *Proc. SPIE - Int. Soc. Opt. Eng.* **4783**, 123–130.

Sinn, H., Sette, F., Bergmann, U., Halcoussis, C., Krisch, M., Verbeni, R. and Burkel, E.: 1997, Coherent dynamic structure factor of liquid lithium by inelastic x-ray scattering, *Phys. Rev. Lett.* **78**, 1715–1718.

Skold, K. and Price, D. L. (eds): 1986, *Neutron Scattering, Part A*, Vol. 23 of *Methods of Experimental Physics*, Academic Press.

Smirnov, G. V. and Shvyd'ko, Y. V.: 1986, Experimental study of the time evolution of the decay of Bragg modes of the collective nuclear excitations in a crystal, *Pis'ma. Zh. Eksp. Teor. Fiz.* **44**, 431–435. [(1986) *Sov. Phys. JETP* **44**, 431-435].

Smirnov, G. V. and Shvyd'ko, Y. V.: 1989, Measurements of the lifetime of a nucleus excited near a resonance, *Zh. Eksp. Teor. Fiz.* **95**, 777–789. [(1989) *Sov. Phys. JETP* **68**, 444-450].

Smith, S. T., Badami, V. G., Dale, J. S. and Xu, Y.: 1997, Elliptical flexure hinge, *Rev. Sci. Instrum.* **68**, 1474.

Snigirev, A., Kohn, V., Snigireva, I. and Lengeler, B.: 1996, A compound refractive lens for focusing high-energy x-rays, *Nature* **384**, 49–51.

Sortais, Y., Bize, S., Abgrall, M., Zhang, S., Nicolas, C., Mandache, C., Lemonde, R., Laurent, P., Santarelli, G., Dimarcq, N., Petit, P., Clairon, A., Mann, A., Luiten, A., Chang, S. and Salomon, C.: 2001, Cold atom clocks, *Physica Scripta* **T95**, 50–7.

Stenger, J., Binnewies, T., Wilpers, G., Riehle, F., Telle, H. R., Ranka, J. K., Windeler, R. S. and Stentz, A. J.: 2001, Phase-coherent frequency measurement of the Ca intercombination line at 657 nm with a Kerr-lens mode-locked femtosecond laser, *Phys. Rev. A* **63**, 021802(R).

Stepanov, S. A. and Ulyanenkov, A. P.: 1994, A new algorithm for computation of x-ray multiple Bragg diffraction, *Acta Cryst.* **A50**, 579–585.

Stepanov, S. A., Kondrashkina, E. A. and Novikov, D. V.: 1991, X-ray surface back diffraction, *Nucl. Instrum. Methods Phys. Res. A* **301**, 350–7.

Stetsko, Y. P. and Chang, S.-L.: 1997, An algorithm for solving multiple-wave dynamical x-ray diffraction equations, *Acta Cryst.* **A53**, 28–34.

Stetsko, Y. P., Kshevetskii, S. A. and Mikhailyuk, I. P.: 1988, Bond-method determination of the lattice parameters of single crystals under conditions of extremal diffraction ($\theta \approx \pi/2$), *Soviet Technical Physics Letters* **14**, 13–14.

Steyerl, A. and Steinhauser, K.-A.: 1979, Proposal of a Fabry-Perot-type interferometer for x-rays, *Z. Phys. B* **34**, 221–227.

Sturhahn, W. and Kohn, V. G.: 1999, Theoretical aspects of incoherent nuclear resonant scattering, *Hyp. Interact.* **123/124**, 367.

Sturhahn, W., Toellner, T. S., Alp, E. E., Zhang, X., Ando, M., Yoda, Y., Kikuta, S., Seto, M., Kimball, C. W. and Dabrowski, B.: 1995, Phonon density of states measured by inelastic nuclear resonant scattering, *Phys. Rev. Lett.* **74**, 3832.

Suorti, P. and Freund, A.: 1989, On the phase-space description of synchrotron x-ray beams, *Rev. Sci. Instrum.* **60**, 2579–85.

Sutter, J. P.: 2000, *Applications of special x-ray diffraction cases in silicon crystals*, PhD thesis, Purdue University.

Sutter, J. P., Alp, E. E., Hu, M. Y., Lee, P. L., Sinn, H., Sturhahn, W., and Toellner, T. S.: 2001, Multiple-beam x-ray diffraction near exact backscattering in silicon, *Phys. Rev. B* **63**, 094111.

Sykora, B. and Peisl, H.: 1970, A high resolution x-ray diffractometer, *Z. angew. Physik* **30**, 320–322.

Tanaka, M.: 1983, The dynamical properties of a monolithic mechanism with notch flexure hinges for precision control of orientation and position, *Japn. J. Appl. Phys.* **22**, 193–200.

Tanaka, T. and Kitamura, H.: 2001, SPECTRA: a synchrotron radiation calculation code, *J. Synchrotron Rad.* **8**, 1221–1228.

Teworte, R. and Bonse, U.: 1983, High-precision determination of structure factors F_h of silicon, *Phys. Rev. B* **29**, 2102.

Thompson, A. C. and Vaughan, D. (eds): 2001, *X-Ray Data Booklet*, Lawrence Berkeley National Laboratory, Berkeley, California 94720. LBNL/PUB-490 Rev. 2, second edition.

Thurn-Albrecht, T., Steffen, W., Patkowski, A., Meier, G., Fischer, E. W., Grübel, G. and Abernathy, D. L.: 1996, Photon correlation spectroscopy of colloidal palladium using a coherent x-ray beam, *Phys. Rev. Lett.* **77**, 5437–5440.

Toellner, T. S.: 1996, *High-Resolution X-Ray Probes Using Nuclear Resonance Scattering*, PhD thesis, Northwestern University.

Toellner, T. S.: 2000, Monochromatization of synchrotron radiation for nuclear resonant scattering experiments, *Hyp. Interact.* **125**, 3.

Toellner, T. S. and Shu, D.: 2001, private communication, *unpublished*.

Toellner, T. S., Hu, M., Sturhahn, W., Bortel, G., Alp, E. E. and Zhao, J.: 2001, Crystal monochromator with a resolution beyond 10^8, *J. Synchrotron Rad.* **8**, 1082–1086.

Toellner, T. S., Hu, M. Y., Sturhahn, W., Quast, K. and Alp, E. E.: 1997, Inelastic nuclear resonant scattering with sub-meV energy resolution, *Appl. Phys. Lett.* **71**, 2112.

Toellner, T. S., Mooney, T. M., Shastri, S. D. and Alp, E. E.: 1992, High resolution, high angular acceptance crystal monochromator, *Proc. SPIE - Int. Soc. Opt. Eng.* **1740**, 218–223.

Toellner, T. S., Shu, D., Sturhahn, W., Alp, E. E. and Zhao, J.: 2002, private communication, *unpublished*.

Tolansky, S.: 1960, *Surface Microtopography*, Interscience, New York.

Touloukian, Y. S., Powell, R. W., Ho, C. Y. and Klemens, P. G.: 1970, *Thermal Conductivity - Nonmetallic Solids*, Vol. 2 of *Thermal Properties of Matter*, IFI/Plenum, New York - Washington.

Trammell, G. T. and Hannon, J. P.: 1978, Quantum beats from nuclei excited by synchrotron pulses, *Phys. Rev. B* **18**, 165–172. Erratum: *Phys. Rev. B* **19**, 3835 (1979).

Tuomi, T.: 2002, Synchrotron x-ray topography of electronic materials, *J. Synchrotron Rad.* **9**, 174–178.

Tuomi, T., Naukkarinen, K. and Rabe, P.: 1974, Use of synchrotron radiation in x-ray diffraction topography, *Phys. Stat. Solidi (a)* **25**, 93.

Tuomi, T., Rantamäki, R., McNally, P. J., Lowney, D., Danilewsky, A. N. and Becker, P.: 2001, Dynamical diffraction imaging of voids in nearly perfect silicon, *J. Phys. D: Appl. Phys.*

Udem, T., Holzwarth, R. and Hänsch, T. W.: 2002, Optical frequency metrology, *Nature* **416**, 233–237.

van Bürck, U., Mössbauer, R. L., Gerdau, E., Rüffer, R., Hollatz, R., Smirnov, G. V. and Hannon, J. P.: 1987, Nuclear Bragg scattering of synchrotron radiation with strong speedup of coherent decay, measured on antiferromagnetic ^{57}FeBO$_3$, *Phys. Rev. Lett.* **59**, 355.

van Bürck, U., Siddons, D. P., Hastings, J. B., Bergmann, U. and Hollatz, R.: 1992, Nuclear forward scattering of synchrotron radiation, *Phys. Rev. B* **46**, 6207.

Vaughan, J. M.: 1989, *The Fabry-Perot interferometer*, Hilger, Bristol.

Verbeni, R., Monaco, G., Krisch, M. and Sette, F.: 2003, unpublished.

Verbeni, R., Sette, F., Krisch, M., Bergmann, U., Gorges, B., Halcoussis, C., Martel, K., Masciovecchio, C., Ribois, J. F., Ruocco, G. and Sinn, H.: 1996, X-ray monochromator with 2×10^{-8} energy resolution, *J. Synchrotron Rad.* **3**, 62–64.

von Laue, M.: 1931, Die dynamische Theorie der Röntgenstrahlinterferenzen in neuer Form, *Ergeb. Exakt. Naturwiss.* **10**, 133–158.

von Laue, M.: 1960, *Röntgenstrahl-Interferenzen*, Akademische Verlagsgesellschaft, Frankfurt am Main.

Weckert, E. and Hümmer, K.: 1997, Multiple-beam x-ray diffraction for physical determination of reflection phases and its applications, *Acta Cryst.* **A53**, 108–143.

White, G. K. and Roberts, R. B.: 1983, Thermal expansion of reference materials: tungsten and α-Al$_2$O$_3$, *High Temperatures - High Pressures* **15**, 321–328.

Wille, H.-C., Gerken, M., Gerdau, E., Shvyd'ko, Y. V., Rüter, H. D. and Franz, H.: 2003, Excitation of the nuclear resonance in ^{61}Ni at 67.4 keV by synchrotron radiation, *Hyp. Interact. (C)* **5**, 1–4.

Wille, H.-C., Shvyd'ko, Y. V., Gerdau, E., Lerche, M., Lucht, M., Rüter, H. D. and Zegenhagen, J.: 2002, Anomalous isotopic effect on the lattice constant of silicon, *Phys. Rev. Lett.* **89**, 285901.

Windisch, D. and Becker, P.: 1990, Silicon lattice parameters as an absolute scale of length for high precision measurements of fundamental constants, *Phys. Stat. Solidi (a)* **118**, 379–388.

Wong, J., Krisch, M., Farber, D. L., Occelli, F., Schwartz, A. J., Chiang, T., Wall, M., Boro, C. and Xu, R.: 2003, Phonon dispersions of fcc δ-plutonium-gallium by inelastic x-ray scattering, *Science* **301**, 1078–1080.

Xiaowei, Z., Yoda, Y. and Imai, Y.: 2000, Precision wavelength measurement of the 14.4 keV Mössbauer photon, *J. Synchrotron Rad.* **7**, 189.

Xu Shunsheng and Li Runsheng: 1988, The three-dimensional dynamic DuMond diagram for x-ray diffraction analysis of nearly perfect crystals, *J. Appl. Cryst.* **21**, 213–217.

Yabashi, M., Tamasaku, K. and Ishikawa, T.: 2002, Measurement of x-ray pulse widths by intensity interferometry, *Phys. Rev. Lett.* **88**, 244801.

Yabashi, M., Tamasaku, K., Kikuta, S. and Ishikawa, T.: 2001, An x-ray monochromator with an energy resolution of 8×10^{-9} at 14.4 keV, *Rev. Sci. Instrum.* **72**, 4080.

Zachariasen, W. H.: 1945, *Theory of X-Ray Diffraction in Crystals*, John Wiley & Sons, Inc., New York. Reprinted by Dover Publications, New York, 1967.

Ziman, J. M.: 1969, *Principles of the theory of solids*, Cambridge University Press, Cambridge.

List of Symbols

A_{FP}	absorption in a Fabry-Pérot resonator
$\boldsymbol{a}_1, \boldsymbol{a}_2, \boldsymbol{a}_3$	basis vectors for the direct crystal lattice
a,b,c	crystal lattice parameters: $a \equiv |\boldsymbol{a}_1|,\ b \equiv |\boldsymbol{a}_2|,\ c \equiv |\boldsymbol{a}_3|$
a_{Si}	lattice parameter of silicon at 22.500 °C
$\boldsymbol{b}_1, \boldsymbol{b}_2, \boldsymbol{b}_3$	basis vectors for the reciprocal crystal lattice
$\boldsymbol{B}$	reciprocal lattice vectors associated with the Bragg-case Bragg reflections
b_H, b	asymmetry factor (for the reflecting atomic planes perpendicular to the reciprocal vector $\boldsymbol{H}$)
c	speed of light in vacuum
$\boldsymbol{\mathcal{D}}(\boldsymbol{r},t)$	electric vector of the total radiation field inside a crystal
$\boldsymbol{\mathcal{D}}(\boldsymbol{r})$	spatial part of $\boldsymbol{\mathcal{D}}(\boldsymbol{r},t)$
$\boldsymbol{\mathcal{D}}_\nu(\boldsymbol{r})$	wave field vector corresponding to the eigenvector ν
$\boldsymbol{D}_H(z)$	net wave field propagating in the crystal with the wave vector $\boldsymbol{k}_H$
$\boldsymbol{D}_0(\nu), \boldsymbol{D}_H(\nu), \ldots$	components of the eigenvector ν
$d, \tilde{d}$	thickness of a crystal plate
d_{A}	thickness of the spherical crystal analyzer segment
d_H	distance between the reflecting atomic planes associated with the reciprocal vector $\boldsymbol{H}$
$d_e(0)$	extinction length
d_{ph}	photo-absorption length
d_{g}	gap width between the mirrors of a Fabry-Pérot resonator
$\boldsymbol{\mathcal{E}}(\boldsymbol{r},t)$	electric vector of the radiation field in vacuum
$\boldsymbol{\mathcal{E}}_0(\boldsymbol{r})$	vacuum forward transmitted radiation field component
$\boldsymbol{\mathcal{E}}_H(\boldsymbol{r})$	vacuum Bragg reflected radiation field component
$\boldsymbol{\mathcal{E}}_{\text{i}}$	amplitude of the front surface incident monochromatic wave
$\tilde{\boldsymbol{\mathcal{E}}}_{\text{i}}$	amplitude of the rear surface incident monochromatic wave
E	photon energy
E_{M}	energy of Mössbauer photons

390 List of Symbols

E_a	atomic resonance energy
E_c	center of the region of total Bragg reflection on the energy scale
E_H	Bragg energy
E_f	free spectral range on the energy scale
E_{fo}	uncorrected free spectral range on the energy scale
E_R	center of the region of total *exact Bragg back-reflection* on the x-ray energy scale
$E_R^{(s)}$	center of the region of total *exact Bragg back-reflection* on the x-ray energy scale (in the symmetric scattering geometry)
E_P	x-ray photon energy at which the Bragg reflectivity reaches its peak value
E_γ	nuclear transition energy
F_H	structure factor of the crystal unit cell
$f(\boldsymbol{H})$	atomic scattering amplitude
$f^{(0)}(\boldsymbol{H})$	atomic form factor
f', f''	real and imaginary parts of anomalous scattering corrections to $f(\boldsymbol{H})$
f_s	frequency of an iodine stabilized He-Ne laser
$\mathcal{F}$	finesse of a Fabry-Pérot resonator
$G^{ss'}_{HH'}$	scattering matrix
$g(\boldsymbol{H})$	Debye-Waller factor
h, k, l	Miller indices
h, k, i, l	Miller indices in a hexagonal lattice
$h, \hbar = h/2\pi$	Planck's constant
$\boldsymbol{H}, \boldsymbol{N}, \boldsymbol{B}, \boldsymbol{L}$	reciprocal lattice vectors, diffraction vectors
$\tilde{\boldsymbol{H}}$	reciprocal lattice vector perpendicular to the virtual reflecting planes - total scattering vector
$\boldsymbol{K}_0$	vacuum wave vector of the incident plane wave
$\boldsymbol{K}_H$	vacuum wave vector of the Bragg scattered wave
K	magnitude of the vacuum wave vectors $\boldsymbol{K}_0$, $\boldsymbol{K}_H$, etc.
K_u	deflection parameter of an undulator
$\boldsymbol{k}_0$	in-crystal wave vector of the forward scattered wave
$\boldsymbol{k}_H$	in-crystal wave vector of the Bragg scattered wave
$\boldsymbol{L}$	reciprocal lattice vectors associated with Laue-case Bragg reflections
L_{g_o}	*half* of the optical path length for successive reflections in a Fabry-Pérot resonator (uncorrected)
L_g	*half* of the optical path length for successive reflections in a Fabry-Pérot resonator
l_x	linear dimension of the resonator surface illuminated with the incident x rays

M	figure of merit
$\boldsymbol{N}$	reciprocal lattice vector associated with normal incidence Bragg reflection
$\boldsymbol{N}^*$	back-diffraction vector conjugate to $\boldsymbol{N}$: $\boldsymbol{N}^* = \boldsymbol{L} - \boldsymbol{B}$, $\boldsymbol{N} = \boldsymbol{L} + \boldsymbol{B}$
N_a	number of accompanying Bragg reflections excited in a crystal at normal incidence to the (hkl) atomic planes
N_e	number of reflecting planes throughout the extinction length
N_d	number of reflecting planes throughout the crystal thickness
N_g	effective number of interplanar distances d_H fitting into the gap of an x-ray Fabry-Pérot resonator
N_{go}	number of interplanar distances d_H fitting into the gap of an x-ray Fabry-Pérot resonator, uncorrected
N_u	number of magnetic periods of an undulator
N_s	number of scatterers (atoms) in a crystal
N_0	number density of atoms
$\mathcal{N}_r$	effective number of reflected waves required for a perfect interference pattern in a Fabry-Pérot resonator
$\mathcal{N}_s$	effective number of reflections in a Fabry-Pérot resonator required for achieving a perfect interference pattern
n	index of refraction
n	number of beams in multiple-beam diffraction
n_g	refractive index of the medium filling the gap between the mirrors of a Fabry-Pérot interferometer
$\boldsymbol{Q}$	momentum transfer
P	tie point - the common origin of the wave vectors
P, P_H^s, $P_{HH'}^{ss'}$	polarization factors
P_e	parity of a nuclear excited state
P_g	parity of a nuclear ground state
p_A	transverse dimension of the analyzer crystal segments
R	reflectivity
R_A	radius of the spherical crystal analyzer
R_ν	ratio of $D_H(\nu)$ to $D_0(\nu)$ in the 2-beam case
R_{FP}	reflectivity of a Fabry-Pérot resonator
$\mathcal{R}$	effective reflection coefficient of the crystal mirrors of a Fabry-Pérot resonator
$\boldsymbol{r}, x, y, z$	space variables
r_e	classical electron radius
r_{0H}	reflection amplitude at the front crystal surface as measured inside the crystal
r_{H0}	reflection amplitude at the rear crystal surface

$r_{\scriptscriptstyle FP}$	as measured inside the crystal amplitude of the radiation reflected off a Fabry-Pérot resonator
S	source size
$\boldsymbol{s}, \boldsymbol{s}'$	generic notation for the unit polarization vectors $\boldsymbol{\sigma}$ and $\boldsymbol{\pi}$
$\boldsymbol{s}_0$	polarization of the incident wave
$\boldsymbol{r}_{\mathrm{n}}$	radius vector of an atom in the crystal unit cell
T	transmissivity
T	throughput
T	crystal temperature
$T_{\scriptscriptstyle H}$	Bragg temperature, crystal temperatures, at which the Bragg back-reflection takes place
$T_{1/2}$	half lifetime of an excited nuclear state
$T_{\scriptscriptstyle FP}$	transmissivity of a Fabry-Pérot resonator
$\mathcal{T}$	effective transmission coefficient of the crystal mirrors of a Fabry-Pérot resonator
t	time variable
t_{00}	transmission amplitude at the rear crystal surface
t_{HH}	reflection amplitude at the front crystal surface
$t_{\scriptscriptstyle FP}$	amplitude of the radiation transmitted through a Fabry-Pérot resonator
t_{f}	time lag between successive scattering events in a Fabry-Pérot resonator; quantum beat period
$\boldsymbol{U}, \boldsymbol{u}$	translation vectors
$\boldsymbol{u}_{\scriptscriptstyle H}$	unit vector parallel to $\boldsymbol{K}_{\scriptscriptstyle H}$
V	volume of the unit cell
$W(\boldsymbol{H})$	exponent of the Debye-Waller factor $g(\boldsymbol{H})$
$w_{\scriptscriptstyle H}$	Bragg's law correction
$w_{\scriptscriptstyle H}^{(\mathrm{s})}, w^{(\mathrm{s})}$	Bragg's law correction in the symmetric Bragg scattering geometry
$w'_{\scriptscriptstyle H}$	correction to Bragg's law for the reflected wave
$\boldsymbol{x}', \boldsymbol{y}', \boldsymbol{z}'$	basis unit vectors of the coordinate system determined by the diffraction vector $\boldsymbol{H}$ and the inward normal $\hat{\boldsymbol{z}}$
$x, y, z, \boldsymbol{r}$	space variables
y	reduced deviation parameter (deviation from Bragg's condition)
Z	number of electrons in an atom
Z	normalization factor
$\hat{\boldsymbol{z}}$	internal normal to the front crystal surface
$\hat{\boldsymbol{z}}_{BL}$	critical surface normal to the front crystal surface, for which $\hat{\boldsymbol{z}}_{BL} \perp \boldsymbol{K}_{B}$ and $\hat{\boldsymbol{z}}_{BL} \perp \boldsymbol{K}_{L}$
$\alpha_{\scriptscriptstyle H}, \alpha$	parameter of deviation from Bragg's condition (associated with the atomic planes perpendicular to $\boldsymbol{H}$)

List of Symbols   393

β_H	linear thermal expansion coefficient of the crystal in the direction of $\boldsymbol{H}$
β_g	linear expansion coefficient of the material filling the gap of an x-ray Fabry-Pérot resonator, or of the material of the spacer
Γ	energy width of a resonance
γ_0, γ_H	cosines of the angles between the crystal surface normal and the wave vectors $\boldsymbol{K}_0$ and $\boldsymbol{K}_0 + \boldsymbol{H}$, respectively
γ'_H	cosine of the angle between the crystal surface normal and the wave vector $\boldsymbol{K}_H$ of the vacuum reflected wave
$\boldsymbol{\Delta}_H$	momentum transfer due to scattering at the vacuum-crystal interface
$\Delta\theta$	angular width
$\Delta\theta^{(s)}$	angular width of the Bragg reflection in the symmetric Bragg scattering geometry
$\underline{\Delta\theta}$	angular spread of the incident x-rays
$\delta^{ss'}, \delta_{HH'}$	Kronecker delta symbols
$\delta\theta_\parallel$	deviation from normal incidence in the direction parallel to the basal plane
$\delta\theta_\perp$	deviation from normal incidence in the direction perpendicular to the basal plane
$\Delta\Upsilon$	angular dimension of the spherical crystal analyzer segment
ϵ	relative deviation of the photon energy E from Bragg's energy E_H
ϵ	relative deviation of the radiation wavelength λ from d_H
ϵ_H	intrinsic relative spectral width of the Bragg reflection with diffraction vector $\boldsymbol{H}$
$\epsilon_H^{(s)}, \epsilon^{(s)}$	intrinsic relative spectral width of the Bragg reflection with diffraction vector $\boldsymbol{H}$ in the symmetric Bragg scattering geometry
ϵ_{FP}	relative spectral width of the Fabry-Pérot resonances
$\varepsilon = 2\gamma_0 \varkappa/K$	dimensionless refraction correction to vacuum wave vector $\boldsymbol{K}_0$
η_H, η	asymmetry angle: the angle between the diffracting atomic planes normal to $\boldsymbol{H}$ and the entrance crystal surface
υ_H	displacement of an atom from the equilibrium position in the unit cell projected on the direction of the diffraction vector $\boldsymbol{H}$
Θ	angle of incidence (the angle the incident ray makes with the surface normal)
Θ_c	the center of the region of total Bragg reflection

	on the scale of the angle of incidence
Θ_{cr}	critical angle of incidence
Θ_D	Debye temperature
Θ_H, Θ	angle of incidence with respect to the reflecting atomic planes perpendicular to $\boldsymbol{H}$
Θ'_H, Θ'	reflection angle with respect to the reflecting atomic planes perpendicular to $\boldsymbol{H}$
$\Theta_\perp$	incidence angle of the wave reflected perpendicular to the atomic planes
Θ_m	incidence angle for reflection with greatest wavelength
Θ_R	incidence angle for *exact Bragg back-reflection*
θ_H, θ	glancing angle of incidence with respect to the reflecting atomic planes perpendicular to $\boldsymbol{H}$
θ'_H, θ'	glancing angle of reflection with respect to the reflecting atomic planes perpendicular to $\boldsymbol{H}$
θ_B	Bragg angle, the angle defined by Bragg's law
θ_c	the center of the region of total Bragg reflection on the scale of the glancing angle of incidence
θ_{cr}	critical glancing angle of incidence, or critical angle of total reflection
θ'_c	the center of the region of total Bragg reflection on the scale of the glancing angle of reflection
θ_R	center of the region of total *exact Bragg back-reflection* on the scale of the glancing angle of incidence
$\varkappa, \varkappa_H$	refraction correction to vacuum wave vectors $\boldsymbol{K}_0$ and $\boldsymbol{K}_H$
Λ_ν	excitation amplitudes of the wave fields $\boldsymbol{\mathcal{D}}_\nu(\boldsymbol{r})$
λ	vacuum radiation wavelength
λ_c	the center of the region of total Bragg reflection on the wavelength scale
λ_f	free spectral range on the wavelength scale
λ_{fo}	uncorrected free spectral range on the wavelength scale
λ_M	wavelength of Mössbauer radiation
λ_R	center of the region of total *exact Bragg back-reflection* on the wavelength scale
$\lambda_R^{(s)}$	center of the region of total *exact Bragg back-reflection* on the wavelength scale (symmetric scattering geometry)
λ_s	radiation wavelength of an iodine stabilized He-Ne laser
$\boldsymbol{\pi}_H, \boldsymbol{\sigma}_H$	two mutually orthogonal unit polarization vectors related to the plane wave with wave vector $\boldsymbol{k}_H$
ρ_{0H}	reflection amplitude at the front crystal surface as measured in vacuum
ρ_{H0}	reflection amplitude at the rear crystal surface as measured in vacuum

ρ_d	dislocation density
$\boldsymbol{\sigma}_H, \boldsymbol{\pi}_H$	two mutually orthogonal unit polarization vectors related to the plane wave with wave vector $\boldsymbol{k}_H$
τ	lifetime, decay time
$\tau_1, \tau_2, \tau_{23}$	abbreviation for different algebraic combinations of $\tan\theta_{c_1}$, $\tan\theta_{c_2}$, and $\tan\theta_{c_3}$
τ_{00}	transmission amplitude at the rear crystal surface as measured in vacuum
τ_{HH}	transmission amplitude at the front crystal surface as measured in vacuum
Υ	angular acceptance of analyzers
Φ	angular deviation from parallel alignment of the mirrors of a Fabry-Pérot resonator
Φ	scattering angle
Φ	the angle, which the wave vector of the incident radiation makes with the dispersion plane
ϕ_H, ϕ	azimuthal angle of incidence with respect to the plane built up of the vectors $\boldsymbol{H}$ and $\hat{\boldsymbol{z}}$
ϕ'_H, ϕ'	azimuthal angle of reflection with respect to the plane built up of the vectors $\boldsymbol{H}$ and $\hat{\boldsymbol{z}}$
$\phi_\mathcal{T}$	effective transmission phase of the crystal mirrors of a Fabry-Pérot resonator
$\phi_\mathcal{R}$	effective reflection phase of the crystal mirrors of a Fabry-Pérot resonator
φ_A	Airy phase
$\chi(\boldsymbol{r})$	electric susceptibility of a crystal
$\chi_0, \chi_H, \chi_{\overline{H}}$	Fourier components of $\chi(\boldsymbol{r})$
$\chi'_0, \chi'_H, \chi'_{\overline{H}}$	real parts of $\chi_0, \chi_H, \chi_{\overline{H}}$
$\chi''_0, \chi''_H, \chi''_{\overline{H}}$	imaginary parts of $\chi_0, \chi_H, \chi_{\overline{H}}$
χ_g	zeroth Fourier components of the electric susceptibility of the material filling the gap of an x-ray Fabry-Pérot resonator
Ψ_H	angle between the virtual and the real reflecting atomic planes (angle between $\tilde{\boldsymbol{H}}$ and $\boldsymbol{H}$)
ξ	component of the reduced phonon wave vector $\xi = Q/(2\pi/a)$
$\xi^{(n)}$	angular corrections in the modified Bragg's law for the exit wave
ω	radiation circular frequency
Ω	angular range of backscattering

Index

$(+, +)$ multiple-crystal configuration, 34, 144–153, 155, 158–161, 163, 166, 217, 241, 242, 245, 246, 281
$(+, +, +)$ multiple-crystal configuration, 165, 166, 168–170, 217, 270, 271
$(+, +, -)$ multiple-crystal configuration, 159–164, 166, 167, 169, 170, 217, 271
$(+, +, -, -)$ multiple-crystal configuration, 242, 246
$(+, -)$ multiple-crystal configuration, 34, 144–147, 151–155, 159, 160, 218, 223–225, 229, 243, 246, 247, 262, 322
$(+, -, -, +)$ multiple-crystal configuration, 155–159, 260, 264, 265, 268
α-Al_2O_3, 6, 7, 9, 30, 32–34, 73, 76, 106–109, 112, 116, 117, 121, 131–135, 137, 141, 185, 192–194, 196, 200, 204–207, 209, 219–221, 224, 226–234, 238–241, 273, 296, 306, 316, 317, 326–328, 337–341, 350, 358–360, 362–367
^{119}Sn, 248, 257, 359
^{121}Sb, 359
^{149}Sm, 248
^{151}Eu, 248, 359
^{161}Dy, 33, 228–230, 248, 252, 359
^{28}Si, 26
^{30}Si, 26
^{57}Fe, 19–22, 28, 29, 33, 134, 227, 228, 243, 248, 256–258, 262, 359
^{67}Ni, 232, 359
^{83}Kr, 262
n-beam diffraction, 114, 140, 197, 203
– degenerate, 121
étalon
– Fabry-Pérot, 171, 178
– x-ray Fabry-Pérot, 34

2-beam diffraction, 37, 38, 59, 63, 65, 114, 119, 122, 124–126, 134–136, 237
24-beam diffraction, 32, 235, 236
4-beam diffraction, 32, 38, 126, 127, 131, 134, 137–139, 141, 236, 284
– degenerate, 127, 131
6-beam diffraction, 32

absorption edge, 13, 111, 225, 306
accompanying reflection, 8, 9, 34, 38, 114–120, 122, 123, 125, 127, 131, 132, 134–136, 140–142, 203, 235, 236, 273, 284, 296, 297, 303, 342, 391
Airy phase, 176–179, 184–192, 197, 199, 201, 207, 210
analyzer, VIII, 1, 9, 16, 38, 215, 287, 288, 293, 295, 296, 300, 302, 304, 306–308, 313, 315
– $(+, +, \pm)$, XIV, 304, 307, 309, 313, 314
– backscattering, 16, 18, 33, 289, 290, 306
– built-in, 21, 215
– exact backscattering, 300, 302, 303
– flat crystal, 292, 294–296
– high-resolution, 303, 313, 314
– multi-bounce, 304, 307, 309, 313, 314
– nuclear resonant, 22–24
– single-bounce, 302, 304, 306
– spherical, 13, 15, 17, 33, 288–290, 295, 297, 299–302, 304, 306, 309
angle of incidence, 2, 3, 39, 48, 317, 370, 394
– azimuthal, 49, 54, 72, 74, 80, 81, 83, 89, 91–93, 97, 100, 165, 278, 368, 395
– critical, 3, 394
– glancing, 3–7, 48, 54, 68, 70–74, 76, 77, 81, 83, 88, 89, 91–95, 100, 105, 107, 108, 113, 146, 147, 165, 166, 184, 186–188, 190, 201, 207, 221, 223, 243, 244, 248, 249, 257, 262, 274–278, 282, 311, 312, 368, 394

398 Index

– – critical, 3, 140, 272, 394
angle of reflection, 48
– azimuthal, 49, 53, 90, 91, 165, 166, 169, 395
– glancing, 48, 88, 90, 92, 93, 95, 146, 147, 394
angular dispersion, 8, 38, 45, 46, 50, 86–88, 90, 91, 93, 100, 101, 111, 164, 169, 217, 269, 284
anomalous
– scattering corrections, 40, 72, 390
– scattering factors, 40, 221, 336, 339, 342, 350
– transmission, 282, 309
asymmetry
– angle, 8, 46, 50, 52, 54, 57, 72–74, 80, 81, 83, 85, 86, 91–93, 97, 98, 100, 102, 140, 144, 151, 164, 170, 213, 216, 244, 249, 253, 257–260, 265, 269, 272, 274–280, 282, 283, 319
– factor, 49, 50, 68, 71–74, 80, 81, 83, 89, 93–96, 120, 130, 140, 141, 151, 159, 163, 213, 241, 244, 246–249, 253, 255, 257, 262, 263, 274, 276, 278, 280, 282
atomic form factor, 40, 219
atomic planes, 3, 4, 6–8, 14, 25, 40, 42–44, 46–48, 50, 61, 70, 72, 76, 77, 89, 90, 100, 106–109, 114, 118, 127, 128, 132–135, 137–142, 180, 181, 185, 188, 189, 192, 193, 196, 204, 207, 208, 210, 218, 219, 223, 232, 237, 238, 245, 246, 250, 254, 256, 267, 268, 317, 327, 331, 332
atomic scattering amplitude, 40
avalanche photo diode (APD), 229, 323

basal plane, 117, 122, 124–129, 136, 138–142, 235, 236, 297, 393
beamline, 249
– exact Bragg backscattering, 303
Borrmann effect, 282
Bragg angle, 4, 31, 34, 68, 70, 113, 119, 258, 312, 394
Bragg back-reflection, 8, 9, 25, 26, 32–34, 116, 125, 127, 173, 220, 223, 228, 230–232, 235, 251, 284, 294, 296, 297, 306
– exact, 8, 84, 141, 368
– multiple-beam-free, 238, 316
– symmetric, 309
– two-beam-case, 117, 350

Bragg backscattering, VII, 1, 9, 23, 26, 31, 33, 78, 114, 115, 118, 120, 172, 229, 332
– exact, 53, 54, 83, 170, 226, 238, 303, 368, 369
– optics, 32, 33
– spectrum, 227
– two-beam, 125
Bragg diffraction, 3, 5, 6, 9, 24, 32, 37, 43, 53, 65, 73, 75, 76, 87, 89, 91, 93, 96, 99, 101, 105, 106, 110–114, 122, 123, 126, 127, 130, 131, 143, 144, 154, 165, 172, 173, 177, 178, 180, 185, 192, 199, 210, 216, 255, 261, 269, 273, 280, 282–284, 312, 350
– four-beam coplanar, 9, 122, 127, 141
– multiple-beam, 8, 114, 122
– – in backscattering, 124
– multiple-crystal, 38, 143, 144
– non-coplanar, 123
– nuclear resonant, 111
Bragg energy, 4–7, 14, 68, 69, 76, 78, 220, 234, 247, 273, 279, 390, 393
Bragg reflection, VIII, IX, 3, 4, 6–8, 14, 16, 34, 50, 59, 66, 72, 74, 77, 81–83, 93–95, 98, 101, 107, 108, 112, 113, 123, 124, 132, 133, 143, 149, 151, 153, 155, 156, 159, 161, 163–166, 170, 185, 186, 191, 195, 197, 201, 203, 206, 210–213, 216–219, 221, 222, 224, 225, 227, 232–234, 239, 243–246, 249, 250, 254, 255, 258, 261, 263, 265, 266, 268–270, 272–274, 276, 279, 281, 282, 284, 288, 291, 293–297, 302–304, 306, 308, 316, 318, 322, 359, 361, 362, 368
– accompanying, 8, 9, 34, 38, 114–120, 122, 123, 125, 127, 131, 132, 134–136, 140–142, 203, 235, 236, 273, 284, 296, 297, 303, 342, 391
– allowed, 342, 350
– angular width, 6, 7, 16, 78–80, 91, 96, 125, 144, 150, 153, 163, 164, 169, 216, 236, 250, 254, 264, 278, 342
– asymmetric, 58, 148, 151, 152, 156, 157, 160, 167, 168, 170, 217, 260, 263, 265, 308
– energy width, 6, 16, 76, 91, 111, 144, 220, 223, 224, 288, 292, 293
– – relative, 6, 7, 76, 77, 79, 106, 164, 216, 217, 223, 241, 247, 269, 273, 293
– forbidden, 121, 131–134

Index   399

- high-indexed, 216, 217, 219, 221, 225, 250, 269, 304
- in backscattering, 79, 114, 116, 117, 121, 127, 269, 273, 288, 289, 296
- low-indexed, 250, 251, 259, 264, 273
- multiple-beam-free, 9, 117, 296, 350
- parasitic, 294, 297
- symmetric, 106, 154, 170, 242, 272, 342, 350
- two-beam-case, 38, 124

Bragg scattering, 3, 42, 44, 54, 111
- geometry, 46–49, 61, 65, 72, 74, 88–90
-- asymmetric, 82, 84, 94
-- symmetric, 75, 82, 84, 94
- multiple, 42
- reversibility, 50, 51
- symmetric, 50

Bragg temperature, 358, 392
Bragg's law, 4, 68, 69, 177, 222, 226, 270, 392, 394
- dynamical, 5, 37, 71, 72, 74, 75, 84, 86, 92, 93, 147, 149
- kinematic, 6, 71, 74, 75, 81, 114, 116, 124, 128
- modified, 5, 37, 71, 72, 74, 75, 84, 86, 91–94, 102, 105, 149, 160, 188, 190, 368, 395
-- for exit wave, 93, 147, 149

Bragg's law correction, 5, 6, 72, 74, 77, 85, 93, 153, 190, 267, 273, 327, 342, 350, 392

Bragg-case
- reflection, 131–134, 137, 235
- wave, 46, 57, 120

Brillouin scattering, 10
Brillouin zone, 70, 71, 115, 116
broadening
- geometrical, 221, 223, 241, 242, 246, 270, 292, 293, 295
- nuclear resonance, 19

capillary optics, 307
cascade
- of asymmetric Bragg reflections, 263
- of Bragg reflections, 263
- of reflecting crystals, 261, 263, 267
cavity, 325, 326
- x-ray, 324–326
channel-cut crystal, 34, 246, 251–253
- artificial, 253–255, 267
coherence
- length, 28, 286
- longitudinal, 286

- spatial, 188, 192, 208
- transverse, VII, 12, 286
coherent scattering, 1–3, 105
collision time, 111
compound refractive lens (CRL), 250, 256, 258, 264, 303, 307, 313
configuration
- dispersive, 153, 155, 158, 241, 244
- double-dispersive, 166
- nondispersive, 153–155, 158, 223–225, 243, 246, 263, 266–268
conjugate pair of reflections, 117–127, 140, 235, 236, 273, 297
conjugate waves, 117, 118, 120, 123
correction
- anomalous scattering, 40, 72, 390
- Bragg's law, 5, 6, 72, 74, 77, 85, 93, 190, 267, 273, 327, 342, 350
critical
- angle of incidence, 3, 394
- angle of total reflection, 3, 140, 394
- glancing angle of incidence, 3, 140, 272, 394
- surface normal, 120, 392
crystal
- ^{57}FeBO$_3$, iron borate, 20, 21
- Be, beryllium, 224, 284, 288
- BeO, beryllium oxide, 224
- C, diamond, 9, 32, 34, 116, 197, 203, 229, 235, 284–286
- centrosymmetric, 66, 104
- GaAs, 116, 235
- Ge, germanium, 9, 32, 33, 116, 127, 235
- sapphire, α-Al$_2$O$_3$, 6, 7, 9, 30, 32–34, 73, 76, 106–109, 112, 116, 117, 121, 131–135, 137, 141, 185, 192–194, 196, 200, 204–207, 209, 219–221, 224, 226–234, 238–241, 273, 296, 306, 316, 317, 326–328, 337–341, 350, 358–360, 362–367
- Si, silicon, 6, 9, 26–34, 72, 73, 76, 109, 116, 117, 121, 127, 172, 223–226, 232, 235, 239, 243, 244, 247–249, 255, 257, 262, 263, 274, 276, 281, 282, 284, 286, 288, 290, 294, 296, 297, 300, 304, 308, 309, 335, 336, 342, 360
- SiC, silicon carbide, 224, 232
- weakly-absorbing, 109, 213
Czochralski method, 360, 362, 364

de Broglie wavelengths, 10
Debye
- model, 336, 340

– temperature, 224, 232, 336, 339, 340, 394
Debye-Waller factor, 20, 40, 195, 219, 224, 232, 336, 339, 340, 390, 392
decay
– constant, 198, 328, 330
– time, 201, 329, 330, 395
deflection parameter, 12, 390
density of states, 10
– phonon, 18, 22
– – partial, 22, 252
– vibrational, 22
deviation parameter, 45, 49, 66–71, 77, 87, 104, 106, 107, 129, 139, 185, 392
– reduced, 67
diamond structure, 203, 235
diamond, C, 9, 32, 34, 116, 197, 203, 229, 235, 284–286
diffraction function, 105
diffraction vector, 43, 46–50, 53, 54, 61, 80, 81, 87, 97, 99, 115–119, 121, 123–127, 131–135, 144, 165, 219, 235, 236, 242, 250, 269, 279, 390–393
diffusion, 10, 19
dislocation, 361–365, 367
– density, 230, 239, 362–366, 395
dispersing
– crystal element, 270–273, 275, 276, 278–280, 307, 310
– element, 164, 170, 217, 269–274, 279–283, 308, 309, 313
dispersion, 2, 14
– angular, 8, 38, 45, 46, 50, 86–88, 90, 91, 93, 100, 101, 111, 164, 169, 217, 269, 284
– plane, 43, 46–49, 51, 53, 54, 80, 81, 87, 89, 97, 99, 145, 146, 241, 266, 267, 269, 278, 279, 309, 368, 395
– relation, 10, 13, 215, 287–289, 340
dispersive configuration, 153, 155, 158, 241, 244, 263, 267
double-dispersive configuration, 166
DuMond diagram, VIII, 74, 79, 143, 145, 148, 152, 156, 157, 160, 161, 167, 168, 222, 241, 243, 246, 247, 260, 270, 278
Dy_2O_3, 229, 230
dynamical theory, VIII, IX, 4, 5, 7, 26, 31, 32, 37–39, 41, 69, 71, 74, 75, 106–110, 112, 114, 126, 135, 136, 141, 144, 149, 172, 173, 181, 196, 219–221, 227, 228, 231, 233, 239, 240, 243, 247, 248, 252, 262, 264, 266, 274,
276, 278, 317, 318, 327, 328, 342, 350, 358, 365, 366

electric susceptibility, 4, 39, 64, 182, 211
– Fourier component, 39, 40, 66, 68, 77, 131, 135, 213
electron-hole pair, 13
elliptical mirror, 309, 310
Ewald sphere, 41, 42, 58, 59
excitation
– charge transfer, 13
– collective, 10, 11, 13, 20
– condition, 41, 44, 45, 59, 114
– – multiple-beam, 123
– core level, 13
– detour (Umweganregung), 121, 131, 134
– electronic, 10, 13, 14
– magnetic, 10
– single particle, 13
extinction length, 66, 76, 77, 103, 106, 107, 109–113, 191, 197, 198, 201, 284, 292, 293, 342, 350, 358, 359, 364–366, 389, 391

Fabry-Pérot
– étalon, 171, 178
– interferometer, 30, 34, 171, 177, 328, 331, 333, 391
– resonator, 171–175, 177–180, 184, 185, 189, 192, 197, 202, 284, 286, 316, 389–392, 395
figure of merit, 216, 233, 268, 269, 274, 276, 280, 283, 286, 391
finesse, 35, 171, 176, 178, 187, 195, 197, 201, 208–211, 284, 286, 317, 319, 328, 329, 390
Fizeau resonator, 174, 180
flux, 57, 65, 216, 287, 304, 306
– density, 57, 176
– spectral, 12, 305
focal
– length, 310–312
– point, 309, 310
four-beam diffraction, 32, 38, 126, 127, 236
free spectral range, 178, 190–192, 198, 201, 330
free-electron laser (FEL), VII, 12, 29, 110, 286, 305
full-width at half-maximum (FWHM), 22, 113, 187, 247, 256, 258, 266, 280, 281, 291, 310, 312, 324

fundamental equations, 41, 54–56, 58–60, 71, 114, 126, 127, 129–131, 141

geometrical broadening, 221, 223, 241, 242, 246, 270, 292, 293
grazing emergence, 45, 52, 56, 59, 87, 89, 91, 142
grazing incidence, 31, 32, 34, 37, 52, 55, 58, 59, 81, 91, 127, 140, 180, 307

heat-exchange method, HEM, 360
HEMEX, 337, 360, 363–367
holography, x-ray, 286
hyperfine interactions, 19–21, 28

incoherent scattering, 21
index of refraction, 1–3, 59, 73, 74, 87
inelastic scattering
– nuclear resonant (NRIS), 20, 22, 229, 230, 252, 256, 258
– of thermal neutrons, 15
– x-ray (IXS), 13–18, 23, 24, 215, 225, 287–291, 299, 302, 304, 305, 307, 310, 313, 314, 403
– – resonant (RIXS), 13, 225
interference, 20, 106, 114
– multiple-wave, 173, 174, 176, 177, 179, 185, 187, 203, 218, 284
– two-wave, 176
interference filter, 23, 24, 34, 171, 174, 179, 197, 218, 284, 286
– Fabry-Pérot, 23, 34, 171, 284
interferometer, VIII, 1
– Fabry-Pérot, 30, 34, 171, 177, 328, 331, 333, 391
– Mach-Zehnder, 187
– Michelson, 27, 187
– triple-Laue-case (LLL), 27
– two-wave, 187
– x-ray Fabry-Pérot, 9, 23, 30, 34, 35, 319, 328, 330–332, 363
interferometry
– intensity, 263
– multiple-wave, 180
interplanar distance, 3, 5, 7, 40, 68, 69, 73, 76, 147, 150, 151, 156, 190, 194, 195, 197, 207, 213, 224, 226, 232, 242, 391
interplanar spacing, 40
invar, 319, 320
isotope
– nuclear, 355, 359
– stable, 19, 355

isotopic effect
– on lattice parameter, 26, 31

kinematic
– approximation, 70, 82, 116, 236
– theory, 4, 6, 70
Kirkpatrick-Baez
– focusing system, 313
Kyropoulos method, 360, 364

Lamb-Mössbauer factor, 20
lattice parameter, 25–29, 31, 221, 224, 226, 229, 335, 337, 338, 342, 350
lattice vibration (phonon), 10, 19
Laue scattering geometry, 46, 47, 112, 229
Laue-case
– reflection, 131–134, 137, 235
– wave, 46, 57, 120
lens
– compound refractive (CRL), 250, 256, 258, 264, 303, 307, 313
lifetime, 19, 355, 395
– half, 355, 392
linac, 12, 286, 304

Mössbauer
– clock, 29
– effect, 19, 20, 28
– radiation, 19, 28–30
– resonance, 22
– spectroscopy, 19
– wavelength standard, 29
Mach-Zehnder interferometer, 187
Michelson interferometer, 27, 187
Miller indices, 40, 219, 226, 335, 337, 350
mirrors
– back-reflecting, 35, 286, 315, 328
– backscattering, 30, 286
– crystal, 34, 184, 187, 191, 192, 195, 285, 286, 319, 322, 332, 333
– elliptical, 309, 310
– flat, 311
– grazing incidence, 307
– multilayer, 312
– – graded, 307
– – laterally graded, 312
– paraboloidal, 307, 309–312
– resonator, 319, 330, 331
momentum transfer, 8, 15, 17, 24, 42–44, 50, 121, 165, 219, 289, 291, 307, 391, 393

- additional, 43–45, 51, 53, 63, 64, 97
- resolution, 287
- total, 43, 44, 46, 50, 51, 53, 89, 99, 165

monochromator, VIII, 1, 12, 16, 33, 38, 151, 215, 216, 221, 241, 249, 252, 256, 257, 259, 260, 263, 264, 267, 269, 272, 283, 286, 287, 289, 302, 306, 308, 309, 315, 316
- $(+,+)$, 241–244, 255–261, 264–266, 268, 269, 273, 280
- $(+,+,-)$, 270
- $(+,+,-,-)$, 244–252, 254–256, 266, 268, 322, 323, 365
- $(+,+,\pm)$, 217, 269–283, 286, 288, 307–310, 313, 314
- $(+,-)$, 322
- $(+,-,-,+)$, 242, 260–268, 280, 281, 308, 313
- backscattering, 15–17, 23, 24, 31, 33, 224, 226, 228, 230, 234, 249, 279–281, 283, 302, 303, 313
- exact backscattering, 303, 313
- four-bounce, 248, 252, 262, 322
- four-crystal, 243, 259–261, 264, 266, 268, 280, 281
- high-heat-load (cooled), 13, 15, 17, 20, 229, 303, 314, 322, 341, 360
- high-resolution, 13, 33, 34, 215, 228, 242, 250, 252, 263, 314
- in-line, 16, 17, 218, 223, 243, 246, 254–256, 304, 313
- multi-bounce, 16, 17, 33, 34, 217, 241–252, 254–283, 286, 288, 304, 307–310, 313, 314, 322, 323, 365
- multiple-crystal, 217, 242
- nested, 16, 244–252, 254–256, 266, 268, 322, 323, 365
- nondispersive, 218
- single-bounce, 16, 33, 218, 222, 224, 228, 234, 242, 249, 279, 280, 283, 302, 303
- single-crystal, 223, 242, 281
- spectral function, 216, 253, 258, 259, 263, 266, 280, 281, 304, 313, 322
- three-crystal, 269, 270, 272, 275, 278, 280, 281, 308
- two-bounce, 13, 218, 280
- two-crystal, 255, 258, 259, 261, 265, 266, 268, 281

multiple resonator, 197
multiple scattering, 111, 114, 143, 181, 182

multiple-beam
- diffraction, VIII, 8, 37, 38, 114–121, 125, 197, 203, 223, 224, 235, 237, 284
- excitation, 123, 124, 136, 236, 237

multiple-crystal
- Bragg diffraction, 38, 143, 144
- configuration, 165

multiple-wave
- fringes, 177, 179
- interference, 173, 174, 176, 177, 179, 185, 187, 203, 218, 284
- interferometry, 180
- resonator, 174

myoglobin, 256

neutrons
- thermal, 10, 15

nondispersive configuration, 153–155, 158, 223–225, 243, 246, 263, 266–268

normal incidence, IX, 2, 7, 25, 31, 32, 37, 54, 69, 72, 77–83, 89, 93, 94, 96–98, 100, 101, 106, 108, 109, 112, 117, 127, 128, 131–142, 166, 172, 178, 179, 192, 193, 196, 197, 200, 202–204, 206, 207, 210, 224, 233, 235, 237, 240, 285, 294, 297, 317, 318, 327, 332, 333, 342, 358, 365, 366, 368, 391, 393

nuclear forward scattering (NFS), 20, 21, 198, 230, 243

nuclear isotope, 355
nuclear resonance, 19, 22
nuclear resonant
- Bragg diffraction, 111
- inelastic scattering (NRIS), 20, 22, 229, 230, 252, 256, 258
- scattering (NRS), 18–20, 22, 33, 198, 215, 225, 228, 229, 252, 265
- spectroscopy, 198

optical path length, 174, 191, 390

paraboloidal mirror, 307, 309–312
parity, 355
Pendellösung
- effect, 239
- fringes, 364
phase
- diffractive, 188
- refractive, 188
phonon, 10, 18, 20, 22, 33, 252
- acoustic, 17, 18
- density of states, 18
-- partial, 22, 252
- dispersion, 18

- dispersion relations, 18, 215
photo-absorption, 5, 65, 76, 104, 106, 108–110, 113, 185, 191, 192, 195–197, 224, 242, 281, 284, 285, 342, 350
- length, 110, 284, 389
plane
- atomic, 3, 4, 6–8, 14, 25, 40, 42–44, 46–48, 50, 61, 70, 72, 76, 77, 89, 90, 100, 106–109, 114, 118, 127, 128, 132–135, 137–142, 180, 181, 185, 188, 189, 192, 193, 196, 204, 207, 208, 210, 218, 219, 223, 232, 237, 238, 245, 246, 250, 254, 256, 267, 268, 317, 327, 331, 332
- basal, 117, 122, 124–129, 136, 138–142, 235, 236
- dispersion, 43, 46–49, 51, 53, 54, 80, 81, 87, 89, 97, 99, 145, 146, 241, 266, 278, 279, 309, 368, 395
- virtual reflecting, 44, 50, 213, 390, 395
plasmon, 13–15
- dispersion, 15
Pointing vector, 57
polarization, 1, 39, 273
- component, 54, 55, 58, 129, 130
- effect, 122
- factor, 55, 59, 60, 68, 123, 129, 130, 136, 213, 391
- state, 122, 129, 136
- state π, 55, 122, 123, 130, 136, 138, 140
- state σ, 55, 122, 130, 132, 133, 136–138
- vector, 55, 59, 122, 297, 392, 394, 395
- vector π, 129
- vector σ, 129
protein, 256

quantum beat period, 198, 200, 201, 329, 392
quantum beats, 20, 198, 201

Raman scattering, 10
reflection cone, 123–125
refraction, 1, 2, 6, 44, 54, 70, 74, 114
refractive index, 2, 6
region of total reflection, 5–7, 66, 67, 71, 73–75, 78–81, 83, 96, 109, 188, 190, 191, 199, 201, 207, 208, 368
resonance
- Fabry-Pérot, 187, 190, 195–197, 199–202, 206–208, 210–212, 316, 318
- condition, 171, 185–187, 189, 190

- nuclear, 19, 20, 22, 33, 34, 198
-- long-lived, 111
-- low-lying, 33
- transmission, 192, 195
- width, 187, 192, 194, 195, 316, 317
resonator, VIII, 1
- Fabry-Pérot, 171–175, 177–180, 184, 185, 189, 192, 197, 202, 284, 286, 316, 389–392, 395
- Fizeau, 174, 180
- multiple, 197
- multiple-wave, 174
- thermal neutron, 172
- x-ray, 9, 23, 24, 198, 315, 322, 323, 326, 329, 330
- x-ray Fabry-Pérot, VII–IX, 38, 60, 64, 110, 144, 171–173, 180–188, 190, 192–194, 196–201, 203–211, 284, 286, 315, 317, 318, 391, 393, 395
reversibility of Bragg scattering, 50, 51
ring detector, 294, 300–303
Rowland's circle, 298

scattering geometry, 43, 46, 49
- asymmetric, 46, 51, 53, 80, 81, 83–85, 87–89, 94, 99, 216
- asymmetric Bragg, 82, 84, 94
- backscattering, VIII, 7, 16, 25, 37, 78, 96, 101, 106, 166, 223, 224, 226, 240–242, 270, 288, 291, 292, 295, 299, 300
- Bragg, 46–49, 58, 60, 61, 65, 67, 72, 74, 88–90, 282
- coplanar, 146
- exact backscattering, 25, 53, 279, 293–295, 297, 299–302, 309
- Laue, 46, 47, 50, 58, 229, 239
- symmetric, 50, 52, 72, 73, 77, 81, 390
- symmetric Bragg, 5, 46, 72, 73, 75, 76, 79, 81–85, 91, 94, 111, 219, 392, 393
- symmetric Laue, 46
scattering matrix, 55, 129–131, 135, 136, 141
self-amplified spontaneous emission (SASE), 12
slope error, 253, 310
space group, 116, 117, 284, 335, 337
spatial coherence, 188, 192, 208
spectral flux, 12, 305
spectrometer, 18, 22, 215, 225, 303
- inelastic x-ray scattering, 13, 14, 16, 17, 23, 33, 215, 272, 287–289, 302, 304, 305, 313

– high-resolution, 16, 177, 223
– x-ray, 11, 13–15, 306
spin, 19, 355
– relaxation, 19
state
– excited, 10, 19, 355
– ground, 355
Stokes, anti-Stokes components, 17, 22, 290
storage ring, 12, 215, 304, 330
structure factor, 39, 219, 221, 239, 339
surface normal, 44, 46, 47, 49, 87, 119, 130, 132, 133, 135, 137, 140–142, 213, 393
– critical, 120
surface roughness, 210, 211, 253
– admissible, 211

thermal
– conductivity, 340, 360
– contraction, 207
– expansion, 207, 319, 335, 339, 360
– – coefficient, 69, 207, 208, 213, 226, 279, 335–337, 339, 393
– neutrons, 10, 15
– parameters, 221, 342, 350
– vibrations (phonons), 40, 219, 232
thickness oscillations, 106, 190, 239, 292
Thomson scattering, 225
throughput
– of a monochromator, 242, 243, 245, 248, 258, 265, 268, 272, 274, 276, 280, 283, 285, 286, 302, 304, 313, 392
– of an analyzer, 287, 289, 302, 306
– peak, 215, 216, 242, 250, 251, 256–259, 262, 264, 265, 268, 281, 308, 309
tie point, 42, 115, 116, 119
time-of-flight, 113, 316, 323, 326, 329
total
– momentum transfer, 43, 44, 46, 50, 51, 53, 89, 165
– reflectivity, 74, 144
total reflection, 66, 77
– region, 5–7, 66, 67, 71, 73–84, 93, 94, 96, 97, 109, 147, 161, 162, 166, 188, 190, 191, 199, 201, 207, 208, 368
– – center, 70–72, 74, 76, 79, 93, 103, 108, 390, 393, 394

– – width, 67, 68, 76, 77, 79
two-beam
– approximation, 109, 110, 132–135, 141, 142, 219, 233, 236
– diffraction, 37, 38, 59, 63, 65, 114, 119, 122, 124–126, 134–136, 237

Umweganregung (detour excitation), 121, 131, 134
uncertainty relation, 19, 110, 111, 113
undulator, 10–12, 24, 303, 305
– spectrum, 12, 304
unit cell, 39, 219, 226, 339

virtual planes, 44, 50, 213, 390, 395

wave vector, 41, 42, 63, 70, 87, 89, 115, 116, 118–122, 124, 126, 131, 135, 138, 140, 142, 278
– in-crystal, 41, 43
– reduced, 18, 395
– vacuum, 38, 41, 43, 90, 97
wavelength meter, 9, 25, 215
wavelength selector, 164, 165, 217, 269–273, 276, 277, 307
weak-link
– mechanism, 254, 255, 267, 268, 319

x-ray
– cavity, 324–326
– resonator, 9, 23, 24, 198, 315, 322, 323, 326, 329, 330
x-ray Fabry-Pérot
– étalon, 34
– interferometer, 9, 23, 30, 34, 35, 319, 328, 330–332, 363
– resonator, VII–IX, 38, 60, 64, 110, 144, 171–173, 180–188, 190, 192–194, 196–201, 203–211, 284, 286, 315, 317, 318, 391, 393, 395
x-ray holography, 286
x-ray laser
– free electron (X-FEL), VII, 12, 29, 305
x-ray photon correlation spectroscopy, (XPCS), 23
x-ray topography, 361
– white beam, 360–362

Springer Series in
OPTICAL SCIENCES

New editions of volumes prior to volume 70

1 **Solid-State Laser Engineering**
By W. Koechner, 5th revised and updated ed. 1999, 472 figs., 55 tabs., XII, 746 pages
14 **Laser Crystals**
Their Physics and Properties
By A. A. Kaminskii, 2nd ed. 1990, 89 figs., 56 tabs., XVI, 456 pages
15 **X-Ray Spectroscopy**
An Introduction
By B. K. Agarwal, 2nd ed. 1991, 239 figs., XV, 419 pages
36 **Transmission Electron Microscopy**
Physics of Image Formation and Microanalysis
By L. Reimer, 4th ed. 1997, 273 figs. XVI, 584 pages
45 **Scanning Electron Microscopy**
Physics of Image Formation and Microanalysis
By L. Reimer, 2nd completely revised and updated ed. 1998,
260 figs., XIV, 527 pages

Published titles since volume 70

70 **Electron Holography**
By A. Tonomura, 2nd, enlarged ed. 1999, 127 figs., XII, 162 pages
71 **Energy-Filtering Transmission Electron Microscopy**
By L. Reimer (Ed.), 1995, 199 figs., XIV, 424 pages
72 **Nonlinear Optical Effects and Materials**
By P. Günter (Ed.), 2000, 174 figs., 43 tabs., XIV, 540 pages
73 **Evanescent Waves**
From Newtonian Optics to Atomic Optics
By F. de Fornel, 2001, 277 figs., XVIII, 268 pages
74 **International Trends in Optics and Photonics**
ICO IV
By T. Asakura (Ed.), 1999, 190 figs., 14 tabs., XX, 426 pages
75 **Advanced Optical Imaging Theory**
By M. Gu, 2000, 93 figs., XII, 214 pages
76 **Holographic Data Storage**
By H.J. Coufal, D. Psaltis, G.T. Sincerbox (Eds.), 2000
228 figs., 64 in color, 12 tabs., XXVI, 486 pages
77 **Solid-State Lasers for Materials Processing**
Fundamental Relations and Technical Realizations
By R. Iffländer, 2001, 230 figs., 73 tabs., XVIII, 350 pages
78 **Holography**
The First 50 Years
By J.-M. Fournier (Ed.), 2001, 266 figs., XII, 460 pages
79 **Mathematical Methods of Quantum Optics**
By R.R. Puri, 2001, 13 figs., XIV, 285 pages
80 **Optical Properties of Photonic Crystals**
By K. Sakoda, 2001, 95 figs., 28 tabs., XII, 223 pages
81 **Photonic Analog-to-Digital Conversion**
By B.L. Shoop, 2001, 259 figs., 11 tabs., XIV, 330 pages
82 **Spatial Solitons**
By S. Trillo, W.E. Torruellas (Eds), 2001, 194 figs., 7 tabs., XX, 454 pages
83 **Nonimaging Fresnel Lenses**
Design and Performance of Solar Concentrators
By R. Leutz, A. Suzuki, 2001, 139 figs., 44 tabs., XII, 272 pages
84 **Nano-Optics**
By S. Kawata, M. Ohtsu, M. Irie (Eds.), 2002, 258 figs., 2 tabs., XVI, 321 pages
85 **Sensing with Terahertz Radiation**
By D. Mittleman (Ed.), 2003, 207 figs., 14 tabs., XVI, 337 pages

Springer Series in
OPTICAL SCIENCES

86 **Progress in Nano-Electro-Optics I**
Basics and Theory of Near-Field Optics
By M. Ohtsu (Ed.), 2003, 118 figs., XIV, 161 pages

87 **Optical Imaging and Microscopy**
Techniques and Advanced Systems
By P. Török, F.-J. Kao (Eds.), 2003, 260 figs., XVII, 395 pages

88 **Optical Interference Coatings**
By N. Kaiser, H.K. Pulker (Eds.), 2003, 203 figs., 50 tabs., XVI, 504 pages

89 **Progress in Nano-Electro-Optics II**
Novel Devices and Atom Manipulation
By M. Ohtsu (Ed.), 2003, 115 figs., XIII, 188 pages

90/1 **Raman Amplifiers for Telecommunications 1**
Physical Principles
By M.N. Islam (Ed.), 2004, 488 figs., XXVIII, 328 pages

90/2 **Raman Amplifiers for Telecommunications 2**
Sub-Systems and Systems
By M.N. Islam (Ed.), 2004, 278 figs., XXVIII, 420 pages

91 **Optical Super Resolution**
By Z. Zalevsky, D. Mendlovic, 2004, 164 figs., XVIII, 232 pages

92 **UV-Visible Reflection Spectroscopy of Liquids**
By J.A. Räty, K.-E. Peiponen, T. Asakura, 2004, 131 figs., XII, 219 pages

93 **Fundamentals of Semiconductor Lasers**
By T. Numai, 2004, 166 figs., XII, 264 pages

94 **Photonic Crystals**
Physics, Fabrication and Applications
By K. Inoue, K. Ohtaka (Eds.), 2004, 214 figs., IX, 331 pages

95 **Ultrafast Optics IV**
Selected Contributions to the 4th International Conference
on Ultrafast Optics, Vienna, Austria
By F. Krausz, G. Korn, P. Corkum, I.A. Walmsley (Eds.), 2004, 281 figs., XIV, 506 pages

96 **Progress in Nano-Electro Optics III**
Industrial Applications and Dynamics of the Nano-Optical System
By M. Ohtsu (Ed.), 2004, 155 figs., XIV, 226 pages

97 **Microoptics**
From Technology to Applications
By J. Jahns, K.-H. Brenner, 2004, 303 figs., XI, 335 pages

98 **X-Ray Optics**
High-Energy-Resolution Applications
By Y. Shvyd'ko, 2004, 181 figs., XIV, 404 pages

Printing: Saladruck, Berlin
Binding: Stein+Lehmann, Berlin